Community Informatics:
Enabling Communities with Information and Communications Technologies

Michael Gurstein
Technical University of British Columbia

IDEA GROUP PUBLISHING
Hershey USA • London UK

Acquisition Editor:	Mehdi Khosrowpour
Managing Editor:	Jan Travers
Development Editor:	Michele Rossi
Copy Editor:	Maria Boyer
Typesetter:	Tamara Gillis
Cover Design:	Connie Peltz
Printed at:	Sheridan Press

Published in the United States of America by
  Idea Group Publishing
  1331 E. Chocolate Avenue
  Hershey PA 17033-1117
  Tel: 717-533-8845
  Fax: 717-533-8661
  E-mail: cust@idea-group.com
  http://www.idea-group.com

and in the United Kingdom by
  Idea Group Publishing
  3 Henrietta Street
  Covent Garden
  London WC2E 8LU
  Tel: 171-240 0856
  Fax: 171-379 0609
  http://www.eurospan.co.uk

Library of Congress Cataloging-in-Publication Data

Community informatics : enabling communities with information and communications technologies / [edited by] Michael Gurstein.
        p. cm.
    Includes bibliographical references and index.
    ISBN 1-878289-69-1 (cloth)
        1. Community life--Technological innovations. 2. Computer networks--Social aspects. 3. Electronic villages (Computer networks) 4. Information society. I. Gurstein, Michael.

HM761 .C65 2000
306.4'6--dc21   00-025310

British Cataloguing in Publication Data
A Cataloguing in Publication record for this book is available from the British Library.

NEW from Idea Group Publishing

Community Informatics:
Enabling Communities with Information and Communications Technologies

Table of Contents

CI APPLICATIONS

CI AND DEVELOPMENT

Preface

This book is both very new and somewhat old. It is new in that it represents the most current thinking/research/practise in applying Information and Communications Technologies (ICTs) to community needs. It is old in that many of the theories and approaches which have gone into the writing of most of these papers represents continuities and extensions of existing activities and efforts.

What is particularly striking however, is the degree of convergence which can be seen in these papers around questions of community use of ICTs. This book includes the work of information scientists, computer scientists, political scientists, sociologists, community planners, social and regional development specialists, urban planners, and practitioners of rural development, journalists, environmentalists, political activists and odd hybrids of many of the above categories.

ICTs are becoming ever more pervasive in reshaping the economic and social landscape and in shifting our attention to the virtual and the "cyber." However, at the same time our interests and engagements are continuously being drawn back to our most basic of connections—to our own bodies, to our families, and to our communities.

The term "informatics" is a slightly exotic one for North Americans who are more used to the narrower terminology of "computer" or "information scientists." But the term "informatics" implies something that is lost in the terminology of science, that is the capacity to act on and through the technology with which one is working. Where computer "science" suggests the dispassionate gaze and the formal engagement of the scientist, "informatics" looks towards the applications of the technology, towards its use in and on the world in which we are living.

Equally, some prefer to use the term "community networking" as in Doug Schuler's seminal book *New Community Networks: Wired for Change* (Addison Wesley, 1996), making the direct link between the technology and how it is being used in a community context. For some, the community networking terminology is sufficient. However, as one can readily see from this book, the diversity of applications and areas of application for technology enabling communities perhaps is too broad to be encompassed by the language of "networking," however suggestive the term is of "process" as well as "product."

There seems now, moreover, to be somewhat of a convergence around the term "community informatics." This perhaps is an extension of the broad-based acceptance of the terminology of "health informatics" among practitioners and academics alike in the study and application of how ICTs are being used in the sphere of health practise.

In a sphere of activity as individualized or single-user oriented as "personal" computing, it may seem anomalous to be developing approaches and strategies supportive of "community" computing applications. However, as will be seen below, this shouldn't be surprising when it is realized that for most, and even in the most advanced societies, communities are still at the centre of the human systems which provide us and our families with many of the basic components from which construct our physical, social and cultural well-being.

Communities provide us with the opportunity to mature and educate our children and increasingly to train and retrain ourselves as adults. They give us the means for sharing the burdens and opportunities of our physical well-being with our neighbours and allow us to participate in the shaping of the immediate components of our daily lives. They are the means through which we participate in our culture as producers and not simply as consumers. Communities for many are the essential framework through which they can be productive participants in society.

Increasingly communities are the contexts within which we can find ways of intervening in and responding to some of our modern dilemmas and critical problems in the environment, in the bridging of social and economic divides, in maintaining the kind of physical surroundings in which we wish to live. Thus while the computer and the "virtual" seem to be ever more pervasive, the role of the "community" as the crucible in which we can create a shared and meaningful future is, if anything, increasing in importance. And along with this goes the need and opportunity to enable those processes with whatever technical or human means that are available.

I have chosen to order the chapters into several sections. The "Introduction" is meant to provide a broad overview of Community Informatics (CI) as it might currently be understood and to suggest some terminology that might be useful in discussing CI research and practise.

The section on "Background and Issues" presents papers which give an introduction to certain of the issue areas with which CI is engaged, "access" (Clement and Shade), "environment and sustainable development" (Beale), and "community development" (Loader, Eagle and Hague). The next section presents various approaches to understanding and theoretically framing CI including from the perspective of socio-logical theory (Pigg), political theory (Baker), regional development (Simpson and Cawood) and empirically (Bruce). The section which follows presents the variety of experiences and approaches to community or civic networking from the U.S. perspective (Schuler), the perspective of the Community Technology Center Move-ment (Miller), Milan which is Europe's oldest civic network (de Cindio), a civic networking project in Argentina (Finquelevitch), a civic democracy networking project in Antwerp (Pierson) and a review of the experiences with telecentres in Europe and Africa (Falch).

The next section presents a variety of CI application areas including cybercafes (Stewart), cultural survival for Australian Aborigines (Turk and Trees), local govern-ment policy discussions (Ranerup), local tourism support (Agostini et al.), and university-community relations (Collins). The following section presents experiences in using CI approaches to support social and economic development in developing countries (Colle), a critique of recent efforts in this area (Fortier), the use by the United Nations and related non-governmental organizations (Lawrence and Brodman), in Los Angeles neighborhoods (Pitkin, Richman and Krouk), and in a severely disadvantaged outer suburb of Edinburgh, Scotland (Slack). The concluding section presents a number of more in-depth CI case studies including in a school management process in the USA (Halaska), in community-based planning efforts in Germany (Pipek), in creating an on-line news source in the former Yugoslavia (Hamman), and by national not-for-profit Internet service providers globally.

Acknowledgments

The chapters in this book draw from a wide variety of disciplines and reflect the academic and professional interests of a singular range of very talented people. I'd like to thank all of them for contributing to this volume and in this way helping to define Community Informatics as an area of research and of practise. On behalf of these authors and myself, I'll thank the range of research and other sources of funding of which the work in this volume is a product.

The bulk of my work in preparing this book was undertaken while I was most fortunate to have been appointed to a chair in the Management of Technological Change at the University College of Cape Breton funded by Enterprise Cape Breton Corporation, the Canadian Natural Sciences and Engineering Research Council and the Social Sciences and Humanities Research Council. Without the opportunities for research and reflection afforded by this research chair, my own work and the bringing together of the essays in this volume would not have been possible. I hope that each of these feels that by supporting, if indirectly (and perhaps unexpectedly) the development of this book they have made a useful contribution to the broader issues of the "management of technological change."

A number of people provided critical help at various stages of the project. An excellent team of reviewers, almost all of whom were also authors of the book, made an outstanding contribution. I'd also like to thank Stephen MacLean of Sydney, Cape Breton, who provided key administrative assistance and Bob Morgan of the University College of Cape Breton who encouraged me to pursue this enterprise. Jan Travers and Mehdi Khosrow-Pour of Idea Group have provided steady and patient support throughout the ups and downs of putting the volume together. Mostly I'd like to thank my wife without whose encouragement none of these endeavors would have come to a successful conclusion.

Michael Gurstein
Technical University of British Columbia
January 2000

Introduction

Community Informatics: Enabling Community Uses of Information and Communications Technology

Michael Gurstein
Technical University of British Columbia

There is an emerging need for all sectors of society to find ways to optimize the opportunities which information and communications technologies present. Research and development work in Information Systems and information technology has accepted a model of computing where the individual interacts directly with the computer and, through the computer and communication systems, with other individuals. Thus the objective of IT research and development has been to continuously enhance and extend the capabilities of **individuals** *within the context of the corporations, organizations, and governments for which they work. However, ICT also can be used to support* **communities** *in their efforts for social and economic development. Community informatics is a technology strategy or discipline which links economic and social development efforts at the community level with emerging opportunities in such areas as electronic commerce, community and civic networks and telecentres, electronic democracy and on-line participation, self-help and virtual health communities, advocacy, cultural enhancement, and others.*

As you read this chapter, the world is being transformed by information and communications technologies (ICTs). From scarcely a million users of the Internet in 1990, the current estimate is 125 million users and growing exponentially.[1] Where ICT-enabled commerce was unknown 10 years ago, it is estimated that $1.5 trillion annually of transactions will be undertaken via the Internet by 2005. Some tens of thousands of discrete Web sites are being created daily, and it is estimated that there are currently several billion discrete "pages" on the World Wide Web. Internet companies have surpassed in paper value entire conventional industries. Normally conservative commentators are arguing that Internet-enabled education will put in jeopardy the current tertiary educational systems of the world and have a transformative impact on all other levels of education.

Much of the research and development work in the area of ICTs has been focused on pushing the frontiers of the hardware or the software—to make it faster, smaller, cheaper, and more functional. The model implicit in this is of the individual directly interacting with the computer and, through the computer and communication system, with other individuals building "virtual" relationships in a "virtual world." IT research and development has been directed to continuously enhance and extend the capabilities of individuals working with these machines, and in this way enhance the activities of the corporations, organizations, or governments in which they work.

But many applications and application areas are not accommodated within this schema; for example, ICT[2]-enabled activity also can be focused on "physical communities" as well as on "virtual"[3] ones, and on those currently at risk of being excluded from participating in an ICT-enabled world and the opportunities which it presents, alongside the rather narrow demographics of current users. The technology juggernaut is moving forward, and increasingly, segments of society find themselves displaced or simply left behind as a consequence. "Community informatics" (CI)[4] is concerned with carving out a sphere and developing strategies for precisely those who are being excluded from this ongoing rush, and enabling these individuals and communities to take advantage of some of the opportunities which the technology is providing. It is also concerned with enhancing civil society and strengthening local communities for self-management and for environmental and economically sustainable development, ensuring that many who might otherwise be excluded are able to take advantage of the enormous opportunities the new technologies are presenting.

Community: Virtual and Physical

Community informatics pays attention to physical communities and the design and implementation of technologies and applications, which enhance and promote their objectives. CI begins with ICT, as providing resources and tools that communities and their members can use for local economic, cultural, and civic development, and community health and environmental initiatives among others.[5]

CI includes the technology/ICT and the "user" (and the "uses"),[6] and is as concerned with community processes, user access, and technology usability as it is with systems analysis and hardware or software design.[7] CI accounts for the design of the social system in which the technology is embedded as well as the technology system with which it interacts.[8] Thus CI is an extension from "organizations" to "communities" of the "socio-technical" approach to systems design, and reflects the increasingly ubiquitous distribution of personal computers and Internet access to communities and individual end users as well as corporations and governments.[9]

Discussion on the impact of ICT has, to date, concentrated on the "virtual" world—"virtual communities" and the "virtual relationships" which are, for example, enabling on-line electronic commerce (e-commerce). But for most people their "physical" relationships with their "physical" communities remain of more significance. Physical communities continue to provide the context in which people raise their families, educate their children, ensure their health and well-being, conduct their businesses, tend their surroundings, and influence the ongoing management of their civic affairs. Attention therefore needs to be paid to how communities, community affairs, and "civil society" in general are interpenetrated, enhanced, and enabled through the use of ICTs. These are areas of concern for CI.

Community Informatics

CI studies how ICT can help achieve a community's social, economic, political, or cultural goals. Fundamental to this is "access" to the technology, since without at least minimal access, little can be accomplished.

Clement and Regan[10] identify an "Access Rainbow" with seven discrete levels: governance and policy, literacy and social facilitation, service providers, content and services, software tools, devices, and carriage facilities. This includes "technical" (telephone connections and computers), "economic" (the cost of using and maintaining these systems), "social" (cultural, educational/literacy, and social barriers limiting use of the systems), and "physical" access (as for the physically disabled).

Also of interest is how to manage and situate the organization providing access in the community;[11] how to organize a technology context (institutional, organizational, training, etc.) which optimizes the use of the technology and the related opportunities;[12] and how public or community-access opportunities are to be linked into ongoing nontechnical service or other organizational structures as, for example, linking public access sites into existing public facilities in the local community.[13]

CI may also include distinctive software, hardware, and applications design (C-Suite, on-line voting, community Web sites);[14] specialized approaches to automated information processing and management;[15] the development of community-oriented ICT training, education, and organizational design;[16] or management approaches.[17] Insights on how communities are organized, pursue their common objectives, and manage themselves internally to develop and process information, and how they govern themselves, are all elements in a CI analysis and design of oriented ICT applications.[18]

History

The Internet was developed in the 1960s as a private network for facilitating communication within small scientific communities, particularly those engaged in defence-related research. Over several years these connections spread to link scientists from several disciplines and communities throughout the United States. From there, as graduate students looked to maintain this type of electronic contact, the network extended even further into the nonscientific community and in the early

1980s linked several thousand computers interacting as a single telephone-based network.[19]

Freenets were founded and maintained by volunteers—computer professionals, professors, and others—who made the Internet resources of the university available to community groups and members of the general public. The Cleveland Freenet, arguably the world's first, was created in 1986 at Case Western Reserve University. It offered free dial-up access to a university server for local community members who had a computer, a modem, and an inclination to connect to the Internet (or anyone else who did not mind paying long-distance calling charges).[20]

Freenets built on the experience of pre-Internet public computer networks. The Community Memory Project at the University of California at Berkeley, for instance, had installed a networked system of public-access computer terminals in laundromats and libraries in 1976. Through the 1980s, "basement" computer hobbyists set up their own dial-in Bulletin Board Systems (BBSs) accessible to anyone with a computer and a modem, hosting discussions on a range of topic areas.

Quickly, the Freenets began to build content on their servers by and for their members. When logging on to the Cleveland Freenet, for instance, you would see a (text-based) "map" of the city, with an option to "go" to the courthouse, library, or post office by typing a certain key. Community groups posted newsletters and events listings, or individuals posted classified ads. Freenet members could communicate with one other or with anyone else in the world with an e-mail address.

Freenets generally survived on volunteer labour and donations. As it grew, the Freenet movement came under pressure. Volunteers were burning out, since it was extremely labour intensive to maintain the growing computer networks. At the same time, university administrators and government officials began to take notice of the increasing human and capital resources at stake. Some Freenets have restructured and continue to grow and operate, maintaining a principled commitment to free public access to information and electronic networking. Others have evolved to "community networks," charging for service (often with a sliding rate scale), while often expanding from providing raw service to playing a development role in the community.[21]

Apart from semantics, the change from "Free" Net to "Community" Net has also meant a broader focus on community development issues as they relate to technology. Schuler argues that community networks can and do take action at any point where ICTs intersect with what he calls the "core values" of the community: education, culture, communication, democracy, individual health and well-being, and economic equity and opportunity.[22]

Most CNs offer dial-up access to the Internet. However, as commercial Internet service providers (ISPs) continue to prove they can offer high-speed, low-cost Internet access, some CNs have been happy to abandon server maintenance and the resale of dial-up accounts, which has proven to be bothersome for volunteer-based organizations. Newer CNs are also increasingly taking on the role of advocate for the broader interests of communities facing technological change.

Community Access

In Canada, telephone access has been almost universally available for decades. However, in the early stages of the Internet, accessing the Internet (or as it came to be termed, the "Information Highway") was a problem for many in rural and remote areas.

To respond to this, the Canadian Government in 1995 launched a *Community Access Program* (CAP)[23] with the objective of ensuring broad-based low-cost access to the Internet from even the most rural or remote locations. Within the first year of its operation, the program was overtaken by events, in that commercial ISPs found that they could profitably provide Internet service to almost all Canadians (including those in rural and remote areas) at competitive rates.

CAP then shifted from being an ISP of last resort to providing public access sites where the non-computerized public might gain access to the Internet. Where once CAP had been oriented towards providing *technical* access, the modified program and related initiatives have been slowly evolving toward broader issues of providing *social* access including to the un- and under-employed, to those lacking in computer and literacy skills, and to the physically disabled. CAP evolved and continues to develop in the direction of providing throughout the country a network of these sites.[24] Similar efforts elsewhere in the world go by other names, as for example CINs (community information networks under the Missouri Express Program),[25] in the UK under the name of telecottages or community resource centres;[26] or what are known in Africa and elsewhere in less developed countries as telecentres.[27]

In this context, certain technologies and technological approaches will be more supportive of community access than others. For example, some applications (e.g., videoconferencing) require Internet-access capability (bandwidth capacity) unavailable to all but the very wealthy or the institutional user. In other cases, the continuous upgrading of hardware and software to handle ever more elaborate processing requirements may be a major burden for communities with few resources for this type of ongoing expense. Similarly, technologies which require extensive maintenance or installation support may not be feasible for communities where such skills are unavailable or are very costly.

"Access," of course, is more than simply "technical" access. It includes how to provide widespread "public" access where human, financial, and technical resources may be scarce. This is currently an issue in many countries where the cost of access by individuals is prohibitive (the term "digital divide" is now being widely used in this context), or where there may be cultural or other reasons for having "community" access in addition to individual "in-home" access. Providing an appropriate physical facility for such "community" access is often an issue, as is managing and sustaining the institution or organization through which the access is being provided, organizing the facility to optimize the use of the technology and the opportunities which it provides, and linking this into ongoing local service delivery and other institutions. In Canada, for example, many of the community access sites have been in schools and libraries, while others have been in community centres, fire-halls, and even in commercial premises. The strategies for use and for ongoing sustainability of these centres are similarly varied.

A 'Community Informatics' Approach

(The following section is adapted from the paper "A 'Community Informatics' Approach to Health Care for Rural Africa" with Bruce Dienes, presented to *The Africa Telemedicine Project: CONFERENCE '99 "The Role of Low-Cost Technology for Improved Access to Public Health Care Programs Throughout Africa,"* Nairobi, Kenya, February 19-21, 1999.)

CI is an approach to ICT, which includes a concern for the accessibility of the hardware, the software, the connectivity, and the information; and for the use and user to which the technology is being applied, particularly within the context of the user's physical community. Incorporating the user and his community into the system design process introduces new elements and new "stakeholders" into an extended approach to ICT design, development, and implementation.

Access Facilities

How the user gets access to the technology is of particular interest. For many and particularly in less "connected" regions and countries, this will be through public or community-access facilities, i.e., telecentres, CAP sites, "cybercafes," etc. These centres, in addition to providing communications and small business support, also may become centres for the delivery of electronically mediated health, training, and public information services. This presents both responsibilities and opportunities— the responsibility to design activities so as to effectively provide these services, and the opportunity to help communities while (and not incidentally) developing sufficient revenues to ensure sustainability.

The Design of the Service

Central to the success of the activity will be the information or service being provided. There is a vast amount of information and services available on the Internet. However, relatively little of it is appropriate or useable in contexts where environmental conditions, resource scarcities, skill deficiencies, and cultural expectations and practices are different from those in wealthy areas or the developed countries and particularly the U.S. (the source of most of this information).

For services to be widely useful, information providers must design and provide services of specific interest to the end user, and particularly which takes into account the specific contexts of the various regions and cultural and linguistic groupings. What would be best is if information service providers were widely distributed and close to those using the information so as to localize information from other regions and to develop information of special interest within local contexts.

Design of the Telecentre

The telecentre as the site of community access is central to the impact of ICT in many local communities. The range, number, and distribution of telecentres will determine whether services are available to the few or to the many. The effectiveness of the centre will determine the effectiveness and success of the service delivery. The planning of the telecentres will need to be as nodes in a service delivery network. The design of the operations of the telecentre should be as an intermediary between the information (or service) provider and the information (or service) user. For this the telecentre will need a different model than currently, so that the Internet-enabled service is designed to be provided through the *telecentre* rather than directly to *individuals* as is the conventional approach.

The physical and organizational design of the telecentre should also reflect its likely use as a service facility and should include among others:

- connectivity;
- a paraprofessional staff with on-line and information management skills;
- translation facilities for the key languages served in the community;
- multiple uses—education, extension, small business support, communications;
- sources of revenue to ensure sustainability; and
- links to an established physical institution.

Design of the Community System

There is a tendency to think that the mere presence of electronic resources will meet the requirements of a "community" without the need for further intervention or leadership. The design of the community system into which the ICT-enabled service or information will be transmitted will be particularly important for a CI project working through public access telecentres.

In many contexts, the on-line information or service may be provided to **groups** rather than to **individuals**. In some cases these groups will already have been formed; in other cases they may need to be formed specifically to take advantage of the opportunities presented. Thus, for example, a prenatal group might be developed which meets regularly to obtain advice on nutrition and related matters by means of the Internet. The group would identify its information needs, the request or search would be undertaken by a paraprofessional trained for such activities, and he or she would in turn pass the information back to the group where it would be assimilated/processed/applied.

It will, in this context, be necessary to look at the entire service process as a system, including the information or service provider/designer, the paraprofessional intermediary, the professional, and the group or community information user/recipient.[28] Effective planning and development for all stages of the process will be needed for the service activity to be successful. It may also be desirable to establish a process of information-sharing between **groups** with similar concerns, as a parallel to the useful and beneficial interactive communication processes which have developed between **individuals.**

On-Line Service Delivery

CI approaches will be central to using ICTs linked into community, institutional, and social systems for delivering community-based services through telecentres and networked PCs.

Among the services that are currently being provided through the Internet are:

- Information
- Education and training
- Mentoring and consultation
- Self-diagnosis/self-monitoring

- Transaction processing

Some examples of these kinds of resources in the health services area include Med Help International, a virtual medical center for patients, and the Global Health Network, available in eight languages.[29] Similar such resources (although perhaps not as extensive or sophisticated) are available for the variety of other services being made available for community support.

Information as a CI Service

Useful and useable information is at the core of ICT-enabled services. For this information to be retrievable, understandable, and relevant to the consumer there needs to be "mediating structures." These link the electronic "service" with the end user. For example, a technologically trained paraprofessional would translate the needs of a community support group into appropriate Internet search criteria, and then sift, interpret, and translate the returned information and put it into a form that is useable by the "client" community.

A useful model of an on-line community information resource can be found in the Community Toolbox[30] whose mission is to "promote community health and development by connecting people, ideas, and resources." The site includes on-line training for developing community leadership, an index of health resources, both physical and virtual, local and national, and on-line discussion and chat to facilitate users to share information or to organize community health initiatives.

The Internet is uniquely equipped to be a low-cost, high volume, very fast deliverer of this type of information. (It is clear that one major role that an organized network of telecentres could play is to create and maintain index Web sites and search engines that meet the specific needs of information consumers in specific regions.)

On-Line Support

"On-line support" is the mechanism whereby individuals provide information, comfort, and mutual assistance through the medium of the Internet. This can be done by e-mail, newsgroups, or Web conferences (asynchronous), or by chat (synchronous), although in most cases it is done asynchronously.

Support of this type is now available for a variety of conditions, circumstances, and activities; has become one of the most active "application" areas for the Internet; and is a crucial component of most successful e-commerce developments. In CI applications, on-line support has led to real-life meetings and the development of formal face-to-face organizations. In other cases, it has led to the development of information repositories in the form of archives or Web-sites. It has even led to the development of advocacy movements and campaigns.

The challenge is to ensure that the considerable benefits available to those participating in these on-line processes can also be made available in the public access/telecentre context as well. To achieve this it will be necessary to develop a means to identify and implement innovative on-line practices and evolving technologically enabled service support opportunities.[31]

The Technology

The technologies of interest from a CI perspective are evolving very quickly as with all areas of ICTs. Communities were first introduced to the value of using ICT with the availability of low cost and easy-to-operate personal computers and particularly the early Apple Computers which targeted community (and educational) users. Community organizations, as others, became adept at using word processing, spreadsheets, and databases to support the variety of their activities, closely following developments in other sectors including the SME and the public sector.

With the development of the Internet, however, a new horizon of opportunities has arisen for using ICT in support of community goals. As communities in both developed and less developed countries have come increasingly under threat from the cultural power of the media and the economic power of large and centralized corporations, ICT can assist them in reasserting their significance in contemporary social, economic, and political life.[32]

At least in theory, ICT can help to overcome the isolation and the sense of being separated from the mainstream which many communities in rural and even in some urban environments experience. ICT is coming to be seen as a potential "leveler," allowing those previously made marginal to participate more actively and effectively.

Hardware

Among the hardware issues of concern from a CI perspective are the overall cost of hardware purchase, maintenance, and replacement. The current minimum requirement for individual access to the Internet is a personal computer (PC) with a modem and a telephone line. While the cost of this equipment is declining rapidly, the machine power required to handle normal applications is continuously increasing, as version after version of software with additional resource requirements ("bloatware") comes into general use. This has led to an extraordinary cycle of computer obsolescence/necessary replacement while the technology itself is still in perfectly good running order. Though the hardware has been declining in cost, the pace of change and increase in functionality has been even faster, so that there is a continuous need for financial expenditures for technology upgrading.

Equally wastefully, PCs that may be used only for particular applications sit idle for much of the day. Developing ways of sharing access to the hardware, whether through telecentres, other community facilities, or through low cost commercial services (cybercafes), is one response to this; another possibility is the development of hardware solutions which allow users to run certain applications on otherwise obsolete (and thus very low cost) equipment or to share certain facilities through networked systems or through the sharing of applications software.[33]

Software

While the cost of ICT hardware continues to fall in a dramatic fashion, the cost of software, perhaps because it is in a much less competitive environment, has not kept pace. The cost of applications and operating system software (which is often built into the purchase price of the machine) adds significantly to the cost of ICT access. As with the hardware, the continuous development of new or more elaborate (and frequently not required) functionality and upgrades for existing software has driven the need for continuous replacement. Not incidentally, this has added significantly to ICT operating (installation and maintenance) costs while creating the risk of "orphaned" hardware and data as the ability to work with information in different formats becomes lost over time.

For operating systems and an increasing number of applications, an alternative to "commercial" software is emerging. This not only provides access to functional and robust software at affordable prices (in some cases "free"). Its development is also providing a broad (and alternative) model of appropriate and effective organizational structuring for information-intensive activities and for communities which wish to increase their level of information absorption and utilization. This model, often referred to as "open source," is derived in part from the "hacker" culture which provided much of the software for the "Freenets" and is based on Linux, an alternative operating system to the dominant Microsoft Windows.[34]

The "open source" model is an approach to software development based on the free distribution of the essential codes around which the software is written, and the development of an informal, Internet-mediated community of interest around the "debugging" activity which is supporting the improvement of this software. The result is an extension of functionality for the benefit of the "virtual community" working on the software and general low-cost software distribution to the user community at large.

The software developed using this approach is rapidly gaining market-share as it has been found to be technically superior to the competing products. Similar models of software development are being undertaken in a variety of application areas. Of particular interest from a CI perspective is that the resulting software is distributed free to users, with the only cost being in the time required for installation and maintenance.

Connectivity

"Connectivity" is the term used to describe the link between the local computer and the other computers and networks which comprise the Internet. Connectivity is available by means of conventional telephone lines at various speeds. As with the other areas discussed above, the applications being developed for the Internet are such as to require increasing speeds of access/connectivity for their effective operation. In most cases, the cost of connectivity increases with the speed of the connectivity being provided, since additional hardware (to the conventional telephone line) may be required to achieve the faster connectivity speeds. In addition to access by conventional telephone lines (twisted pair copper wires), connectivity is increasingly and competitively being provided by fibre optic cable, coaxial cable (cable television), satellite, and even cellular telephone. As with other ICT components, the quantity (speed of access) has been continuously increasing while the cost of access has been in more or less continuous decline in most parts of the world.

From a CI perspective the applications of most interest are those requiring either minimal or maximum bandwidth; for example, e-mail requires the least amount of connectivity. However, applications such as Telehealth and ICT-enabled distance learning often require very fast connections with a very large throughput sufficient to support voice, video, and graphics.

Bulletin Boards

Among the first of the ICT technologies to attract the attention of the nonspecialist was the "bulletin board." The bulletin board system (BBS) is a dial-in electronic space which could store transmitted electronic messages. Individuals could dial-into the board to retrieve the range of messages placed there, including those that might have been specifically left for them. These types of systems also are known as "store and forwards" in that messages could be sent to the BBS for storage and then forwarded on to their intended recipients or even to other BBSs thus forming a primitive electronic network. Communities were interested in BBSs since they could provide a common focal point for maintaining—in electronic and reasonably accessible form—local information (databases) concerning such matters as events, volunteer availability, organizational capacity, and so on.

With the development of the Internet as a kind of network of these localized BBSs, the individual BBS receded into disuse and obsolescence, except in areas where the Internet was unavailable or where access was prohibitively expensive, such as parts of Africa. GreenNet, a store-and-forward BBS, still allows for dial-up message forwarding for certain parts of Africa where long-distance access to the Internet is the only available form of e-mail and where the relatively rapid message download from the electronic BBS is a much cheaper, if more cumbersome, alternative.

Community Networks

In several instances community BBSs which began as individual (and hobbyist) efforts evolved into larger organizations with a broader array of activities and objectives. In a few instances these linked into community organizations and became the basis for community information networks. These CNs provided Internet access and particularly dial-up Internet accounts to the larger community when such access was restricted, as initially, to those with an affiliation to a university or, as later, when such access was both costly and restricted to major centres. The evolution from BBS to CN/local ISP often was only through the common thread of a shared interest in electronics and communications technologies.[35]

A particularly interesting example of this is provided by the Chebucto Community Network in Halifax, Nova Scotia, which evolved from a local BBS into an active community network and local voluntary organization. Chebucto combined links to the university, the local hobbyist community, and to the voluntary sector. What is particularly interesting about Chebucto is that they were sufficiently technologically adept to develop software which transformed the major elements of a BBS into the backbone for supporting a CN (particularly the methods for handling local information and voluntary participation).[36]

UseNet, News, and E-Lists

In addition to e-mail access, CNs typically provided discussion lists (Usenet or news lists) dealing with topics of local interest generally including discussions of local civic government and larger political issues. In some instances, they were used by community groups to maintain contact with their members or to recruit new members. These lists provide a significant and more sophisticated service to that provided by the BBS and have proven to be a means for a local community to take advantage of the ease of use, asynchronous communications capacity, and information storage and retrieval facility of on-line communication to pursue their objectives.[37]

The World Wide Web

Closely linked to the evolution of community networks and to other types of community-oriented public access has been the development of "community Web sites." Initially, these sites were meant to be "places" where local communities could present themselves to themselves and to the world, and their development was linked into the ongoing framework of "local access." In communities with voluntary CNs, these often took responsibility for the development of the initial community Web site. In other communities, universities with the skills and the interest took this upon themselves.[38] In still other communities where access was provided by a local private sector ISP, they would undertake to develop the site either as a contribution to the community or as a commercial venture, renting out "space" or advertising on the site. In many instances, the development of the local Web site was seen as a way of promoting the local economy and particularly local enterprises.[39]

As the Net and particularly the technology of the World Wide Web has evolved, however, this approach has rather gone out of fashion as search engines and common entry points ("portals") into the electronic jungle of the Web have become the centre of attention. Static versions of community Web billboards apparently have not been effective in generating interest in local communities, particularly since it is difficult to capture the attention of the casual Web "surfer." The serious user of the Web looking for particular products or locations for settlement or investment is going to be interested in the more sophisticated sites which would differentiate themselves from their thousands of electronic competitors.

Community sites, sometimes with local government or chamber of commerce or other voluntary organization sponsorship, are now very common, with similar formats presenting local statistics, tourist highlights, links to local organizations and businesses, and occasionally forums or discussion groups for local residents or those with an interest in the local area. Parallel to this is the more recent phenomenon of well-capitalized commercial operations looking to become the common entry point or "portal" for particular communities, especially larger and wealthier ones. Often these are associated with local newspapers or even with newspaper chains with local outlets, who are concerned to protect their role as providers of local information and particularly as the primary outlets for local display and classified advertising.

This marketplace is in something of a flux, with major and very well-capitalized efforts like Microsoft's effort going out of business. Franchises [40] have been springing up and looking to achieve a critical mass quickly and are competing with a large number of more or less well-financed local efforts[41] also vying in the marketplace. How this will evolve is not yet clear. An early strategy of attempting to become the dominant local site by making arrangements with local government to put local ordinances and civic information up on the Web seems not to have been effective. Alternative approaches such as putting funds into local news gathering, specifically for Web distribution, is very costly except for local newspapers who are able to re-purpose existing news-gathering facilities.

One approach which seems to be having some success is to attempt to become the most popular forum and chat group for a local area—capturing the attention and involvement of those with an ongoing interest in using the Net for discussing local issues as well as nonresidents with an interest in the local area (e.g., possible in-migrants, those with a family connection, and so on).

Perhaps the most interesting and comprehensive development in this area is the "Local Ireland" project undertaken by the Irish consulting firm NUA in conjunction with Irish Telecom and with funds from the Government of Ireland. This is a plan to develop sites including forums and chat lines on a county-by-county basis, and to link this in with the very considerable international interest among Irish expatriates and Irish descendents in exploring their Irish roots. There is an attempt as well to link this with local tourism and local economic development.[42]

Band Width—Broad Band

A key concern in many communities is "the size of the pipe" which is entering into the community. This refers to the "bandwidth" or carrying capacity of the telecommunications connection into the community through which access to the Internet and to other electronic services must pass. The understanding is that if the "pipe" is too small, the community will be unable to access the range of services (such as full motion video, videoconferencing, and very large file transfer for graphics or animation) on which the future economic well-being of the community is assumed to depend, as well as providing opportunities for such value-added services as on-line training, telemedicine services, and video on demand.

A major telecommunications supplier has even developed an integrated "community networking" package with training, facilitation, and community development components to assist communities in acquiring, developing, and managing the range of applications which would be available in a "fat pipe"/broad bandwidth context. [43]

Geographic Information Systems (GIS)

A final technology of CI interest is GIS or geographic information systems, which is a way of developing integrated databases linked to maps. GIS provides quite sophisticated tools for mapping local geographical or other (environmental, for example) information so as to display, manipulate, and manage this as a support for local self-management. This kind of graphical display of information, which previously had only been accessible to professionals, puts a new and empowering tool into the hands of communities.[44]

Community Informatics Applications

CI is concerned with a wide range of ICT activities and applications, but particularly those that link the technology with community interests and objectives.

Community Internet Access
"Public" access to ICT is being made available within communities through a variety of government and not-for-profit community access sites, telecentres, and civic networks; and through for-profit cybercafes and Internet-enabled telephone centres among others.[45] Initially a number of CNs were established as a means to provide Internet access and particularly e-mail accounts to those unable to acquire them in other ways. However, with the very rapid proliferation of low-cost e-mail and even free e-mail accounts by companies such as HotMail, this role for CNs has diminished very significantly. This has had a negative impact on a number of North American CNs and forced them to rethink their strategies and objectives, including those towards becoming the local providers/managers of "public Internet access."[46]

Community Information
The provision of "community information" includes a range of information of local interest such as local listings, directories, a local calendar/schedule of events, and so on. In many cases, social service entitlement and referral information (as in earlier non-computerized library-supported community information centres) and public health information is provided.[47] In some cases this is being done through the local CN which, on a voluntary basis, is maintaining a community database, now generally in the form of a community Web site. In other instances, this is being done on a competitive and commercial basis, often by the local newspaper which is providing free listings for voluntary organizations, and even in some cases by civic governments in order to build traffic to the site (and thus potential or actual advertising revenues).[48]

Civic/Community Participation Online
ICT is being used to enhance civic and civil society participation through nonpartisan electronic democracy projects, through party-sponsored civic forums, and through government-sponsored public consultation initiatives.[49] Such "electronic democracy" is providing to citizens the initial means to obtain and circulate information of local political/civic interest. In addition, it is able to support electronic interaction with local officials and politicians concerning this information and issues of local concern in those few instances where public officials are willing to enter into this type of interaction. Ultimately the on-line expression of public opinion will be linked in some manner directly to the formal discussion of ongoing local issues, even to the extent of formal electronic "voting" on issues and direct participation in decision making.[50] A major challenge for CI is developing the means to manage this type of interaction, both enabling an enriched electronically enhanced democracy while ensuring that such an interaction does not disintegrate into democracy by plebiscite or by instant public opinion polling.[51]

Community Service Delivery Online

ICT is being used as a means for providing public services, including information and registration concerning entitlements and certification, health information and counseling, and employment information and small business support (including mentoring).[52] The direct provision of services to individuals in their local communities or at home through ICT is only in its infancy. However, one can expect that this will grow dramatically in the very near future as it comes to be realized how cost effective this approach may be in a number of sectors and particularly in those areas which are highly information intensive such as information provision, training, registration/licensing, and so on.[53]

Community E-Commerce

Both commercial and noncommercial agencies are making efforts to ensure that some of the opportunities emerging through electronic commerce are being made available to geographic communities, (alongside virtual communities), as for example through e-malls, community Web sites, links between SMEs and on-line commerce and others.[54] A number of initiatives are currently underway to find ways of linking local commercial and production activities with the facilities offered by e-commerce, both directly to retail and trade purchasers and to suppliers.[55] An interesting example developing in several rural Atlantic Canadian communities is linking community access sites to genealogy research, part of which may now be done best through the Net and part of which may involve local (and locally hired) researchers.

Education/Training/Community Learning Networks

A major and rapidly emerging CI application area is in education, training, or lifelong learning. Increasing areas of education and training are being provided on-line, including the on-line distribution of course material in text, oral, and even video format along with an asynchronous or synchronous interactive component through e-lists or forums or chat facilities. The medium is still in its early stages and techniques and curricula incorporating some of the unique opportunities which the technology affords are just now being developed. Also the methods for linking on-line facilities with training and lifelong learning needs, and with existing community organizational and institutional structures, are currently being created along with the means to link this into ongoing and emerging opportunities for on-the-job training in a variety of work contexts.[56]

Community and Regional Planning

Much more sophisticated community involvement in local land use and environmental planning than had previously been possible is resulting from the application of GIS technology.[57] New developments in GIS include four-dimensional models that track and project impact of management strategies over *time*, predicting, for example, the impact on forests of various forest management strategies, or the impact on business and communities of various development plans.[58]

Telework

ICTs may support local economies by allowing for work to be done remotely from the workplace, or "telework." Decentralized computing linked to a communications capacity as, for example, through the Internet or dedicated data-lines, allows work to be done from any connected remote location. Skills and training can be provided remotely. The playing field for technology-enabled education, training, and other work-related services is being leveled. Some have suggested that technology would allow for certain enterprises or their employees to be located anywhere so long as they were tele-connected. This notion has, at least to date, proven more vision than reality, however, with the difficulties of organizing remote production, remote management, and the overall conservatism of organizations in how they undertake their activities inhibiting a significant uptake of these opportunities, at least as yet.[59]

A CI Model for an Integrated Service Delivery System

ICT gives the promise that information-intensive services and applications, which had previously been limited or inaccessible to many because of physical isolation, disability, or the costs associated with access, will now be much more widely available. CI is concerned with how to design electronically enabled "services" or applications so that they are as widely available and useable as possible.

In the above we have begun to discuss CI-enabled services and application delivery systems. A preliminary model of this type of service delivery system would include:

- *A community-based technical capacity to receive services and information* in a manner which can be made accessible and disseminated to a wider population, as for example through the network of telecentres that are being established throughout many rural and urban areas.

These telecentres will then need to have a minimum level of connectivity and processing power and the technical skill for ongoing management and maintenance.

- *A social/organizational capacity to receive and redistribute services and information.*

This will require that the telecentres have the local capacity to identify service and information needs that can be provided electronically. Additionally they will need to have the skills to access and process the information and services as they are made available. They will also (and crucially) need the means to interpret and translate that information and those services from and to the local populations, who in turn may be from a variety of linguistic and cultural groups and with widely varying levels of literacy and education. There will also be a need for the capacity to copy and distribute the information in paper form and orally through presentation to community meetings.

- *An organizational capacity* to localize external information of use to specific services and, where necessary, to compile and develop its own information resources.

This capacity would probably be best achieved though a semi-independent centre linked to, but organizationally distinct from, an institution such as a university or library (ensuring that the services and information are accurate and reliable but not so esoteric that lay people cannot understand them).

There will need to be a capacity to design and develop information that is useful to and useable by a highly varied population with limited education and literacy skills and a range of cultural and linguistic backgrounds. This information need not be redesigned specifically for each subpopulation, but it must be presented in a manner such that it can be translated and redistributed by local information paraprofessional intermediaries as well as professionals in certain instances. A means will also be needed to identify what information or service can and would be of most local benefit given the limited resources available for these activities.

- *A technical capacity to mount and deliver this information* via the various modalities of the Net, including and particularly e-mail, the WWW, and chat.

The service provider should be capable of keeping abreast of ongoing technology development and to undertake limited research and development activity in support of their work. This will ensure that the activity moves both at a pace appropriate to the expanding capacity of the overall delivery system as well as to its users. Particularly in the context of multimedia, the Web is moving towards streaming audio and video as a major delivery vehicle. Similar increases in delivery capacity are likely to continue (satellite service delivery is one reason) and thus the opportunity to integrate audio and video into ongoing programming may present itself sooner rather than later and any program developed should make accommodation for this.

- *A human capacity to mobilize resources* in this sector, including leadership and vision.

Many communities lack the skill to bring together the resources that might be available with the need, in a sector as new and as rapidly evolving as this one. Traditional leadership in communities may be ill-equipped, feel themselves to be inadequate, or lack the information to move forward with ICT, and those with the skills may only be interested in pursuing individualized efforts. Thus communities need access to appropriate leadership and leadership training to assist in pursuing technology-enabled opportunities as they emerge.

- *A social and organizational capacity to utilize and implement the services and information* which are being provided through the facility.

An organized development of groups to work through and with the telecentres and the telecentre paraprofessional worker, and to receive and implement the information and service in their own lives, and that of their families and communities, will be needed. These groups will need to be planned for and developed as part of the overall program, since without these groups the program will have little resonance or effect in the communities to which the programs are being provided.

- *An overall program development and management capability* to integrate and expedite the various elements of the models, to ensure funding as required, and to ensure that the various components of the system work together smoothly. The program development capacity will also need to establish service delivery priorities to ensure that efforts are directed towards areas where the most cost-effective intervention can be obtained.

- *Linkages between the formal system and the informal system:*
  - At the initial stage of identifying priority areas, done in full knowledge of the possible cost avoidance benefits of investment in the telecentre rather than the more expensive professional service.
  - At the information and service design level, to ensure the reliability, appropriateness, and effectiveness of the information and service being provided.
  - At the telecentre, where it is likely that a cohabitation will occur between professionals and the paraprofessional and the community organizations. It is likely that the telecentre will be used for professional purposes in addition to CI uses. There could be significant synergies between the two as, for example, in the joint use of technology for the paraprofessional, technical support for both activities, and so on. Thus, protocols for appropriately handling the relationship between the two types of service will need to be developed.
  - Ensuring an appropriate level of integration between information measures and measures to support individual, family, and community service use, assisting in the local assessment of information usefulness and reliability.

- *A means to ensure evaluation and feedback* from the service and information back to the information source and development group. Assistance in undertaking evaluation and feedback procedures will be desirable.

Features of a CI Application

This section will begin to identify those elements or features of a CI strategy which would distinguish it from other types of applications. The key characteristic of a CI application, of course, is that it is meant to support community objectives and to be useable by community actors or agents. This is what gives a CI application its identifying characteristics.

a) *Ease of use*: A CI application will be designed so as to optimize its ease of use, including installation and ongoing maintenance, where possible by the community or end user. While the "back end" of the application may be complex (they often seem to be based on UNIX/LINUX architecture which is famously not user-friendly), the "front-end" should be transparently easy to use and to implement.

b) *Training*: A CI application will necessarily include a significant training component, not as an afterthought or as an "add-on," but as an integral part of the system. The training should be directed to allow for the nonprofessional user to manage the system being operated.

c) *Mediation*: A CI application will likely be designed so as to accommodate a "mediator" between the application or service provider and the end user, since the end use of the service is meant to be available to the general community member and not to specific or specialized users.

d) *Applied technical system*: A CI application is for the most part identified by what it does, not how it operates. Its support for community health, for example, is assessed by its effectiveness in contributing to community health, not by the elegance of its software design. The particular way in which the contribution is made or its technical features (for example, design elegance) will, from a CI perspective, be of lesser interest or relevance.

e) *Governance structures*: The manner in which a CI application is managed or "governed" will be distinctive in that particular attention will be paid to supporting community processes and community decision making. Thus the manner in which access is provided, the management of the information provision activity, and the broad service system which is developed incorporating the use of ICT, are all of significance in implementing a CI application.

f) *Sustainability*: Key to CI is the ongoing sustainability of the various applications and services. How public access facilities, community networks, community Web sites, or other CI applications can be sustained after a transition from volunteer management and/or public sector support is key to a CI success. Thus identifying an appropriate "business model" for an ICT-enabled service is of crucial importance.

Issue Areas

CI is an emerging "discipline," but it is also a creative and practical response on the part of communities to the difficulties which they and their members are facing in the modern world. In practice, CI is one of the ways in which communities are responding to the variety of crises which they are facing with the decline of local economies, the loss of the means for local decision making, the cutting back (in most developed societies) of the "social safety net," and the imposition of "cost discipline" on the range of public services.

Issues being addressed in the context of CI include:

- Public and community access to the range of opportunities which ICT is presenting, given the "rainbow" of access limitations facing many communities, including physical, training costs, and other limitations on universal or even general access.
- Public and community access to the opportunities and rewards which are emerging through electronic commerce. As increasing elements of economic activity migrate to the electronic sphere, there is the danger that only those with access to the technology, and particularly to sophisticated knowledge and skills, will be able to take advantage of e-commerce, inevitably resulting in an

increase in the social barriers to economic advance.

- Community initiatives to resolve "digital divide" issues. The "digital divide" refers to the division between those who have access to and are capable of taking advantage of the digital sphere and those who, because of financial status, education, skill levels, and so on, are not. Indications are that the width of this divide may be increasing with relatively fewer people on the positive and more on the negative side. Related to this is an increasing danger of economic polarization both within a society and between societies.
- Linking community "values" and content on the Internet. Of particular interest and ongoing concern to "physical" communities in the context of the Net is interposing their fundamental concerns and values into the virtual space of the Net in such areas as pornography, "hate" speech, and promoting or providing tools for violence. Issues of information control or censorship, particularly for children, are among the most contentious currently affecting the Net. These arise precisely because the Net exists not only in virtual space, but that at each juncture and for each user it transects physical or domestic space. How this can be managed and "domesticated" from a "community" perspective is one of the most pressing issues affecting the Net.
- The internationalization/globalization of the Internet and the services and products whose production and distribution are enabled by the Net. There appears to be a declining role for the state in maintaining control over the technology and for ensuring a degree of equity in the distribution of technology's benefits (as, for example, resulting from a decline in the capacity for taxing transactions). There is a further danger from uncontrolled access to information and manipulation of information, with the attendant risks to personal privacy and financial integrity in the absence of effective means of Internet regulation and supervision.
- "Who owns" the information, both in terms of copyright/distribution permissions and of authentication.
- The various political/cultural/religious biases or "agendas" which underlie or are perceived to underlie the information being provided. (How this is handled, and the widespread legitimacy and neutrality of the information being provided, may be a determinant of the longer term usefulness and acceptability of the service.)

Conclusion

To date the firestorm of development and implementation of ICTs, and particularly the Internet and the World Wide Web, has been linked almost exclusively to a limited range of activities in the information and research intensive sectors, certain highly entrepreneurial technology intensive enterprises, and to those with a strong interest in and knowledge of information and information technologies. Other sectors including the not-for-profit and the public sectors have lagged behind the leading edge of development. Also many without a specific interest linked to information technology or information-intensive activities—particularly those in manual and resource-intensive employment, those with difficulties with literacy, certain ethnic groups, those with lower incomes, and so on — all have been slower to embrace the technology.

The pace of change and the opportunities and risks which are being presented are of such potential significance that efforts need to be made to ensure as broad a base of participation as possible. As an example, as considerable segments of certain aspects of economic activity are transferred to the on-line sphere, if those who are at risk from these changes are not "on-board," their economic and social position will be jeopardized; if elements of the society are unaware or unable to access Internet-delivered training opportunities in a timely and effective fashion, then their relative position within the society could further erode, and likewise for informal or even formal participation in the political sphere. If access and use of these technologies is effectively available only to those who are already advantaged, then the degree of disadvantage of the already disadvantaged will increase.

The reality is that many of those who are being left behind in the technology race are those most economically and socially vulnerable. While the better educated and more advantaged are able to derive considerable benefit from their participation in a "virtual world" as accessed via ICT and to shift a variety of their activities and commitments to this sphere, those less advantaged or simply less able to effectively participate (such as the elderly, the physically disabled, the geographically isolated, as well as the economically marginal) may not have the means to individually undertake the same shift. They may thus find themselves further disadvantaged and at risk whether from unemployment, lack of education opportunities, or lack of political power where these have effectively shifted to the on-line mode.

There is thus the need to pay attention to the means by which those at risk of being left behind may be provided with support to overcome these limitations. This would appear to be a role for the "community sector" and a mission for CI.

Alongside this, CI can be supportive in ensuring that there is an ongoing place for "physical" communities within the context of the "virtual" cyber-spatial world of electronic commerce, virtual communities, and virtual service delivery. While all of these necessarily have a physical component, the tendency appears to be to ignore the physical community and the needs of those who rely on the physical community for their economic and social well-being, while moving to reinforce and transfer resources to the "virtual." Responding to these pressures is a further role for CI and a CI approach.

I would like to thank those who have read and commented on this chapter including my ICT-4-LED/MBA class and Dr. Bruce Dienes.

References

Aikens, G. S. (1997, April). *American Democracy and Computer-Mediated Communication: A Case Study in Minnesota.* Unpublished PhD thesis. Cambridge University. <http://aikens.org/phd>.

Bibby, A. (1995). *Teleworking: Thirteen Journeys to the Future of Work.* London:

Calouste Gulbenkian Foundation.

Clement, A., & Regan, L. (1996, June). "What do we mean by 'universal access'? Social perspectives in a Canadian context." Paper presented at INET96: The Internet–Transforming Our Society Now. Montreal. [Online]. <http://www.crim.ca/inet96/papers/f2/f2_1.htm>. [1999, February 9].

Day, P. and Harris, K. (1997). *Down-to-Earth Visions: Community Based IT Initiatives and Social Inclusion.* International Business Machine Corporation (UK).

Dienes, B. (1997). *WiNS '97 Final Report.* [Report]. Sydney, NS: Centre for Community and Enterprise Networking, University College of Cape Breton.

Dienes, B. and Gurstein, M. (1998). "Remote Management of a Province-wide Summer Employment Program Using Internet/Intranet Technologies". *Annals of Cases on Technology Applications and Management in Organizations,* 1, Hershey, PA: Idea Group Publishing.

Gurstein, M., & Andrews, K. (1996, Oct.). *Wire Nova Scotia (WiNS): Final Report.* [Report]. Sydney, NS: Centre for Community and Enterprise Networking, University College of Cape Breton.

Gurstein, M., & Dienes, B. (1999, Feb.). "A 'Community Informatics' Approach to Health Care for Rural Africa." Presented to The Africa Telemedicine Project: CONFERENCE '99 "The Role of Low-Cost Technology for Improved Access to Public Health Care Programs Throughout Africa," Nairobi, Kenya, February 19-21.

Gurstein, M., & Dienes, B. (1998, June). "Community enterprise networks: Partnerships for local economic development." Paper presented at the Libraries as Leaders in Community Economic Development conference, Victoria, BC. [Online]. <http://ccen.uccb.ns.ca/flexnet/CENs.html>.

Gurstein, M., Lerner, S., & MacKay, M. (1996, November 15). *The Initial WiNS Round: Added Value and Lessons Learned.* [Report]. Sydney, NS: Centre for Community and Enterprise Networking, University College of Cape Breton.

Gurstein, M. (1999, forthcoming). *Burying Coal: The Centre for Community and Enterprise Networking and Local Economic Development.* Vancouver, BC: Collective Press.

Gurstein, M. (Ed.). (1999, forthcoming). *Community Informatics: Enabling Communities with Information and Communications Technologies.* Hershey, PA: Idea Group Publishing.

Gurstein, M. (1999a). "Fiddlers on the Wire: Music, Electronic Commerce and Local Economic Development on a Virtual Cape Breton Island," C. T. Romm and F. Sudweeks (Eds.), *Doing Business on the Internet: Opportunities and Pitfalls.* Berlin: Springer Verlag [Excerpted in Internet Intelligence Bulletin, (UK) Issue 74, April, 1999].

Gurstein, M. (1999b, Feb.). "Flexible Networking, Information and Communications Technology and Local Economic Development." First Monday. <http://firstmonday.dk/issues/issue4_2/index.htm>.

Gurstein, M. (1999, forthcoming). *The Net Works: The Internet, Local Economic Development and the Future of Work in a Global Economy.*

Gurstein, M. (1998a). "Information and Communications Technologies and Local

Economic Development". In G. MacIntyre (Ed.), *A Roundtable on Community Economic Development*. Sydney, NS: University College of Cape Breton Press.

Gurstein, M. (1998b, May 31). *Applying the Concept of "Flexible Networks" to Community Access Computing*. [Online] <http://ccen.uccb.ns.ca/flexnets>.

Gurstein, M. (1996, June 24-26). "Managing Technology for Community Economic Development in a Non-Metropolitan Environment". International Conference on Technology Management: University/Industry/Government Collaboration." Istanbul, Turkey. UNIG: UNESCO.

Hagel, J., and Armstrong, A.G., (1997) *Net Gain: Expanding Markets Through Virtual Communities*, Boston, MA: Harvard Business School Press.

Kanfer, A.G. (1994, Oct). "The role of the Web in connecting a community: The development of the Champaign County Network (CCNet)". Paper presented at the Second International World Wide Web Conference, Chicago, IL.

Landry, J. (1997). "Negotiating the Forum of Electronic Public Space: The Battle Among Community Computer Networks Constituency Groups." Unpublished Master's thesis. Montreal: Concordia University.

Leiner, B.M., Cerf, V.G., Clark, D.D., Kahn, R.E., Kleinrock, L., Lynch, D.C., Postel, J., Roberts, L.B., Wolff, S. (1998, Feb), "A Brief History of the Internet" for the Internet Society (ISOC), <http://www.isoc.org/internet/history/brief.html>.

National Working Party on Social Inclusion (INSINC). (1997). *The Net Result: Social Inclusion in the Information Society* (Report), IBM (UK).

Pigg, K. (1999, April) "Community Networks and Community Development." Paper presented at the International Association for Community Development Conference, Edinburgh, Scotland. <http://www.ssu.missouri.edu/faculty/kpigg/IACD99.html>.

Pigg, K. (1998, May). *Missouri Express: Program Implementation Assessment.* [Report]. University of Missouri.

Piore, M. & Sable, C. (1984). *The Second Industrial Divide: Possibilities for Prosperity*. New York: Basic Books.

Raymond, E.S. (1998, Nov.). "The Cathedral and the Bazaar". <http://www.tuxedo.org/~esr/writings/cathedral-bazaar/cathedral-bazaar.html>.

Rheingold, H. (1993). *Virtual Community: Homesteading on the Electronic Frontier*. Reading, MA: Addison-Wesley.

Rice, R. E. (1993). "Using Network Concepts to Clarify Sources and Mechanisms of Social Influence." W. Richards Jr. and G. Barnett (Eds.), *Progress in Communication Sciences*, vol. 12 (pp. 43-62). Norwood, NJ: Ablex.

Rice, R. E., and Aydin, C. (1991, June). "Attitudes toward New Organizational Technology: Network Proximity as a Mechanism for Social Information Processing." *Administrative Science Quarterly*, 36, 219-244.

Rice, R. E. (1987). "Communication Technologies, Human Communication Networks and Social Structure in the Information Society." In J. Schement and L. Livrou (Eds.), *Competing Visions and Complex Realities: Social Aspects of the Information Society* (pp. 107-120). Norwood, NJ: Ablex.

Romm, C. T. and Pliskin, N. (1997a). "Electronic Mail as a Potent Coalition Building Information Technology." *The ACM Transactions on Information Systems* (TOIS).

Schuler, D. (1996). *New Community Networks: Wired for Change*. Reading, MA:

Addison-Wesley.

Schwartz, E. (1996). *NetActivism: How Citizens Use the Internet*. O'Reilly & Associates.

Scott, M., Diamond, A., & Smith, B. (1997). *Opportunities for Communities: Public Access to Networked IT*. Canberra, Australia: Department of Social Security.

Sproull, R., and Kiesler, S. (Eds.). (1991). *Connections: New Ways of Working in the Network*. Cambridge, MA: MIT Press.

Waterloo-Wellington Information Network for Employment and Training. (1998). WWINet Operating Committee: Terms of reference. [Online] http://wwinet.hrdc-drhc.gc.ca/wwinet/termref.html [1998, May 20].

Winner, L. (1992). "Silicon Valley Mystery House". In Sorkin, M. (Ed.), *Variations on a Theme Park: The New American City and the End of Public Space*. New York: Noonday Press.

Zuboff, S. (1988) "The Panopticon and the Social Text". In *Age of the Smart Machine*. New York: Basic Books.

Endnotes

1 The most generally available and widely accepted compilation of Internet user statistics is that provided by the Irish consulting firm Nua, <http://www.nua.ie>.

2 The term ICT is often used to cover a similar area as is IT or IS in North America, with the inclusion of "communications" technologies and particularly the Internet.

3 Much of the discussion concerning "communities" in the context of IT/ICT in recent times has been concerned with "virtual communities" rather than with "physical" communities. In this paper we are concerned exclusively with the use of IT/ICT by *physical* communities. Cf. Rheingold (1993) and Hagel and Armstrong (1997) for discussions of *virtual* communities and their use of IT/ICT.

4 Others who use the term include the Community Informatics Research and Applications Unit (CIRA) at the University of Teesside, Middlesborough, UK, <http://wheelie.tees.ac.uk/cira/index.htm>. See also the use of the term "Social Informatics" at the Center for Social Informatics at Indiana University <http://www-slis.lib.indiana.edu/SI/>.

5 "Community Informatics" draws heavily from the on-going practical and research work linked to the area of "community networks" and "community networking", cf. Schuler (1996).

6 This approach is often associated with those adopting what is called a Participatory Design (PD) Methodology, and concerned with the study of the Computer Human Interface (CHI) and with development of approaches to Computer Supported Collaborative Work (CSCW). The literature on all of the above is very large and can be easily explored on the World Wide Web by employing any available search.

7 This area among others is discussed at some length in the various chapters in Gurstein (Ed.) (1999). *Community Informatics: Enabling Communities with Information and Communications Technologies*, Hershey, PA: Idea Group Publishing.

8 Cf. Pigg, 1999, April.

9 This approach is often linked with Enid Mumford and the ETHICS Method-ology (see Mumford, E. [1996] *Systems Design: Ethical Tools for Ethical Change.* Macmillan).

10 Clement & Shade, 1996, June. See also The Net Result: Social Inclusion in the Information Society, Report of the National Working Party on Social Inclusion (INSINC), IBM (UK), 1997 (Chapter 4).

11 Gurstein, 1998a.

12 Gurstein & Dienes, 1999, Feb.

13 Pigg, (1998, May 15) and Gurstein, & Andrews, (1996, October 31).

14 See for example <http://www.chebucto.ns.ca/CSuite/AboutCSuite.html>, which discusses the Chebucto Suite software for community networks.

15 One of the more interesting areas in which this has developed is in the community use of GIS software as a support for community-based planning.

16 Gurstein, M., Lerner, S., & MacKay, M. (1996, November 15). Gurstein, & Andrews (1996, October 31). Dienes, B. (1997).

17 Dienes & Gurstein, 1998.

18 Gurstein. (1996). "Managing Technology for Community Economic Devel-opment in a Non-Metropolitan Environment" International Conference on Technol-ogy Management: University/Industry/Government Collaboration" June 24-26, 1996, Istanbul, Turkey UNIG: UNESCO.

19 Leiner, Cerf, Clark, Kahn, Kleinrock, Lynch, Postel, Roberts, & Wolff, S. 1998, Feb.

20 Cf. Schuler (1996). Kanfer (1994).

21 Cf. A very interesting discussion of "communities" and "networks" by Phil Agre, "Rethinking Networks and Communities in a Networked Society," Red Rock Eater Service, June 1, 1999, via e-mail.

22 Cf. Schuler op. cit.

23 There is extensive information available on this through the WWW; cf. <<http://cap.unb.ca/>> and links.

24 The Canadian Government has continued with a number of additional programs designed to stimulate broad access to and use of the Internet as through Volnet ,for voluntary organizations <http://www.volnet.org>, a program to provide an equivalent level of access to urban communities to that being made available to rural ones through CAP, and most recently what is being called the "Smart Communities Initiative" <http://smartcommunities.ic.gc.ca>.

25 Cf. Pigg , 1998, May.

26 Day & Harris, 1997.

27 There is a very extensive set of documentation on telecentres, particularly in Africa, available through the International Development Research Centre's ACACIA project, < http://www.idrc.ca/acacia/>.

28 In a private communication, Liz Rykert, one of the more experienced practitioners in on-line group facilitation argued that the notion of the "mediator" between the information and the end user was precisely the kind of mediation which many currently receiving public services, as for example social support services or employment counseling services, might wish to avoid and that the "disintermediating"

effects of ICT was precisely one of the most beneficial results for many of those on low or fixed incomes. While agreeing in general with her point in certain instances, the need for "mediation" will remain for many given the complexities of the technology, the complexities of the information being provided, or the sheer lack of individual access which many will continue to have for the foreseeable future.

29 <http://www.medhelp.org/> and <http://www.pitt.edu/HOME/GHNet/GHNet.html>.

30 <http://ctb.lsi.ukans.edu/>.

31 See Gurstein & Dienes, 1998, <http://ccen.uccb.ns.ca >, on the benefits of combining telecentres with existing organizational structures, such as a library system and see overall the approach undertaken by C\CEN as presented through the Web site.

32 Cf. for example, Gurstein & Dienes, 1998, June.

33 The Canadian Government through its Community Access Program is currently exploring approaches to this type of solution, which if it is successful should be a benefit to the educational, library, and other "public" ICT-accessed community worldwide. Cf. <http://www.pc.ibm.com/us/products/netfinity/3500tour/>.

34 For very useful discussions of this, see the work of Eric Raymond and particularly his essay, "The Cathedral and the Bazaar," available on his Web site <http://www.tuxedo.org/~esr/>.

35 Leiner et al., 1998, Feb.

36 <http://www.chebucto.ns.ca/>, In this see also the work of the The National Center for Supercomputing Applications University of Illinois at Urbana-Champaign; Kanfer, 1994, Oct.

37 One of the more interesting and well-documented instances of this is Minnesota E-Democracy, cf. Aikens, G., 1997, April.

38 A very interesting, if slightly dated example of this can be found in the following project developed by the University of Illinois—Chicago <http://www.uic.edu/~schorsch/goals.html>, whose goals were:

a. *Provide Internet Access, Electronic Mail Addresses, Communication Software, High Speed Modems to 50 Community Organizations, and Subsidized Access for an Additional 100 Nonprofit Organizations.* The UIC Academic Computer Center will provide remote dial-up access for the core 50 community organizations. The subsidized Internet access for the additional 100 nonprofit users will be negotiated with private local suppliers of remote dial-up Internet access.

b. *Provide Telecommunications Training and On-line Information Education.* The curriculum will focus on the basics of effective telecommunications and on adding value to on-line information through collaboration and application to user needs. Training will include lecture-based demonstrations on the concepts and potential of the system and hands-on training to teach the specific skills required. ITRC's regular offerings in computer literacy and application training will also be available. All appropriate training documentation will be developed and provided by UIC's Academic Computer Center and the ITRC.

c. *Promote Innovative Interaction between Information "Haves" and*

"Have Nots" and Cross Cultural Communications. Outreach and education to community organizations will emphasize the need for providing information content while encouraging faculty, staff, and community development professionals to interact with on-line community-based information in their research, writing, and planning. Further, the early findings that electronic networks help "level the playing field" suggests that information interaction between the Mexican-American, African-American, and academic communities may lead to broader community development coalitions organized around information interaction. In turn, cross-cultural information interaction has the potential to raise a community's understanding of itself, its relationship to the outside world, and the outside world's understanding of Chicago's under-represented communities.

d. *Seed the Development of Youth-Oriented Electronic Grassroots Think Tanks.* In two youth-oriented, community-based computer education labs—one in both adjacent neighborhoods—multimedia computers will be provided to allow for hyperlinked multimedia (Netscape) access to the Internet. Students will be encouraged to develop World Wide Web content relevant to on-line discussions regarding community development.

e. *Provide Gopher and File Transfer Protocol (FTP) Clearinghouse Services for the UIC Neighborhood Initiative Program.* A UIC gopher server and FTP site will be established to provide on-line access to relevant documents, resources, and scheduling data to facilitate information dissemination between UIC, community organizations, nonprofits, and donors. Each program area will have its own FTP directory and gopher menu subsystem.

f. *Provide Gopher-Based Pointers to Other Internet Information Relevant to Housing, Economic Development and Jobs, Community Education and Health, and Nonprofit Management.*

g. *Provide a Searchable On-line Version of the Donors Forum's Philanthropic Database.*

h. *Provide Wide Area Information Services (WAIS) for Selected Documents on Chicago Area Empowerment Zone/Enterprising Communities, Affordable Housing, Community Planning, and Nonprofit Management.* WAIS capabilities allow for full-text searches. This service will facilitate quick research on Chicago-area community and nonprofit development.

i. *Establish two Listservs:* UIC Neighborhood Initiatives Community Partnership and Nonprofit and Donor Community Listservs. Listservs will be used for frequently updated notices and lists such as calendar events, government and media listings, request for proposals and awards, jobs and internships, and technical assistance needs and availability.

j. *Establish three Online Conference Forums*: UIC Neighborhood Initiatives Community Partnership, Nonprofit and Donor Community, and Network Evaluation and Comment Conferences. Conference forums will be provided for ongoing, issue-oriented dialogues. This allows for long-term online community building around critical thinking about public policy and opinion, program development, management and training issues, collaboration, and outreach. A conference dedicated to the evaluation of this network provides a forum for daily feedback on this project. It also allows project staff to respond to user needs and concerns in a timely manner. Further, conference forum archives will help new users become quickly familiar with the nature of this on-line community.

39 For a useful if slightly dated discussion on Community Networks and Community Web-sites see Kanfer, A.G. "Ghost Towns in Cyberspace", 1997, <http://www.ncsa.uiuc.edu/edu/trg/ghosttown/>.

40 Cf <http://www.neighborhoodlink.com/> and <http://www.mytownnet.com/>.

41 E.g., <http://www.newyork.com>.

42 <http://www.local.ie/> Another instance of the creative use of a Web portal to support local community objectives is the Cape Breton Music site <http://cbmusic.com>. For a background discussion of this site and its links to the local community and to a virtual community of Cape Breton music lovers, see <http://ccen.uccb.ns.ca/articles/E-CommPaper.html>.

43 This is how Nortel describes their "Integrated Community Network (ICN)." There have been hundreds of community information networks, which started out as Freenets and have failed, because there was no business rationale supporting the concept of free. As we often say in ICN "free is not sustainable." Integrated community networks cannot possibly duplicate existing services, because ICNs are not physical networks. ICN is a vision and a process that enables a community to understand the implications of a networked society for changed behavior and changed spending in order for them to succeed in this new economy. The ICN process allows collaborative coalitions of community leaders to better understand the role of network connectivity as the fundamental enabling resource that allows them to act on their new understandings. Such changed action usually leads to a much higher priority of spending to support connectivity applications. ICN also enables network providers in existing communities to partner with their existing customers to understand that the manner in which they currently offer telecommunication services needs to change, because it typically supports an industrial rather than an information age economic model. Those network providers that understand this fundamental difference will thrive, and those that don't will see their customers go elsewhere. Those network providers that do participate in ICN initiatives invariably make significant changes in how their services are priced, packaged, and delivered.

Concerning a specific ICN being implemented in Tasmania, Australia, Building a People Network, workshops continue to be staged statewide, marketing materials to inform and persuade the evolution of a true "people network." Priority applications are being developed and implemented, and the infrastructure and networking issues are, in turn, evolving to meet the everyday needs. Several public access programs have been utilizing the foundational strategies of the Tasmanian Community Network and Networking Tasmania, a government services WAN. Service Tasmania's one-stop shops for all government transactions and services are extremely popular. Within the next two years, over 60 centers will bring communities online. The Computers in Schools program and Access to Learning program are bringing together youth and aged people, with students teaching elders Internet skills, and the entire community enabling a job-ready workforce. Remote sites such as King Island and Flinders Islands are preparing for infrastructure upgrades. Community and business development strategies have led to over 2,000 new jobs created through six call centers in 18 months. The government's Industry Development program helps businesses engage in information technology. The Electronic Showcase tells the world about the quality products and services that can be obtained here in Tasmania. Business and professional services have also emerged. Remote diagnostics, administrative support and training are among those telehealth applications moving to pilot. The Electronic Commerce Center is training for awareness and generating funding. LIST (Land Information System Tasmania) integrates over 300 types of information relating to land, so that investors, real estate professionals, emergency personnel and the general public can obtain information from one place. Project Tiger builds on LIST and is about layering information concerning mineral deposits and mining sites. EnAct allows anyone wanting to know about the latest legislative developments in Tasmania to access free but important and relevant information. <http://www1.nortelnetworks.com/bbprods/icn/QandA/>.

44 See for example the use of GIS by the City of Seattle to assist home owners in locating their property, <http://www.ci.seattle.wa.us/maps/>.

45 For additional resources on "access" issues, see <http://www.benton.org> and <http://www.fis.utoronto.ca/research/iprp/ua/>.

46 This issue has been and continues to be very extensively discussed on the e-mail list Communet-L, which functions as the internal communications vehicle for much of the Community Networking "movement" in the United States.

47 The provision of public access has in some instances flowed out of these community information centres, as for example in Australia. See <http://www.vicnet.net.au/vicnet/contents.htm>.

48 See for example the range of cities identified through <http://www.citylink.com>.

49 Schwartz, (1996) and http://www.e-democracy.org/mn-politics-archive/. which is the "classic" electronic democracy experiment.

50 Although this is not yet happening it is being widely discussed as for example Kaczmarczyk, A. Perspectives of Cyberdemocracy http://www.imm.org.pl/mat/AcbdemA.html and the Dutch Ministry of the Interior and Kingdom Relations, *Electronic Civic Consultation: A Guide to the Use of the Internet in Interactive Policy Making,"* nd.

51 Cf Gurstein, M. "A Cathedral" of Public Policy to a Public Policy "Bazaar", <http://ccen.uccb.ns.ca/articles>.

52 Gurstein and Dienes, (1999).

53 An interesting example of this can be found at the very extensive Web site created by the Canadian Department of Industry, <http://www.strategis.ic.ca>.

54 Gurstein, (1999b), (1998b).

55 See for example the Government of Canada's Mainstreet program, the World Bank's Virtual Souk, the U.S.-based WebMarket, and similar initiatives elsewhere.

56 See for example <http://www.handsnet.org/information1241/information.htm>. A very interesting experiment is underway in rural Nova Scotia where a commercial ICT-enabled training company (MacKenzie College) is col-

laborating with the local telephone company (MTT) and local community access sites to provide the means for local access to sophisticated training software.

57 Bruce Dienes in a private communication provided me with several interesting examples of this from East St. Louis, Illinois. "The UIUC—University of Illinois, Urbana/Champagne, prepared data that supported a change in the proposed path of a public transit light rail system. They also use it to track land usage, match potential building projects with flood plains, etc. The use of GIS requires a university or similar resource to crunch the data and prepare the presentation, but the final product is a nice balance between being visually easy to comprehend (so the local citizens can interpret it and make their own decisions) but also steeped in statistics and data (so the local Planning Board will take notice of it.)."

58 Dienes, private communication.

59 Telework has become a major theme of Government activity in support of the information society and particularly in Europe and with the European Union; cf. <http://www.eto.org.uk/>.

Community Informatics:

Background and Issues

Chapter I

The Access Rainbow: Conceptualizing Universal Access to the Information/ Communications Infrastructure

Andrew Clement
University of Toronto, Canada

Leslie Regan Shade
University of Ottawa, Canada

This chapter presents the Access Rainbow, a seven-layer conceptual model of access to the information/communications infrastructure intended to strengthen public interest perspectives in current 'information highway' policy discussions. In particular, it aims to provide the basis for a workable definition of 'universal access' and point to concrete steps for achieving this ideal.

A prime impetus for the discussion of universal access is the rapid incorporation of digital networking into the information/communication infrastructure (ICI) for conducting a widening range of social, economic, educational, and political activities. While the benefits are mixed, powerful forces are propelling the shift toward network-based transactions. Acknowledging that in many cases narratives about the advent of network technologies are simplistic, deterministic and mythical in scope (Mosco, 1998), this work assumes that digital networks will continue to be increasingly central to daily life and anticipates a time when they are regarded as a mundane, but vital part of the social infrastructure. We consider how the promise of network services may be achieved in a socially equitable and productive fashion. We seek to develop and apply a pragmatic model of access to the ICI which respects and embraces public interest perspectives. We discuss several issues central to defining access, present a seven-layered model that addresses the key requirements of an access definition, and, finally, show how the model has been applied in policy discussions in Canada.

Defining Access

Defining access to the ICI is difficult for several reasons. While access is consistently identified as a key principle in policy discussions, it is not an end in itself. Access simply enables further activities that can only partially be specified beforehand. There are three main questions to address: 1) Access for what purposes?; 2) Access for whom? and 3) Access to what?

Access for what purposes?

In broad terms, enhancing the ICI holds the promise of enabling all citizens to participate more fully in all aspects of economic, social, cultural and democratic life (Clement, 1998; Karim, et al., 1998; Schon, et al., 1998; Dutton, 1999). A central notion is the possibility of participative interaction with others. In contrast with existing electronic media, digital networks allow people to be creators as well as recipients. In many situations, computer-based information and communications technologies (ICTs) offer significant advantages over conventional media for accessing, creating, exchanging and sharing information in the conduct of daily affairs, thus benefitting the social individual in each of his or her major roles as consumer, producer, caregiver and citizen. There are myriad possibilities: exchanging gossip, buying goods, checking the weather forecast, making a living offering information services, playing games, learning a new language, building a community resource file, assessing medical treatments for an ailing parent, contributing to civic debates, and so on. This list could be very long, and ultimately the purposes of access can never fully be defined because citizens should be free to invent their own uses and hence find new value in the infrastructure.

Access for whom?

A brief answer to this question is easy—access for all, at least for all citizens who need and wish to make use of the ICI. To the extent that network services are valuable, no one should be excluded from the opportunity of participating in their advantages. Furthermore, the benefits for everyone expand as more people become reachable through the network. As Borenstein notes, "The utility of [digital] networks appears to rise exponentially with the number of interconnected users" (1998, p. 6).

However, not all citizens are alike, and we need to recognize the diversity of people and their particular access needs. In part this involves recognizing the obstacles to access that are characteristic of various 'populations,' most notably age, gender, income, education, disability, language, ethnicity, geographic location (urban vs. rural and remote), and nationality (developed vs. developing countries) (Castells, 1996; Golding, 1996; Ebo, 1998; Loader, 1998; Shade, 1998). Not

surprisingly, there are clear indications that various aspects of the information/
communications infrastructure are less accessible to individuals already disadvantaged according to these characteristics. While the development of a major new area
of social infrastructure provides an opportunity to redress long-standing disparities,
unless these differences are taken into account during this crucial formative period,
there is a serious risk that the future society will be more widely divided along these
distressingly familiar lines. As Hirschkop (1996, p. 93) warns, unless all people have
access to a range of network services, "instead of extending access to the community,
the new technology will install a new form of communication apartheid."

At the same time it also true that not everyone will find network access
advantageous all the time. For a variety of reasons, such as to find solitude or simply
avoid being overwhelmed with e-mail, people may not want to use network services.
Promoting universal access is not equated with making it imperative that everyone
be "plugged in" as much as possible. The overriding objective is to ensure that
everyone enjoys a range of communicative options suited to their particular life
circumstances.

Access to what?

Until recently, models of information/communications infrastructure emphasized the purely technical aspects, notably physical connectivity. In order to define
more fully what access to the ICI encompasses, and to account for the intricate
interplay of its social and technical aspects, a broader model highlighting multiple
dimensions of access needs to be delineated. While broadening the focus on access
beyond narrow digital connectivity issues makes the architecture more complicated,
it also makes it easier to take fuller advantage of long-standing media practices and
policy understandings for providing socially desirable and universally accessible
services.

In spite of rapid convergence to binary digital representations for many overtly
distinct information forms, new networked media will not soon make obsolete the
existing sophisticated infrastructure based on paper and electronic media. While the
Internet is currently the focus of most of the attention, it is important to see it and
other data communications developments as additional components of the wider
infrastructure with a long history of evolution. Undoubtedly, digital networks will
displace many activities currently conducted via highly evolved conventional media
such as the postal service, document publishing and distribution systems, telephone
services, TV and radio broadcasting, but are unlikely to render any of them
completely redundant. Rather, digital networks will be increasingly interwoven with
these other networks in complex and mutually redefining ways. Thus, any access
model needs to address the full range of conventional and new media.

At a time when political and economic support for many public media is
diminishing (e.g., postal and public library services, and, in Canada, national public
broadcasting), it is important to challenge prevailing market-oriented rhetorics
which undermine the principle of universality of services benefitting everyone.
Citizens should not face a degradation of existing services until they have superior
alternatives to choose from. Since the new technologies offer greater capabilities at
lower costs, significant benefits can be achieved without anyone being worse off.

Those who promote new services which potentially undermine existing widely used services should bear the onus for ensuring that current users will not be disadvantaged.

Canada has a unique history of ensuring national communication links (postal, telecommunications, broadcasting, and satellite) through federal government intervention (including subsidies, content quotas, and principles of universality). With the development of the ICI, however, the government has opted to let market-led forces and industry initiatives predominate. Amidst an environment characterized by competition, telecommunications deregulation, and increasing cuts to social services, issues of social cohesion, national and cultural sovereignty and notions of citizenship remain crucial. This leads to the question of what services should be considered essential to the maintenance of good societal functioning (Reddick, 1998). Ensuring these services are available may then require specific protections or promotion via collective public initiatives. An access model should assist in identifying where and how public action is required.

In summary, a model of access needs to:
- include support for a multiplicity of usage roles involving creation and dissemination as well as retrieval of existing information;
- address the full range of possible users and the diversity of their life situations;
- recognize the interplay of social and technical dimensions in the development of infrastructure;
- encompass both conventional and new media;
- highlight "access gaps," areas of social need likely to be "left out" by market forces acting alone; and
- help identify what services should be considered "essential."

The Access Rainbow: a Socio-Technical Architecture

In light of the various considerations identified above, we developed an integrated model for analysing and discussing access to network services (Clement and Shade, 1996). A key feature of this model is that it illustrates the multifaceted nature of the concept of access. Inspired by the layered models used for network protocols, the lower layers emphasize the conventional technical aspects. These have been complemented with additional upper layers emphasizing the more explicitly social dimensions. The main constitutive element is the service/content layer in the middle, since this is where the actual utility is most direct. However, all the other layers are related to each other and are necessary in order to accomplish proper content/service access (see Figure 1 Access Rainbow).

The layers also reflect the important regulatory separation of *carriage* and *content*. This distinction has a long history, going back to the regulation of railroads, but remains relevant as we witness the rapid growth of huge, vertically integrated, global media conglomerates that threaten to impose monopolistic or oligopolistic control over vital societal infrastructure. Keeping the layers distinct but interoperable

Figure 1.

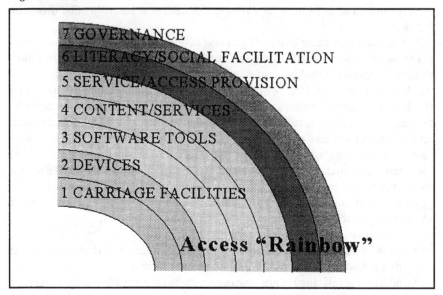

7 GOVERNANCE
6 LITERACY/SOCIAL FACILITATION
5 SERVICE/ACCESS PROVISION
4 CONTENT/SERVICES
3 SOFTWARE TOOLS
2 DEVICES
1 CARRIAGE FACILITIES

Access "Rainbow"

to allow competition within sectors is seen by some as a key step in providing consumer-oriented service (Sirois, Forget, 1995).

We have adopted the rainbow metaphor because it embodies several concepts central to public interest ideals. It simultaneously suggests unity and diversity. Although the coloured layers can be distinguished from each other, there are no definitive boundaries between them. They are intrinsically related to one another and integral to the whole. No single strand is sufficient; all are necessary. There are, of course, limitations to such a metaphor. The broad arch of a rainbow as well as the spectrum of colours usually conveys harmony and inclusiveness, but in the case of the ICI this is a romantic ideal. The current situation is just the opposite, fraught with contradictions and enduring tensions. The rainbow thus ironically directs our attention to the many difficult challenges we face in achieving the ideal of an inclusive and supportive societal infrastructure.

We begin our exploration of the model with the overview found in Table 1 : Access Rainbow Overview Table. For each layer of the model, we provide:
- a brief working definition;
- a description of its key features, including public interest desiderata;
- illustrative examples; and
- current candidates for designation as 'essential' for achieving universality, with emphasis on those aspects most likely to require further public initiatives.

The following discussion revisits each of the seven layers in turn, expanding the description by highlighting likely gaps and obstacles to achieving universal accessibility and identifying key policy questions. While the model is generally applicable to conventional and digital media, the latter are highlighted due to space

Table 1: Access Rainbow Overview Table

Layer	Description	Examples	Essential aspects
7. Governance How decisions are made concerning the development and operation of the infrastructure.	Information/communication infrastructure (ICI) development implicates a very wide range of stakeholders who are differentially placed in terms of their ability to contribute effectively to the decision-making process. The central challenge of governance is to foster a democratic process that allows all the stakeholders to become informed of the issues and participate equitably in choosing among alternatives.	Legislation, regulations and regulatory bodies (e.g. CRTC, FCC), public enquiries, policy task forces, advocacy campaigns, on-line forums, local civic bodies, 'access' boards/councils, markets	Public consultation process Research and social impact assessments New institutions (e.g. national and local "Access Councils") Conception of the "Electronic Commons"
6. Literacy/ Social facilitation The skills people need to take full advantage of ICTs, together with the learning facilitation and resources to acquire these skills.	ICTs are complex and still immature technologies requiring a range of skills to use effectively, especially when creating new content. Acquiring these skills is largely a social process involving a combination of formal and informal methods within the context of supportive learning environments. The means for acquiring networking skills need to be affordable, readily available, attuned to the learners' varied life situations and sensitive to language, cultural and gender differences.	Keyboarding skills, knowledge of diverse software packages, World Wide Web navigation and search strategies, WWW site design, creation and updating, database and spreadsheet manipulation, systems purchasing and configuration, troubleshooting, effectiveness assessment	Basic literacy, numeracy, media 'savvy' Computer literacy (keyboarding, web navigation) "Local experts" (i.e. in workplace or neighbourhood)
5. Service/access provision The organizations that provide network services and access to users.	Most users gain network access through their employers or educational institution providing a wide range of necessary access services. Even individual subscribers need affordable, ongoing relations with network service organization(s) capable of quickly responding to varied needs.	Employers, educational institutions, Internet Service Providers (ISPs), telcos, cablecos, community nets, libraries, schools and other public facilities, community organizations.	Local public access point (e.g. schools libraries, hospitals clinics, day care centres, post offices, community centres) Community networks
4. Content/ Services The actual information and communications services people find useful.	The central role of the ICI is to facilitate access to a wide range of information and communication services that people find valuable in their daily lives as citizens, producers, consumers and caregivers. They need to be able to interact meaningfully with others, obtain useful information easily without being overwhelmed and contribute creatively to the store of available information. Content and services should be affordable, reliable, usable, diverse (culturally/linguistically/politically), secure, privacy enhancing, individually filterable, and free of censorship.	Telephone enhancements (e.g. 911, call answering, caller ID), radio/television programming, electronic mail, newsgroups, the World Wide Web, databases	Electronic mail Newsgroups Emergency services World Wide Web databases.(e.g. environmental, job banks, health) Government information, library holdings, political process, civic/local events
3. Software tools The programs that operate the devices and make connections to services.	Software is the critical new ingredient that extends the conventional ICI to include digital networking. These tools are under rapid development and are being embedded in a growing range of devices. They need to be affordable, multilingual, interoperable, privacy enhancing and easy for everyone to use.	Browsers, e-mailers, search engines, authoring and editing tools, distribution list servers, groupware, operating systems, wordprocessing, multimedia, spreadsheet and database packages.	Web browsers, e-mailers and authoring tools that are usable by everyone across a range of technical platforms, natural languages and human (dis)abilities. They should include encryption and other privacy enablers.
2. Devices The actual physical devices that people operate	Contrary to the general trend of 'convergence' seen in carriage media, we are witnessing a proliferation of device forms, with a widening mix of capabilities, prices and sizes. Wireless connections are increasingly a feature of these devices. They need to be affordable, not rapidly obsolete, easy to use (especially by those people with disabilities), interoperable, and close at hand to where needed.	Telephone terminal equipment, TV and radio receivers, pagers, modems, cable modems, set-top boxes, Net PCs, Web TVs, kiosks, printers, scanners, workstations, PDAs	Workstation Set-top box Public kiosks Universal Design
1. Carriage The facilities that store, serve or carry information.	There are several types of digital information infrastructures developing, with the Internet being the most prominent. Networks previously based on analogue transmission media are also being converted to digital. Carriage media need to be widely available, affordable and interoperable.	Telephone (ISDN, xDSL) and cable transmission, radio/television broadcast, Internet, WWW server space. Satellite and other wireless transmission.	Telephone: affordable single party service, digital dial tone (e.g. DSL) phone number portability Cable (with modem?) Internet connection locally

limitations. Given the broad scope of the model, the details provided are necessarily suggestive rather than comprehensive.

Carriage Facilities

The facilities that store, serve or carry information are undergoing rapid change with the expansion of digital techniques. The Internet Protocol is clearly emerging as the dominant standard. As the pace of technological development accelerates, digital networking is being extended into areas previously based on analogue transmission media, such as telephony and radio and television broadcast. Accompanying digitalization is the growth of very high bandwidth media such as fibre-optic cable and wireless transmissions in the gigahertz range. The growing importance of networked repositories means that such server-space (e.g., for WWW service, distribution lists, and e-mail store and forward) also needs to be considered as part of the carriage layer.

While these technical advances in carriage facilities promise greater capacities, wider availability and reduced unit costs, there are well-founded concerns that some people, notably those with low incomes or living in rural or remote areas, will actually be worse off. The most prominent access gap results from the expense of providing terrestrial-link carriage services to those living in rural and remote areas. Currently in rural Canada there are still several hundred thousand party lines which are therefore incapable of computer or even fax usage. With the erosion of regulatory mechanisms to subsidize service to these areas from urban revenues, there is a real prospect that penetration rates may actually drop, even for conventional telephone services. Satellite-based telephony may invert the economics of distance thus making remote areas relatively cheap to serve. However, so far these satellite services are generally aimed at high-income markets. Income disparity is an issue even in areas that are relatively cheap to serve. Some telephone and cable companies have begun withdrawing investment from inner-city areas since they do not anticipate sufficient economic returns from their low-income inhabitants. Prices could also rise as these companies consolidate their hold in particular markets and return to monopolistic practices following the bout of competition and mergers triggered by the U.S. Telecommunications Act of 1996.

The growing popularity of data-intensive applications (e.g., graphics and especially video) means that the current widely available analogue standard is becoming obsolete, while adopting higher speed digital services (e.g., via ADSL or cable modems) incurs a significant monthly premium beyond the reach of many subscribers. This raises the question of what constitutes the minimum "essential" bandwidth and hence what should be guaranteed in some way. Given that most useful digital services are still text-based and therefore not dependent on broadband capacity, and that upgrading local infrastructures would incur enormous costs, it appears premature to declare high-speed access (e.g., >56K baud) as "essential." However, this will likely change over time.

Policy questions: What funding mechanisms (e.g., a universal access fund) should supplement or replace traditional internal cross subsidization to ensure universal carriage service? Should carriage providers be required to meet public accessibility criteria to operate? Are penetration rates suitable measures of universality, or should it also be based on proportion of income required to access basic services? Who will ensure the interoperability of networks? What is the minimum "essential" bandwidth?

Devices

Many access and display devices already exist, and more are being readied for market in the near future. Contrary to the trend of "convergence" seen in carriage media, we are witnessing a proliferation of devices, with a widening mix of capabilities, prices and sizes. Wireless connections are increasingly common and there is a trend to mobile, hand-held networked devices (Chapman, 1998).

The main issues from an access point of view are the high cost of current workstations and their rapid rate of obsolescence. The annual ownership costs are thus much higher than comparable equipment costs for other major electronic media (radio, TV, telephone). The emergence of cheaper Net PCs or Web TVs together with rising economies of scale will only partially ease the situation as long as the rate of technological change remains very high. Again it is those with lower incomes who are at greatest risk.

Another key issue is usability. Many devices are difficult to use and not readily adaptable to individual differences, especially for people with disabilities. Principles of "universal design," which aims for accessibility by everyone, including those who are disabled in various ways, are only just beginning to be followed and need to be adopted more widely (Shneiderman, 1999). As Perry, et al. (1997) note, "Designed access is preferable to retrofitted access."

Where people cannot afford their own equipment, there should be public access facilities close at hand that are suited to a wide range of users. This will be especially challenging away from urban centres.

Policy questions: Are the ICT devices located close to where people need them? Are they affordable? Are they easy to use, especially for people with disabilities? Are they adaptable to human variations? Are they interoperable? Are the devices designed to be augmented with technical refinements, or do new designs render old designs obsolete?

Software Tools

Software is the critical new technical ingredient which expands the capabilities of the ICI. Until recently, software tools have only been related to computers, but they are now becoming incorporated into many common appliances. While there are attempts to make software easier to use, the widening range of products and features adds to their complexity and usability difficulties. Even GUI (graphical user interface) interfaces using icons, pull-down windows, and other non-textual elements can present additional problems for the visually impaired. Monopolistic practices in the software industry, as revealed in the Microsoft antitrust trial, contribute obstacles to the smooth interoperability of software packages from competing vendors. All these factors make software more expensive to buy and learning to use it effectively a major challenge. Even more than with devices, there needs to be greater adoption of "universal design" principles and the participatory design practices involving diverse user communities (Greenbaum and Kyng, 1991). Furthermore, software is largely written with English speakers in mind, thus disadvantaging other language groups. The general lack of encryption and other privacy protection features (e.g., in Web browsers) represents another inhibition to the widespread usefulness of software.

Policy questions: Are major software tools easy for everyone to learn and use? Are they affordable, interoperable, privacy protective and platform independent? How can a wider range of user constituencies participate in software design? How can software tools be developed in an open, nonproprietary fashion? Are they available in languages other than English?

Content/Services

Improving access to the information and communications services people find valuable in their daily lives provides the central rationale for ICI development. Such services must include the ability for users to interact in a creative and participative fashion as well as simply to receive stimuli. People should find them as readily available resources that contribute to fulfilling their roles as citizens, producers, consumers and caregivers. To achieve this goal, content and services should be affordable, timely, reliable, and easy to use. They should cover a wide range of offerings that reflect social diversity in terms of cultural identity, linguistic prefer-ences and political views. People should be freely able to express their observations and opinions without fear of censorship or sanction. While enjoying unrestricted access to what others have contributed, they should be able to avoid full exposure to materials they find offensive or disturbing. Interaction between individuals and organizations should be based on the assurance of privacy, security and authentica-tion.

Essential on-line databases potentially include those providing access to information about environmental conditions (e.g., weather, pollution), employment opportunities (e.g., job banks), health matters, emergency services, library holdings (e.g., via OPACs), political processes and civic/local events. Many of these are currently available via the Web without charge to the user. However, commercial influences could undermine some of the diversity and quality of the information. For example, some health-oriented sites are supported through advertising, including from pharmaceutical firms, raising questions of potential bias. A significant and growing proportion of Web activity goes through large commercial portal sites (e.g. AOL, MSNBC, Yahoo, Sympatico, Disney). While in important ways these make finding materials easier for many people, they also channel attention to those that are most lucrative to the portal host. This tends to marginalize content providers that cannot afford to pay for visibility on the portal site or that appeal to relatively small market segments. This has national and cultural sovereignty implications, leading in Canada to calls for policies that ensure prominent "shelf space" for Canadian content in on-line cultural emporia.

Government-held information is another area requiring special attention. Large amounts of information collected in the name of the public interest with public funds are stored in computerized databases. Over time, it should be relatively inexpensive to make much of this information available via the WWW with no personal privacy violation. While some useful steps are being taken in this direction, they are so far haphazard. Much more needs to be done to fulfill the public's right to know. A similar principle applies to information of public concern held by private organizations, such as health and safety records, public legal filings, licensing violations, environmental activities, and privacy practices. As with government information, this would be relatively cheap and technically easy to accomplish with network facilities. However, the institutional obstacles are formidable and this form of public accountability may require legal mandating.

Electronic mail is the main candidate for an essential universal digital network service (Anderson, et al., 1995; Markle Foundation, 1998). It demonstrably fills an important communications niche as witnessed by the rapid growth in the volume of transactions and number of subscribers. Partly this has been enabled by several companies, such as Hotmail/Microsoft, and Juno, offering it free of charge to anyone in exchange for carrying advertising on each message. Whether or not this practice is a short-term phenomenon that will end once the market has stabilized, it does at least indicate that the costs of e-mail are sufficiently low, that ensuring it is available to everyone would not be prohibitively expensive.

As with software tools, content/service design must accommodate a broad range of users (including people with disabilities), cultures, and languages. Services should be available in text-only form so they are accessible by older, low-end devices and software. Mechanisms should allow users to filter out offensive materials, individually and for their children. Privacy-enhancing technologies (e.g., encryption) should be available for individuals to maintain control over their personal information. In terms of content, not all service needs of various groups are represented. The overwhelming dominance of the English language is also a major impediment for global access (Fishman, 1998). In order to meet the needs of diverse cultural groups, a range of material must be created by these peoples with respect to their particular cultural priorities and heritage.

Policy questions: Which of these services are actually basic/essential? Are they free or very affordable, authentic, and reflective of the diversity of the community? How can one ensure participative decision making and design regarding essential content for various diverse communities? Which essential content/services are unlikely to be offered universally by market forces, and are hence worthy of public support?

Service/Access Providers

The organizations that provide network access to users are a vital but relatively neglected aspect of the overall ICI. Recent statistics indicate that users access the Internet mostly from their places of employment. But workplaces offer much more than just free access to equipment and connection—they provide a range of services that in combination (ideally at least) provide a supportive use environment. These services include purchasing, configuring, upgrading, maintaining, troubleshooting, repairing, and documenting systems as well as staffing help desks and training sessions. All of these are skill demanding and expensive but vital to routine, unproblematic network use. As network access becomes commodified and sold to individual subscribers some of these support functions become easier, but they don't disappear completely. Individuals have come to rely on an expensive and sometimes unresponsive collection of Internet service providers (ISPs), telephone or cable companies, and hardware and software vendors to maintain minimal network access. While the number of people who have Internet access as individuals is growing rapidly, it is a long way from being the norm. In Canada, domestic access to the

Internet was recently just 28% (Ekos, 1998).

We can therefore expect that for years to come, the many people who do not have access through their workplaces nor can afford domestic purchase of networking services will have to depend on public access facilities. Consideration should be given to implementing public access points in public, nonprofit, volunteer, community organizations. These should provide access to computer terminals, software, applications, and broadband access to the Internet should be provided (OECD, 1997, p. 58). With schools and universities increasingly supplying students with public access workstations and network accounts, educational institutions are becoming prime providers of such public network access. However, although both the Clinton and Chretien governments have advocated K-12 access to the Internet in public schools, this does not necessarily mean that every child will have ready and useful access to the Internet. In many cases the emphasis is on getting the wires to the school, with little support for other aspects of the learning and use environment. Teachers in particular are often not consulted, nor provided with the time to modify curriculum and offer their students the assistance they need (Shade, 1999). Similarly, although public libraries and other community organizations, such as hospital clinics, daycare centres, post offices, and community centres, are well positioned to provide access to digital network technology, an overall climate of fiscal restraint threatens to diminish the viability and availability of such services. Taking on the role of public access provider can become a major burden that draws resources away from other service priorities, thus weakening the institution.

Nonprofit, volunteer-run community networking organizations have grown up in many towns and cities specifically to provide various forms of public access. Although in many cases they began simply as Internet service providers, they are becoming increasingly important as sources for both community-based content and training. While their primary focus on digital networking gives them some advantages over more conventional public sector organizations, they so far generally lack reliable access to funding sources. In Canada, the federal government's Community Assistance Program (CAP) is helping thousands of local organizations establish their own networking facilities, but the long-term sustainability of these promising initiatives is a thorny and so far unresolved issue. Incorporating local, nonprofit and civically responsive organizations into the ICI is a central policy challenge for infrastructure development.

Policy questions: How accessible and sustainable are public access sites and projects? How can existing public institutions, such as public libraries and community centres, be supported in taking on responsibilities for public network access? How can sustainable models of community networks be built? Given a widening gap between education, income, and employment levels, how can community organizations and sites effectively support citizens in the use and development of the ICI? How can commercial Internet service providers participate in public policy discussions accompanying technological design and implementation?

Literacy/Social Facilitation

New digital media place significant learning burdens on users. Even when the design improvements identified in the discussion of the Devices and Software and Content/Services layers become commonly adopted, people will find mastering the inherent complexities of the infrastructure a continuing challenge. Achieving widespread effective use of network technologies must therefore be supported by a variety of formal and informal learning facilities. According to Neice (1996),

> A person's relative degree of digital literacy will enormously influence their participation in and access to the information techno-structure. A complete understanding of the socially included and the excluded, in the information techno-structure, will depend on a much richer and more detailed understanding of the nature and form of digital literacy skills than is available today.

Digital, or network literacy, is too often treated in practice as being mainly about keyboard and menu navigation skills. It is more usefully viewed as encompassing a broad range of knowledge and skills. Knowledge includes an understanding of the various types, sources and uses of global networked information; the role of networked information in research and problem solving; and systems whereby information is stored, managed, and transmitted. Skills include the ability to retrieve specific sorts of information using a range of tools such as search browsers and on-line databases; the ability to manipulate, enhance, or increase the value of information; the ability to purchase and configure local systems and then troubleshoot them when they (inevitably) don't work as expected; and the ability to analyze and resolve both professional and personal services that increase the quality of one's life through actively creating as well as passively consuming information.

While good materials (of the right level and language) in combination with self-study are important in acquiring the necessary skills and knowledge, the social aspects of learning are often overlooked. Much learning relevant to accomplishing practical tasks occurs informally "on the job" with the assistance of "local experts." These are friends or colleagues close at hand who are approachable and knowledgeable about both the technological capabilities and the task requirements. The access/service-providing organizations mentioned above are clearly a principal source for such learning support. However, those who will be most in need of learning assistance will be those who can't attend educational institutions or have jobs in workplaces where network services are readily available. Local, community-based support for learning therefore become essential.

Policy questions: How can and should digital literacy be defined? What are the basic levels of digital literacy required to obtain good employment, become informed civically or otherwise play an active role in society? How can one foster public support of education/training and a climate of experimentation and diffusion of innovation?

Governance

Governance is about the ways in which decisions are made concerning the development and operation of the information/communication infrastructure. Current governance and policymaking is centred in the major developed (OECD) countries, with corporate and industry stakeholders playing leading roles. Privatization and the deregulation of global telecommunications markets are seen as the way to foster competition and technological innovation. The principal role for individuals in this approach is as consumers, choosing from among the growing array of digital goods and services. This represents an overly narrow range of actors and interests. Important issues, such as public accountability, cultural identity, social inclusiveness and cohesion, and national sovereignty, are being threatened by giving such a priority to market forces. A major challenge during this crucial formative period is to broaden the range of participants and ensure that decision-making is democratic at the local, national and global levels (Herman, McChesney, 1997).

The general citizenry has much at stake in the current phase of ICI expansion, but faces formidable difficulties in effectively shaping the development process. Social priorities need to be established, technological and economic alternatives examined, issues clarified, potential "impacts" assessed and policy options debated. This requires a combination of research projects, experimental trials, education campaigns, expert testimonies and public hearings that engage the interest of a wide range of citizens and their organizations. All of this is feasible and there is considerable experience to draw upon. However, what appears to be the main obstacle to greater public involvement is the unwillingness of the leading actors to broaden the discussion.

Canada has had a long history and some notable achievements with active public involvement in decision making over complex infrastructure initiatives (e.g., Aird Commission, 1926, and the MacKenzie Valley Pipeline Enquiry, "Berger Commission," 1974). In the case of the ICI, the most publicly visible policy discussion involved the Information Highway Advisory Council (IHAC). In a break with tradition, it did not conduct significant research nor hold public hearings. Its final reports provide a useful overview of the many issues involved in the information highway discussion, but the recommendations largely reflect the dominance of the telecommunications and media industry representation on the Council (IHAC, 1995, 1997). The Council did recognize the need for a comprehensive strategy for achieving universal access, and in response the Canadian Government committed itself to:

> "By 1997, ... develop a national access strategy ... to ensure affordable access by all Canadians to essential communications services. ... Developing this strategy will involve widespread consultations with all interested parties." (Government of Canada, 1996, p. 24)

Unfortunately, public consultations did not occur and the goal of universality has been replaced by a much narrower and vaguer ambition to make "Canada the most connected nation in the world." However, several public interest organizations are continuing to pursue the access issue (Buchwald and McDowell, 1997; Clement and Shade, 1998). Through a series of workshops, funded in part by the federal government, they have begun formulating their own national access strategy (Clement, 1998). Central to the strategy is declaring cyberspace as an 'electronic commons' embracing a variety of public and private spaces for a range of profit and noncommercial exchanges. The electronic commons is a shared resource vital for supporting the varied activities of daily life, one that needs to be equitably apportioned and managed in the public interest by carefully balancing the contending legitimate demands for its use. In keeping with the rainbow model, access to electronic network services is treated as a complex social/technical phenomenon, and not mainly about connection to high-speed transmission facilities.

Some of the key governance principles include:
- giving priority to community-based initiatives;
- redressing existing inequities which inhibit access;
- ensuring there will be no decline or degradation of existing information/ communication services;
- using commercial expansion to fund broadening of access; and
- affirming the public's right to full participation in decision-making concerning development of the communications/information infrastructure.

The strategy proposes establishing "Access Councils," broadly representative and publicly accountable bodies at the local and national levels mandated to pursue the goal of universal accessibility. Among its functions would be to define an evolving set of 'essential network services' and establish priorities for a Universal Access Fund.

Policy questions: How can one establish a participative public policymaking process? What new governance institutions are needed? What are the prospects for "digital democracy"—i.e., what role can the Internet itself play in involving citizens in ICI development and other policy areas? (Alexander and Pal, 1998). How can community-based and public interest organizations be supported in taking an active part in ICI governance? How can one govern locally or nationally within an increasingly global arena? How do various trade regimes (e.g., WTO, NAFTA, GATT, OECD policies, and the proposed Multilateral Agreement on Investment - MAI) affect governance? (Clarke, Barlow, 1997).

Applying the Model

This section shows how the Rainbow model has been used in two different scenarios: in examining how new technologies might be implemented and accessed in a nonprofit organizational setting and as part of a recent policy discussion surrounding the development of a Canadian health information infrastructure. The Rainbow model has also been applied in policy discussions concerning a gender-based analysis to the needs of women and women's groups within Canada with respect to access to information and communication technologies (Balka, 1997).

Non-Profit Organizational Setting: The Social Development Network

The Social Development Network (SDN) (http://www.web.net/sdn) comprises a broad cross section of 8,000 nonprofit organizations in Ontario. SDN members believe that accessible and affordable information and communications technology should be used to achieve their individual and shared goals; therefore, SDN encourages broad-based participation which flows from shared commitment to this mission. SDN's goals are to provide an electronic forum for not-for-profit organizations to work together in support of social development, promote inter-organizational capacity-building, make technology easier to use for organizations and people in the nonprofit sector, add value with information and communication technology, and promote cooperation among organizations and people involved in social change.

Mielnizcuk (1996) has adapted the Rainbow model for the development of SDN, which is described as "a human network supported by a technical one." While the uppermost layers represent the organizational and social considerations for a working system, the middle layers consider process and service content and the lower layers address the technical components. Each of these layers interact and "in order for a system to be responsive, participants, stakeholders and users must have explicit opportunity to shape all levels." The Rainbow model was particularly helpful in providing a comprehensive and coherent framework around which a disparate and far flung collection of small organizations could coalesce.

Of special importance is SDN's emphasis on facilitation and training to develop lead users at each location of member organizations. These lead users are supported by readily available shared resources and skilled support staff. Organizational members and sectors commit to ensuring that local innovations and challenges are broadly shared with others through designated facilitators, and they will identify specific equity problems between information haves and have-nots.

(See Figure 2 "Social Development Network: Information Infrastructure" found at http://www.web.net/sdn/sociotec.gif. Also see http://www.web.net/sdn/govserv.gif)

Developing Canada's "Health Info-Structure"

New information and communications technologies hold considerable promise for the general public to play a larger role in making informed decisions about their health and that of those they care for. However, there has so far been relatively little attention given in Canada to the complex issues of accessibility—how members of the public may readily obtain relevant, reliable and timely health information and communicate with others about health matters. There are significant economic, technical, and social barriers to access that can prevent many citizens from benefitting fully from the health info-structure, and which may further increase the disparities in health status among citizens.

Federal government planning for incorporating digital networking into Canada's health system has been centred in the Advisory Council on Health Info-Structure which reports to the Minister of Health (Health Canada, 1998). A recent study commissioned by the Advisory Council examined accessibility issues with a view to recommending steps towards the elimination of access barriers for Canadian citizens and suggesting a design framework for the development of a Canadian health infrastructure (Clement et al., 1998).

The study identified a major gap in the networked info-structure as the lack of a definitive electronic repository of basic health information resources and linkages to the wide and growing range of Internet-based information/communications services. This gap could be overcome by developing the Canada Health Space (CHS), "a universally accessible health information/communications 'commons.'" In keeping with the foundational principles of Canada's public health system, it would be owned and governed by Canadians and operated in the public interest. It would make available to all Canadians an integrated set of electronic resources with multiple providers and entry points. As a complement to independently maintained resources and by providing an effective means for enabling Canadians to obtain health relevant information and to communicate with each other about health-related matters, it should develop a popular reputation as the principal place for Canadians to begin looking for health related information and communications services. At the same time it should become the preferred means of publicity for those offering health services, information, or opinion. In time, it should also become a mandatory reporting vehicle for every individual or organization whose products or activities are intended or likely to affect significantly the health of Canadians.

The Accessibility study drew upon the Rainbow model in several ways. Foremost, it highlighted the notion of accessibility as a complex multimodal and multilayered notion in which the social aspects of the health information/communication infrastructure play a prominent role. This is in contrast to treating access mainly as connection to high-speed networks linking databases of health information with individual health consumers, a view which dominates the current policy discussion. A consequence of this shift in attention is the need to keep the development of the social infrastructure in balance with technological investments. In particular, policy must recognize and support the ongoing role of health information intermediaries in providing, interpreting, and disseminating information.

The Access Rainbow was also useful in articulating specific aspects of the Canada Health Space and its overall architecture.

Carriage

Given the variety of health information users and use circumstances, the health info-structure must offer multiple carriage modes. Integrated with the direct, in-person communication between an individual and front line health (information) provider, there needs to be a coordinated array of transmission options, including via:

- telephone (e.g., 1-800 numbers, involving a flexible combination of skilled human agents, and Interactive Voice Response technologies);
- Internet (e.g., WWW, e-mail, chat, etc.);
- fax (e.g., fax form entry and fax-back services); and
- standard mail (e.g., brochures and other bulky or glossy-printed materials).

In so far as digital modes are used to carry health information for the general public, the initial priority should be on taking full advantage of the currently available voice grade network to extend the reach of the CHS rather than promoting applications requiring broadband facilities available to the fewer already relatively advantaged users.

Devices

Health information services should be designed to work well with widely available "off the shelf" hardware components. Given that the users of health information services are more likely than the public generally to be physically disadvantaged in some way, it is especially important that public access devices meet the highest standards of usability. The priority locations for such devices should include health information centres, clinics and hospital waiting rooms.

Software Tools

Again, as with devices, the emphasis should be on taking advantage of the software tools that are already widely used. Software for public access points should be carefully chosen for wide usability. Given the often sensitive nature of health matters, privacy and even anonymity protection features are especially important.

Content/Services

Health information and communication services that many people will find useful include: preventative health information, medical treatment options, health system performance statistics, health promotion/advocacy organizing information, community health indicators, environmental hazards, health and safety records of employers and product manufacturers and disease/disorder-related discussion groups. The information needs to be timely and reliable. The source of information should be clearly identified, especially if it is commercially sponsored, so its credibility can be assessed. Information should be available from alternative sources, including from nonprofit, public or community-based organizations. Making queries and obtaining information should not compromise privacy rights.

Since users of a health info-structure (e.g., the lay public, experienced health intermediaries, frontline health providers) will reflect differing degrees of health expertise, the information will need to be presented at several levels of sophistication. In particular, information geared to the nonmedical specialist should be available.

People should be readily able to find others sharing similar health concerns and freely discuss these matters. They should have an opportunity to learn of other people's experiences, opinions and ratings of particular health services as well as contribute their own publicly accessible reviews.

Service/Access Provision

Health information/communication intermediaries (e.g., librarians, frontline health providers, medical clinics, voluntary and community health promotion organizations) are an established part of the existing health info-structure. Far from digital networks rendering them increasingly obsolete, they have a vital and ongoing access role to play. Since they are already experienced in providing information services, they are well positioned to host public access sites, evaluate software tools, develop digital materials, assist users in learning to interpret health information, and contribute to decision making around health info-structure development. They thus need to be better supported in these roles and the networks that link them strengthened. Digital networking can help in this regard.

Literacy/Social Facilitation

Beyond conventional and computer literacy, users of the CHS will need to develop skills in making sense of information involving unfamiliar medical terminology and reflecting widely differing perspectives. People need to be able to judge the relevance, reliability and sources of information. The network of health information/communication intermediaries mentioned above will be especially valuable in helping people learn to navigate health databases and interpret the information they find.

Governance

The good functioning of the health info-structure is a matter that concerns everyone, in a variety of ways. Decision making around development of the health info-structure should therefore reflect this range and diversity of stakeholders. This will require ongoing public participation in all phases of design, implementation, and operation. Large medical institutions and suppliers are already well positioned to play an active part in this process. It is the participation of smaller, resource-weak health organizations which will need facilitation. In particular, health information intermediaries and community-based health promotion/advocacy organizations will need support in this.

Conclusion

This chapter has sketched a holistic model for defining access to the information/communications infrastructure that addresses universality and other key public interest concerns. While much work needs to be done in testing, refining and elaborating the model, probably the most important task is to apply it to current policy debates. One possible such application of this model is to provide the basis for defining essential aspects of access at every infrastructure level which in turn could guide the expenditures from a possible Universal Access Fund.

A major design challenge now facing policy makers is to specify a multifaceted architecture for developing network technologies that takes all seven elements of the Rainbow model into account and affords access to everyone by virtue of their membership in society. The design process must be broadly participative and dynamic. It must be carried out in the face of strong pressures from rapid technological change, ideological opposition, ignorance of technical possibilities and social implications, strained public resources and societal instability.

Acknowledgments

We appreciate the helpful suggestions for revision provided throughout the years, including comments by Harmeet Sawhney and Lucy Suchman.

References

Alexander, C.J. and Pal, L.A. (1998). *Digital Democracy: Policy and Politics in the Wired World.* Toronto: Oxford University Press.

Anderson, R. H., et al. (1995). *Universal access to e-mail: Feasibility and societal implications.* Santa Monica, CA: The Rand Corporation. URL: http://www.rand.org/publications/MR/MR650/

Balka, E. (1997). "Viewing the World Through a Gendered Lens." Prepared for Universal Access Workshop: Developing a Canadian Access Strategy: Universal

Access to Essential Network Services. February 6-8. URL: http://www.fis.utoronto.ca/research/iprp/ua/gender/balka.html

Borenstein, N. (1998). One Planet, One Net, Many Voices. *CPSR Newsletter*, 16(1), 1, 5-8.

Castells, M. (1996-98). *The Rise of the Network Society*, Vol 1-111. Blackwell.

Chapman, G. (1999). "Digital Nation: The Future Lies Beyond the Box." *The Los Angeles Times*, Monday, January 4.

Clarke, T. and Barlow, M. (1997). *MAI: The Multilateral Agreement on Investment and the Threat to Cultural Sovereignty*. Toronto: Stoddart.

Clement, A. (Ed.) (1998). Key Elements of a National Access Strategy: A Public Interest Proposal, Information Policy Research Program, University of Toronto, August 5, 18. http://www.fis.utoronto.ca/research/iprp/ua/

Clement, A., D. Chan, W. Duff, C. Marton, L. Shade and P. Stoyanova (1998). Accessibility to Canada's Health Info-Structure. A paper submitted to the Advisory Council on Health Info-Structure, Office Health and the Information Highway, Health Canada, October, http://www.fis.utoronto.ca/research/iprp>

Clement, A. and L.R. Shade (1998). "Kanadische Burgerinitiativen gestalten ein Netz fur alle." pp. 354-365 in *Internet & Politik*, Claus Leggewie und Chirst Maar (Hrsg). Bollman.

Clement, A. and Shade, L. (1996). "What Do We Mean By Universal Access?": Social Perspectives in a Canadian Context, presented at the Internet Society conference (INET96), Montreal, June. http://www.fis.utoronto.ca/research/iprp/dipcii/workpap5.htm/

Dutton, W.H., (Ed.) (1999). *Society on the Line: Information Politics in the Digital Age*. Oxford University Press.

Ebo, B., (Ed.) (1998). *Cyberghetto or Cybertopia?: Race, Class and Gender on the Internet*. NY: Praeger Press.

Ekos Research Associates, Inc. (1998). Information Highway and the Canadian Communications Household. February 28. URL: http://www.ekos.ca/FEB98.HTM

Fishman, J.A. (1998-99). "The New Linguistic Order." *Foreign Policy*, No. 113 (Winter), 26-40.

Golding, P. (1996). "World Wide Wedge: Division and Contradiction in the Global Information Infrastructure." *The Monthly Review*, (July-August), 70-85.

Government of Canada (1996). *Building the Information Society, Moving Canada into the 21st Century*.

Greenbaum, J. and Kyng, M. (1991). *Design at Work: The Cooperative Design of Computer Systems*, Lawrence Erlbaum.

Health Canada. (1998). *Connecting for Better Health: Strategic Issues*. Interim Report, Advisory Council on Health Info-Structure, September 30. http://www.hc-sc.gc.ca/ohih-bsi/menu_e.html

Herman, E.S. and R.W. McChesney (1997). *The Global Media: The New Missionaries of Global Capitalism*. London; Washington DC: Cassell.

Hirschkop, K. (1996). "Democracy and the New Technologies." *The Monthly Review*, (July-August), 86-98.

Information Highway Advisory Council (IHAC) Phase II Report, Preparing Canada for a Digital World. URL: http://strategis.ic.gc.ca/SSG/ih01642e.html

Karim, K.H., Smeltzer, S. and Loucher, Y. (1998). "On-Line Access and Participation in Canadian Society." Paper presented to the Knowledge-Based Economy and Society Pilot Project, Hull, Quebec, March. Hull: International Comparative Research Group, Strategic Research and Analysis, Department of Canadian Heritage.

Levy, D. (1994). "Fixed or Fluid? Document Stability and New Media." *ACM European Conference on Hypermedia Technology*. Edinburgh, September 18-23, 24-31.

Loader, B., (Ed.) (1998). *Cyberspace Divide: Equality, Agency, and Policy in the Information Society*. London: Routledge.

The Markle Foundation. (1998). E-mail For All Outreach Campaign. URL: http://www.iaginteractive.com/emfa

Mielniczuk, S.A.(1996). Information infrastructures: An integration of the social and technical. URL: http://www.web.net/sociotec.html

Mosco, V. (1998). "Myth-ing Links: Power and Community on the Information Highway." *The Information Society* (January-March).

Neice, D. (1996). Information Technology and Citizen Participation. Prepared for Canadian Heritage. URL: http://www.fis.utoronto.ca/research/iprp/ua/neice.html

OECD. (1997). *Towards a Global Information Society: Global Information Infrastructure, Global Information Society: Policy Requirements*. Paris: OECD.

Perry, J., E. Macken, N.Scott, and J.L. McKinley (1997). "Disability, Inability and Cyberspace," pp. 65-89 in *Human Values and the Design of Technology*, B. Friedman (Ed.). Stanford,CA: CSLI Publications, Cambridge University Press.

Reddick, A. (1998). *Criteria for Defining Essential Communication Services*. Ottawa: Public Interest Advisory Council (PIAC), April.

Schon, D.A., B.Sanyal, and W.J. Mitchel (1999). *High Technology and Low-Income Communities: Prospects for the Positive Use of Advanced Information Technology*. Cambridge, MA: MIT Press.

Shade, L.R. (1999). "Net Gains? Does Access Equal Equity?" *Journal of Information Technology Impact*, 1(1), 23-39. URL: www.jiti.com

Shade, L.R. (1998). "A Gendered Perspective on Access to the Information Infrastructure." *The Information Society* (January-March), 33-44.

Shneiderman, B. (1999). "Universal Usability: Pushing Human-Computer Interaction Research to Empower Every Citizen." Position Paper for National Science Foundation & European Commission meeting on human-computer interaction research agenda, June 1-4, Toulouse, France.

Sirois, C. and C.E. Forget (1995). *The Medium and the Muse: Culture, Telecommunications and the Information Highway*. Montreal: The Institute for Research on Public Policy.

Status of Women Canada. (1996). Gender-Based Analysis: A Guide for Policy-Making. Ottawa: SWC.

Universal Access Workshop. (1998). *Call for a National Task Force on Universal Access*. URL: http://www.fis.utoronto.ca/research/iprp/ua/tfcall.htm

Chapter II

Requirements for a Regional Information Infrastructure for Sustainable Communities — The Case for Community Informatics

Thomas Beale
Mooloolah, Queensland, Australia

Although many community informatics (CI) efforts to date have been successful (see e.g., Beamish, 1995), few appear to have been part of a strategic response to the "big" problems of the environment and society; typically they address specific needs of particular communities, and have limited influence outside their scope. Nevertheless, they have created a wealth of experience which may be used in new endeavors, since they have exercised numerous technologies and economic models. However, if the concept of "community" is going to play a more important role in establishing sustainable economic, ecological and cultural systems for the future, community informatics will need to develop not just better solutions, but a *strategy* to support the community ideal.

This chapter argues that the object of such a strategy should be a *regional information infrastructure* (RII), whose purpose is to provide a *knowledge-processing* context for physical communities. The definition of requirements for an RII is effectively an expression of a community ideal, or vision, in terms of a technological strategy. Technical architectures stemming from the requirements represent a repository of solution patterns which can be reused and evolved. Of course, any such strategy is predicated not only on a *certain kind* of community vision, but the idea that "community" is indeed an effective response to global and local problems; thus the themes of community and knowledge are firstly explored as a background to the discussion of the RII.

A strategic approach offers a number of advantages over formulating *ad hoc* solutions in response to specific problems: firstly, it provides a thinking framework within which CI can more effectively operate as a discipline; secondly, it constitutes a more holistic approach to using information technology (IT) for community development, and should result in more effective solutions.

With an information infrastructure in place, physical communities would have a place to express and explore their own identity, history and interests; as a result, self-aware communities would emerge where once there were disconnected individuals. Once on the path of self-development, communities could use the RII to support regional agriculture, trade, local governance, environmental monitoring, basic health and education, recycling and waste disposal, and social and cultural activities.

Self-management holds the promise of reducing unemployment, reversing rural decline, and greatly improving the prospects for the environment. Furthermore, a well-designed infrastructure might provide the means of better focussing government social expenditure at points of need, leading to better outcomes.

Most of the qualities of the RII presented here are essentially precepts for using existing information technologies, such as:
- Being knowledge-oriented rather than technology-oriented.
- Enabling geographical region-specific data-gathering and knowledge management.
- Emphasizing participatory rather than unidirectional communication.
- Close integration into the community rather than elitist IT centres.
- Having an inclusive user-base and removing barriers to access.
- Encouraging interdisciplinary construction and use of knowledge.

Correctly implemented, the RII should provide a means of reestablishing the interplay between local environment, economy and society in the minds of people locally, consequently encouraging more sustainable and diverse social structures, in which better management of human impact on the environment will be possible.

Background

The Role of Information Technology
There are two important landmarks in the social history of information technology: the advent of the affordable computer, and the Internet. The first put information processing and knowledge work within the economic reach of both corporations, with their hundreds of desktops, and home users. The second has provided the means for connecting users, regardless of geographical location.

Although there are possibly more technically important events in the history of computing, these are now primary drivers in the "information revolution," in the sense that they have made the information-related activities of communications and symbolic processing part of our daily lives; many people, in their *private* capacity, are now active users, transformers, and producers of information.

Prior to the Internet, the information revolution predominantly consisted of computing in corporations, where its main effect has been to make existing processes far more efficient, and mass media, where it is characterized by *unidirectional* communication; overall, it tended to reinforce existing societal control structures.

The advent of the Internet and cheap access has changed the landscape considerably: it has provided a context for *participatory* communication and enabled the establishment of *new kinds of community*. For this reason, it has enjoyed enormous popularity; users can publish "alternative" material, and be read by millions. Despite this, it too has attracted criticism, including that it creates a new frontier, or "digital divide" between the connected and the unconnected, thus exacerbating the marginalization of the poor.

Although a milestone in history, the internet directly encourages only certain kinds of community, and tends to lean more toward the globalization rather than the localization of information. Any strategic CI endeavour needs to take account of these factors.

The Concept of "Community"

There are essentially two commonly accepted definitions of the term "community" (taken from the *Concise Oxford Dictionary*), which are adapted for the purposes of this discussion, as follows:

- *Geocommunity*: local community, body of people living in the same place. The term "geocommunity" is used here to emphasise its geographical character.
- *Community of interest*: body of people having a common interest such as profession, religion, area of academic study, ethnicity.

To this we can reasonably add a third:

- *Community of commerce*: people working together in the same company, or other companies linked in a supply chain.

This last type is really a type of community of interest, but is sufficiently different to be considered separately.

In all cases we can add to the definition the idea that members of a community of any kind *identify* themselves as such, and perform at least some activities in a *collaborative* manner.

Communities of Interest

Historically, much interaction among members of communities of interest has occurred via the manual technologies of letters, journals, books and physical travel, particularly for communities which are very specialised, such as academic groups.

All this has changed in recent years, starting with the early electronic infrastructure of the telegraph, radio and telephone, and culminating in the Internet. The Internet has vastly increased the creation and development of communities of interest, in the form of *virtual* or *on-line* communities, due to the fact that it possesses four attributes not previously available in the one infrastructure: it is (nearly) global, it is (nearly) real-time, it is interactive, and it (mostly) has high bandwidth. (The lack of bandwidth prevented the earlier electric and electronic technologies having the same effect as the Internet; likewise the lack of interactivity in broadcast television). Today even the most far-flung members of a virtual community can transmit large amounts of information quickly and communicate in real-time.

Some "cyberutopians" (e.g., Negroponte, 1995) claim that the virtual world contains the solutions to today's problems of poverty and inequality, in the form of communities independent of place. While digital communities clearly have a lot to offer, it is unlikely that they will replace geographical communities: it is an undeniable fact that humans like to interact with each other physically. Not only do we communicate on an individual and family level, but most fulfilling social behavior occurs in the real world, in music concerts, dance classes, farm days, football matches, pubs and cafes.

Communities of Commerce

Communities of commerce are a prime example of existing communities which have been empowered by information systems and Internet technology. In today's world, these are networks of firms concerned with various phases of production, distribution, and sales. Electronic communication may take the form of semiprivate shared Internet sites called *extranets*, or simply swapping of private e-mail addresses and documents.

Inside corporations, the ability to perform projects using teams has been vastly improved with the advent of *intranets* (private Internet-like networks) which offer better internal communication (via e-mail) and information dissemination (via a web). More sophisticated tools are used for *project management*, and the building of *corporate memory*, i.e., the history of projects and people in a business, giving the corporation life beyond its employees. E-commerce enables large and small companies alike to accept financial transactions online, avoiding some of the significant costs of marketing and retailing. Exploiting these new technologies has enabled large conglomerates to completely globalize their manufacturing, distribution and marketing.

Some of the key lessons which have been learned from the use of IT in companies are:

- Intranets enable *collaboration* between groups of individuals working on projects.
- IT solutions greatly improve the efficacy and possible scale of activities, such as *planning and project management*.
- Computer systems can be used to implement *corporate memory*.

Geocommunities

The physical community, or *geocommunity*, is quite different from its virtual cousins. Consider some of the sociological differences between geocommunities and communities of interest:

- The geocommunity usually has a system of governance and a *common economic framework*. Members are obliged to behave and operate according to certain rules to a much greater extent than in on-line communities.
- The geocommunity as a whole is characterised by many and *diverse interests* rather than a specialised interest. For example, education, health, agriculture, local arts, security and employment all concern most individuals in a geocommunity.
- Most geocommunities in the developing world, and some rural ones in the

industrialised world, are *directly involved in production* activities relating to their own sustenance and shelter.

- Some of the primary concerns of the geocommunity are *subject to the constraints of real-world resources*, including water availability, arable land, environmental pollution, urban development, and transport. Solutions to problems in these areas require widespread cooperation, and often have important consequences for health and survival, unlike issues in the virtual world.
- Members *do not typically choose* to be part of the community—they were either born into it, or arrived due to a particular circumstance, e.g., to be with relatives or following work. Consequently, there is no *a priori* reason for members to agree with each other on many or even any issues. Geocommunities thus require mechanisms for both *cooperation* and *conflict resolution.*
- Membership is *socially diverse*, including the wealthy, poor, sick, disabled, elderly, and immigrants.
- Members do not just identify with the community, they make *sacrifices for it*, in the form of mutual obligation, taxes, sharing goods, or in extreme cases, by fighting for it. Thus, the geocommunity usually has a much higher level of importance to an individual than a on-line one.
- Geocommunities are the *focal points for culture*. Universities, sport and ethnic activities all centre on particular places.
- There is a high level of *face-to-face interaction* in geocommunities.

Overall, geocommunities are far more *complex* than communities of interest: they are a highly connected, interdependent *web of relationships*. Getting on together is ultimately mandatory for survival in a geocommunity, whereas in a virtual community, all one has to do is log out.

Finally, it should be recognized that, at an abstract level, *active processes* are the primary concern of geocommunities, whereas virtual communities are primarily concerned with *ideas*; put another way, physical communities are about *doing* things, not just thinking about them.

Nothing has been said so far about how communities are arranged in the real world. Two things are generally true: a) communities exist *wherever people identify them* to be, and b) they exist in a hierarchically *nested* fashion according to the issues they deal with. People typically identify with a number of communities, starting with their village or neighbourhood, progressively including the city or region, and finally a language or ethnic community.

The Community Landscape

It is interesting to map the various types of community in the two dimensions of *globalness of constituency* and *specialization of interest*, as per Figure 1.

Solid areas represent communities in which the use of information technology is well established and sophisticated; crosshatched areas indicate minimal or growing areas of IT use.

Communities of interest are shown in the two areas at the top of the figure. The solid area at the top represents very global academic, technical and recreational communities, some of which have existed on the Internet for years, using Usenet

Figure 1. Types of Community

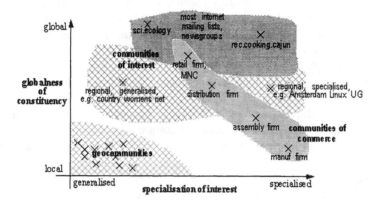

news or mailing lists, and more recently, the Web.

The upper crosshatched area represents a new area of Internet use by regional groups. Most members of these groups meet online for the first time, but due to living in at least the same general area, do in fact have some real-world concerns in common. A rural women's network which began recently in outback Queensland, Australia, for example, has grown from a modest research project to an active group with hundreds of members (Simpson, 1999). This type of group is worthy of further study, because it may be an instance in which a community which starts out online gives identity and shape to an underlying geocommunity.

Communities of commerce are represented in the elongated solid area starting on the lower right. In large supply chains, the various industries which contribute to a product or service tend not to be very local, and as they get larger and more generalized (e.g., multinational conglomerates, large retail companies), they become more globalized.

Geocommunities appear in the lower left-hand corner of the figure, according to their generalist, local nature. This is the area least well served by new technology, due to its complexity and non-specialisation.

How Important are Communities?

If community informatics is to have a major corrective impact on societies and the environment, the question must be raised of how relevant both the ideal and reality of communities is.

While most people in the industrialized world probably belong to "communities" of work or leisure, they often do not identify themselves strongly with a community which relates to where they live, or to satisfying their basic needs. Indeed, a majority of industrialised people probably do not belong to what is arguably the most important community—their *home* community; further, there may not even be a cohesive home community for them to be part of. This is the reverse of the situation in most of the developing world. As a result, most people in the industrial world (who are generally the heaviest consumers on the planet) do not have a local "interaction context," and thus do not see "community" as important.

How has this situation arisen? The replacement for many of the original functions of geocommunities in the developed world comes from the matrix of industrial production in which modern consumers live. Most individuals and families in the industrialised world satisfy all their needs with goods and services bought from the outlets of corporations or by government services. They are typically employed by large organisations, their agents, distributors, or suppliers. In a world of increasingly globalized production, it is recognised that this vast sea of consumers is the indirect cause of most environmental problems around the world, due to the simple fact that almost all resource exploitation is carried out on behalf of consumers (see e.g., Jacobs, 1991, Ch. 3). Worse, people living in developing countries with a previously healthy community culture are being indoctrinated to aspire to the industrial paradigm (see e.g., Helena Norberg-Hodge in Mander and Goldsmith, 1996).

If we reflect on the highly dispersed nature of today's industrial systems of production, distribution, consumption and waste disposal, the cause of many problems is clear: the pattern of increasing globalization has led to a *delinking* of production and consumption activities. As both a cause and result, consumers tacitly agree to have minimal involvement in the production of their needs, including the management of their wastes, and instead simply purchase products and waste disposal using monetary exchange. The list of undesirable outcomes resulting from this situation is too long to mention here, but a crucial aspect of the process is that *informational and tangible feedback between producers and consumers is lost*. The result is that a great deal of resource exploitation is not limited by any controlling (negative) feedback, and proceeds as if there were no limits to the resource base, leading to the environmental predicaments of our time. Secondly, due to the illusion of infinite production, consumption itself has almost no limits placed upon it, and is growing out of control. In this process, the marketplace itself has become a globalized mechanism encouraging endless environmental destruction.

The question of interest here is: what can the community strategy offer to improve this situation, considering that most people in the industrialised world do not strongly identify with a geocommunity, and are (supposedly) happy consumers in the world market economy? Note also that one of the goals of some of many admirable community programs (e.g., in Schön, Sanyal and Mitchell, 1999) is to raise the standard of education and living of people in disadvantaged urban areas; if successful, the beneficiaries will eventually join wealthier echelons of society, and ultimately become part of the larger problem.

We could translate the above question as: what does *mainstream* society respond to? The answer has to do with the core mechanism common to all economic systems—the market. Although lower income people are mostly price-responsive, and wealthier people tend to take into account quality, and some ethical concerns, it can be said in general that people in the consumer economy respond to *better market alternatives*. This means that if better alternatives (as judged by the big-picture concerns above) are to be offered, *they must be visible in the market economy*, and they must be on average, *at least as attractive* as their competitors.

Many better alternatives do exist: regional produce, ethically produced imported goods (e.g., under fair trade programs), high-quality services providing employment in the home community and so on. We can now see two ways in which the community ideal, by taking into account the fact that most people today act as *consumers*, can become an important instrument of environmental and societal change:

- By introducing better or *distinctive* local alternatives into the marketplace, consumers are brought into the local economy, and can become part of a growing community.
- As a consequence of the shift to more local consumption, the global marketplace itself can be broken down into information-rich, regional marketplaces.

The above argument is essentially an economic one. It shows that the community paradigm can *become* an important antidote to the industrial economy by, perhaps cynically, assuming that most people can only be "converted" by appealing to them through the market.

A more profound argument is worth touching upon. The economic system in which we live can be understood as a *self-preserving organism* (see e.g., Ch. 8, Ayres, 1994):

- It is self-organising, in that it has a structure and behavior which continues to replicate, governed in this case by the relatively simple rules of corporations, governments and stock markets.
- It has a continuous need for resources and energy for its internal metabolism, satisfying its prime directive, "production."
- It grows and evolves, both in size and complexity.

The above is not merely an analogy, since, *in reality*, the industrial economy is not controlled by any of its constituent parts, and has only two limiting mechanisms: absolute resource exhaustion and loss of consumer base. In this analysis, the economic system is a self-replicating pattern, which absorbs resources for its own growth, and whose emergent properties are fairly independent of its microscopic makeup (i.e., of particular firms or governments). This tells us that trying to change it as if it were a controllable linear system is probably futile, unless catastrophic means are employed.

We can now begin to see the community paradigm as a *planetary survival strategy*: it is an alternative self-replicating pattern, also using resources to achieve its ends, but with two crucial differences. Communities operate in a sea of other communities, like the inhabitants of a forest. Instead of growing uncontrollably, their resource usage is limited by a local resource base, and they tend to evolve into *efficient users of limited resources*. Secondly, communities don't exist to produce, but to survive.

One can see fairly quickly that healthy communities could start to act against the destructive global economy by diverting its resources, and starving it of its need for endless production. At a very deep level, therefore, the concept of community is worth preserving and developing.

IT and Geocommunities

What about already functioning, cohesive communities? Can IT support for existing or emerging geocommunities help with the really pressing problems experienced by people in the real world? In principle, the answer should be "yes," primarily for the reason that real-world communities are where *resources can be focussed directly onto solving problems*. Most of the big problems—environmental degradation, social dislocation, unemployment, and cultural homogenisation—are essentially problems of resource management and social structures in the tangible world; in fact in many cases, they are simply the large-scale consequences of local problems.

Under what conditions can IT provide solutions? The first is that any on-line expression of a physical community must be *representative of both the people and the problems*; in other words, *maximum "reflection"* is required—not only maximum user participation, but a maximal virtual reflection of internal community processes. If this is not the case, on-line activities risk marginalizing certain community members, ignoring significant problems, and failing to provide good solutions.

Next, we must understand the types of interactions which characterize communities. In general terms, these are:
- Trade
- Collaboration
- Decision-making and conflict resolution
- Cultural and recreational

We have already mentioned trade in the context of bringing mainstream consumers into the community. Obviously marketplaces fulfil an essential function in healthy communities as well. IT can also provide on-line support for local produce markets, LETS (local exchange trading systems; see e.g., Lang, 1994) and other local services.

At the core of healthy communities are active projects implementing change. Effectively implementing projects requires the existence of *collaborative structures* to establish and manage real projects at the point of need, i.e., the capability for groups of people working in teams to perform tasks.

An important corollary to collaboration is *spheres of interest*. Not everybody is involved in the same project or group, and not everyone needs access to the most detailed information of other projects. As in corporations, private working contexts which export a certain amount of information are essential to managing the work of the community as a whole.

At a deeper level, communities need to be able to *learn* about themselves, their world, and from their mistakes and successes. As in corporate contexts, IT can provide support for:
- *Self-knowledge*: of people, place and culture, and of resource usage.
- *Community memory and history*: knowledge of events and activities over time.
- *The learning function*: the ability to construct meaningful knowledge from data, facts and events.

As communities progress, decisions are taken, and conflicts inevitably arise. Moving forward in such situations requires mechanisms for discussion, polling, and different kinds of voting, according to the model of democracy (or otherwise) used by the community. Mainstream IT is already being used for opinion polls and user feedback, and community-oriented solutions such as the on-line "preferendum" described in Newman and Emerson (1999) are beginning to emerge.

One theme which is much more important to geocommunities than their virtual cousins is culture. It should be remembered that information technology in its centralized and globalized form is already dominating the thought space of most of us through television, print, *and a significant part of the Internet*. Reestablishing local cultures means fighting for "mindshare," and without the regional use of IT, this is unlikely to happen.

Local versus Global Information

Since information and knowledge are basic to social development, it is worth considering their properties. Here, we are interested in the properties of information found in local versus global contexts.

Most sites on the Internet are maintained either by communities of interest or businesses, and have little or no relationship to geography, appearing the same to all comers. The lack of relation with a geographical location is of course their strength: it enables them to bring together people from all over the world to discuss a common interest. Naturally, there are exceptions, for example newsgroups for local Linux users, mailing lists discussing particular environmental problems or political repression. But even here, discussions are fairly general or universal in nature, due to the geographically disparate constituency.

In general, therefore, most virtual communities do not really address the specific needs of people in a local context.

Consider the many types of information which are specific to a locale, be it village, catchment area, ethnic region or nation, including:

- *Wilderness & biodiversity*: endemic and endangered species; records of remnant and fragmented ecosystems.
- *State of the environment*: state of the local environment in terms of land, water quality, pollution, erosion, salinity problems and so on.
- *Climate and microclimate*: climate information is geographically specific; microclimate is particularly important, since it has a significant effect on architecture and agriculture.
- *Agriculture and forestry*: information on species, forest management, horticultural techniques, successful local techniques, land history.
- *Ecological footprint*: region-based resource audit statistics, e.g., footprint data, input/output tables.
- *Local economy*: local businesses, banks, societies; local trading.
- *Infrastructure services*: council information, health, education, political issues.
- *Community services*: general discussions, announcements, special interest groups, support groups.
- *Language, arts and culture*: education, discussions, announcements and events relating to regional artistic, cultural and ethnic activities.
- *Local knowledge*: experience, stories, local science and folklore.
- *Demographics*: underlying data relating to people, organisations and places in

local communities.
- *Geographical information*: geographical data used as a basis for visualising other data.

Clearly this kind of knowledge is intimately representative of and connected to local social activities as well as the land. Whether it is qualitatively different from "global" or "general" information is another question.

Psychological Importance

If local information were widely available on a local basis, it would be possible for people to become more *psychologically connected* to their region, as have been all long-lived regional cultures. An intimate knowledge of the local context paves the way for more sustainable resource usage, and acts as a deterrent to polluting activities.

Most Data is Local

The simplest form of information is data. Most data that is gathered relates to particular places or people, and is by definition local. This includes raw data for ecologies, soils, climate, people and economies. A minority of data used in computing systems is not local, but this is usually synthesized or averaged, as in the cases of global atmospheric temperatures or GDP.

The fact that most interesting data are local implies that local people can be involved in gathering them and using them to learn about themselves. However, for many categories of data this isn't currently the case, since data gathering is mostly carried out by central agencies (e.g., the census bureau), or else specific research groups concentrating on one place only. Older sources of reliable data such as tribal peoples are disappearing, to be replaced by people whose industrialised lifestyles do not include intimate knowledge of their context.

Local Knowledge is Empirical

Knowledge at the local level is often *applied* ("how" rather than "why"/ "what") and *empirical* rather than general or theoretical; it may be an *interpretation* of general principles into specific procedures and designs, or data gathered by observation. Due to the myriad factors and unknowns which influence concrete experiences, such knowledge may take a great deal of time and effort to be assembled. For example, while the principle of "mutually beneficial subsystems" (e.g., in permaculture, a sustainable agriculture system) is easily stated, it has taken many years of trial and error, experiment, and innovation, to translate this into practical guidelines for differing climates, biomes, cultures, and other regional contexts (and in fact, much of the knowledge is centuries old). Clearly the results of such work should be recorded and maintained as an ongoing knowledge asset.

Local Knowledge is Holistic

Local knowledge of all kinds is linked to the *common physical context* of people, culture, and land. Consequently, numerous *relationships* exist *between* "facts," which are not present in theoretical knowledge. For example, techniques originating in agronomy, hydrology, entomology, and economics are all intimately linked at the point of application on a large farm, in the guise of soil management, cultivation, water management, dam-building, pest management and business management, even though the separate disciplines are not linked at the theoretical level. Local knowledge is *holistic*, and should be understood more in terms of *networks of ideas* rather than discrete items.

Local Knowledge is an Immediate Learning Resource

Community repositories of *locally relevant* applied knowledge constitute an *immediate learning resource*, particularly for community newcomers starting with minimal knowledge or experience of any sort.

In both rural and urban areas, many activities cause significant modifications to the environment, using tangible resources in the process, including construction, farming, road-building, and forestry. In all such activities, basic mistakes are often repeated, due to the failure to apply preexisting knowledge. The main stumbling block is *accessibility* of relevant, empirical knowledge. Enhancing the accessibility of proven local experience could greatly reduce costly mistakes, and contribute to a more efficient use of resources.

Local Information is Empowering

Some measure of local *autonomy* is a general goal of all self-aware communities. This extends to information as well: control over information sources and participatory communication are preferable to centralized ownership and one-way communication. As a simple example, a community which has its own demographic database detailing skills and availability clearly has more control over the local employment situation than one which relies on the standard mechanisms of government employment agencies and job advertisements.

A proven example of local economic autonomy is the well-known LETS concept, in which local trading is enabled with the use of an alternative currency of exchange.

A Regional Information Infrastructure

Requirements

The discussion so far suggests that information and communications technologies used in a *regional information infrastructure* could provide numerous benefits to community development, to society and to the environment.

We are now in a position to draw together the qualitative properties—the requirements—of such an infrastructure. They are summarized in the following list.

1. Regionally embedded.
2. Maximize participation
3. Participatory rather than one-way communication.
4. Integrated into the community; non-elitist.
5. Knowledge bases and community memory.
6. Learning function.
7. Spheres of interest.

8. Support for collaboration and project management.
9. Decision making and conflict resolution.
10. Marketplace mechanism.
11. Ecological footprint auditing.
12. Agent of global change.
13. Global networking and Internet access.
14. Cooperative rather than corporate software development.
15. Economically sustainable.

In order to facilitate the following discussion, two new terms will be used. The infrastructure itself will be denoted as the RII, or Regional Information Infrastructure while nodes (servers, points of presence) will be denoted as RICs, or Regional Information Centres.

Regionally Embedded

Ideally the infrastructure is regionally embedded, consisting of computer systems and network nodes (RICs) which correspond approximately to the placement of physical communities. This is not so much a technical requirement, since many of the information resources of a community—including its e-mail, Web, discussion groups and documents—can be "housed" anywhere on the Internet, allowing members of a community to log on from home and interact with their on-line resources via a standard Web browser. A number of companies provide exactly this kind of service today (see e.g., Smith and Barty, 1999). The real reason is to allow *autonomous control* over the information resource.

Autonomy is important for a number of reasons. Firstly, the software solutions constructed will depend heavily on the particular community—both in the functional sense (different communities have different problems), and culturally (computer per house is fine for better off people; computer per village may be the case in poor rural areas).

Secondly, as a community's on-line needs become more sophisticated, the technical limits to the outsourced web-housing solution become apparent. For example, it may want to integrate demographic and geographical information systems and other specialist databases. It may also want to have its own personnel administer parts of the computer system, and to be able to experiment more flexibly with the system as a whole than the outsourced interface will allow. Problems of unreliable Internet connectivity may even arise (as would be typical in much of Africa, due to poor telecommunications), in which case the community completely loses access to its information. Eventually it becomes mandatory to set up local servers.

The following goals should be aimed for in developing the regional nature of the infrastructure:

- *Each community* should have its *own on-line area*, which identifies it, and provides it with a private context. In other words, every community should have a presence in the infrastructure.
- Similarly, *each region*, whether defined as a biome, ethnic or language area, should be represented.

- At least some *people from the region are employed in the local RIC*, helping to prevent technical elitism, and aiding communication between technical people and users.
- RICs are *close enough to the people they serve* that the people or company running them have at least some real-world concerns in common with users; ideally they are part of their own user base, ensuring that the providers identify with the real-world priorities of the users.

Maximize Participation

The motivation for maximizing participation in the RII is clear: to faithfully represent the whole community and its activities. The principle to be observed is: *no person should experience barriers to participation*. In the RII this translates into understanding the user base in terms of financial resources, technical skills, literacy, disabilities, language and cultural norms.

Social Reach versus Digital Reach

One way to think about RII accessibility is to define its *social reach* as the total extent of people it influences, and its *digital reach* as those directly connected by computer. Digital communications is accordingly understood as just one means of information transport, with other mechanisms such as paper, telephone, demonstration, and classroom teaching being used as required.

Such an approach not only starts to take account of differences between individual abilities and cultural preferences, it also *enables the RII to be developed economically*, because there is no "big-bang" requirement for an instant high-bandwidth digital infrastructure to every home and business before anything can be achieved. Thus the RII in both a poor L.A. neighborhood and a village in Bangladesh might consist of a single computer and modem in a meeting hall, with demonstration and word-of-mouth making the link between operators of the computer and the rest of the group.

Working on this basis puts a slightly different cast on the subject of *access*, which has been the subject of much CI work to date (see e.g., Beamish, 1995, and Mitchell, Ferriera and Shiffer in Schön et al., 1999). The assumption by many authors that every person in a community needs to be digitally connected may be reasonable for islands of urban poor in wealthy cities, but not for people in poor countries.

Bridging the gap between the digital and social reaches of the RII may not be so difficult in the technical sense (some people may simply wish to use the telephone to ring an RIC operator), but may create requirements in the areas of language (small language groups are not well represented on the Internet), English proficiency, and comprehension of visual rather than verbal forms of representation. Content should be designed to be usable by the *final* audience, in terms of language, presentation and complexity.

The important point is that digital communication is not confused with the real goal of social communication.

Social Defrayment of Costs

To ensure inclusiveness, it must be guaranteed that poorer users are not marginalized by the costs of access. Although there may be various mechanisms by which users pay individually, community-supported or -subsidized access to the RII should always be available, for example in the form of free RIC cafes (like Internet cafes), low-price computer rentals and so on.

Engagement

Solving the problem of access does not guarantee participation, of course. The RII must appear relevant and useful to prospective users, and needs to provide ways of introducing them to the on-line community environment.

Various strategies for engagement are possible. General approaches include "soft" introductions to IT via community events such as Internet cafe nights, interactive guided training sessions with small groups and so on, with a background of community marketing and advertising of the facilities and events. Possibly the two most important aspects of managing engagement, going by experiences in companies and institutions, are a) that *motivation is born of need*, and b) to make new users feel they are *choosing* to use something which helps them, and which they find enjoyable. Users should never feel as if they are having technology imposed upon them.

One of the precepts of CI is that information systems *enable* and support human endeavors, rather than replace them. In concrete terms, this means supporting projects or initiatives devised by people in the community, rather than defining them.

An effective strategy therefore suggests itself: identify *existing initiatives* and *proposals* in the community, and provide IT support for starting and/or managing project groups accordingly. The key is to find group leaders or *facilitators*, and provide IT training relevant to the job at hand. These people become "seed" people, helping to educate those around them in the use of information systems, while retaining their focus on their projects. The job of engagement is then far simpler than if generic IT training programs were supplied, or IT-led projects were devised. Having said this, there are some circumstances in which the RII would need to provide facilitators, for example in villages in poor countries and rural households, where there may be absolutely no initial understanding of using computer technology.

As a result, rather than being faced with the problem of instantly training a whole community, an RIC can spend more resources on training fewer project leaders, whom they can rely on to introduce and familiarize others with the technology.

Participatory Communication

In order for the RII to encourage participatory rather than one-way communication, a simple precept can be adopted. By way of illustration, consider the following example of non-participatory communication in a community context.

Many councils already have their own Web presence. Unfortunately, their sites are commonly deficient in that they broadcast out, but only allow minimal communication in the other direction (such as a single e-mail address for "comments"). This phenomenon may stem from a deeply ingrained cultural assumption that the work of representative bodies, once elected, is to be done without reference to the community, resulting in a non-participatory style of presentation. Ironically, the lack of efficient communication between councils and citizens is often cited by both sides as a reason for the failure of programs, poor service and so on (Beamish, 1995, Ch. 4, p. 2), describes this phenomenon in North America). Many councils have excellent resources and committed people, but are often as frustrated as the people they serve, since their genuine desire to understand what people want is not matched by the logistic ability to actually discover these needs. Lack of two-way communications also exacerbates less desirable aspects of such organisations, namely lack of transparency and accountability, and ultimately *relevance*.

One way to improve the situation is to make participatory communication a requirement for all civic bodies represented in the RII, as follows. Any body seeking to represent itself electronically must not only provide material for dissemination, but also discussion and other appropriate communication channels in which *their own personnel participate*. This is technically very simple to achieve even with basic Internet software, and could vastly improve the level of direct engagement between councils and citizens. This precept could be taken further, by requiring all bodies to *demonstrate a minimal level of responsiveness*, or face removal from the system. Such a measure is not as extreme as it sounds: it simply allows the community to enforce the level of involvement it desires from its representative bodies.

Generalizing, the same precepts can be applied to local and regional organizations of any kind, including businesses, when they go online in the RII. By this mechanism, for example, companies could be held accountable by community members for inappropriate advertising or poor quality goods.

Integration into the Community

One of the most important principles is that RIC personnel are not an elitist, specialized group of computer experts. Computer experts will of course be needed, but the likeness to corporate IT environments need go no further.

Psychologically speaking, the RII should encourage solutions for some of the more negative dynamics found in corporations and communities:

- Improving self-confidence, removing fears associated with ignorance of technology or being shown up as incompetent. This is particularly important for older users, people with low literacy and some disabled people.
- Reducing the potential for the RII to be used as a power base by particular individuals.

Principles of management at RICs which can ensure community engagement include:

- No professional may work in their specialization 100% of the time. The intention is to ensure that all personnel retain a holistic, current picture of the community as a whole.
- Personnel swaps should be arranged between RICs, in order to spread experience and ideas, and add to job interest.
- Personnel are required to interface to the community regularly, whether it be giving presentations on new services, training, helpdesk or other functions.
- Management of funding and budgets should be done with inclusive input from the community as well as the RIC, based on the vision the community has for itself. In other words every expenditure should be justified and prioritized as

part of a holistic community strategy, rather than the corporate situation of internal empires competing for a given budget.
- All RIC initiatives are requested by the community, not imposed upon it.
- There is a mechanism for community feedback on the performance of the RII.

One of the attractions of an RIC is that "real" technical jobs become available to university graduates from rural areas who would normally find a corporate city job. The employment of originally local people in skilled jobs in the local area would contribute to the quality of service, since local people know the problems and interests of their own community, and are better equipped to communicate with their own people.

Knowledge Bases and Community Memory

Support for knowledge bases is a core RII requirement: they form the *content* of the representation of communities and regions online. Two of the important knowledge resources in an RII have already been alluded to—demographic and geographic information; these are discussed in more detail below.

The basic scheme of information processing in a knowledge base is as follows:
- Information is *captured* from the real world and stored as data, which may be structured according to an *information model*.
- Higher levels of information may be *abstracted* or *synthesized* from stored data, e.g., by a process of review and editing.
- Information is made available by *dissemination*, or on request.

Abstraction and Synthesis

Abstraction are strategies for turning large amounts of data into usable knowledge. Primary data is captured from external processes, such as the numerous articles from an on-line discussion or field data about insects. It can then be distilled or edited into more abstract levels of information. To obtain a quality result, human intervention is necessary. For example, an on-line discussion of 350 articles about orchard-planting techniques might yield a five-page document explaining the essential ideas, if edited by an experienced writer. Clearly for the purposes of learning, this will be far more accessible and efficient than the original material.

Synthesis is another approach to dealing with large amounts of data. Internet search servers implement a simple synthesis function, using queries based on keywords to find Web pages. More powerful methods are used in corporate situations, and should be applied in the RII. In general, a synthesizing function requires that the data be structured according to an underlying information model.

Holistic Knowledge, Information Models, and Interoperability

The RII is likely to host knowledge which is both inherently complex (e.g., ecological models), and holistic (i.e., containing relationships between diverse elements). A lesson that has been learned the hard way in corporate contexts is that neither of these can be ignored. The temptation is to build information models which are too simple, without internal relationships, and systems which are not interoperable, preventing the representation of holistic knowledge.

For most people who never see the inside of database systems, the costs of bad information modeling are often experienced as repeated requests for the same information and an inability to correlate simple facts which are related in the real world.

To prevent the same errors in the RII, the following technical approach should be used:
- A common information modeling methodology should be used for all databases, ensuring interoperable information.
- Modeling should take account of relationships.
- Databases should be interoperable at the system level.

Demographics

Two basic types of self-knowledge are arguably essential to a good community information system: *demographic*, or information about people and organisations, and *geographic*, or spatial information describing the physical landscape. Almost all other information becomes more valuable if contextualized with respect to one or both of these.

Demographic data is not conceptually complex—it consists of names, addresses, contact numbers, and additionally for organisations, roles and places of business. However, the importance of a *reasonably complete* demographic information model should not be underestimated, since the various databases referring to demographic information often require different details to be recorded. The typical result of not making demographics a central concern of an information facility is numerous separate databases containing very similar information, creating a maintenance nightmare for administrators.

The other requirement of a demographic system is that it *integrates easily* with other databases. Prudent choice of implementation technology, in particular software component technology, can ensure this.

Geographical Information Systems

Geographical information systems, or GISs, provide a way of spatially mapping a great deal of information, including demographics. A GIS enables three-dimensional representations of the landscape to be created, on top of which layers of other information can be placed, and with which some very sophisticated analysis can be performed. For example, starting from a contour map of a region, a GIS can create an internal representation which can be used to calculate the locations and volumes of water runoff due to rain, determine the effects of building dams, or evaluate shadow formation due to sun movement. Additional layers containing representations of man-made structures such as buildings, water mains, electricity distribution networks and so on can be overlaid. Other useful layers include the biological: biodiversity data, wilderness areas, crop types, and the cultural: cinemas, clubs, cafes, galleries. The ability of the GIS to generate any combination of base maps and overlays, with numerous graphical display options, is what makes it such a powerful tool—it effectively transforms what may initially be a mass of numbers or text items into a graphical presentation quickly comprehensible to the human eye.

GISs are thus extremely valuable in agriculture, conservation and infrastructure development. Popular access to GIS-generated maps and statistics would be of inestimable value in educating local people about their area, and presenting other RII information. (See Burrough and McDonnell, 1998, for a good introduction to GIS.)

Learning Function
Information Sources

In practice, information is entered into systems via a user interface of some kind, but the important issue is its *source*. We can identify two broad classes of information source, namely, specialist users and the community itself.

Information systems have historically been concerned with specialist users, and a significant legacy of IT design exists in this area. In the RII context, specialist users might be naturalists entering wildlife data into a biodiversity database, or RIC employees managing the demographic system.

In the second category, one of the unique functions which the RII can perform is the *gathering of data directly from the community*. In some cases, user submissions to discussion forums, such as mailing lists and local discussion groups constitute a ready-made set of data which can be regarded as community property, and be used for any appropriate purpose. However, there are other circumstances in which users might supply data. Every person in the community, from amateur naturalists to cooking enthusiasts, represents a potential source of data. (In fact, in some disciplines such as entomology, much of the theory is built upon data gathered by amateurs in the field.) Community-sourced data could be provided by phone, e-mail or a form-based Web system.

With a community-based data-sourcing facility, the RII would give the community as a whole *access to its hidden expert knowledge*—the tipi builders, natural healers, birdwatchers, and local historians—a complete reversal of the science-based culture in which nothing that has not come from professional scientists can be accepted. Once community-sourced information bases are established, both experts and audience will materialize and self-organise.

The possibilities here are clearly immense; the only criterion for community-sourced information is that it be of potential interest to the rest of the community.

Quality of Ad Hoc Information

There are a couple of requirements for handling community-sourced information, since there is an inherent lack of uniformity and standardization in how it is provided. The first is that all items are attributed to a person or organisation, are timestamped, and can be verified to be the work of the stated author. The second is that some judgement of quality or veracity must be made with respect to each item. For example, an experienced farmer may contribute a design for the irrigation of avocado trees, which has proven itself over many years in the area. This item would be accorded a high level of general applicability. By contrast, a farm researcher might submit an untested suggestion for companion planting. This item might be marked "untested."

In general, some way of tagging all items with a level of quality must be used, ensuring that the content of the resulting databases can be used appropriately. This is likely to require human expert intervention. Note that the aim is not to discard suggestions or untested hypotheses; on the contrary, input from all quarters is encouraged. The quality-tagging system is designed to make sense of large amounts of data of inevitably variable quality and bias.

The RIC as a Learning Device

In the long term, users might try to submit information which already exists. This is as useful as first-time submission, since an identical or similar submission can be used to modify existing ones, improving their quality. For example, all field observations can be added to a database to build up both dynamic (time-based) and statistical pictures of an ecology, simply by aggregating identical observations appropriately. In design-related areas, users can review and modify existing submissions, or just indicate whether the knowledge is useful for their circumstances. (A successful precedent for just this dynamic is open source software development, described below.) Users could also provide their own quality ratings of services and products described in RIC databases.

At another level, RII information systems can be constructed to allow user customization of interfaces, remember traversal paths taken, and ultimately be an active and cooperative party in the user's exploration of community knowledge.

Active Knowledge

The worth of any knowledge resource is shown by its usage. To prevent RII knowledge bases becoming passive or elitist, they need to have a "live" feel. A simple way to achieve this is for experts to make themselves available via a discussion group (or even phone, if they prefer) to answer questions relating to each area. Useful exchanges would go into the knowledge base. From the point of view of the RII, a "home design centre," for example, might thus be: a home Web page; a learning knowledge base; a discussion group and a phone number. For the user it is no longer a "database," it is an expert service.

In summary, community information-sourcing leads to a number of benefits:
- The community remains actively engaged in using the RII.
- The comprehensiveness and overall quality of databases grows over time.
- It transforms the RIC from a normal database into a learning community memory.

Spheres of Interest

Spheres of interest define *private working contexts* within a larger information base, i.e. a place to create and modify information of any kind, including e-mail, discussions, documents, plans, and financial information.

A context could correspond to a project, such as the "alternative energy group" in a village, an organization, such as a local council or club, or a whole village. Access would be controlled on the basis of membership in the appropriate group. Nonmembers could be granted read-only access, or no access; however particular visitors (e.g. specialists from another country) might be given special rights. The intention is not to create empires of secret knowledge, but to impose some organization on what would otherwise become chaos.

Any person can belong to more than one sphere of interest. Some members of a group take on specific roles, such as "treasurer," "secretary," "chairperson," "administrator" and so on. It should be possible to establish identities for such roles in their own right, thus allowing, for example, e-mail to be sent to the treasurer, or the secretary to own certain document files.

An advantage of defining separate contexts to work in is that both what is inside the context and what is exported must be defined and maintained. While the interior may be "well-organised chaos," the exported view may be a concise presentation of the group's work, facilitating efficient communication between groups and the outside world.

Collaboration and Project Management

Collaboration is about *focussing resources* onto managed, cooperative activities to achieve a desired result. Processes exist at various levels, from the whole-of-community, such as a harvest, to the project group, such as planning and executing a street festival, to the individual, such as building a house.

Corporate experience has shown that useful tools for collaboration include e-mail, discussion groups, and mailing lists, as well as advanced planning, estimation and management tools; they create information *within* the contexts provided by spheres of interest.

In order to use such tools effectively, the general evolution of successful collaborations needs to be taken into account. This could be characterised as follows:

- Recognition and communication of shared interests/vision.
- Definition of shared frame of reference, or "operating rules," of which there are typically formalized and unstated components.
- Definition of goals to be met in achieving vision.
- Application of common effort (may require formal project management).
- Ongoing construction of shared body of knowledge.

A crucial requirement is the existence of a *facilitator*: someone who mediates and directs communication activities, manages engagement, and interprets both formal and cultural (unstated) rules during the process.

The above evolution could thus be supported with e-mail, discussion groups, an explicit facilitator role, and connections to RII knowledge bases.

Depending on the complexity of the intended endeavor, project management (PM) may be required. Most complex endeavors (e.g., farming, software construction) can be characterized as engineering processes, consisting of the activities: requirements definition, design, using existing patterns, implementation, testing, deployment and maintenance. Engineering project management is about the efficient organisation of resources during the stages of this process, within overall time and cost constraints.

As in corporate contexts, IT support for such processes includes document repositories and PM tools; however, the integration of these into other knowledge bases, community histories and learning facilities in the RII is also essential.

Decision Making and Conflict Resolution

The model of decision making in various communities around the world varies greatly, although various flavors of democracy are generally preferred. Whatever the case, the RII should not dictate a particular model, but should support the community's own model.

Take, for example, the case of participatory democracy. In order for the process to be meaningful, voting needs to occur on an *informed* basis. The RII can support this by providing a means of *information dissemination*, which must be simple enough to guarantee accessibility for all participants. Community discussion and debate are often required, especially for contentious issues. This may occur both online, via discussion groups, and face to face, in which case the RII may simply be used to organise meetings.

Actually making decisions may require simple voting software, or something more sophisticated, depending on the type of decision, the number of alternatives, and how much compromise is likely to be needed. An implementation of the on-line preferendum concept mentioned earlier appears to offer powerful possibilities.

Marketplace Mechanism

The activity of any economy revolves around trading, i.e., the exchange of goods and services. Information systems are an excellent means to model markets, since they can perform the searching, matching, and the subsequent exchange transactions.

The RII can provide marketplaces in a similar way to the large on-line stores on the Internet today. The principles of abstraction described earlier can be applied: a detailed but easy-to-use search engine makes the difference between each buyer spending hours browsing separate sites on the Web, or finding a product or service in minutes.

The Informed Marketplace

There may be profound implications of using local IT systems for marketplace trading. Consider trading in the globalized economy: most goods are sold to consumers with almost no information about how, where and at what environmental and social costs they were produced. Whether it be reconstituted orange juice ("produce of many countries") in a supermarket or a timber table (plantation or forest?), the typical consumer has very little hope of learning this information; in fact, beyond a cursory understanding of what the product is, most consumers decide to buy using a single datum: price. *Most consumption occurs on an uninformed basis.*

In order to redress this situation, a requirement could be devised which prevented vendors from being represented in any on-line marketplace unless they provided not only the "specification" of their product, but also a *minimum of mandatory data* on circumstances of production and distribution. A second require-ment could be that this kind of information is *as clearly displayed* as the basic data of price and availability.

These deceptively simple precepts make for an *informed marketplace*, and offer the possibility for RII to help bring about the age of the *informed consumer*, and the environmentally and *socially responsible producer*.

In an on-line produce market, for example, required information could include producer locations, plant and animal species and varieties, quality standards used, organic ratings, and kilometers travelled. Armed with such information, buyers would be able to exercise a much more informed choice than in the globalized supermarket paradigm; further, they would be able to source produce from closer to home, enabling them to buy fresher produce picked in its ripe state, with minimal levels of spoilage. On the production side, it would be possible to reduce the number of intermediate agents who buy, store and distribute produce, removing a significant cost component of today's food. In a market where produce can be pre-sold before it is picked, producers obtain a more reasonable price, and consumers obtain a better product.

A Level Playing Field

The on-line regional marketplace can make a further contribution to the community. In the current mainstream economy, large companies can afford expensive broadcast time and newspaper space for advertising, an impossibility for small companies. This occurs to such an extent that many small operators are prevented from entering the market; others are driven out, regardless of the quality of their product, simply because consumers do not know they exist.

However, in an RII, ethical and technical precepts can come together to improve equality. An ethical guideline may be stated which ensures that all companies, regardless of financial resource, have an equal *minimum amount of on-line "real estate" in the system*, above which they may purchase space as desired. The technical aspect is that the initial representation of a company seen by consumers, in a well-designed on-line marketplace, will not be the company's own Web pages, but a synthesized index designed by an RIC, under which each company will be given a few lines of information. Based on these summaries (approximately equal for all companies), users may then choose to jump to a Web site. Under this system, the marketplace is effectively owned by consumers, and companies are its clients.

Alternative Exchange Systems

An obvious function for an RII is to implement LETS for communities using or wanting to use local exchange. The first requirement is that the software developed reflects the model desired by the community (there are numerous alternatives under the LETS banner), and not an externally imposed one. Choices include: whether to publish accounts and whether to link the exchange unit with the national currency.

The other main requirement is that the LETS system be completely integrated with the information base of the RII marketplace, so that when agreements are struck, payments are correctly made.

Ecological Footprint Auditing

The problem of loss of informational feedback in the globalized economy was alluded to earlier. A unique function which regional, community-based information systems can perform is the auditing of local resource usage, enabling people to directly *see the extent of their own resource impact*.

To date, no serious attempt has been made to audit resource usage other than on a national basis, which is in any case, more part of the ideology of GDP economics than concern with the environment. But most environmental problems are local, albeit globally pervasive; to name a few: forest clearing, salinization, biodiversity loss and chemicalization of waterways. Consumption patterns are also quite regional, since they are culturally determined (although they are becoming more globally homogenized).

Auditing should try to identify three types of impact, namely, the local and remote impacts of the community, and the local impact of exportation. If some of these statistics could be gathered at household, community and regional levels, communities would become more aware of their total impact, and could begin to alter their consumption patterns.

While some resource impacts would admittedly be logistically difficult to calculate, resources for which usage can usually be estimated, and over which most communities can exercise some influence, include water, petroleum products, electricity, timber, and waste and pollution absorption amenity. Energy balance and resource input/output tables as used at the national level (see Boyden, Dovers and Shirlow, 1990) could be developed for regions or communities using RII systems, and publicized prominently to those same communities. Footprint calculations as described in Wackernagel and Rees, (1996), could also be used.

If studies in regional sustainability were able to suggest *sustainable targets* for usage of each resource, the publication of these targets, along with appropriate educational material, might form a basis for *self-limiting resource usage* by communities.

Agent of Global Change

One possibly surprising area where the combination of the community ideal and information technology could have a positive impact is in addressing poverty in the developing world. It has long been recognized that government aid programs and even the best efforts of non-governmental organizations (NGOs) have had at best, only marginal impact. A major obstacle to progress is that whatever small successes are achieved are eroded by the legacy of colonization: the tide of debt, wars financed by corruptly diverted money, and the interests of foreign corporations. In the end the resources arriving at the doorstep of the impoverished are pitiful.

Perhaps we should not be surprised that institutional "help" does not often work. The best source of help for a person in need is another person; by extension, the best help for a poor community could be a better off one, if only the means of contact existed.

The RII can help in a number of ways. If communities in wealthy countries voluntarily chose sister communities in poor countries (or vice versa), they could provide material and technical help, while benefiting from cultural interactions themselves. The RII would provide the communications support for programs such as:

- Provision of education and training materials; on-line education.
- Donation of "old" computer hardware and software, and other useful goods.
- Funding of scholarships and exchanges.
- Cultural programs.

The key requirement is a facility for developing communities anywhere to find helper communities elsewhere on the Internet; to this end each community should publish a short list of major skills, financial and material resources, required skills and knowledge and any other facts relevant to this process.

Global Networking and Internet Access

Access to the internet is a clear requirement of an RII. RICs in a regional infrastructure can perform the same function as existing Internet service providers (ISPs), but with a number of advantages. Firstly, since RICs are content-rich and provide numerous community-related functions, users are more likely to use them than generic ISPs for Internet access. Secondly, since RICs can be run on a not-for-profit basis, they may be able to provide cheaper Internet access.

With RIIs offering Internet access, two useful results follow:

- Users originally just wanting Internet access are exposed to their community, online.
- Existing ISPs would be motivated to become RICs in the RII.

Ultimately, a successful RII could have a "civilizing" effect on the Internet as a whole, by encouraging existing ISPs to become part of the infrastructure, and possibly even leading them to less profit-oriented business models, improving the level of inclusiveness. Given that it should also lead to the establishment of numerous RICs, each of which acts as a new ISP point of presence, the total number of Internet access points would grow, bringing information technology to new urban and rural communities alike.

Cooperative Model of Software Development

Software development will be one of the primary technical activities in the RII. There are two levels of software: *infrastructure*, which includes operating systems, networking, and security, and *application*, which includes all software produced to meet specific community programs. Infrastructure software is fairly generic, and much of it can be shared between RICs, while application software may be sharable between communities with similar needs.

Software development can be a complex process, and the culture adopted by RICs may have lasting effects on the RII as a whole. Most software in the world has been developed by companies, and is sold as binary executables, on a per-unit basis. The source code usually remains the property of the company, and users have little say in the development process.

However, some software has been developed under alternative models, and is made available in the public domain, or as shareware (voluntary payment of small sum), or under what are now known as "open source" licenses, which give users and other developers access and use rights of the source. Historically, some well-known tools for software developers were produced under this mechanism, known as "GNU" software. Very recently, industry interest in open development has exploded, due to the success of the free unix operating system, Linux, and some influential articles, such as Eric Raymond's "The Cathedral and the Bazaar" (1998).

The most interesting lesson for community informatics from this new movement is that good software can be developed in a situation of "controlled anarchy," whereby developers volunteer to cooperate on a particular project, and publish the results as freely available source. The effect is that skills are pooled, results are shared, and effort is not wasted on replicating solutions, indicating that such a model should be considered for the RII.

It should be noted that freely available software does not mean that no fee can be charged; there is no reason why RICs cannot be remunerated for development and support services.

Economic Sustainability

The overriding economic concern for an RII is to exist as a *sustainable* enterprise, that is to say, one whose long-term financial viability is taken into account at the start, and is always maintained. At the same time, the RII must be able to serve as broad a user base as possible, without marginalizing those with limited ability to pay.

The best way to serve the community interest economically is usually on a *not-for-profit basis*, since no further income is required above the expenses of the business.

To make the RII a viable business, outgoings have to be matched by capital and income. Possible sources of establishment capital include direct shareholding, one-off government funding, and secured loans from ethical investment funds.

Direct sources of income are likely to include:
- User fees.
- Transaction commission: a small fee (e.g., 2%) is deducted from every on-line marketplace transaction sell-price and it goes to the RIC.
- Local advertising (preferential rates to local producers?).
- Ongoing local, state or federal government funding.
- Local *ad hoc* funding on a per-project basis from businesses and governments.

Peripheral income might come from:
- Teaching exchanges with other RICs: learners coming to an RIC might pay something for training (or their RIC might do this).
- Larger RICs could perform accreditation services for products and services.
- Sale/swap of software solutions developed at an RIC.

The revenue generation strategies used by existing community networks, such as the FreeSpace model, described in Beamish (1995, Ch. 2, p. 29), would be worth investigating.

Making the RIC work as a viable economic entity requires achieving a balance amongst these sources, subject to ethical requirements already stated. In particular, the ubiquity and economic affordability of RIC services must be guaranteed; any user scheme must support this.

Whatever details are applicable to a particular community, it is essential that a viable financial plan for the RII be developed as early as possible, which is revisited often.

Technical Architecture

The technical design and implementation of a regional information infrastructure demands a book in its own right, but it is worth providing a brief overview,

Figure 2. RIC Software Architecture

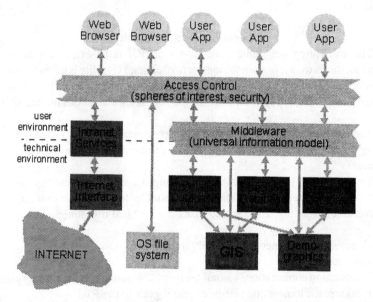

mainly to show that everything described so far can be implemented with existing technology.

Technically speaking, an RII would exist as a logical network of RICs, each constructed in much the same way as an ISP. Much of the software and *technical* knowledge is the same at all RICs, which means that early RICs could be used as design templates, allowing many development tasks to be done once and reused. With existing worldwide availability of cooperatively developed software, there should be no impediment to equipping even a very small RIC with sophisticated software.

The principle elements of an RIC IT architecture are illustrated in Figure 2.

From the user point of view, the RIC is presented through the use of standard browser applications, and in some cases, custom-designed applications. An access control layer implements both security and spheres of interest for all underlying services. In this model, it would be possible to define a particular group having its own e-mail identities, a specified membership, a group-private Web site, discussion groups, document area, database access, as well as an exported view of the Web site and some discussion groups.

The intranet provides a community-private e-mail, discussion groups, mailing lists, and Web servers, while the Internet connection connects the community to the outside world via similar services.

A middleware layer allows users to see information, not in terms of which database it comes from, but according to an integrated, logical information model.

Conclusion

The concept of community is profoundly important both for our development as self-aware individuals and members of society, and as an environmental survival strategy, since it offers an alternative pattern of resource usage to the growth economy.

As a survival strategy, its success is mandated upon widespread replication of the community pattern, in a manner not dissimilar to the replication of the corporations which control the industrial economy. This means that, like corporations, communities must have their own operational knowledge contexts, and must share their patterns. The regional information infrastructure offers a mechanism for doing this. Its own pattern must therefore be *defined* (the main aim of this chapter), including development and replication mechanisms, such as software and knowledge sharing. To achieve sufficient visibility, it must employ information technologies of *similar power* as in today's corporations. Once established, RIIs would allow physical communities to become actors in the world economy, and provide an operational matrix in which consumer power and choice can actually be exercised.

On the individual and community levels, an RII is a means of creating and evolving communities *in situ*, because it provides a *community mind* for group thinking and learning. As with the human organism, the community mind aims to be a reflection of the thinker's true nature and a tool for self-realization—not a means of controlling it.

For this reason, the RII requirements described here include no rules for enforcing social equality, diversity, or other moral value; instead they propose simple precepts which may be applied in order to encourage values cherished by a community.

References

Ayres, R.U. (1994). *Information, Entropy and Progress*. American Institute of Physics Press.

Beamish, A. (1995). *Communities Online: Community-Based Computer Networks*. Master in City Planning, Massachusetts Institute of Technology. Available at: http://alberti.mit.edu/arch/4.207/anneb/thesis/toc.html.

Boyden, S., Dovers, S., and Shirlow, M. (1990). *Our Biosphere Under Threat— Ecological Realities and Australia's Opportunities*. Oxford University Press Australia.

Burrough, P.A. and McDonnell, R.A. (1998). *Principles of Geographic Information Systems*. New York: Oxford University Press.

Daly, H. and Cobb, Jr., J. (1994). (2nd Ed.). *For the Common Good—Redirecting the Economy Toward Community, the Environment, and a Sustainable Future*. Boston MA: Beacon Press.

Dobson, A. (1995). (2nd Ed.). *Green Political Thought*. London: Routledge.

Jacobs, M. (1991). *The Green Economy—Environment, Sustainable Development and the Politics of the Future*. London: Pluto Press.

Lang, P. (1994). *LETS Work—Rebuilding the Local Economy*. Bristol: Grover Books.

Mander, J., and Goldsmith, E. (Eds.). (1996). *The Case Against the Global Economy and for a Turn Toward the Local*. San Francisco: Sierra Club Books.

Negroponte, N. (1995). *Being Digital*. Cambridge MA: MIT Press.

Newman, Dr. D. R. and Emerson, P. J. (1999). *The on-line preferendum: a tool for voting, conflict resolution and decision-making*. School of Management, Queen's University Belfast. Available at: http://www.qub.ac.uk/mgt/papers/prefer/

Raymond, E.S. (1998). *The Cathedral and the Bazaar*. Available at: http://www.tuxedo.org/~esr/writings/cathedral-bazaar/.

Schön, D.A., Sanyal, B., and Mitchell, W.J. (Eds.). (1999). *High Technology and Low-Income Communities—Prospects for the Positive Use of Advanced Information Technology*. Cambridge, Massachussets: MIT Press.

Simpson, L. (1999). *The New Pioneers: Rural Women and Communication Technology*. Keynote paper at Community Networking 99: Engaging Regionalism conference, University of Ballarat, Australia, October 1999.

Smith, R.W. and Barty, G. (1999). *HarvestRoad WebPOWER—Distributed Publishing & Management System*. Harvest Rd Communications, Pty., Ltd. Available at http://www.harvestroad.com.au.

Wackernagel, M., and Rees, W. (1996). *Our Ecological Footprint—Reducing Human Impact on the Earth*. Canada, BC: New Society Publishers.

Acknowledgments

Robert Tap, BSc, of Crystal Waters international community, for valuable insights on the community learning function; Launz Burch, BSc, writer and on-line educator for insights into the evolution and facilitation of groups and communities; James Beale, BSc, entomologist, for general review.

Chapter III

Embedding the Net: Community Empowerment in the Age of Information

Brian D. Loader, Barry Hague and Dave Eagle
University of Teesside, United Kingdom

Introduction

Throughout the world millions of people are getting online to the Internet to exchange information and communicate with each other to form what Howard Rheingold has famously described as 'virtual communities' (1994). The revolutionary potential of the new Information and Communications Technologies (ICTs), currently epitomized by the Internet and other Web-based technologies, to transform social relations has not surprisingly grasped the imagination of the media, academics, politicians, businesspeople and members of the public more generally. It has produced an extensive and often fierce debate about the possible beneficial consequences of such technological developments for social interaction which is based more around common interests rather than spatial proximity. Such optimistic visions have also been matched by alternative dystopian depictions of the new media facilitating the emergence of surveillance societies (Lyon, 1994; Davies, 1996). Yet, in whatever form the arguments are couched, their emphasis on remote communication often acts to disassociate individuals from the everyday experience of the communities they live in. It is as if there is no place for localized face-to-face interaction between people in the Information Age. Whilst we do not preclude 'communities of interest' and recognize that the term community itself can be used in many ways, our own approach to community informatics (CI) has been shaped by the desire to reconnect locally spaced communities to the wider electronic network of cyberspace.

Community informatics is an approach which offers the opportunity to connect cyberspace to community place: to investigate how ICTs can be geographically embedded and developed by community groups themselves to support networks of people who already know and care about each other. It thereby recognizes both the transforming qualities of ICTs as well as the continuing importance of community as an intermediate level of social life between the personal (individual/family) and the impersonal (institutional/global). The numerous community enthusiasts, some of whom are mentioned in this book, who are building interactive Web sites, virtual chat rooms and electronic-lists as tools to support local communication between their members, are a striking testament to the value of a CI perspective.

This chapter attempts to explore a model of CI which has as its primary focus the empowerment of individuals within local social networks by means of the adoption and exploitation of ICTs. It is an approach which emphasizes the need for the new technologies to be shaped by the human aspirations and desires emanating from existing community social structures rather than the expectation of new modes of social intercourse required to meet the technical, commercial and political designs of 'outsiders.' It seeks to challenge the often well-intentioned but flawed policy initiatives to impose technological solutions to perceived social problems from above. Instead it is an approach which stresses that technologies should be embedded within existing cultural and social relations. This is a perspective which requires the development of policy strategies aimed at stimulating the stock of what Robert Putnam has described as 'social capital' (Putnam, 1995): the strength of social networks, the level of trust between members, the degree of collaboration, the extent of participation, and the experience of negotiation which enables communities to control their own economic, cultural, and social development. It thereby proposes a new form of community development for the information age.

The chapter is structured first to provide a background to the difficulties of defining the notion of community, its continued importance for understanding social life, the limitations of cyber visionary approaches to CI and the need for an 'empowerment' perspective, and the essential requirements for such an approach in terms of access and awareness raising, skill development, and support. The final section of the chapter provides a more speculative discussion of an approach which addresses the problem of sustainability through the suggested inversion of current public service organization and flows of public resources.

Context

As in previous periods of economic and social restructuring the place of communities in our societies and their potential displacement has been a significant talking point. The dramatic social and cultural upheavals arising from the development of what Manuel Castells describes as global informational economies (Castells, 1996) and the concomitant processes of post industrialization experienced in many economically advanced societies are in large part driven (but not determined) by the transforming qualities of the new ICTs. Community structures built around modernist cultural identity and architectures, standardized work practices and related social discipline are frequently depicted in the popular media as in the process of terminal decay. Rising crime levels, the breakdown of family structures (particularly in western society), the persistence of high unemployment figures, significant educational underachievement and growing health inequalities are all common traits said to arise from the collapse of community living. Its replacement is portrayed as the emergence of atomized and fragmented social structures characterized by widening disparities of wealth and power which act to exclude large numbers of people from the economic and social benefits of society.

It should perhaps come as little surprise therefore that reforming social democratic politicians faced with widening social fissures and mounting social problems, should turn to the regeneration of community structures as a policy remedy for such social blight. In the UK, for example, the New Labour administration has placed community development as a cornerstone of its social and economic policies. It features significantly in debates about 'the third way' (Giddens, 1998a; Hargreaves, 1998). Predating such discussion, the role of local social networks in social and economic restructuring has been a significant aspect of USA policy initiatives. The theoretical underpinnings of such initiatives can be found in the work of "communitarian" movement, where scholars such as Amitai Etzioni (1995, 1997) Michael Walzer (1992) and Charles Taylor (1993) have all challenged the individualizing nature of free market liberal economics and its detrimental consequences for social cohesion and economic well-being.

Such concerns for the diminution of communities and the perceived decline of social life in late modern societies finds a strong resonance in the writings of Howard Rheingold whose championing of the Internet as a means for rebuilding social relations is already legendary (1994). Enthused particularly by his experiences as a participating member of the WELL, Rheingold was one of the first to highlight how the new 'many-to-many' communication capability of ICTs could change "the level of person-to-person interaction where relationships, friendships, and communities happen" (Ibid, p.12). The complex computerized communications networks, such as the Internet, were facilitating 'virtual communities' arising from the continuous interaction of millions of people around the globe. Such connectivity provided, for Rheingold, the prospect of countering "America's loss of a sense of a social commons" (Ibid) and rebuilding collective social structures. The advantages of virtual community structures could be demonstrated by the very experiences of the rapidly growing numbers of on-line participants. Such electronic networks acted to facilitate social support between members, enabled access to a wealth of information and advice often altruistically provided by other participants, provided a basis for enhanced political deliberation and democratic activity and offered the potential for mutual respect of personal development and expression.

A clear affinity is thus discernible between those who favour a communitarian approach to policy-making and the prophets who envisage cyberspace as a new electronic social commons. Both perspectives share a *fin de siecle* foreboding for social and economic development and each proposes the search for mechanisms to reestablish community social structures as a means of redemption. The convergence of these approaches also provides the foundation for the development of CI which draws heavily upon the theory and practice of each. But before we can go on to explore the themes and issues raised by the emergence of this new field of inquiry it is necessary to question, albeit briefly, two significant predications. First, that communities are beneficial for social, economic and political development. Second, that ICTs are indeed likely to foster greater community cohesion.

A Common Sense of Community?

The concept of community is notorious for being commonly understood by the majority of people whilst being universally indefinable by academics. It is imbued with a halo of comforting optimism which politicians find hard to resist when appealing to the electorate. Yet it remains stubbornly difficult to find an acceptable concord about a single meaning. For Raymond Plant all definitions of community remain 'essentially contested' (Plant et al., 1980, Ch. 9). It can be spoken of as a geographical phenomena where individuals share a common sense of territory or place. However, this tells us little about whether the local residents interact or even know of each other's existence. An alternative use of the term usually refers to communities which have a common interest. These are used to describe a social structure based around shared characteristics other than location, such as ethnic origin, religion, leisure or occupation. Again, this does not imply that members of such social networks have much commitment to the local social systems of which they are a part. Consequently, yet a further categorization of community can be found in the literature which relates to the sentiments often associated with community action or pride. Wilmott, for example, refers to the idea of 'community of attachment' (1986, Ch. 6) to capture this understanding, whilst Lee and Newby offer a similar notion of shared identity which they call 'communion' (1983, Ch. 4).

Each of the above concepts can be seen as conceptually distinct from each other. All contain explanatory uses and limitations. At times there may be significant synergy between them—with shared locality, interests and communion converging. The contention that this represents an idealized form from which late modern society is diverging is, however, probably both historically incorrect as well as unrealizable in contemporary society. It is not only misplaced to confuse these different variants and create unrealistic expectations, but it also fails to acknowledge the negative aspects to be experienced from being a part of some communities. The strength of social ties can result as much from bigotry against other cultural groups as from people caring for each in a particular vicinity. Notions of 'us-and-them' generated from perceptions of 'outsiders' and the development of personal identity which arises from them are surely capable of creating a shared and mutually reinforcing oppressive environment. Moreover, even the most idyllic country village setting may be perceived quite differently by some of those who are its inhabitants. Fiona Williams suggests, for example, that the experience of many women living in such commuter belt communities is of a daily imprisonment whilst their spouses are empowered and stimulated through engagement with the wider world (Williams, 1993). From these angles, it is not even clear that close-knit communities are desirable at all.

Yet the potential stifling and one-dimensional cultural qualities of local social relations no more deserve to be exaggerated than their apocryphal counterparts. More valuable, we would assert, is to keep very much in mind the recognition that geographical location does not naturally lead to social relations and that what is of interest are the circumstances which produce such a variety of strong, weak or nonexistent social networks. A methodology which is likely to inform such inquiry and which can help lay the foundations for a CI perspective is social network analysis (Scott, 1991; Wellman, 1988). Although still relatively new, it has made significant progress in recent years as a means of understanding communities as networks of social relationships which can encompass all three of the above conceptions of community. It is an approach which suggests that community structures can be heterogeneous, capable of a great deal of change, and subject to a complex range of phenomena shaping their construction (Suttles, 1972). Such a framework has the potential advantage of offering CI a modus operandi which avoids naive normative entrapment while enabling the critical exploration of the social shaping capabilities of ICTs for community empowerment.

Whatever the difficulties of finding an agreed meaning of the term, community life remains very important as what Bulmer (1989, p.253) describes as 'intermediary structures' which bridge the social divide between personal and family relations and the wider societal institutions which people often find impersonal. In its various forms it provides a public space where individuals can ground their moral code through interpersonal experience. Somewhere that offers the possibility of developing a shared sense of obligation and mutual dependence, a social context for individuals to feel 'rooted.' "A human being has roots by virtue of his real, active and natural participation in the life of a community which preserves in living shape certain particular treasures of the past and certain particular expectations for the future" (Weil, 1952, p.41). Yet also a public domain where moral and political values can be contested and debated in a condition of equality and mutual respect between members.

Such a space is likely to remain crucial for social and economic development for some time to come. What remains to be clarified is the possible effect of ICTs on the development of such intermediary social networks.

Enabling Technologies?

It is something of a paradox that the very technologies that are being heralded as providing the basis for a resurgence of community life may also be the medium for introducing greater social discord and individual isolation. Communications media which are capable of connecting more people together around the globe and transmitting an ever greater volume of multifarious information are also capable of making people more remote from each other (Haywood, 1998). Indeed, recent research has even suggested that "greater use of the Internet was associated with declines in participants' communication with family members in the household, declines in the size of their social circle, and increases in their depression and loneliness" (Kraut et al., 1998, p.1). Similarly, the poor quality of deliberation and discussion on the Internet has also called into question the value of electronic forums as public spaces (Wilhelm, 1998).

Such analyses of computer-mediated communication (CMC) provide a sharp contrast to the earlier utopian prophesies of cyber visionaries such as John Perry Barlow (1996a, 1996b). Barlow and others believed that they were "creating a world that all may enter without privilege or prejudice accorded by race, economic power, military force, or station of birth" (Barlow 1996b). The Internet could produce a space unfettered by the social, political, cultural and economic encumbrances of the modern world. Only here could individuals shake off the identities bestowed upon them by government agencies, corporate institutions and social conditioning.

Liberation would be found through the unconstrained free expression of the individual mind. The limitations of this vision have been dealt with elsewhere (Loader, 1997), but its attraction remains powerful to those disenchanted with current social trends and represents, we suggest, a misleading view of the relationship between the new media and community development.

Instead, a picture begins to emerge which suggests growing divisions between social strata (UN Human Development Report, 1996) whose fault lines are increasingly marked by the economic opportunities afforded by access to the new technologies and the new modes of culture arising from them. Indeed, many of the most recent additions to the world's top wealthy people derive their position from the new technologies and their associated knowledge industries (Bill Gates currently occupies pole position of the *Forbes* magazine's ranking of the world's billionaires). Anthony Giddens suggests these recent trends in inequality give rise to two forms of social and political exclusion. "One is the exclusion of those at the bottom, cut off from the mainstream of opportunities society has to offer. At the top is voluntary exclusion, the 'revolt of the elites': a withdrawal from public institutions on the part of the more affluent groups, who choose to live separately from the rest of society" (1998b, p.27). What Giddens does not make explicit, however, is the relationship between this social cleavage and the contention that the processes of inclusion and exclusion may be increasingly structured by the social and economic construction of the new technologies and their diffusion.

For many of the world's most affluent members new ICTs do indeed appear to offer access to a global information economy, most notably the financial markets, which enable the transcendence of time and space boundaries. Network connectivity thereby provides the potential for such an elite to evade the clutches of democratic representational intrusion. Such 'voluntary exclusion' by contrast is not available to the majority of the world's population whose life opportunities are more closely associated with spaces which are places: bounded, culturally specific and historically defined (Graham & Marvin, 1996). Even leaving aside the stark disparities between countries (Holderness, 1998), whole areas of cities and rural communities in advanced societies are being excluded from the economic and political opportunities to interact online. The increasingly liberalised telecommunications markets in many countries have led to the uneven development of the infrastructure supply. Distorted by 'cherry picking' of the most profitable geographical locations, it leads to the likelihood of existing marginalised social areas becoming more excluded as 'virtual ghettos.' Urban neighbourhoods and districts where a high proportion of the inhabitants do not even have a telephone in their home are not likely to be on the threshold of entering Barlovian cyberspace.

What is becoming seen by some as the new orthodoxy (Webster & Robins, 1998) is the contention that global economic restructuring is spawning a new digital elite: the 'digerati.' As yet such formulations are somewhat ill defined but they have in common the notion that a new global social strata is emerging which is capable of successfully interacting with and shaping the global informational economy. Robert Reich for example identifies the emergence of 'symbolic analysts' within a tripartite division of America's labour market. "Symbolic analysts solve, identify, and broker problems by manipulating symbols. They simplify reality into abstract images that can be rearranged, juggled, experimented with, communicated to other specialists, and then, eventually, transformed back into reality" (1992, p.178). They are an increasingly rootless group who depend for their well-being upon "the value they add to the global economy through their skills and insights" (1992, p.196). It is precisely this group which is more likely to 'voluntarily exclude' themselves from the reciprocal moral confines of geographically grounded democratic communities by their cultural inclusion into the informational elite. Ironically, in many respects John Perry Barlow personifies this new privileged caste.

In its most dystopian vision, this new electronic elite is able to gain advantage through its domination of the cultural attributes associated with CMC. "Increasingly," maintains Castells, "CMC will be critical in shaping future culture, and increasingly the elites who have shaped its format will be structurally advantaged in the emerging society" (1996, p.360). Knowledge capital, and its perpetual conversion, becomes the resource which differentiates life opportunities. The creation of cultural codes, symbols and language enable the digerati to fix the perimeters of discursive interaction. Control over their wider social assimilation is the foundation of political power. "The price to pay for inclusion in the system is to adapt to its logic, to its language, to its points of entry, to its encoding and decoding" (Castells, 1996, p.374).

The new technologies cannot therefore be easily divorced from relations of power based upon existing social and economic structures. ICTs are socially constructed and it is therefore of vital importance to know how such tools are developed. How democratic, we might ask, is the process of the social shaping of the new ICTs? (Sclove, 1995). Or as Castells cogently remarks "Who are the interacting and who are the interacted in the new system...largely frames the system of domination and the processes of liberation in the informational society" (Castells, 1996, p.374). What appears to be emerging to-date is a fragmented pattern of access to networks dominated by a relatively small, well-educated interactive elite whilst the majority are encouraged to 'lock-into' a prefigured network of 'video-on-demand,' tele-gambling, electronic plebiscites and home-shopping facilities.

Technologies, then, are not value free but neither do they determine our future. They are shaped and developed through social relations and thereby offer the potential of being used as a tool by the disadvantaged and excluded to challenge entrenched positions and structures. A great deal of research suggests that feelings of powerlessness can be detrimental to people's health, educational achievement, and employment opportunities. Robert Putnam, conversely, has emphasized the importance of the level of 'social capital' within a community for individual fulfillment. Such analyses place the collective empowerment of people at the forefront of the economic and social development of communities. The manner in which this may be realized, in our view, is a central concern of CI. Its facilitation we suggest is dependant upon three interrelated themes—access and raising awareness, skill development and support. It is to each of these and the issues they raise that our attention will now turn.

Access and Awareness

If CI initiatives are to deliver the empowerment of socially and economically disadvantaged individuals and communities on which, we argue, they are premised, then it is vital that key issues concerning the broadening of access to ICTs are identified and addressed. We have argued elsewhere that, to-date, government sponsored initiatives to address the question of access have focused largely on broadening access to ICT hardware and software and providing widely available basic training in their use. We have further argued that, laudable as these initiatives are, there is a need to broaden the debate to include consideration of access to information, access to 'community networks,' access to decision-makers, and access to a basic source of income (see Hague and Loader, 1999, Ch. 1). Working within this broader framework, what follows is an attempt to raise some key issues concerning access to electronic networks and to describe the way in which these issues have been addressed and the lessons that have been learned in relation to our own work in the North of England.

Cost, Ownership and Control

Broadening physical access to ICTs clearly involves substantial costs in terms of hardware, software, technical support and telephony (the latter being particularly important in a UK context where currently relatively high local call costs can make Internet access prohibitively expensive). To some extent, those for whom personal or community group ownership is not a viable option can make use of existing and growing free or low-cost public access points sponsored by national and local governments, voluntary organisations and, in some cases, the commercial sector. These are usually situated in public libraries, town halls, high street locations and the like. However, praiseworthy and important as such provision may be, we would argue that it is no substitute for the provision of ICT facilities that are owned and managed by and for local community groups. Community ownership and control of the physical kit, its location, the uses to which it is put, and the way in which access is broadened and coordinated are essential components of the approach to CI we advocate. In short, we would contend that the development of an ICT literate and skilled workforce is but a small part of a more fundamental objective to foster more motivated and independent citizens by giving them ownership of key resources through which they can control and shape the production, retrieval and dissemination of information.

From our experience, within a UK context, there are two distinct but related types of resource bases upon which community groups can draw in attempts to establish and maintain community-owned and -controlled ICT facilities. The first is the range of grant-awarding bodies including the EU (e.g., ERDF and ESF funds), national government (e.g., SRB), Quango's (e.g., National Lottery Charities Board) and local authorities. The fact that there are so many sources of grant aid for ICT-related community projects is something of a double-edged sword. On the one hand, it might be argued that there are plentiful supplies of money out there awaiting the enterprising community group. On the other hand, it must be acknowledged that the plethora of potential sources and their multifarious bidding procedures present a potential barrier to such groups. In recognition of both the importance and complexity of the fund-raising process, many large organisations, such as local authorities and universities in the UK, now employ specialist staff to advise on funding sources and put together bids. If community groups are to effectively locate and secure the grant aid moneys that are out there, then they too must have access to such expertise. Those concerned to promote CI initiatives should be mindful of such a need and seek ways to make the necessary expertise available.

The second type of resource base can be found closer to home and constitutes the range of organisations, groups and individuals who are natural 'stakeholders' in local ICT-related community development initiatives. Thus cash or in-kind contributions might be sought from all those who have an interest in the social and economic well-being of the local community generally and/or a specific interest in the provision and uptake of ICT-related products and services. This would include, for example, local public service organisations and information providers (local governments, health authorities, the Benefits Agency, etc.), educational establishments, hardware and software manufacturers, telecommunications providers, and other local commercial interests. Of course, identifying those who may have an interest in the development of a particular project is one thing and getting them involved in a partnership for its achievement is quite another. The latter, like any relationship, involves the cultivation of a situation where each partner feels they can both give and receive and that the balance between these two is agreeable.

A concrete example of at least the beginnings of the kind of 'stakeholder' partnership described above can be found in the Trimdon Digital Village project (see Loader, Hague and Brookes, 2000). Here, the University of Teesside recruited local residents onto tailor-made, free-of-charge at point of access, ICT related 'local learning' courses (see 'Developing Skills,' this chapter). This enabled core funding from the Higher Education Funding Council (HEFCE) to be drawn down which helped finance the installation of the necessary hardware and software through the university acting as a stakeholder. Thus the facilities will remain in the community not only for the continuation of courses delivered within the local community but also to provide access to a much wider range of applications. British Telecommunications also provided some free telephony to facilitate Internet access, in return receiving unquantifiable benefits in terms of public relations as well as building a future market for its products. In addition, community members were assisted in creating their own presence on the WWW (www.btinternet.com/~trimdon2000-trimdon2000/) by a local Web-authoring business (Daelnet: www.daelnet.co.uk/) and the TDV project has benefited from countless hours of voluntary time committed by community members towards project development, fund-raising and training.

Two notes of caution should be added to the above tale. First, attempts to bring other significant stakeholders on board have, to date, proved problematic. In particular, public service information providers have shown a reluctance to commit resources to the project. This may, in part, be explained by the inherently conservative nature of bureaucratic organisations combining with the fragmentation within and between these organisations to mitigate against a swift and positive response to such a new and innovative project. It may also, however, provide evidence of a reluctance to cede power to local community groups in the manner required by CI projects based upon community ownership facilitated by a partnership of stakeholders. The second note of caution concerns the sustainability of the project and others like it (see 'Sustainability,' this chapter). If experiments like TDV are to endure, then ways must be found to meet long-term costs including kit replacement and updating , and ongoing costs for technical support and telephony.

Location, Location, Location

As any good store keeper will tell you, location matters. Many initiatives aimed at improving access to ICTs, both in the UK and elsewhere, have involved the development of purpose-built 'electronic village halls' or community access points at huge expense and often with very little in the way of prior community consultation. It is of little surprise to us that some of these facilities very quickly become as undervalued and undesirable to potential 'buyers' as would a luxury mansion located on the hard shoulder of a motorway.

It is our contention that decisions concerning the physical location of a kit that is being installed with the purpose of broadening community access should be guided by one overarching principle. This is that, consistent with a raison d'être of empowerment, community members themselves are best placed to decide where the kit should be located for maximum accessibility. Of course, a basic grasp of marketing principles would tell us this. Yet it is surprising how many large-scale projects are embarked upon with inadequate or nonexistent community consultation or 'market research.' CI is clearly as much to do with developing techniques and methodologies for involving community members in collective decision making as it about hardware and software development.

From the starting point of the above principle, our research with local communities would suggest two further rules of thumb as a guide to location. The first is that the introduction of ICTs should be done in a way that is sensitive to: existing information requirements and patterns of information retrieval; existing vehicles and locations for communication and interaction within and beyond the community; and existing social, economic and political networks. In short, those attempting to establish electronic community networks should, wherever possible, work with existing local social networks, facilities and gathering places. Of course, this existing infrastructure may be fraught with economic, social, political and cultural divisions. Of course, it may be a source of oppression and exclusion. But it is our belief that CI initiatives must be introduced in a way that is sensitive to such realities and attempt to work with and not round them. To locate a new purpose-built ICT centre in the middle of a community and declare it open to all is not to overcome these hurdles, it is to ignore them and, potentially, fall at the first.

The second rule of thumb, which follows from the first, is that multiple sites are preferable to one central site. Attempts to accelerate the diffusion of ICTs and to encourage new and innovative uses must recognize the heterogeneous and contested nature of communities. Different individuals and groups within any community use established information and communication tools in different ways, for different reasons, at different times, and in different places. CI initiatives driven by a commitment to community empowerment ought to recognize and encourage such diversity, whilst also recognizing the existence of underlying inequalities that might be either challenged, reinforced or even exacerbated through the introduction of ICTs. Again, an 'open to all' community ICT centre is not offering equal access to the whole community if, as is likely in practice, it systematically attracts certain groups and individuals and discourages others.

The approach to location of ICTs in the community advocated here can be illustrated with two brief examples. The first, again, comes from TDV. Following extensive consultation with the local community (based on discussions with the Trimdon 2000 community regeneration group and questions in a Community Appraisal survey conducted by this group), it was decided initially to locate on-line machines at three sites distributed throughout the village: the Community College, the Library, and the Labour Club (the network has since been expanded to include the village primary schools and a local pub). Evidence from ongoing survey-based evaluation of the project provides a ringing endorsement for the 'multi-site' design of TDV. Eight months after public access to the TDV computers was established, some 37% of survey respondents reported that someone in their household was making use of them. The figures for the respective sites are 24% making use of the Community College computers, 16% making use of the Library and 14% making use of the Labour Club. Clearly, it is different people who are making use of the computers at the respective sites. This survey evidence, backed by evidence from focus groups run with local residents, suggests that people will use the computers at a location that they feel comfortable with and already make use of for other purposes, and that these locations will vary from individual to individual, and group to group.

The second example comes from the Hendon Hub project, an SRB- and ESF-funded project involving the development of an electronic network of five community groups in the east end of Sunderland (an area in the North East of England formerly associated with ship building and now experiencing high levels of socioeconomic deprivation). The groups comprise three youth and community projects, a community care project and a women's centre, and each has had an on-line PC with videoconferencing facilities, networked to a central server, installed on their premises. Again, focus group evidence suggests that people are making use of these computers who would simply not have considered accessing computers at a purpose-built location, or indeed any location other than the one that they visit to 'do other things.' In other words, the kit is being used because it is available to complement existing patterns of social support and networking. The Hendon Hub project is still in its infancy and subject to ongoing evaluation. However, there is room for cautious optimism in terms of its impact on community access to ICTs. Some of the caution may dissipate if and when the number of sites can be expanded to more adequately reflect the diversity that exists within even a small and relatively tight-knit community such as Hendon.

Content: What are We Selling?

Borrowing from the language of marketing again, in seeking to 'sell' a particular product or service, we follow a process of identifying the customer's needs, matching these needs to product features, and selling these features as benefits to the prospective customer. Let us consider the diffusion of ICTs to socially and economically disadvantaged individuals, groups and communities within this framework.

Nobody can have escaped the hype surrounding ICTs in general and the Internet in particular that has emanated from a variety of sources (including politicians, public servants, academics, and the computer and telecommunications industries) in recent years. In policy terms, the selling proposition adopted by many politicians and public servants has been as follows:

ICTs are a good thing per se. Those who can access and have the skills to utilise these ICTs will gain obvious advantages (primarily economic) for themselves and will be more useful (primarily economically) to society. Those individuals and communities who do not 'skill up' for the Information Society lay themselves open to economic and social marginalisation.

The above scenario is open to question on at least three counts. Firstly, it is far from clear that ICTs are, or are considered to be by most people, a good thing per se. ICTs, and the uses to which they are put, are subject to social shaping and as such lend themselves to the achievement of a variety of possible outcomes, the desirability of which is contestable. What is clear, it seems to us, is that for all but a small minority of people, ICTs have little or no intrinsic interest. There is little needs based demand for ICTs per se. The needs-based demand that is out there in abundance is a multifarious demand for distinct types of information and communication-based content to which ICTs might facilitate access. It follows from this that the features that should be sold to the ICT uninitiated are not the bells and whistles of the technology but specific, tailor-made packages of content to which it might afford access.

Secondly, it might be argued that the diffusion rate of ICTs is not best served by emphasizing negative benefits in terms of the fate awaiting those who do not skill up for the Information Society. Research on the diffusion of innovations consistently shows that 'preventive innovations,' i.e., those designed to ward off some undesired future state or event, have a relatively lower rate of adoption than innovations that are designed to emphasize positive benefits to the adopter's lifestyle (Rogers, 1983: Ch. 5).

Thirdly, the link between the feature of ICT skilling and the purported benefit of economic prosperity is highly contentious, particularly in relation to geographic communities. The empirical evidence to back such a claim is simply not yet available. Furthermore, a selling proposition linking ICT skilling to employment prospects is likely to prove very resistible to individuals and communities experiencing third-generation structural unemployment. This is not to dismiss the possibility of an important role for ICTs in economic regeneration. It is simply to urge caution in terms of assuming a simple causal link and to point out the hollowness of the selling proposition within disadvantaged communities.

The above issues, we feel, are best addressed through a 'bottom-up' approach to the diffusion of ICTs through CI projects. Such an approach starts from the premise that, whilst for many, ICTs have little intrinsic interest, a 'hook' can be found for all or most people to make them consider engaging with new technology. Again, the approach advocated here foregrounds existing behaviour and needs in relation to work, leisure and community life and targets 'awareness raising' activities here rather than promoting any intrinsic value of ICTs.

Many active and dynamic groups, clubs and societies exist in the community. One approach to targeted awareness raising is to tailor activities to these existing interest groups and capitalise on the enthusiasm already present. An advantage of this approach is that interest groups tend to exist across gender, age, and ethnic barriers.

Let us briefly consider a couple of examples. Local historians abound in the North of England. The processes of the local historian can be greatly aided by the use of ICTs, with multimedia databases, scanners, and the Internet having obvious potential in gathering, storing, and disseminating information. Offering general IT training courses to local historians may be attractive to some. Offering local history courses, which use ICTs and include specific training to develop the required skills will be likely to be attractive to many more. Electronic music, particularly its generation with easy to use software packages, appeals to a very wide cross-section of the community. A cheap and cheerful low-end package for the creation of dance tracks can be an excellent icebreaker for youth groups in particular, opening up opportunities for broader awareness raising and more substantial projects.

In the borrowed language we have thus far adopted, the above is the equivalent of the established practice of 'niche marketing.' What moves it beyond marketing and associations with market based exchange are, we would argue, the philosophical underpinnings that provide justification for community informatics as advocated here. To locate these we turn to political theory and, in particular, to David Held's (1996) concept of 'democratic autonomy.' For Held:

...politics is a phenomenon found in and between all groups, institutions and societies, cutting across public and private life. It is expressed in all the activities of cooperation, negotiation and struggle over the use and distribution of resources. It is involved in all the relations, institutions and structures which are implicated in the activities of production and reproduction in the life of societies...(1996, p.310).

The principle of democratic autonomy therefore requires that:

"[people] should be able to participate in a process of debate and deliberation, open to all on a free and equal basis, about matters of pressing public concern" (Ibid., p.302).

It is for this reason, rather than any intrinsic qualities of ICTs, that citizens have a right to access the new information and communication channels made possible by new technologies and to have a say in decisions about their use and development. Furthermore, it is our contention that such citizen involvement must include collective involvement at the level of the local geographic community. For it is here, to a large extent, that moral identity is shaped and a frame of reference within which to interpret the world is formed (for 'communitarian' arguments along these lines, see e.g., Bell, 1993).

Community informatics should, then, involve inviting community groups to explore the potential for ICTs to assist in meeting their needs and to become involved with the development and use of applications or features designed for that purpose. In this way, the benefits that are being sold relate to an enhancement of existing social and political activity and interaction, and ICTs are required to play a facilitating role. This does not preclude the possibility, and exciting prospect, of ICTs opening up new forms of activity and interaction. It simply recognises that the value of new ICTs may be most effectively introduced and understood in relation to the familiar.

A crucial aspect of need related to access to ICTs concerns access to high quality and relevant information. From the CI perspective advocated here, this means not only equipping people with the skills to effectively navigate their way around the sea of information available on the WWW, but also encouraging the development and provision of locally specific electronic information. As an example, our research with communities in the North of England continually provides evidence of a demand for, and difficulty in obtaining, the same kinds of locally based information. Prominent among these are information on local council services, health services, welfare benefits and leisure services. Of course, some of the organisations responsible for providing such information have established or are in the process of establishing a WWW presence. However, such Web sites generally do not currently provide an efficient vehicle for the specific single user enquiry. More specifically they don't tend to be interactive. For example, it is not currently possible, in the UK at least, for an individual to be able to look for information on local dentists, select their preferred dentist, check surgery times, register their details, seek advice on their condition, and make a convenient appointment if necessary, all within one easy to navigate on-line session.

Of course, the technology itself presents no insurmountable barriers to the provision of the kind of facility described above. A range of other factors mitigate against such provision, including the aforementioned fragmentation of public service information providers, the lack of a Web presence for some, and the marked lack of interactivity built into the Web sites of others. What is required is a willingness on the part of public service information providers to commit resources to the provision of high quality on-line information, to coordinate this on-line provision with that of other information providers, and to open themselves up to electronic dialogue with service users.

Broadening the context slightly, the interactive capabilities of ICTs open up all kinds of possibilities for electronically facilitated 'digital democracy' based on citizen deliberation and direct access to the decision-making process. However, again, to-date in the UK the reality lags significantly behind both the rhetoric and technological capability (see Hague and Loader, 1999).

The organisational, financial and technological barriers to the achievement of electronic service delivery and digital democracy are not insignificant. It is our contention, however, that the major obstacles are cultural and political in nature. Quite simply, it may be too much to expect politicians and professionals to cede power to people through facilitating electronic interactivity. Again, the value of a CI approach as advocated here comes into focus. By raising awareness of the capabilities of new ICTs; developing the confidence, skills and imagination to think about novel applications; and encouraging collective community action (see 'Access to Community Networks' below), CI initiatives may provide a catalyst for empowering the community to exert pressure from below for new kinds of electronically assisted public service and political activity.

A final, and fundamental, point to note about access to information is that surely one of the most exciting and potentially empowering aspects of new ICTs is the opportunity created for individual citizens and community groups to cease being simply consumers and become producers, controllers and manipulators of information. Many people who were previously economically and culturally disbarred from such activities are now producing their own electronic publications and have a global presence on the WWW. In terms of content, this is clearly an area where CI initiatives should be concentrating efforts to widen and deepen access.

Access to Community Networks

From our perspective, one of the most interesting questions relating to access to ICTs concerns the extent to which they can be geographically embedded and developed by community groups to support networks of people who already know and care about each other. The numerous community enthusiasts who are building interactive Web sites and electronic lists as tools to support local communication between their members are testament to the value of this aspect of CI. However, there are a number of questions concerning the effects ICTs might have on access to such networks that are open to empirical investigation but that, to our knowledge, remain largely unanswered. Many of these questions relate to the level and distribution of 'social capital' within communities (see Loader, Hague and Brookes, 1999, for further details of our view on the relationship between CI and social capital):

Is a high level of social capital a necessary prior condition for establishing a successful electronic community network?

From our limited experience with the communities we have worked with to date, we would not go so far as to claim a high stock of social capital as a necessary prior condition for successful CI. However, we would postulate a positive relationship between the level of social capital and such factors as initial acceptance of electronic networking, diffusion rates within the community, quality of interaction, and sustainability. Certainly, the most successful experiment in geographically based CI with which we have been involved is taking place in a community with a strong collectivist ethos surviving from the old traditional industries around which it grew up, and a high level of participation in a wide range of community activities (see Loader, Hague and Brookes, 1999).

Can CI Activities Facilitate the Building of Social Capital?

In assessing the extent to which CI initiatives might empower local communities, we should consider the extent to which they encourage people to engage in collective activity, foster local connectivity and help to build the feelings of solidarity, reciprocity, trust and altruism on which social capital is founded. Again, our limited experience has thrown up some encouraging examples. One of these concerns the aforementioned Hendon Hub initiative. Here, key 'access workers' within the five projects involved have made face-to face contact through a series of ICT training workshops and are using the Hub network to exchange information and offer mutual support (largely on ICT related matters such as software applications, technical problems, and skills) using e-mail. At the time of writing, this contact is being extended to the use of video conferencing facilities and the projects are collaborating on the establishment of a common presence on the Web. Despite the fact that they are literally only a couple of miles distant from one another, there had been little or no contact between the projects prior to the establishment of the Hub.

What are the Characteristics of Network Users vis-à-vis Non-Users?

Given the heterogeneous makeup of even the most tight knit community, it can be assumed that access to social capital will be unevenly distributed. In other words, some community members will be relatively well integrated and connected within the local community and relatively rich in terms of their ability to draw upon social capital. The prospect then arises that it is these already advantaged groups and individuals who will have most to gain from the establishment of an electronic community network. Quite simply, it may be the case that those who are well networked in 'real' space, will be those who are well networked in 'cyberspace.' Our experience certainly lends weight to this argument. For example, survey evidence from TDV suggests that the likelihood of an individual making use of the Trimdon computers is positively related to their level of involvement with the Trimdon 2000 community regeneration group.

Each of the above questions justifies further empirical research (including the measurement of social capital itself) and we would urge all those involved in CI experiments to consider them in evaluating individual projects. In the longer term we would contend that, if CI is to establish itself as a distinct discipline or field of study, then some common methodological tools should be developed.

It is our hope that the above treatment has encouraged a broadening out of the debate surrounding access to ICTs. That said, it should be pointed out that one crucial aspect of access is missing from it—namely, the development of skills—and it is to this that we now turn.

Developing Skills

Clearly, a key challenge in relation to broadening access is addressing the huge gap between the potential of ICTs and the skills necessary to exploit that potential which exists throughout society and to varying degrees of severity based on such factors as geography, age, socioeconomic status and educational attainment. Policy makers have not been slow to realise this and government sponsored initiatives abound. In the UK, for example, we have witnessed 'IT for All,' 'Computers Don't Bite' and currently, 'Web Wise' (www.bbc.co.uk/education/webwise/) programmes aimed at awareness raising and skills development. Such initiatives are, of course, timely and praiseworthy. That said, consistent with our previous argument, we would suggest that a CI approach to skills development ought to differ from them in terms of both the emphasis in its key objectives ('community empowerment' as opposed to 'economic necessity') and its mode of delivery ('bottom up and bespoke' as opposed to 'top down and off the shelf'). What follows is an outline description of the approach to skills and training that has informed our work at CIRA. It is not intended as a 'how to do it' guide. Rather, we aim to share the evolution of our thinking on skills development and offer insights from mistakes made and lessons learned in the field.

Where, When and What?

A key plank of CIRA's approach to skill development has been the development of what we term a 'local learning' strategy. On the basis of focus group work with several local communities, we determined that those unfamiliar with ICTs are looking for training which is delivered in their own time, at their own pace and in a familiar environment. Our initial thinking was that the technology lent itself very well to this need for flexible delivery. We thus considered developing on-line and CD-ROM-based training materials to facilitate learning from the premises of the community groups with which we worked. However, further consultation quickly revealed the folly of such an approach. The very real uncertainty, apprehension and fear of unfamiliar technologies experienced by our target groups clearly necessitated face-to-face, one-on-one tutor support. Staff in the University of Teesside's School of Computing and Mathematics therefore developed some 'Introduction to ICTs' short courses for delivery on the premises of community groups by trained local learning tutors, supported by 'work at your own pace' work books (the pilot runs of these courses impressed upon us the vital importance of selecting the right kind of tutor for such courses and providing them with the necessary support).

Using the above model, the University has been able to provide a large number of people, who reported that they would not consider a more formal course at a designated training centre or college, with some basic ICT skills. Indeed, many of these 'trainees' have gone on to more advanced training and formal qualifications. It is planned to extend the local learning model to include a range of non-ICT courses (including e.g., 'Community Appraisal' and 'Local History') which it is hoped will encourage practical application and development of the ICT skills picked up in the introductory modules.

The latter point above highlights another key lesson we have learned concerning ICT skill development. This is that, consistent with the view that ICTs are simply facilitating tools, skills are often more readily acquired as a by-product of work on some other objective than they are as a direct result of isolated ICT training. Many of us who consider ourselves relatively ICT literate picked up our skills in pursuit of other work or leisure interests. Why should we expect the communities we work with to be any different? Acceptance of this reality has informed the design of the aforementioned ICT courses where, whenever possible, skills are introduced through real projects (e.g., job searching, fund-raising, organising community events). Furthermore, there are endless examples of such skilling taking place in a more informal, non-course based context. The people of Trimdon, for instance, picked up some basic computing skills and familiarised themselves with some statistical software (with the minimum of guidance from CIRA staff) in order that they might conduct and analyse their own Community Appraisal survey. Many of the people who took part in this exercise had never used a computer before. As previously mentioned, members of the same community, with some expert guidance, designed their own Web site without any previous experience. The likelihood that many of these people would have enrolled in courses on 'computing for statistics' or 'WWW authoring' is slim indeed. Yet their desire to achieve other objectives led them to acquire some basic skills in both.

The above highlights the importance of informal training in ICT skill development. Indeed, our own experience highlights time and again the importance of a 'cascade' effect in the dissemination of ICT skills. Many members of the Trimdon community who have not had any formal ICT training, or have gained little benefit from that they have received, have acquired useful skills through one-on-one support from other community members who possessed ICT skills (further evidence of the existence of a high level of social capital?). In the case of the Hendon Hub project, the entire training strategy is built around 'cascade theory.' A key member of each of the five projects has been designated a 'community access worker,' and these workers are undergoing a programme of formal training. In return, they have been given the brief of disseminating the skills they acquire to other members and users of the project. In this way, focus group evidence suggests, community members who would not have considered more formal training (even of the local learning variety) as an option are gaining exposure to ICTs and developing some basic skills.

It must be noted that there are many problems and potential pitfalls with the cascade approach to skill development. Not least of these is the onus placed on the skills, abilities and motivation of the informal trainers. Those receiving training through "sitting by Nellie," as the management literature would have it, are very prone to picking up Nellie's bad habits, being limited by Nellie's sphere of competence and coaching ability, and being included or excluded in terms of Nellie's willingness to pass on her expertise.

A further problem with the cascade approach concerns the lack of formal accreditation for skills acquired in this manner. Whilst, for some, formal recognition and a piece of paper to demonstrate competence may not be important objectives, for others they may open doors to job opportunities and educational courses that would otherwise remain firmly shut. The answer here must surely be to combine formal and informal training and establish clear links between the two for those who wish to follow them. CI practitioners might usefully promote the establishment of a seamless route through nonaccredited training to postgraduate level for ICT training. Of course, such a route would involve a wide range of organisations including Universities, colleges of further education, schools, commercial training organisations, professional associations, voluntary organisations and community groups. With the involvement of such a diverse range of organisations comes inevitable problems of coordination, clashes of perspective and competing interests. Arguably, the solution to such problems lies in the notion of partnership that underpins CI as described in this chapter.

Providing Support

ICTs are not always user friendly. This will come as no surprise to most of us. For a willing user with something significant to gain it is reasonable to expect that the effort required to overcome some of the difficulties will be expended. For many, this is not the case. Expecting this group of users to move beyond the basic procedures learned during initial training is unreasonable unless they are supported. Users require ongoing support if they are to continue and extend their use of technology.

Most users require technical support to help them with upgrading hardware and software and installing new or additional hardware and software. They also require technical support to help them to deal with faults and routine maintenance. This much is obvious, as the general lack of user friendliness apparent when the technology is working and stable is nothing compared to the totally alien look and feel 'under the hood.' Even the language used is alien to most users. How many PC users know what a DLL is? How many want to know? The simple solution when faced with a machine which is telling you something that you don't want to hear, in a language that you can't understand, is to walk away from it. The number of PCs on desks in small businesses and voluntary sector organisations which rarely get switched on, or used for anything beyond basic word processing, is evidence of this.

In addition to this need for technical support many users also need other types of support. Moral support is often achieved quite simply by an 'expert' being available at the end of the telephone. This is important for users who have yet to reach that point where they have the confidence to experiment with the technology without fear of breaking it, and to try unexplored features just because they are there.

Access to ongoing training is important, as is guidance on training needs and the existence of a coherent programme of training. Language can be a problem here too. Users and groups of users are quite capable of identifying the sorts of activities that they would like technology to assist them with. They may not be able to put this into the same words as those that are used in catalogues of training modules. A need for help in the production of posters advertising community events does not intuitively indicate a need for 'Introduction to Pagemaker 5.00.'

Even relatively simple and commonly used terms are often not understood. The difference between a database and a spreadsheet seems obvious to ICT professionals, but in small businesses and community groups, we regularly find database applications implemented as spreadsheets.

Community groups often request advice on defining what are appropriate uses for general purpose ICTs such as personal computers. What they really want to know is what are inappropriate uses. We endeavor to leave this decision to the community. Although there is a common aversion to allowing people to 'waste time' playing games on PCs we generally advise communities to allow game playing as way to build familiarity with the technology. In one rural community a group of young game players have joined an international community of players of a multi-user Internet-based game. This led them to developing plans to attend a national and an international conference.

We see the provision of ongoing support as a significant part of any community informatics project. This implies a very long-term relationship with communities and great difficulties in finding sources of funding for CI projects in a world where funding is based so much around well-defined endpoints and exit strategies. The ongoing support requirement of a community group or individual is no different to that of a corporate user. This has significant implications in terms of sustainability.

Sustainability

If the CI experiments being conducted around the world today are to have a lasting impact on the social, economic, political and cultural landscape of local geographic communities, then answers must be found to some vital questions concerning sustained development. Ways must be found to renew or replace the short-term 'pump priming' finance that is often provided to launch projects and meet initial capital and revenue costs. Typically, a CI experiment may initially raise the necessary finance to install machines and meet telephony and training costs for a given period. Arguably, far greater challenges lie ahead in terms of meeting ongoing costs for equipment update and renewal, continuing technical support, training requirements and telephony.

Sustainability, of course, is about far more than just fiscal support over time. It will crucially depend, we have argued, upon embedding the technology in the fabric of communities and providing good reasons to continue to engage with it. In short, sustainability will require the development and continuing provision of high quality, locally relevant, up-to-date and interactive content in terms of access to information, services, etc. The responsibility for the creation and maintenance of this locally orientated content must lie with a partnership of local public service organisations and community groups and members themselves.

The empowerment of communities through the use and adoption of ICTs crucially depends upon this sustained development. The resource implications of an approach to CI which is partnership orientated have been highlighted through the foregoing discussion. Awareness raising, skill acquisition, and support can only partially be provided on a remote basis in cases where the levels of social capital are still insufficiently high. The prospects for the future sustainability of many CI initiatives, we believe, are likely to depend upon the realization by public service organizations that they may offer a significant return on investment. Improvements in health, education, employment opportunities and crime prevention which may accrue from strong social networks facilitated by CI are an important incentive for policy-makers. The most significant challenge is nothing less than the reconfiguration of rationally administered state provision, paternalistic professionally determined needs, and bureaucratic organizational delivery systems towards an emphasis upon partnership, decentralized support, and the facilitating of self reliance (Loader, 1998). Whilst the task remains a highly demanding one, it is far outweighed by the possible benefits in terms of better quality public service provision, citizen partici-pation and social inclusion.

Conclusion

We neither adopt a pessimistic perspective that the technology will act to fragment and disempower community organization or that it will necessarily produce a virtual panacea. Such technologically deterministic approaches fail to consider, in our view, the value of social agency, the ability of individuals and groups to resist certain technologies of domination and their capacity to shape the technologies for their own purposes. Consequently a CI approach acknowledges that community structures are (and always have been) heterogeneous and that the diffusion of technologies and their use will reflect such variation. It recognizes that ICTs can as easily act to reinforce existing patterns of exclusion as to provide opportunities for global networking. But it does not preclude the possibility that such recognition can itself act to stimulate supportive action on the part of agencies (public and private) to enable communities to contest such inequalities. Such a perspective is indeed consistent with the reconfiguration of governance in western societies unwilling to support collectively financed and organized welfare states and increasingly intolerant of the limitations of its democratic institutions.

References

Barlow, J.P. (1996a) Thinking locally, acting globally, *Time*, 15th Jan.
Barlow, J.P. (1996b) A cyberspace independence declaration. *Cyber-Rights List*, Feb. 8th.
Bell, D. (1993) *Communitarianism and its critics*, Oxford: Clarendon Press.
Bulmer, M. (1989) 'The underclass, empowerment and public policy,' M. Bulmer, J. Lewis and D. Piachaud (Eds.) *The goals of social policy*, London: Unwin Hyman
Castells, M. (1996) The information age: Economy, society and culture, Vol. I. *The rise of the network society*, Oxford: Blackwell.
Davies, S. (1996) *Big brother: Britain's web of surveillance and the new technological order*. London: Pan Books.
Etzioni, A. (1995) *The spirit of community*. London: Fontana.
Etzioni, A. (1997) *The new golden rule: Community and morality in a democratic society*. New York: Basic Books.
Giddens, A. (1998a) *The third way: The renewal of social democracy*. Cambridge: Polity Press.
Giddens, A. (1998b) Equality and the social investment state. In. I. Hargreaves & I. Christie (eds.) *Tomorrow's politics: The third way and beyond*. London: Demos.
Graham, S. & Marvin, S. (1996) *Telecommunications and the city: Electronic spaces, urban places*. London: Routledge.
Hague, B.N., & Loader, B.D. (1999) *Digital democracy: Discourse and decision making in the information age*. London: Routledge.
Hargreaves, I., & Christie, I. (1998) *Tomorrow's politics: The third way and beyond*. London: Demos.
Haywood, T. (1998) Global networks and the myth of equality: Trickle down or trickle away? B. D. Loader (Ed.) *Cyberspace divide: Equality, agency and policy in the information society*. London: Routledge.
Held, D. (1996) *Models of democracy* (2nd edition). Oxford: Polity.
Holderness, M. (1998) Who are the world's information poor? B. D. Loader, *Cyberspace divide: Equality, agency and policy in the information society*. London: Routlege.

Kraut, R., Lundmark, V., Patterson, M., Kiesler,S., Muk opadhyay, T., & Scherlis, W. (1998, September). Internet paradox: A social technology that reduces social involvement and psychological well-being? *American Psychologist*, 53(9), 1017-1031. Available: http://www.apa.org/journal/amp/amp5391017.html.

Lee, D., & Newby, H. (1983) *The problem of sociology*. London: Hutchinson.

Loader, B.D. (1997) *The governance of cyberspace: Politics, technology and global Restructuring*. London: Routledge.

Loader, B.D. (1998a) *Cyberspace divide: Equality, agency and policy in the information society*. London: Routledge.

Loader, B.D. (1998b) Welfare direct: Informatics and the emergence of self-service welfare? J. Carter (Ed.) *Postmodernity and the fragmentation of welfare*. London: Routledge.

Loader, B.D., Hague B.N. & Brookes, P. (2000) Trimdon digital village: A search for civil democracy in an information age'. In S. Coleman & A. Jones (Eds.) *Citizenship in a wired world*. Rowman & Littlefield (forthcoming).

Lyon, D. (1994) *The electronic eye: The rise of the surveillance society*. Cambridge: Polity Press.

Plant, R., Lesser. H., & Taylor-Gooby, P. (1980). *Political philosophy and social welfare*. London: RKP.

Putnam, R. (1995) Bowling alone: America's declining social capital. *Journal of Democracy*. 6(1), January, 65-78.

Reich, R. (1992) *The work of nations: Preparing ourselves for 21st century capitalism*. New York: Vintage.

Rheingold, H. (1994) *The virtual community*. London: Secter & Warburg.

Rogers, E.M. (1983) *Diffusion of innovations*. New York: Free Press.

Scott, J. (1991) *Social network analysis: A handbook*. London: Sage.

Sclove, R. E. (1995) *Democracy and technology*. London: Guilford Press.

Suttles, G. (1972) *The Social construction of Communities*. Chicago: University of Chicago Press.

Taylor, C. (1983) Rationality. In M, Hollis. and S, Lukes. (Eds.) *Rationality and relativism*. Oxford: Blackwell.

Walzer, M. (1992) The civil society argument. In C, Mouffe. (Ed.) *Dimensions of radical democracy: Pluralism, citizenship and community*. London: Verso.

Webster, F., & Robins, K. (1998) The iron cage of the information society. *Information, Communication & Society*. 1(1), 23-45.

Wellman, B. & Berkowitz, S.D. (Eds.) (1988) *Social strucures: A network approach*. Cambridge: Cambridge University Press.

Wilhelm, A. G. (1998) Virtual sounding boards: how deliberative is on-line political discussion? *Information, Communication & Society*, 1(3), 313-338.

Williams, F. (1993). Women and community, in J. Bornat, C. Pereira, D. Pilgrim and F. Williams. *Community Care: A Reader*. Basingstoke: Macmillan.

Wilmott, P. (1986) *Social networks, informal care and public policy*. London: PSI.

Understanding the Context of CI

Chapter IV

The Role of Community Information in the Virtual Metropolis: The Co-Existence of Virtual and Proximate Terrains[1]

Paul M.A. Baker
Georgia Institute of Technology, USA

Community Informatics and the Virtual Metropolis

Traditionally communities have been linked to the underlying geography, so that the identity of a community, for instance a neighborhood in a city, was linked to an underlying physical place, as part of a legal jurisdiction. A different kind of community is made possible by the self-identification of individuals with a common interest. In defining the concept of community informatics, Michael Gurstein in his preceding introductory chapter, makes a distinction between the type of "virtual community" made possible by the use of information and communications technologies (ICTs), and the augmented communication that ICTs can facilitate in a physical community. Thus the term connotes at least two different kinds of aggregate relationships, the first primarily physical (proximate), and the second, primarily conceptual (virtual). An example of this would include, for instance, alumni of the hypothetical Prestigious University who, while no longer physically present on campus, maintain strong identities as *alumni*, which can be thought of a part of the conceptual space defining "the University." Initially they were part of a physical community, but ultimately they are part of a virtual community. Another variant of this would be primarily virtual, citizens who consider themselves part of a large metropolitan area, for instance, Washington, DC, and refer to themselves as *Washingtonians* even if they might live in an adjacent jurisdiction in the neighboring state of Virginia. In this sense we could say that in either case we had a virtual (or conceptual) relationship that bears only a symbolic connection with the underlying "place."

The widespread deployment of ICTs has made relatively common a kind of nonphysical "space" constructed of digital bits that Manuel Castells has referred to as a "space of flows" (Castells, 1989). At first glance, an abstract place, it can seem more familiar if thought of in terms suggested above, in which identity is related to a virtual (or conceptual) space rather than linked to a purely physical one. As usage of ICTs have become commonplace, this virtual space has gone from being an exotic place, inhabited by a few intrepid souls, to a teeming bazaar of information driven activity.[2] And, in many respects, activities common in the physical world have been replicated in the virtual world. This has interesting ramifications for proximate, geographic communities.

The increasingly widespread availability of Internet access provided by commercial Internet service providers (ISPs), at least in U.S. urbanized areas, has undermined one of the early *raisons d'être* for community based information networks,[3] establishment of access (in this case, originally simply connectivity) to the Internet. This raises an interesting question for some of these early colonizers of the virtual world — what is the role for community networks in the Brave New Wired World? It has become apparent that mere connectivity to the Internet *does not* equal access, as the entire concept of access to information infrastructure has become a more complex and multifaceted proposition. The conceptual approach of community informatics emphasizes the need to address the larger contextual issues encompassing access, including access to equipment, understanding of the use of information, motivation to achieve access to these infrastructures, and inclusivity and diversity of participants. Finally, and most problematic, what sort of parameters do we use to measure the boundaries of "community" in the virtual metropolis, or gage the efficacy of ICTs in geographic community development efforts?

This chapter explores some of the structural and policy issues related to the implementation and operation of community-based information networks and examines three different types of community networks (CNs), drawing on theories of diffusion of innovation to provide a framework for analysis. The three cases presented here focus primarily on the implementation of CNs, that is, the combination of ICTs (hardware/software), the organizational structures, and the participants (or users) designed to facilitate information flows in several types of proximate communities. Conclusions are then drawn as to role of virtual interactions in the operation of CNs. For the purposes of analysis, the critical components of the framework include:

- *Key Actors*, or participants involved in the community network (ICT framework);
- *Innovation Factor*, or what action was critical in the initiation of the community information system;
- *Opportunities and Barriers*, variables influencing the implementation and development of the community information systems;
- *Policy Outcomes/Assessment*, what outcomes occurred as a result of the implementation of the systems.

The analysis further explores the relationship of these information-related communities to the underlying political and physical geography, or alternatively, explores the relationship of proximate space to virtual space.[4]

One form of these information-related communities is generally referred to as a *community network*,[5] which at least conceptually, contains some interesting contradictions. Communities such as traditional neighborhoods or villages are geographically based. In contrast, a network is an interlinked set of nodes, such as computers, interconnected to allow for the transfer of information from one physical location, to another. A 'virtual community' then is an abstract information-theoretic construct generated by the use of ICTs to store and construct an information framework that contains linkages to the geographic world. Yet the most readily defining characteristics of these community-based information networks is their construct in a virtual space, the Castellsian "space of flows" liberated from the locational ties of place, even as they are physically linked to a substrate of computer chips, wires, and software. A CN therefore can, to varying extents occupy "space" in both the virtual and physical world. In fact the most effective CNs seem to manage to draw on the strengths of the respective realms to weave "community" in the geographic world.

While the literature on computer mediated communication (CMC)[6] has richly described the "on-line communities" that have emerged in cyberspace realm, in terms of proximate communities, what does community, in a sense of neighborhood, mean if cyberspace offers the possibility to substitute an electronic flow of information for backyard fences and town squares? Is it possible to have a "here" when there is no "there"? If the geographic link is weakened, then what sort of connection exists among members of a community?

If, on the other hand, a community is defined by relationships rather than place, then what is the nature of relationships? What linkages connect people in an analogous manner to proximate or geographic place? Galston (1999), building on Thomas Bender's (1982) definition of community,[7] suggests that at least four key structural components comprise a virtual community in contrast to a physical community:

- *Limited Membership*—A typical feature of on-line groups is weak control of admission and participation of members. The rapid turnover tends to dilute the sense of intimacy and community from a point of stability. Membership limitations are much more easily exercised when physical space is occupied, either by preventing access to a given community space, or by other restrictions to the space in terms of meeting times, etc.
- *Shared Norms*—Virtual groups appear to develop protocols for behavior in response to three kinds of imperatives: promoting shared purposes, safeguarding the quality of group discussion and managing scarce resources in the virtual commons. In this respect physical communities are similar with the norm reference frequently related to underlying geographic considerations (e.g., a school, a park, a historic monument, traffic flow, and so forth).
- *Affective Ties*—As an explanatory aspect is complicated by debates among experts as to whether genuine community is facilitated by ICTs or is merely a type of "pseudocommunity."[8] The intensity of "flaming" and the rather emotional language that can occur in on-line communications, however, might serve as an indicator that some type of emotional attachments are possible, even in non face-to-face settings. In physical communities affective ties might

also be inferred by frequency of participation as gauged by actual physical presence.

- *Mutual Obligation* – While a sense of mutual obligation to the other members may occur, it seems that face-to-face contact strengthens subsequent on-line interactions. Therefore it might not be unreasonable to speculate that an on-line community with a geographic identity (and the possibility of further face-to-face re-enforcement) would be more likely to have a greater sense of mutual obligation develop.

The analysis of the community aspect of ICTs examines the nature of these various types of linkages, or relationships made possible by ICTs, and speculates on how these innovative modes of communication impact the various types of community.[9]

Information, Networks and the Rise of the Virtual Space

The rapid spread of ICTs has had an immediately observable effect — the production of a large number of articles speculating about the extraordinary changes that will result from the widespread deployment of the Internet. Early advocates of cyberspace seized upon potential benefits generated by ICTs and network effects spinning out a future where there is no "there"—anywhere can be everywhere, or nowhere. Not without some irony, proponents of these viewpoints can be seen at conferences where they cluster in hallways, rather than in on-line chat rooms, or engaged in other computer mediated forms of communication—an unintentional assessment about both the value of in person communication as well as computer mediated communication (CMC). This speaks to the importance of the physical world in the maintenance of various types of community.

According to some of these wired pundits, distance, and to some extent, its fellow traveler, time, have been vanquished in the information society.[10] The "End of Place" has been proclaimed, and we are loosened from the shackles of geography. An alternative conceptualization of the effects of ICT adoption is not that these technologies replace physical space, but constitute a new overlay in the fabric of communication, rather than a substitute for proximate interaction. The power of ICTs initially seems to be in their ability to reduce the friction effect of distance, the amplification of flows of information, allowing communities that can be everywhere and anywhere, sustained by information connection rather than grounded in proximate space. This latter effect leads to the observation that the nature of community relates to a connection, or interest, with the substitution of an interest in "place" with an interest in an conceptual construct. We are no longer restricted to communications with people we are in proximate connection with, but rather those with whom we share a similar **intensity** of interest, shared norms or elements of belief, and to some extent, affective ties. Paradoxically, this linkage effect also can strengthen extant bonds of interaction in a physical community.

Observation of these effects has led some proponents of CMC to speculate that the acceleration and increase in density of information and communication flow offer solutions to many of our social problems, particularly those related to sense of loss of physical community and decrease in social capital. However, given that the use of (or nonuse of) such extant forms of ICTs as the telephone has not "created" an ideologically desirable geographic sense of community, the odds that increased flows of information and communication in and of themselves will generate a sense of community are slim. The use of telephones for instance, can facilitate communication, but would not necessarily *create* a community of "phone users" analogous to "Internet users." We do not pick a telephone up and randomly dial other users without some reason, be it a particular interest, or some other pointer that would lead us to expect that some sort of prior existing connection exists. Without the aggregating effects of geographic proximity, some other sort of cohesive community force needs to be present. Placed in the broader framework of community informatics, rather than as an isolated dependency on the technology per se, the community building possibilities of CMC and ICTs appear to be a reasonable assumption.

This leads to construction of a counter-hypothesis, that ICTs *cannot* create community, virtual or physical, but operate by facilitating communication of individuals with common interests, who might not otherwise be able to congregate in a single physical location. The aggregate of these individuals with common interest or shared beliefs, which can be said to constitute latent or potential communities, can therefore interact virtually via information networks. Thus "if you build it, they will come" seems less true than "where two of three are gathered together, there will be community." ICTs allow those two or three individuals to be anywhere, not necessarily physically co-located. What ICTs do *not* do is create a community if no commonality of interests exists. Further extending this thesis, once a virtual community becomes large enough to be self-sustaining, in that enough individuals participate to generate ongoing dialogue, new ideas, or other reasons to continue communications, and to allow normal maintenance functions to occur including dealing with community conflict, developing communication protocols, and sustaining the infrastructure" (in this case the electronic or software requirements), it can be said to "exist." If the common focus of this community deals substantively with a physical community (say a town or neighborhood), then the technologies offer new and increased opportunities for interactions of the members of the virtual community. In the case of a community network, one in which a geographic identity also exists, it may overlay geographic space in unexpected ways. Thus while they *may* map onto locations (places, towns, cities, etc.) they may also theoretically *compete* in several regards. This chapter draws upon several cases of community networks to illustrate aspects of this viewpoint, where a physical community can be reinforced by "wires," but "wires" cannot "create" a community that does not exist in one sense or another. Community informatics as a community building approach recognizes that ICTs offer tremendous potential in terms of reinforcing community interaction, specifically in physical communities.

Conceptual Distinctions of Virtual and Geographic

A metropolis is generally considered to be a large urban center of culture and trade, or an incorporated municipality with definite geographic or legal boundaries. The term "virtual," as used in discussions of information infrastructure, generally implies a non-geographic location represented by an electronic coordinate system used for routing communication protocols. Thus, re-conceptualized, it is possible to think of a "metropolis" as an abstract, physically distributed but informationally linked group of individuals and organizations sharing common interests and objectives. Thus from the viewpoint of the individual, embedded in a variety of information, social, and neighborhood networks, linked by personal and ICT connections, "community" might encompass a variety of different types of relationships depending on the context (Wellman, 1997). It follows that for the purposes of the present discussion, a virtual metropolis can be thought of as an additional information overlay of the extant geographic space made possible by the information infrastructure of the region.

This is not a novel idea, as analogues currently exist that reflect a disconnect between conceptual (or *virtual*) location and a physical location. Consider again, the construct "Washington, DC," the capital of the United States. Geographically, "District of Columbia" (the "actual" Washington) is a small area of not much more than 500,000 inhabitants, but conceptually **WASHINGTON** is a megalopolis of 4.5 million people with multiple (and quite independent) political jurisdictions covering three states and a radius of as much as 100 miles. A further extension of the concept of "Washington" goes beyond the geographic into the symbolic, and carries quite a difference sense in that it is the seat of power and influence of a global superpower. Returning to the purely proximate, while the physical region encompassed by the metropolitan Washington area would be well served by working together, in reality the independent communities of interest compete and promote their particular agendas.

Much speculation centers around the possibilities offered by ICTs and virtual spaces, but several concrete observations can be presently made. First, the information infrastructure, which theoretically allows space to be irrelevant, appears to mirror the underlying physical infrastructure. For example, the urban density of physical infrastructure repeats itself in density of information channels (Graham and Marvin, 1996; Rand, 1995). This tendency for the informational infrastructure to mirror the underlying physical infrastructure has some disturbing implications for rural economic development as well as under-served urban areas (Wilson, 1992; NTIA, 1995). Will the information infrastructure, deemed critical to the future of jurisdictions, repeat some of the inequities of existing infrastructures? Will the urban core areas be neglected again? Moreover, the changing form of urban areas seems to reflect the changing patterns of information flow. Garreau (1991), for example, discusses the evolution of the "Edge City" in which suburban centers of information and enterprise act as mini-cities on the edge of the urban core. A key element seems to be the networks of information and knowledge (Glaeser, 1994) or "clustering effect" of high tech industries (Baptista, 1996; Castells, 1996). ICTs increase the velocity of flow of information when a benefit to communications exists.

Community networks, frequently participant-driven efforts to use information technologies to interlink communities, have been cited as having the potential to generate community-building effects. These benefits include greater citizen participation, increased civic interest, and more efficient exchange of local information (Schuler, 1996; NTIA, 1996; Rand, 1995; Katz and Aspden, 1997), and promoting alternative channels for governance (Ghere and Young, 1998; Neu et al., 1998). Extending this possibility, regional information networks, with the geographic variable "region" replacing "community," have been implemented as an overlay to uniting a region (Morino, 1995; PKW, 1996), which can also have community development aspects. While some examples of these types of systems with region-wide orientation exist such as the Blacksburg Electronic Village in Blacksburg, Virginia, and Charlottes Web, in Charlotte NC, extensive, generalizable evaluation studies have yet to be completed, barring some initial reports from efforts (See for instance, Benton Foundation, 1996; Cohill and Kavanaugh, (Eds.) 1997; Guthrie and Dutton, 1992; Rand, 1995; NTIA ,1996).

Hypothetically, the Chamber of Commerce for Neotech City might include an area of 100 square miles, encompassing a variety of physical communities, yet think of itself as a coherent organization representing "Neotech City." Continuing along this line, the telephone network linking Neotech City is part of the geographic space classified as infrastructure, a designed network of wires that crisscross the region, that transcend individual jurisdictional and community boundaries. The wires, and related digital technologies, however, carry information, which provides the substrate for communication across the entire region, essentially instantaneously. In this case the friction of geographic space, that is the time and energy required to traverse a physical space, has been minimized by the flow of information over networks. This idea alters the concept of a physical "there" in that a space may be alternatively defined conceptually by communication, or information flow (Graham and Marvin, 1996; Hiltz, Starr, and Murray Turoff, 1993; Rand, 1995; and OTA, 1995). The same effect that links individual in a virtual space can serve to reinforce the ties of geographic community.

Three Cases of Community Networks

The first case presented is *The Regional Information Infrastructure Policy Project (RIIPP),*[11] funded by the U.S. Department of Commerce's National Telecommunications and Infrastructure Administration, a 12-month long demonstration project of the role, function and benefits of an information and communication infrastructure in community and commercial life. This community network represents an urban-exurban information infrasu.acture linkage between the City of Alexandria and Fauquier County, Virginia, where ICTs were used to extend access to the Internet to an area where it had previously required a long distance call to achieve Net access. The second case, *The Rockville Community Network (RockNet)*[12] in Rockville, Maryland, represents a successful implementation of a local government initiated community network. The third case, the *Potomac KnowledgeWay*,[13] represents a private/public partnership focused on developing a regional community network centered loosely on the Northern Virginia portion of the metropolitan Washington, DC, region.

All three of these community networks are physically located in the Washington, DC, region, where the information infrastructure (specifically, "wires" or telecommunication facilities, switching computers, and access nodes) is so dense that it has been referred to as the *Netplex*[14] through which approximately half of the Internet traffic in the world passes. Fairfax County, Virginia, one of the jurisdictions comprising the Washington, DC, area alone ranked fifth nationally in the number of Internet hosts, or nodes, present on the Internet.[15] The human infrastructure (demographics) of the area itself is somewhat unusual. In the close-in suburbs, the per capita income ranges from approximately $34,023 in Alexandria to $34,414 in Fairfax. For comparison, Virginia as a state has a per capita income of $18,762, the District of Columbia has a per capita income of $24,595, and Maryland $20,552 (all 1993 figures). The mean education levels are also quite high with close to 50% of the residents of most area jurisdictions in possession of at least a BA or BS.[16] As is true of most urbanized areas, the jurisdictions involved are also physically densely populated, which as a rule lowers the cost of extending new information infrastructures.

Case Presentation

Regional Information Infrastructure
Policy Project of Northern Virginia (RIIPP)

Overview

The Regional Information Infrastructure Policy Project of Northern Virginia (RIIPP) was initially organized around community needs for education, training and public information. This joint project focused on community and educational access in Fauquier County, Virginia, a rural ex-urban county of Washington, DC, and more narrowly, on educational Internet access in suburban Alexandria, Virginia. The project was partially funded through the Telecommunications Information Infrastructure Assistance Program (TIIAP), of the National Technical Information Agency (NTIA), U.S. Department of Commerce. The demonstration project linked a variety of institutional actors in the Washington, DC, metropolitan area, designed to demonstrate the role of inter-organizational partnerships in promoting information infrastructure development. The primary institutional participants were: public school systems in two geographically separated jurisdictions, an academic institution (George Mason University) acting as system integrator and policy consultant, and the federal government agency. At the close of the project, results indicated success in implementation of the school based portion of the Internet service in both Fauquier and Alexandria, primarily because of the minimal need of the school based participants for technical assistance and training. In additional, the widespread Fauquier community interest in Internet service, especially World Wide Web service, led to the initiation of five alternative Internet service providers (ISPs) by the project conclusion, which allowed GMU to substitute a locally provided graphical-based ISP for the text based system the project initially started with. Additionally, a self-supporting community of users initially developed around the project based on a "teach the teachers" model. Illustrating the somewhat ephemeral nature of virtual (specifically on-line) nature of communities with little significant affective ties, or mutual obligation, beyond that ascribable to living in the same geographic community, the group faded away as the project wound down, and commercial ISPs provided connectivity.

The geographic variable is particularly of interest here because there is considerable difference between objectives for the two project subsystems. In the urban portion, in the City of Alexandria, a close in suburb of Washington, DC, the project objective was simply to provide Internet connection to two elementary schools, and demonstrate the educational advantages generated by this type of access. In Fauquier County, an ex-urban county part of the Washington D.C. SMSA, the project was more ambitious, in that there were two objectives, 1) provision of educational related Internet access to three public schools in Fauquier, and 2) provision of community access to the Internet as a means of stimulating community and economic development. Given the current arguments that ICTs can be used as a tool for economic and community development, the second objective was included to examine the effects (if any) of a pilot information infrastructure project in an under-served, rural area.

In general, the RIIPP project was not intended to serve as a local ISP, but was mainly concerned with "start-up" issues, such as barriers to entry. The initial overriding theme was to demonstrate the possibility of developing a public-based or -supported community network based in a technology-ready public school system. RIIPP's successful demonstration of the Internet service in both Fauquier and Alexandria schools, as evidenced by the high participation rate of target teachers and the participant feedback, was due in part, to the minimal need for technical assistance, mostly for training. As regards the community access (that is the possibility of the development of an on-line community) portion of the project, the school-site technology and typical school operating hours were, as a rule, inadequate to meet the service demand of an enthusiastic group of community users in Fauquier. Specifically, the need for user support (user "hand-holding"), as well as the system support and maintenance generated by the extension of infrastructure design, was significantly underestimated.

Key Actors

In Fauquier County, units participating in the demonstration project were selected on the basis or readiness, representativeness and a measure of geographic concentration. These included direct access by three schools in Fauquier County Public Schools (FCPS); the Fauquier County branch of the Virginia Cooperative Extension Service (VCES); and selected members of the Fauquier community, with approximately 250 users provided free dial-up Internet service for demonstration and planning purposes. In the City of Alexandria Public Schools (CAPS), the two participating elementary schools were selected because of the significant number of disadvantaged students and the desire of district officials to more rapidly implement plans to network all schools. Aside from the primary participants described above, additional input was provided by other members of the community, including individuals, organizations and governments. This ranged from participation in several focus groups held at several points during the project, for both design and evaluation purposes, to actual measurement of end-use by citizens in Fauquier County, with the related feedback on system operation. The Institute of Public Policy (TIPP), GMU's lead agent for the activity, maintained overall project management and policy evaluation, and additional technical assistance was provided by the Electronic Commerce Resource Center (ECRC) staff, and from the School of Business. The University's Institute for Educational Transformation (IET), under separate contract to Fauquier County Public Schools for the preparation of a school restructuring plan, coordinated teacher training in telecommunications and for the instructional application of telecommunication services. Additional assistance in Fauquier, both in community support and citizen training, was provided by Learning and Beyond, a community-based firm. The project team had initially intended to use George Mason University's computer resources. Subsequent consultation (during September and October 1994) with the computer support staff determined that existing institutional demand would prohibit additional external loads, until completion of improvements to internal infrastructure was finalized. By the end of October 1994, TIPP determined that an alternative access to the Internet would be necessary. This alternative—access through an external service provider—was determined to be appropriate for the duration of the project. Internet service (basic interconnection to the Internet—allowing e-mail, ftp, etc.) was provided by PSI, one of the largest Internet service providers in the United States. Subsequent adaptive changes, as well as the original network system architecture, were completed by PRC Inc., of McLean, Virginia, which prepared the 'scaleable' design linking the Fauquier County and Alexandria architectures.

Technical Considerations

While use of traditional text based Internet services like e-mail, ftp, and telnet continue to grow, the character of the Internet has become more commercially and graphically oriented. As such, there is increasing concern with considerations generally related to commerce, such as privacy, security, access, and who pays and how. The recent development of graphically based interfaces to the Internet has significantly increased both the number of users on the system, and users' expectations, while concurrently increasing demand upon system throughput. Protocols for evaluating the cost and value of digital community-wide communication and information services, however, remains largely unmeasured, an objective of the community informatics approach.

Basic services offered included e-mail, ftp, telnet and gopher—which were used by teachers, students, state and local employees, and individuals after a modest amount of training—and dial-in modem access to an Internet server with a basic terminal device. Demand was anticipated for e-mail within and outside the jurisdictions; telnet to libraries outside the county, like the Library of Congress and the National Agricultural Library; and gopher to federal agencies in the Washington, DC, area. Improved communication between school and home by e-mail was initially planned, with teachers responsive to inquiries from both students and their parents, system implementation problems limited this option. Unlike the two other cases presented in this chapter, at the time of this project, access (or "wires") was still a primary consideration. Thus the system's architecture employed some interesting solutions to the project requirements. The key goal was to establish an Internet connect in a rural area, about 50 miles from Washington D.C. This area, though on the extreme urban fringe of Washington, required a long-distance phone call to reach an Internet service provider. The system design essentially served to stretch the LATA (area in which local calls are "toll free") in order to eliminate the long-distance barrier to Internet service.

The system design served to "extend" the Fauquier School network by connecting it to the Alexandria link to the Internet. A 56kbs Internet connection to Performance Systems Inc. (PSI) linked Polk Elementary (in Alexandria) to the Internet, and another leased line connected the other Alexandria school, Mt. Vernon Elementary, to Polk. From Polk, an additional dedicated line provided the Internet connection to the Virginia Cooperative Extension Service in Warrenton, about 50 miles distant. The original design called for the Extension office to be connected to a hub at Liberty High School in Bealton (about five miles distance from Warrenton), which in turn was connected to two other local schools, all via wireless bridges. Dial-in for the community was to be provided through two access points, one Liberty High School in Bealton, and an additional access point provided in Warrenton. Access to the servers was provided via two Shiva LanRovers (routers) configured to act as terminal servers. The system architecture required several major modifications in order to be functional. Due to unanticipated geographic considerations, the wireless bridge between the VCES in Warrenton and Liberty H.S. in Bealton was replaced by a leased line. A second major problem arose with the dial-in access point. While Internet access was successfully achieved from on-site users, the dial-in capabilities were never reliably implemented. Ultimately, access was provided for the approximately 250 community users through an arrangement with an alternative ISP that had started partially in response to the demand demonstrated by user response to the project.

Assessment

At the time, many proponents of community networks in the U.S. were stepping in to develop community based systems, as many small city and county governments did not believe themselves ready or equipped to deal with the demands and costs of developing an "on-ramp" to the information superhighway. Subsequently, it turned out that many of the necessary skills, and much of the physical resources, were already in place in various public sector organizations, such as the public school, and library systems. In retrospect given the rapid rate of change in computers and telecommunications, the need for communities to provide or encourage actual "wiring" appears to be less important than the provision of content, and system support and training. It may be necessary to change the community's perceptions and thinking about how to leverage the available resources. This can be accomplished through an educational process that presents some of the possible options that are available and readily implemented through creative use of existing resources. In this respect, *access* is less about the wires than the use of information or the motivation to use the technologies.

Setting concerns about technology aside, in the case of RIIPP, it seems reasonable to conclude that local governments in geographic proximity to the community systems as a rule were not concerned with providing connectivity directly, even though there were not (at the time) local commercial alternatives. This is not a condemnation of the local governments, or even unexpected; technology changes so rapidly that even experts in the field had a hard time staying current on the latest wrinkles in information technologies. In this case the use of an objective outside consultant (change agent) appeared to serve as an organizing focal point that sidestepped some problematic local stakeholder interests. This applied to policy as well as in technical areas.

From a technical standpoint CNs multiply the opportunities as well as complexities involved in the delivery of information. While the underlying technologies may be understood by network vendors, it was critical to make sure that they either had had actual experience in delivering access to the Internet in a networked environment, and utilizing the specific type of equipment that is extant. Barring this, the next best option is to ensure that they have the technical expertise as well as a commitment to completing the project in a satisfactory manner.

Another key element that arose was the need to determine the specific goals that were to be achieved. "Access to the Internet" is not a very clear goal, and is analogous to saying that "a library should have books." The community should have clear ideas about what constitutes an appropriate mission statement. For instance: "Citizens should have access to several text based services including, e-mail, remote searches of databases at the library, and updates on local governmental activities through a local dial-up connection." Development of a CN mission can be accomplished through the use of a visioning or focus group process, or other polling or survey type methodology, and has the added advantage of generating the community motivation necessary to develop a viable community information infrastructure.

A variety of options exist for institutional stakeholders to participate. Schools (or other public facilities) with the necessary on-site infrastructure can act as the technology engine for a local community, at least initially, in the form of a pilot or demonstration project, and subsequently generate community support and interest. This has additional policy implications particularly in times of budget shortfalls. Leveraging community resources, including infrastructure, economic or human resources, shows how the entire community can benefit from investment in the public school system, not just those individuals with school aged children.

While ICT "physical infrastructures" (e.g., hardware and software) are generally thought to be the critical entry barrier in implementing a community network, this did not prove to be the case in the subject project. Thus, while the infrastructure may be in place in the school system, systems implementation is a nontrivial additional task that requires human and technical resources for the selected school system. This is not an insurmountable problem, but it does require careful planning and objective setting.

Rockville Community Network (RockNet)
Overview
RockNet is currently an operational community network that was developed to provide a source for community information and a means to inter-link citizens.

According to the mission statement:

> The Rockville Community Network (RockNet) is dedicated to providing free community access to networked community information to promote the spirit of community and meet the information needs of Rockville City residents. RockNet concentrates its efforts on segments of the community and categories of information that have not yet been adequately served by the commercial sector, in much the same way that public libraries and public broadcasting stations operate. RockNet directly publishes information to the people of Rockville City and the Internet via a specific server machine known as [ci.rockville.md.us].

RockNet represents an example of a nonprofit community organization running a community network. What makes RockNet of interest is that the organization was formed as a result of an effort spearheaded by an elected official. The City of Rockville provides server space and resources to RockNet and the City perceives RockNet as an alternative channel for the delivery of municipal information, and as another tool for fostering community development and civic participation.

Community Profile

RockNet is located geographically in Rockville, Maryland, the second largest city in Maryland and county seat of Montgomery County. The city occupies 13.03 square miles within the Metropolitan Washington, DC, area and is located 12 miles northwest of the nation's capital. The City of Rockville operates under the council manager form of municipal government and derives its power from a charter granted by the state of Maryland. Schools and health services are provided by Montgomery County.

The City had an estimated 1995 population of 47,235 and an employment base of 48,900 jobs as of January 1, 1995. The median household income is $52,250, and has a technological penetration of computers of 47.2%.[17] There is no major university located within Rockville, though a community college is located within Montgomery County, and the University of Maryland is located in the adjacent Prince George County. While informally participating in RockNet, the City of Rockville also maintains a separate formal Web site at: [http://www.ci.rockville.md.us/]. The cross-linkages on Web site homepages for RockNet, and the official Rockville home page, reflect the mutually supportive missions.

Key Actors

The RockNet Web site is of particular interest because of its somewhat unusual history, representative of a project championed by an elected official (top-down), rather than by a citizens' group (bottom-up) as is the case with many community networks. In May 1995 Glennon Harrison ran for the Rockville City Council on a platform of increased citizen participation in local government affairs. He recognized that, given the existing connectivity of Rockville, both in terms of citizen access to the Internet as well as by business, the Internet provided a venue for interlinking citizens, and possibly as a tool to strengthen community efforts.

After he was elected he formed a relationship with another city council member, who was the Council's "computer guy"—that is, the individual other council members recognized as having sufficient technical expertise as to inform his observations on the City's use of technology.

By December of 1995 two members of the city council had approached the city administration about what role might be most appropriate for the city in terms of provision of information via electronic means, or in terms of promoting connectivity with and between the citizens. Harrison and Dorsey brought in support material that discussed the possibilities of the Internet and information networks. Following what was referred to as a "buy in" by the city administration, further exploration convinced the key participants (the elected officials as well as the administration) that a broadening of the concept from just municipal information to "community" networking might be a more appropriate objective.

The members of the city council that were championing the project secured the support of council members, the mayor and city administrators for an exploratory public session. Following discussions with several individuals engaged in research in the area of civic and community networks, a presentation and question and answer period was planned for June of 1996. The project champions decided that an open facilitated group effort to generate policy alternatives from citizens would be the most suitable venue given their objectives. There was some indication that the council was somewhat uncomfortable with promoting citizen activities directly, considering potential political risk, but would be willing to provide resources to a community organization that was willing to undertake the effort. The original intent of the meeting then was to serve as a "kick off" to generate citizen interest in community networks. The facilitator had originally planned to provide a briefing on community networking and the opportunities provided by it. However the partici-pants who turned out for the meeting were not only informed participants, they were ready to actually begin design of a community network. The level of interest was such that more than 50 participants showed up to participate on a rainy midweek evening. The facilitator, following a consultation with the project organizers, decided to change the agenda and move from an informational presentation to a facilitated "town meeting approach" in which objectives of a proposed information network were explored.

The results of the facilitated group policymaking effort was sufficiently clear and level of interest high enough that a working committee was formed the same evening and preliminary plans made for an organizational meeting. Given the level of citizen buy-in, members of the city council, the mayor, and representatives of the city administration decided to support the effort. The first organizational meeting was held in mid-July, and agreement reached to proceed with the effort on a more formalized basis. In order to facilitate off-line communication, the group decided to implement a RockNet listserver, which was operational by the end of July 1996. The organizing group recognized that visibility was a key component of developing the networking effort. Glennon Harrison, with support of the group, placed a prototype Rockville Community Network homepage on the World Wide Web in early September 1996.

As of the end of 1999, RockNet had formally incorporated as a nonprofit, and provides information and content over the Internet but does not offer direct dial-in access via modem. While RockNet had originally planned to established several computer outposts in community centers, with one designated as its headquarters and training lab, the limited resources available placed those plans on hold.

Technical Considerations

The Rockville City Council provides server space and resources to RockNet and the City perceives RockNet as an alternative channel for the delivery of municipal information. A working agreement with SAILOR, the Internet service provided by the Maryland State Public Library system, allowed RockNet to establish a direct link to the Internet, rather than merely by placing a Web page on a commercial ISP. This was operational by mid-January 1997, and connected through a Web server located within the City of Rockville's data center.

Assessment

As is true of many community efforts, the RockNet organization continues to struggle somewhat on several fronts. First is the necessity of determining the weight of priorities given limited resources. As one of the goals of the organization is communication, rather than entertainment, RockNet had originally committed to providing e-mail access to citizens. This initially raised the question of who should receive the first accounts, those who could pay (which would help sustain the effort) or those who don't have access to e-mail currently. This replicates tensions that have been evident in other municipal and community efforts elsewhere. While many of these efforts have as a key objective the provision of access to under-served communities, the reality of what amounts to management of expensive technologies is such that support and viability of the organization must temper the objectives of extension of access. Subsequently the availability of free net-based e-mail options (such as yahoo.com or hotmail.com) reduced significantly the priority placed on this feature.

Another issue currently under discussion is the degree of control over the underlying computer servers. While the City has no "official connection" with RockNet, in reality two of the city council members were key proponents of the efforts. In addition the city continues to provide some operating funds, and will allow the CN to operate on city servers. However, as noted previously, the City maintains a separate, rather narrowly focused information server at [http://www.ci.rockville.md.us/]. In this case the distinction between unofficial and official is a difficult one. At present the City has finished a visioning process to help determine direction for the city. As part of the technology plan for Rockville, RockNet has been specifically factored in as one of the city's information outreach efforts.

RockNet has successfully navigated the initial birth pangs of a new organization. A second board of directors has been elected, and plans are underway for securing continuing funding and new resources. RockNet is also finding that the objective of increasing information and communication between citizens requires not only providing access and training, but an outreach and marketing effort as well, to some extent to educate citizens of the advantages of this new community information channel. Thus the organization is working toward training as well as securing funding for its continuance.

Of particular interest is the continued close working relationship between the

City administrators and the RockNet group. This is believed to be the response of championship by several of the city council, as well as "buy-in" by key administrators to the concept of increased citizen participation. It is also believed that successful demonstration of a cooperative arrangement would enhance the visibility of the city administration. In addition, the group has successfully leveraged use of the various publicly provided information infrastructures. This tends to confirm observations made of other systems that while the availability of public resources do not guarantee the success of a community network, the absence of public resources, either through content, provision, information access, or hardware infrastructure significantly reduces the favorable outcome. Finally, the strong geographic community component appears to be one of the factors critical to the continued operation of the network.

Potomac KnowledgeWay

Overview

The Potomac KnowledgeWay (PKW) [http://www.knowledgeway.org/] represents a hybrid of both geographic and non-geographic community types of information networks. While physically based in the Northern Virginia suburbs of the Washington, DC, metropolitan area, its collaborative efforts extended beyond Northern Virginia into Maryland through participation of Maryland Public Library's Sailor Project [http://sailor.lib.md.us], to the north, as far west as Winchester (about 90 miles), south to the Blacksburg Electronic Village (BEV) project [http://www.bev.org], and to Washington, DC,'s CapAccess system [www.capaccess.org]. The objective of PKW was oriented more toward leveraging of existing resources and facilitation of interaction and awareness of other efforts rather than provide "access" per se. PKW described its mission as to promote economic development and the growth of business through connectivity. By analogy, the effort could be considered more akin to an "office" of economic development rather than "community affairs" as is the case with other community networks.

In terms of objectives, one role for these community (or regional) networks, and a key mission for many of the regional efforts, is a focus on addressing issues of access in the broadest sense, both to hardware, and to the information itself (Schuler, 1996). Citizens that lack ready access to information tend to be left out of participation in policy development. PKW's approach was not so much on creating *physical access*, but on facilitating access in its broadest sense—cataloguing information sources, facilitating the cooperation of different types of community resources, and acting as a clearinghouse house for regional information—in short, access to information content, rather than "just" connectivity. Thus PKW was not an ISP—it did not provide direct connection to the Internet through a bank of dial-in modems, but access to Internet resources and information though its indexes of information, and forums for discussion and exchange of ideas and information. The majority of the content provided came from participants in the project, some of which can be described as "community networks."

Community Profile

The geographic community in the case of PKW is more broadly defined than in the other community networks. At its core, it covers the close-in Virginia suburbs of Washington, DC,: Fairfax, Arlington, and Loudoun Counties, and the cities of Alexandria, Fairfax, and Falls Church. The population is an estimated 1.2 million, with incomes and education levels above the average for the Commonwealth of Virginia. The area has a large public research university (George Mason University) present, as well as branch campuses of Virginia Tech and George Washington University, as well as an extensive network of community colleges.

Key Actors

The impetus for development of the PKW arose from stakeholders, with a regional booster orientation, acting as policy entrepreneurs to a perceived need for a plan for Northern Virginia and the Greater Washington area that would attempt to link and unify a region encompassing essentially three states (Maryland, Virginia, and the District of Columbia) in a meaningful fashion. A number of community and business leaders determined that a focus on the information technology and networking expertise of the area would serve as a central orientation. Initially, the primary focus on economic development was expanded to consider how all sectors of the region could adapt to and succeed in an information-based economy. Since the PKW was initially planned, the scope of the project grew to include links to the large portions of the mid-Atlantic area, while still maintaining its initial focus on the Metropolitan DC region.

PKW differed from a typical community network effort in a variety of ways. Again, as in the RockNet case, rather than a grassroots community effort (bottom-up), the origins of the project began with a presentation made to a group of Northern Virginia business leaders (the Northern Virginia Roundtable) in September 1994, by Mario Morino, chairman of the Morino Institute,[18] in the form of a proposal for the creation of a "Networked-Based Information Products and Services Industry" in the region. Ultimately PKW was the result of the Northern Virginia Roundtable's Focal Industries Committee's effort to develop a long-term strategy for core industries in the region, particularly in the information technology and communications sectors.

The resulting organization was a partnership of public, private, and third sector (nonprofit) efforts. Initial members included Virginia's Center for Innovative Technology (CIT), George Mason University, Opportunity Virginia (a group of Virginia business and community leaders), and the Morino Institute, with CIT committing a challenge investment of $150,000. Qualification for this challenge was that the project had to be industry-led and support from the business community had to be present. CIT matched business contributions at a rate of one dollar for every two raised by the Project from regional businesses, indicating that PKW would be one of CIT's major investments in Northern Virginia. With this milestone achieved, leaders of several organizations agreed to move forward on an ambitious plan to incorporate and further the vision.

PKW's Board held its organizational meeting on June 3, 1995. Shortly thereafter, an effort was undertaken to secure the $300,000 in seed funding needed to match the CIT challenge investment. By October 1995, a number of key developments had been completed including the decision to formally broaden the scope of the Project to include not just Northern Virginia but the entire Potomac region, and to reflect that inclusive view in an identity for the region and final name for the project, the Potomac KnowledgeWay. The project' s Web site, *Crossroads*, was named to reflect a "virtual place" where different organizations, individuals or groups could interact or "network" (in this case, literally).

Technical Considerations

The Washington, DC, region is unusual in several ways, and thus PKW's slightly different focus on achieving useful connectivity to information. While not ubiquitous, Internet access is readily available in the area, with more than 90 different[19] private ISPs (Internet service providers) operating. There are also alternative access nodes, including CapAccess [http://www.capaccess.org]; the Maryland Public Library's Internet access project, *Sailor* [http://sailor.lib.md.us]; and access available through terminals at the Alexandria Public Library. Originally, the Potomac KnowledgeWay Project was organized around four programs (from the Potomac KnowledgeWay Web site):

- "The Community Awareness and Education program was a community-based engagement effort geared toward enhancing understanding and awareness of the Knowledge Revolution and building consensus for actions that could advance the region and its constituents. Its components included the Town Hall Series, KnowledgeWay Action Teams, the Net-Worker Program and NewsWay, the Project's newsletter.
- The Regional Collaboration program served as a catalyst for regional action. It brought together institutions and businesses to help them seize key opportunities, forge new partnerships, and stimulate regional change.
- The Information Entrepreneurship program assisted entrepreneurs and intrapreneurs in creating and commercializing information products. The program was designed to encourage innovation and creative growth through an Information Entrepreneurship Coalition, an education series and an Entrepreneurship showcase.
- The Regional Networking program facilitated communications, interaction, and teaming within the region. The program was anchored by The Potomac KnowledgeWay Crossroads, the Web site on the Internet. It provided an "entry point" and road map to understand, locate, and use the region's rich base of resources."

Subsequently, PKW's mission was refocused into three programs (from the Potomac KnowledgeWay Web site):

- "...The **Netpreneur Program** is designed for a new kind of entrepreneur — the netpreneur—that is an entrepreneur whose business would not exist without the Internet or network-based communications. NP is designed to create a social architecture in the region and on the Net that provides venues for matching regional netpreneurs with people who can help them—other netpreneurs, funders, advisors, the media and strategic partners—for information exchange and deal-making. The goal is to help netpreneurs bring net-centric products and services to market faster. The cornerstone of NP will be a Web site, the Netpreneur Exchange, designed specifically for netpreneurs."
- "...**Work Force Enhancement Program** will focus on attracting, developing and retaining the Net-savvy knowledge workforce needed to sustain and expand the industries that drive our economy. A high profile public relations campaign will help attract talent from around the world, while regional

training and education initiatives will develop relevant skills in our existing organizations, individuals and students. Activities will include marketing the region's unique opportunities to college students around the country, promoting the understanding of core skills needed by a Net-savvy workforce, and identifying a virtual university of resources from traditional and nontraditional sources to provide Net-savvy training and education."

- "...**Connected Community** is a community-based engagement program focused on increasing the understanding of the communications revolution, building consensus for actions that can advance our region and its constituents, drawing those parts of our region at risk into opportunities of the communications revolution and advancing an on-line regional information infrastructure that will facilitate communications, interaction and teaming within the region. Programs include the Potomac KnowledgeWay's Web site; a series of sponsored seminars on the communications revolution; and advocacy of easy, affordable Internet access for all segments of the region's population, through libraries, and kiosks in other public or quasi-public places."

This latter program provided the central organizing on-line network effort of Potomac KnowledgeWay (as opposed to outreach efforts). Crossroads, PKW's Web site, and the centerpiece of the Regional Networking program, which was folded into the other programs under the revised programmatic plan. It was designed to help people throughout the region find new ways to communicate, collaborate and spark innovation. At the Crossroads, a *road map* (an additional indexing metaphor) helps individuals and groups locate others in the region with whom they wish to communicate or team. By participating, organizations could promote awareness and understanding of their efforts within the region; receive inquiries that may turn into partnerships, sales, funding or volunteer support; and connect with regional partnerships, research projects and initiatives.

Assessment

The PKW closed down operations during November 1999, with the Web site currently acting as an information archive. While the planning process worked well over a networked medium, demonstrating the communications and collaborative value of ICTs, several factors contributed to its lack of viability. PKW, while providing facilities and electronic support, was not able to provide additional financial or organization support. As long as resources were provided to the group by the energies of the steering committees, planning went ahead. Planning is a good deal less resource intensive than actual implementation of the plans, the point at which the process began to break down. When the key actors began to experience "organizers' fatigue," and began to reduce the level of energy provided, the enterprise began to falter. In a more mature (or less virtually networked) organization, long-term ties might have provided enough slack to carry the organization through the start-up phase. However, as members began to withdrawal, increased loads fell on the remaining members until the effort ran out of willing leaders. In this case, then another failure was of leadership to generate sufficient resources and participants to ensure the continued survivability of the organization.

The community informatics framework suggests that PKW demonstrated the use of ICTs (specifically electronic communication) can be a significant aid in the development of group and community activity. Increased awareness of information, and pooling of resources is vastly improved by taking advantage of information technologies. However it must be noted that while these networks facilitate interaction, they do not serve as a replacement for face-to-face relationships or human interaction.

While PKW was of help in obtaining information and identifying interested actors, in fact it was only after some degree of relationship was established did the electronically facilitated communication prove to be a change agent. Further as Gurstein (introduction) points out, aggregations of static Web pages seem to provide little if any sustainability of "community," particularly if the related geographic community is not clearly delineated. In fact these observations provide several avenues for further research. For instance, what efforts might be undertaken to accelerate and strengthen community-wide relationships, with an objective to utilize a community network to augment communication? How can these systems be most effectively structured to promote citizen-government interaction? And most importantly, over the long term, how do these types of networks facilitate the establishment of or add value to region-wide or cross-community communication?

Case Studies Assessment

The critical components of the three cases presented above, variations on the community network theme, are summarized in Table 1. Each of these cases represents a different organizational structure and set of objectives, relationship between the community network management and users, and connection to a geographic community. In one case it was proactive, in the other cases the public sector involvement is linked to external change agents. They all display to varying degrees components of community: limited membership, shared norms, affective ties, and mutual obligation. In all cases it appeared that once the initial novelty of a community Web site was extinguished (which is simply an interface for ICT operations), unless the Web site carried new and continuously updated information it stood a great chance of being abandoned and becoming a "ghost town in cyberspace" (Kanfer, 1997). A familiar example of the extinction of the novelty effect of technology might be use of the telephone. We rarely phone libraries unless we are looking for specific information or are inquiring as to whether a new book has arrived. After repeated calls and answers of "nothing new," we would expect extinction behavior to set in. Web sites are accessed only as long as they have use. For community networks this means that a commitment must exist to support this part of the "community," or they run the risk of becoming irrelevant and ignored.

Key Actors

The communities represented by selected cases did not offer much variance in socio-demographic characteristics, and were much more similar to each other than to the surrounding regions. Physically close to many of the operating units of the federal government, the Pentagon and the biomedical industries clustered around

NIH; the populations characterized in the cases are generally more highly educated with higher incomes than those of the average jurisdictions in the surrounding states. Given this political and scientific/technical environment, citizens in these areas can be expected to be more innovation tolerant than the average of the surrounding states. There is some indication in the literature on innovation that education and income affect the demand for services.[20] However, as above-average levels of income and education also occur in large population centers, some probability exists that the community development initiatives (that is the implementation of community networks) seen in these cases is as much related to the presence of large population centers as it is to the underlying characteristics of the populations.

Another environmental variable present is the role of centers of "intellectual capital." Apart from the case of generally higher education and income levels in the subject areas, the presence of networks of knowledge, or clusters of firms engaged in similar technological and/or professional activities, may be an influencing factor, particularly in terms of being "early adopters" of technological innovations (Rogers, 1995). There is also some indication that the location of a large university, either in the immediate geographic area or immediately adjacent to the systems discussed in the cases, had an influence on innovative activities. As has been remarked elsewhere (Glaeser, 1994; Castells, 1996), large universities often function as an active agent for change, or as centers of innovation and intellectual resources. In terms of the geographic communities involved in these cases, it appears that the relative homogeneity of the populations did not significantly contribute to any greater identification with the virtual community interactions.[21]

Lastly, community networks, at least those concerned with audiences outside of the core structure, need to take into account the needs and wishes of the users or even potential users of the system. The lessons learned from the RIIPP project clearly indicated that a system optimized for technical objectives (top-down design) might not necessarily be optimized in terms of the demands of citizen users. There is also some evidence that this might ultimately be a factor in the less than successful outcome of the PKW regional network.

Innovation Factors

In reviewing the cases, several different units of analysis seem to appear, depending on the frame of reference that is assumed. If the analysis focuses on the technology (i.e., the ICTs), then the variables seem to relate to system implementation issues. This orientation might be applicable if the community network is viewed as a provider of services or information. In this case, users of a community network are generally characterized as a passive audience for broadcast information, rather than as an active participatory community. It is also important to keep in mind that these systems were at least partly located in urban or urbanized areas. This to an extent is a consideration in terms of system design of the community networks, or what they intend to achieve. This might include a focus on providing pure connectivity to information infrastructure (akin to, say, a rural telephone cooperative), a focus on providing access to community information and linkages that might not be easily obtained elsewhere, or to provide an electronic "place" to act as a center of communication and interaction, rather than as an information utility (Bryan et al., 1998). Given that at least one component of a proximate community is the role of

Table 1: Community Networks Case Study Factor Summary

Variables	RIIPP	Potomac KnowledgeWay	RockNet
Population	Fauquier - 51,300/ Alexandria -115,400	Northern Virginia - 1,200,000 (est.)	Rockville - 47,235
Per Capita Income	25,775 - 34,023	34,023-34,422	27,471
Education Level[1]	48.5%	48.5-53.3%	49.9%
University Influence[2]	Key actor & change agent formed coalition of participants	Peripheral participation	No participation
Locality Type	City and exurban county	Inter-regional network public/private partnership	City (urbanized area)
Connectivity[3]	96.7%/97.8%	98.2%	99.3%
Key Actors	University, city/ county/federal govern- ments, public school system	Non-profit regional group, businesses, local govern- ments	Community network organization, city government
Innovation Factor	Availability of federal grant, public sector en- trepreneur	Private sector policy en- trepreneur, social/business network/technology	Public sector entrepreneur
Opportunities & Barriers	Sufficient capital re- sources, but early on technical learning curve. Training require- ments extensive, gover- nments concerned about political risk, in- itially took caution wait & see attitude.	Initiated by group repre- senting business interests. Capital raised from state & private sources. Key actor was private sector entrepreneur. Low visibility/high profile project. Vision not clear to target community.	Project championed by elected official. Key actor secured support of other el- ected officials. Rock- Net has limited re- sources but broader popular support than other cases.
Policy Outcomes / Assessment	Project demonstrated possibilities of net- worked technologies & indicated rural dem- and for Internet access. Change agent influence replaced by commun- ity/city network devel- opment. Policy innova- tion unlikely absent ex- ternal influence.	Project illustrates that re- sources are not necessarily limiting factor. Focus of project on "information networks" without an un- derlying community seems to explain lack of popular interest. Import- ance of geographic community indicated.	Community partici- pants' efforts indic- ate base of popular support. Local gov- ernment admin. & elected officials allocated resources to a policy innova- tion. Official network followed deployment of community net.

[1] As measured by percentage of population with BS or BA.
[2] The role of intellectual capital (in this case represented by activity of univer- sities) has been noted in the literature on innovation.
[3] Connectivty in this case represents telephone availability to an occupied unit. The presence of telephone represents potential for connectivity.

jurisdiction boundaries, an additional variable that needs to be considered is the role of local government in delivering some of the services or information provided by community-based information networks. In the RIIPP case, there is some evidence that the initial reluctance of the exurban locality to participate in the development of the community network may be related to its being a non-urbanized locality, and hence perceived less need to engage in additional service delivery, particularly without demonstrated demand for these services.

Conversely, the other two systems seem to have been developed as a function of being in an urbanized area. If one of the key functions of a community network is to provide mere connectivity, it will not be able to compete with the resources available to a more established information provider such as the an ISP or local telco. If the design objective of the system (in this case the core community might be thought of as the primary stakeholders, or community "founders") is to be an information provider of community events, activities, and so forth, then it may be out-competed by an established news outlet such as a newspaper. If its objective is to provide local community/governance type information, it may be duplicating the efforts of a locality, unless the locality is a participant in the community information system localities efforts.[22] One of the complexities faced in the analysis of these systems is the need to distinguish the underlying technologies (ICTs) and the actors involved in operating these systems (whether or not the implementers of the community information system constitute a community is an interesting question) from the virtual community that comes into being from the implementation of interactive (participatory) systems. In either case the degree of involvement or participation of the users might be a useful determinant of the nature of a "community" as distinguished from a group of users of services.

Opportunities and Barriers
Internal Organizational Factors

As a generality in terms of organizations, increased availability of personnel, technical, and fiscal resources is linked to higher levels of innovation and the ability to react more quickly to changes in the needs of the networked community. With regard to the operation of a community network, the range of skills and interests is frequently a function of the size of available and interested participants. The larger the population, the more likely the variation in skills, talents, and communities of interest present. Secondly, the presence of slack resources may allow a margin of experimentation, and in this regard, reduce the risk factor. Specifically, in looking at organizations, previous studies on innovation have indicated that organizational slack is related to a higher level of innovation.[23] This was not a limiting factor in the three cases considered above since organizational resources (at least capital resources) were not in critical shortage. An additional consideration is that inefficiently used resources, or internal misallocation of resources in terms of management can result in the same outcome: insufficient resources to complete the necessary organizational objectives.

Certain operational areas, such as technical support for users, exhibited some evidence of strain on resources with resultant system problems. While operation of the systems were generally unaffected, the requirements of the users were not always immediately met. Necessary training or support was initially delayed by limited personnel in either a technical support capacity or training, in the case of the RIIPP. Further, some of the implementation issues, such as inefficient operation of the various task groups, arising in the Potomac KnowledgeWay project may also have been due to insufficient administrative support (human infrastructure) rather than the underlying technological systems. Another element present in all of these cases is the fact that these community networks were the result of multi-organizational collaborations and partnerships, rather than the efforts of a single organization or group of organizers. Thus any analysis of the role of internal organizational factors is confounded by the potentially competing agendas of the various participating entities. In this sense it is difficult to make distinctions between the operations of a community network, the participants involved in the operations, users of services made available on the community network, and of most interest, the virtual community of users that arises. The lack of sufficient resources by the operators of the community network can be offset by a group effort to provide support to other members of the community. For instance, in the case of RockNet, the organization that runs the community network is made up of users with an interest in the operation of a community network, and hence the virtual community and the CN itself overlap. At the other extreme was the PKW which was designed as a clearing house of information, and the users of information had little in common and hence the virtual community in this case was virtually nonexistent.

Political Factors

These networks were chosen for the fact that they all have some type of connection or relationship with a geographic locale. In all three cases one of the participants was a locality, which added political variables to the operation of the systems. A review of the outcomes of these cases leads to the conclusion that political variables, such as the role of individual stakeholders, the political environment (vis-á-vis political risk), and high visibility of the projects, may have affected the project outcomes. In all of the cases presented, there was clear role for policy entrepreneurs or organizational visionaries. This may have been either directly as key actors (system implementers such as the schools) and decision makers, or as change agents (such as a university or elected politician). It is difficult to determine the role of other factors. As noted above, these systems were implemented in relatively resource-rich environments. However there is sufficient evidence from the literature on organizational innovation that these factors may also be of particular importance in resource-constrained environments, where they serve to "marshal resources" or generate sufficient community resources by encouraging collaboration. Community networks need to consider the role and participation of governmental institutions in their operation. While often benefiting from the tremendous benefits from the reduced impact of distance and the boundaries of location, the virtual still can not ignore the proximate.

Alternatively, in environments that are not resource constrained, but ones in which organizational factors are key, the change agents or entrepreneurs may act as risk "buffers" by expending sufficient political capital to absorb the policy risk generated by adopting innovations. In this case, the scenario would be one in which the administrative arm of the locality was reluctant to risk innovation, but under leadership (or pressure) of the policy entrepreneur, would implement the policy innovation. For community networks this offers the possibility that the backing of these types of stakeholders can add to the viability and "legitimacy" of the community part of the network.

Policy Outcomes /Assessment

In terms of initiating community networks, a variety of actors can potentially serve as the key stakeholders, or change agents to crystallize the "wires" part of the systems. School-based infrastructure offered in one case a convenient locus for adding local community information infrastructure to the Internet services, through the use of centralized existing sources. On the other hand, alternative dedicated service providers such as cable-TV operators might be more suited for combining TV and Internet services in a single delivery system.

One problem for certain organizational nonprofit community information systems is that other potential players may suddenly appear when sufficient community interest is generated. These include: locally franchised cable-TV operators, telephone and or wireless communication companies, utilities (there are networks that have been proposed by water, electric and gas companies), alternative Internet service providers, or even other public or public/private sector interests. Communities should consider these groups and organizations an asset to be used in designing a community system. Competition for users may generate offers of technical support and or physical resources that can be used to supplement limited community resources.

Finally, while a variety of successful models exist for developing community-based information infrastructure, in general they have several key elements in common. Typically there is partnership involving several different community groups. The actors/participants with diverse interests ensure that there is a broad base of support, and that the participants will perceive that the system is "theirs." This brings another key component into focus. A well-planned project also looks to the future, and assesses the needs of both the citizens as well as the public sector stakeholders. Flexible planning is also critical. A project "fixed in stone" will find that the rapidly changing world of telecommunications may leave the best planned project "high and dry" when the tides of technology shift.

Conclusion: Questions to Ponder and General Observations

In summation, the determination of community appears to be a function of more than determining boundaries of place and space. As the growing deployment of advanced ICTs allows for alternative modes of communication and information, a entirely new array of relationships emerge. If community can be said to expanded to include the loosely linked networks of interaction, with emphasis shifting from a solely geographic locale to one focused on commonality of interest or purpose, does this eliminate or weaken the role of underlying geographic community? Cautious observers have warned that the use of ICTs will reduce social capital, diminish the nature of geographic community, and weaken community (that is locale based) relationships.

An assessment of the cases presented leaves several unanswered questions that merit more in-depth research:

- If "place" and the friction of distance have a significantly minimized import due to the effects generated by ICTs, then what separates "places" in a virtual context? Does intensity of interest act as an analogue to distance in the "new spaces," in a sense of generation of shared norms, affective ties or even mutual obligations? Could the concepts of "there" and "not there" in cyberspace be more concerned with how frequently a cybercitizen communicates about a given topic (the "there of cyberspace"), with frequency of communication indicating level of interest and hence how more or less "close" they are to the center of a virtual community? How exactly can the virtual connections made possible by ICTs be harnessed to strengthen and develop physical community.

- Centralization vs. patchy-ness—while the decentralizing spatial effect of ICTs has been noted, conversely an opposite effect is noted with respect to concentration of the "physical infrastructure," i.e., the actual wires, servers, and access to bandwidth. ICTs allow a user to be anywhere, but the density of physical infrastructure underlying communication technologies allows for a richer denser flow (in this case, speed) of information. While a physical community may spread over a large area, bandwidth issues, as well as the richness of face-to-face communication, seems to provide a reason to re-centralize.

- Issues of governance have become more complex. Much of the legal operation of the world is geographically based. Taxes and regulation are generally applied at fixed points of place. At what point is information regulated on a global net? Does the process by which a geographic community regulates the interactions of members of the community apply to virtual interactions? What kinds of "community standards" apply when part of the community might not physically be co-located?

The community informatics framework seeks to build on the opportunities made possible by community-based ICT systems as a way of developing a new wave of community participation enabled by the ease of networked based communication. The virtual metropolis presents tremendous possibilities for geographic communities as well as for those charged with governance of the polity. In addition, the potential exists for "virtual groups," that may not have yet realized their commonality of interests to become significant competitors for the attention of citizens. Community development can be initiated by a variety of actors, administrators, elected officials, and other interested stakeholders, any or all of whom can proactively attempt to "colonize" the virtual metropolis or risk being left behind by other "virtual interest groups." Absent participation by this component of the public sphere, one role for communities (as embodied in community networks) is to serve as a forum or conduit for community "objectives," which might otherwise be channeled though the more formal public sector channels.

At least in the United States, the days of the community networks as a provider of connectivity, simple portals to information, pages of links, or listings of government meeting agendas is long past. Commercial ISPs can provide greater connectivity and reliability than can CNs for a significant portion of the population. While areas exist that are of marginal interest to commercial ICT interests, this represents at best a temporary niche activity. The entrance of local newspapers and recognized providers of news into cyberspace has essentially eliminated this role for CNS.

If the current community network structures are at all representative of the possibilities in the virtual metropolis, then we may conclude that while the virtual world is not a competitive threat to the space of places, an adjunct or additional overlay of information and communication, the possibility exists for the virtual metropolis to become a new forum for governance, or at least an augmented or more rich form of community. Even as telephones and automobiles did not eliminate the need for "places," ICTs are unlikely to replace the need to meet face to face. We see that in those places where a geographic sense of community exists, then ICTs serve can serve as intensifiers or more efficient conduits for information flow, be it social, educational or political. In this case community networks amplify the existing geographic community interests, or may enable various types of interests to more readily interact reaching a critical mass within a geographic community that might not have been practical without the networks.

On the other hand, the somewhat disappointing outcomes of the efforts to establish a virtual regional community without an underlying recognized geographic component seems to indicate that a core concept of community must exist, that the successful functioning of a virtual community results from some extant interconnection and not the converse. Building a virtual community simply because it is possible will result in an empty space, unless there is reason to express communication. If you build it, they will come, but only if they have some other reason for going there. Community informatics recognizes that the technology of ICTs does not operate in a vacuum, but in a rich context of social, interpersonal and culture interactions. Implementation of community-based information systems, then, are most likely to be successful when they are used to augment community assets rather than to attempt to create artificial structure where none exists.

Reference

Alexander, J.H., and Grubbs, J.W. (1998). "Wired Government: Information Technology, External Public Organizations, and Cyberdemocracy" *Journal of Public Admin. and Management*: 3(1). [http://www.hbg.psu.edu/Faculty/jxr11/alex.html]

Baker, P. (1997). "Community Networks: an On-line Guide to Resources." [http://ralph.gmu.edu/~pbaker].

Baker, P. (1997). *Local Government Internet Sites as Public Policy Innovations.* Unpublished Doctoral Dissertation, The Institute of Public Policy, George Mason University, Fairfax, VA.

Baker, P. (1996a). Proceedings of Community Networking '96: "Bringing People Together." Taos, NM, May 14-17.

Baker, P. (1996b). "Community Networks: New Tools for Environmental Planning," *Environment and Planning.* May/June, 11.

Baker, P.(1995). "Regional Information Infrastructure Project (RIIPP): The Development of Community Information Infrastructure Through Leveraging of Community Resources." *Proceedings of Telecommunities '95 - Equity on the Internet.* First Annual General Meeting of Telecommunities Canada. Victoria, B.C., Canada, August 21, 1995.

Bauman Foundation (1995). *Agenda for Access, Public Access to Federal Information for Sustainability through the Information Superhighway.* Washington, DC: The Bauman Foundation.

Beamish, Anne (1995). *Communities On-Line: A Study of Community-Based Computer Networks.* Unpublished master's thesis, Massachusetts Institute of Technology. Thesis available: http://alberti.mit.edu/arch/4.207/anneb/thesis/toc.html.

Bender, Thomas (1982). *Community and Social Change in America.* Baltimore: Johns Hopkins.

Benton Foundation (1996). *State and Local Strategies for Connecting Communities.* Washington, DC: The Benton Foundation.

Boncheck, M.S. (1995). Working paper, "Grassroots in Cyberspace." Political Participation Project, Cambridge, MA.

Bryan, C., R. Tsagarousianou, and D. Tambini (1998). "Electronic Democracy and the Civic Networking Movement," in Tsagarousianou, R.; D. Tambini, and C. Bryan (Eds.), *Cyberdemocracy: Technology, Cities, and Civic Networks.* New York: Routledge.

Caircross, F. (1997). *The Death of Distance.* Boston: Harvard Business School Press.

Castells, M. (1996). *The Rise of the Network Society.* Cambridge, MA: Blackwell Publishers.

Castells, M. (1989). *The Information City.* Cambridge, MA: Basil Blackwell, Inc.

Center for Policy Alternatives (CPA) (1996). *State and Local Strategies for Connecting Communities.* Washington, DC: Benton Foundation.

Cisler, S. (1994). "Community Networks: Past and Present Thoughts." Paper presented at "Ties That Bind: Building Community Networks" Apple Conference Center Cupertino, CA, May 4-6, 1994.

Clingermayer, J.C. and R.C. Feiock (1997). "Leadership Turnover, Transactions Costs and External Service Delivery." *Public Administration Review,* 57(3), 231-9.

Cohill, A.M. and A.L. Kavanaugh, (Eds.) (1997). *Community Networks: Lessons From Blacksburg, Virginia.* Boston: Artech House

Dodge, M. (1998)." The Geography of Cyberspace." [http://cybergeography.org/geography_ of_cyberspace.html]

Doheny-Farina, S. (1996). *The Wired Neighborhood.* New Haven, CT: Yale University Press.

Dutton, W.H., Blumler, J. G. and K.L. Kraemer, (Eds.) (1987). "Wired cities: shaping the future of communications." Washington, DC: Washington Program-Annenberg School of Communications, Boston, MA: G.K. Hall.

Dutton, W.H. (1999). *Society on the Line, Information Politics in the Digital Age.* New York: Oxford University Press.

Garreau, J. (1991). *Edge City: Life on the New Frontier.* New York: Doubleday.

Galston, William A. (1999). "(How) Does the Internet Affect Community? Some Speculations in Search of Evidence." In *democracy.com? Governance in a*

Networked World, E.C. Kamarck and J.S. Nye, Jr. (Eds.). Hollis, NH: Hollis Publishing Company.

Ghere, R.K., and B.A. Young (1998). "The Cyber-Management Environment: Where Technology and Ingenuity Meet Public Purpose and Accountability." *Journal of Public Admin. and Management*, 3(1). [http://www.hbg.psu.edu/Faculty/jxr11/gypaper.html].

Glaeser, E.L. (1994). "Cities, Information, and Economic Growth." *Cityscape*, 1(1), 9-48.

Graham, S . (1992). "Electronic Infrastructures and the City: Some Emerging Policy Roles in the U.K." *Urban Studies*, 29(5), 755-781.

Graham, S. and Simon M. (1996). *Telecommunications and the City, electronic spaces, urban places*. New York: Routledge.

Guthrie, K.K and W.H. Dutton (1992). "The Politics of Citizen Access Technology: The Development of Public Information Utilities in Four Cities." *Policy Studies Journal*. 20(4) 574-97.

Hiltz, S. and M. Turoff (1993). *The Networked Nation* (rev.). Cambridge, MA: MIT Press.

Huffman, L. and Talcove, W. (1995). "Information infrastructure: Challenge and opportunity." *Public Management*, 77(5), 9-14.

Jones, S.G. (Ed.) (1995). *CyberSociety: Computer Mediated Communication and Community*. Thousand Oaks CA: Sage Publications.

Kahin, B. and C. Nesson, Eds. (1997c). *Borders in Cyberspace: Information Policy and the Global Information Infrastructure*. Cambridge, MA: MIT Press.

Kahin, B. and J. Keller. (Eds.) (1995). *Public Access to the Internet*. Cambridge, MA: MIT Press.

Kanfer, A., and C. Kolar (1995). "What are Communities Doing On-Line?" Presented at Supercomputing '95, December 7, San Diego CA [http://www.ncsa.uiuc.edu/People/alaina/com_online].

Kanfer, A. (1997). *Ghost Towns in Cyberspace*. National Center for Supercomputing Applications, University of Illinois at Urbana-Champaign, IL. [http://www.ncsa.uiuc.edu/Edu/trg/ghosttown/].

Katz, J.E. and P. Aspden (1997). *Cyberspace and Social Community Development: Internet Use and Its Community Integration Correlates*. New York: Markle Foundation. [http://www.iaginteractive.com/emfa/cybsocdev.htm].

Kingdon, J.W. (1984). *Agendas, Alternatives, and Public Policies*. Boston: Little Brown.

Lohr, S. (1996). "Sizing up the Internet as an Engine of Regional Development." *New York Times CyberTimes*, September 16, 1996. [http://www.nytimes.com/library/cyber/week/0916regions.html].

London, R. (1996). "Checking Perceptions and Reality in Small-Town Innovation Research." *American Behavioral Scientist*, 39(5), 616-628.

National Academy Press (1995). *The changing nature of telecommunications/information infrastructure*. Washington, DC: National Academy Press.

National Telecommunications and Information Administration (NTIA) (1996). *Lessons Learned from the Telecommunications and Information Infrastructure Assistance Program*. Washington, DC: U.S. Department of Commerce.

National Telecommunications and Information Administration (NTIA) (1995).

Survey of Rural Information Infrastructure Technologies. Washington, DC: U.S. Department of Commerce.

Neu, C. R., R.H. Anderson, and T.K. Bikson (1998). "E-Mail Communication Between Government and Citizens: Security, Policy Issues, and Next Steps." Santa Monica, CA: Rand Institute. IP-178. [http://www.rand.org/publications/IP/IP178/].

Nunn, S. and J.B. Rubleske (1996). "Webbed Cities and the Development of the National Information Highway: The Creation of Web Sites by City Governments." University of Indiana Center for Urban Policy and the Environment. [http://www.spea.iupui.edu/dept/urban/rpt-abs.htm].

Office of Technology Assessment (OTA) (1995). *The Technological Shaping of Metropolitan America,* OTA-ETI-643. Washington, DC: U.S. Congress, U.S. Government Printing Office.

Office of Technology Assessment (OTA) (1993). *Making Governments Work: Electronic Delivery of Services,* OTA-TCT-578. Washington, DC: U.S. Congress, U.S. Government Printing Office, September.

Potomac KnowledgeWay (PKW). (1997). *Potomac KnowledgeWay.* [http://www.knowledgeway.org/milestones].

Rand Institute (1995). *The Feasibility and Societal Implications of Providing Universal Access to Electronic Mail (Email) Within the U.S.* Santa Monica, CA: Rand Institute.

Rheingold, H. (1993). *The Virtual Community: Homesteading on the Electronic Frontier.* Reading, MA: Addison-Wesley.

Rocheleau, B. (1998). "Governmental Information System Problems and Failures." *Public Administration and Management: An Interactive Journal.*

Rogers, E.M. (1995). *The Diffusion of Innovations (*4th ed.). New York: The Free Press, Simon and Schuster.

Samaranjiva, R. and P. Shields (1997). "Telecommunication networks as social space: implications for research and policy as an exemplar." *Media, Culture and Society,* 19, 535-555.

Schuler, D. (1996). *New Community Networks: Wired for Change.* NY:Addison Wesley.

Schwartz, E. (1995). "Looking for community on the Internet." *National Civic Review,* 84(1), 37-41.

Stefik, M. (1999). *The Internet Edge: Social Legal and Technological Challenges for a Networked World.* Cambridge, MA: The MIT Press.

Wellman, B. (1997). "An Electronic Group is Virtually a Social Network," in Kiesler, S. (Ed.), *Culture of the Internet.* Mahwah, NJ: Lawrence Erlbaum Associates, Publishers.

Yin, R.K. (1994). *Case Study Research: Design and Methods* (2nd ed.). Thousand Oaks, CA: Sage Publications.

Endnotes

1   I wish to express my great depth of gratitude to comments and suggestions by Andrew Ward, of Georgia Tech, J.W. Harrington, of the University of

Washington, Michael Gurstein, the editor, and two anonymous reviewers.

2 For an insightful take on the social, technical and intellectual changes related to adoption of digital technologies, see Stefik, 1999.

3 These community (in this sense, originally geographically based) information infrastructures have been called by a variety of names including community networks, civic networks, public access networks, and information utilities, depending on the philosophical or disciplinary orientation of the originator of the term. They also convey slightly different meanings depending on the objective of the system.

4 This concept was developed in a paper on the socio-spatial components of telecommunication networks (see Samarajiva and Shields, (1997).

5 It is also interesting to note that the term community network actually precedes the digital sense, for instance the phrase 'a community network of caregivers...' occasionally appears in social work literature. Hence a digital sense is not always implied in common use. For the purposes of this discussion the term will imply the self-identified aggregate of users of a information network with at least a cursory proximate relationship. See Beemish, (1995), Cisler, (1994), Schuler, (1996), for a discussion of community-based information networks.

6 See, for instance, Dutton, (1999), Jones (ed), (1995), Rheingold, (1993), Wellman, (1997).

7 Galston (1999) cites Bender's (1982) definition of community as involving a limited number of individuals in a somewhat restricted social space or *network* (emphasis added) held together by shared understanding and a sense of obligation. Relationships are close, often intimate, and generally involve face-to-face interactions. Individuals are bound together by affective or emotional ties rather than by a perception of individual self-interest. There is a "we-ness" in a community; one is a member or a larger group.

8 Galston (1999) discussing Howard Rheingold's concept of on-line communities.

9 Galston (1999) has written an interesting piece on the changing nature of community as influenced by the Internet.

10 A fairly straightforward exploration of this viewpoint can be found in Cairncross (1997), *The Death of Distance.*

11 The effort did not yield a viable community network, per se, but established the demand for Internet related services, and in the implementation of a local government Web site for Fauquier [http://co.fauquier.va.us], the City of Warrenton [http://ci.warrenton.va.us/], a Web site for the public school system [http://www.libertyhs.com/], and two local newspapers [http://www.citizenet.com/] and [http://www.fauquier.com/index1.shtml], providing community information, but no community network, in a generally accepted sense. Further, the Fauquier County Web site is hosted by a private service provider [http://www.piedmontpress.com/index.html] rather than on a public sector server.

12 [http://www.rocknet.org/]

13 [http://www.knowledgeway.org]. Postscript: PKW suspended active operations in November 1999, but the Web site remains active as of December 1999.

14   Stewart, Thomas A. 1994. "The Netplex—It's a new Silicon Valley." *Fortune* 3/7/94, v129n5, p. 98-104.

15   According to the *New York Times,* Sept. 16, 1996 "Sizing up the Internet as an Engine of Development," Washington, DC, (the city) ranks 20th. If the two are combined, then the area would rank third, after Santa Clara, California, and Middlesex, Massachusetts.

16   Source: U.S. Bureau of the Census.

17   Source is MNPPC Planning Area Profiles URL<http://www.clark.net/pub/mncppc/montgom/factmap/facts/pap.htm>

18   The Morino Institute, a local Virginia-based foundation, fosters the use of networked information technologies to promote regional and economic development. The founder, a local business entrepreneur, is active in community and local development efforts.

19   As of March 1996, though this number is changing rapidly.

20   See for example, Alexander and Grubbs, (1998), Dutton, (1999), Ghere and Young, (1998), Glaeser, (1994), Huffman and Talcove, (1995), Rand, (1995), specifically in reference to information systems; other examples abound in the planning and public administration literatures.

21   That is as measured by level or extent of participation in the on-line activities.

22   Which occurred in both Rockville and Alexandria, where the localities were active participants in the community information systems.

23   Slack resources (i.e., organizational slack resources) refers to resources not necessary for the operation of an organization and hence available to use for other nonessential purposes. Rogers (1995) speculates for instance that one explanation for the importance of organizational size is that a large organization has a greater likelihood of having larger amounts of organizational resources (organizational slack). Conversely, an alternative interpretation might be that limited levels of organizational slack could be a driver for innovation.

Chapter V

Differential IT Access and Use Patterns in Rural and Small-Town Atlantic Canada

David Bruce
Mount Allison University, Canada

Introduction

Popular press and government rhetoric suggest that there has been steady progress in the extent to which individuals, households, businesses, organizations, and communities are using the Internet as part of their daily lives (Bruce, 1998, 1997). However, there is little empirical evidence to support this claim. In this chapter I argue that there has been slow and uneven penetration of Internet use in rural and small-town communities in Atlantic Canada, despite the best efforts of policies and programs. Drawing on evidence from a recent Internet use survey, suggestions are made for improving the performance of policies and programs aimed at increasing Internet access and use.

The purpose of this chapter is to provide an empirical overview of differential Internet access and use patterns, using data collected from a January 1998 survey (Jordan, 1998; Bruce and Gadsden, 1999) of 1501 households in 20 different Atlantic Canadian communities grouped into five distinct "community categories." (Reimer, 1997a, 1997b)

Characteristics of users for purposes of this analysis include age, gender, household income, educational attainment, and employment status. This chapter also explores the extent to which Atlantic Canadians have taken formal or informal courses or training programs related to information technology between 1993 and 1998.

Background

Information technology-related policies, programs, and strategies are being articulated and developed in large part in response to a new economic reality in rural

Canada. Resource-based sectors, such as agriculture, forestry and fisheries, are no longer the main source of wealth generation in an increasing number of rural communities. The service sector has replaced the resource sector as the most important employer of the workforce, accounting for two-thirds of employment. Small businesses, in particular those in the service sector, dominate business operations in rural Canada (BICON, 1998).

In response to this changing economic environment, it has been noted that technology is releasing significant opportunities for marginalized communities, and the implication is that appropriate enabling policies and programs can further those opportunities:

> Technology is not a release from the burden of place, but rather a tool to respond to the challenge and opportunities of the "local." In this context there is an opportunity to be productive and economically vital, not by being "anywhere," but precisely by being "somewhere," and particularly by being part of larger distributed networks where "place" is a "resource" not a burden. (Gurstein, 1999)

The Internet has great potential as a development tool (for connecting people, for electronic commerce, as a medium for distance education, for the delivery of government services, and much more); however, for a community to take advantage of the opportunities presented, many people from all segments of the community must be connected. Likewise, government services delivered online, and electronic transactions of commercial activity, can only be effective and used widely if many are connected.

Further to the issue of "being connected" is the public policy agenda attached thereto. Nationally the federal government has articulated a "Connecting Canadians" agenda to make Canada the most connected nation in the world by the year 2000. Strategic thrusts include establishing community access points across the nation, furthering electronic commerce activities, putting more government information online, and much more. Of particular interest here is the Community Access Program, designed to provide thousands of rural communities across the country with a public location for community members to use the Internet, access training and services, and build interest in the use of Internet for personal and business use. Launched in 1996, the program has made funding available to more than 1,000 communities. Each of the communities involved in the survey which forms the basis of this discussion has such an access centre.

The federal government undertook a major consultation with rural Canadians (Rural Secretariat, 1999) from May to July 1998, to solicit information on the issues and challenges they faced, and their suggestions for appropriate government roles in helping to address those issues. Ten critical issues were identified, one of which was to improve rural telecommunications and increase the of use of the information highway. Other priority areas which could be addressed through increased access to and use of the Internet include developing opportunities for rural youth, improving access to education, improving access to information on government programs and services, and increasing the economic diversification in rural communities.

Within Atlantic Canada each of the provincial governments has developed its own strategic thrusts. For example, the New Brunswick (NB) government has developed a series of funding and support programs to foster increased opportunities for NB-based companies to develop on-line materials, to support staffing at community access centres, to increase information technology training opportunities, to facilitate an increase in the purchase of Internet-connected home computers, and much more (Bruce, 1998). In Prince Edward Island (PEI), a Knowledge Economy Partnership (among government, education institutions, and business) has been formed to facilitate an increase in knowledge economy activities in PEI. In Nova Scotia (NS) the Nova Scotia Community Access Committee (NSCAC) acts as a catalyst and coordination body for provincial initiatives that involve the equitable provision of access to information and technology. Representatives of libraries, education, community networks, government, and industry participate. Also within NS is NovaKnowledge, whose mission is to promote the development of a flourishing, sustainable, knowledge-based economy in Nova Scotia, enabled by information technologies. NovaKnowledge is a unique, multi-stakeholder association that promotes the growth of the knowledge-based economy by forging links between business, government and our educational communities.

All of this information suggests the need to better understand the extent to which people are connected, the characteristics of those who are connected, and their relative location.

Characteristics of Internet Users: Evidence from Other Sources

A number of other studies have highlighted the increases in the number of people using computers and the Internet. Dickinson and Sciadas (1997) pulled together information from the Household and Family Expenditure Survey and identified that in 1996:

- 31.6% of households had a personal computer (up from 10.3% in 1986) and 7.4% of households were connected to the Internet;
- computer ownership and Internet use increased with income (e.g. 13.9% of those in the bottom income quartile and 54.2% of those in the top income quartile had a home computer);
- computer ownership and Internet use increased with education attainment across all income quartiles;
- computer ownership was lower in the four Atlantic Provinces (below 27%) than elsewhere, except for Nova Scotia which was higher than in Quebec and Manitoba only;
- Internet use was lower in the four Atlantic Provinces (below 8%) than elsewhere except for Nova Scotia which was higher than in Quebec and Saskatchewan only; and
- both computer ownership (at 26.5%) and Internet use (at 4.5%) were lower in areas outside of the Census Metropolitan Areas (CMAs), where it was 35.5% and 9.7% respectively. Most residents of Atlantic Canada live outside CMAs.

A 1997 survey of households in nine rural maritime communities showed that 57% of respondents use a computer in the household (Beesley et al., 1998b). The same study showed that almost half of those (41%) use the Internet at home. Home computers were found to be primarily used for word processing (by 68% of respondents), education (by 67%) and entertainment (by 58%). Slightly less than one-third used them for their home office (31%) or for bookkeeping (30%).

Caragata (1998) used AC Neilson Data to describe the level of Internet access and use across Canada. Results showed that:

- Men accounted for 57% of all users in 1996, but by 1998 this number fell to 51%.
- 27% of those aged 55 to 64 use the Internet, an 18% increase over 1997.
- In Ontario and British Columbia, 41% of residents are using the Internet, with the lowest levels of use in Atlantic Canada (36%) and Quebec (36%). (This fact is interesting since *Computing Canada* [1996] suggests that Internet use by businesses is highest in Atlantic Canada.).
- 23% of Canadians do not use the Internet because they have "yet to find a reason."
- 18% of nonusers say the Internet is too expensive. This is confirmed by Beesley et al. (1998b), where price was identified as a barrier to owning a computer among those who did not own a computer.

Drawing on Statistics Canada's 1989 General Social Survey, Norris and Johal (1992) examined knowledge and use of computers in urban and rural areas and found that:

- The incidence of computer use and ownership of personal computers was higher in urban areas.
- Only 21% of the rural population had taken a computer course compared with 31.5% of urban dwellers.
- Computer training and use were higher among those in younger age groups.

Connecting User Characteristics with Business Internet Use

If businesses tap into and use information technology, it is likely to contribute to their economic success and that of their communities. One can look at the rising number of home-based businesses (Bruce, 1993; Jordan, 1998) that operate in Atlantic Canada and not dismiss their role of stimulating local economies. Home-based businesses can provide Web page design, network consulting, telemedicine, real-time consulting, and other information technology services to local and global clients (Mandale, 1998). Small businesses are important to the livelihood of small communities. Information technology can help to create economically healthy communities with opportunities for residents to find satisfying work locally (Gurstein, 1999).

Atlantic Canada, although beginning to build a profitable information technology industry, could be using this industry more extensively to further its economic advantage. Mandale (1998) suggests that the "growth of the IT sector and knowledge-based industries fueled by IT development hinges on one crucial human factor—a supply of workers with new skills" (p.148). It is important that Atlantic Canada formulate policy that ensures skill-building programs are available to residents.

The issues and opportunities identified suggest the need for more information about who is using the Internet so that policy and program development may better target those groups and communities who are currently marginalized and unable to participate in this aspect of community and development.

What follows is a discussion of the extent to which rural Atlantic Canadians are using the Internet, with an emphasis on understanding the different demographics of the users. It also discusses differences among communities with different characteristics. The discussion concludes with some ideas to make stronger connections between people and the Internet for individual and community benefit.

Analysis of Survey Results

Communities were grouped into five distinct categories (CGs) for analysis purposes, based on four different contributing factors. These factors are:

- The extent to which the economy of the community is exposed to **global** market forces (such as the primary resources industries, manufacturing, communications, business services) or to **local** markets (such as construction, transportation, trade, government services, education, health, accommodations).
- The extent to which the economy of the community is subject to more **stable** market conditions (manufacturing, communications, trade, government services, education, health, accommodations) or to more widely **fluctuating** market conditions (primary resource industries, construction, finance, real estate and insurance).
- The extent to which the population in the community is **capable** of supporting itself (more education and skilled, more self-employment, lower age dependency ratio) or is **more dependent** on others (lower education and skill levels, less self-employment, higher age dependency ratio).
- The extent to which economic and social outcomes in the community are positive and **leading** (higher incomes, less poverty, stable families, higher rates of home ownership) or are somewhat negative and **lagging** (lower incomes, more poverty, unstable families, lower rates of home ownership).

The various combinations and permutations of these factors produces many different community groupings; the five discrete community categories chosen for this study are as follows:

- CG1—Characterized by leading social and economic outcomes, with local and stable economic markets, and with a population having somewhat high capabilities.
- CG2—Characterized by leading social and economic outcomes, with local

Table 1: Community Groups

	CG1	CG2	CG3	CG4	CG5
NS	Kentville	Port Hawkesbury	Barrington	Springhill	Guysborough
PEI	Kinkora	Crapaud	Souris	Wellington	Tignish
NB	Woodstock	Clair	Blackville	Campbellton	Pacquetville
NFLD	Gander	Happy Valley-Goose Bay	Anchor Point	Stephenville	Point Aux Gaul

and stable economic markets, but with a population having somewhat low capabilities.

- CG3—Characterized by leading social and economic outcomes, with global and fluctuating economic markets, and with a population having somewhat low capabilities.
- CG4—Characterised by lagging social and economic outcomes, with local but fluctuating economic markets, and with a population having somewhat high capabilities.
- CG5—Characterized by lagging social and economic outcomes, with global but stable economic markets, and with a population having somewhat low capabilities.

The selected communities from each province are shown in Table 1 by community grouping.

The Total Design Method (Dillman, 1978) for self-administered mail surveys was employed. A total of 4,057 surveys were sent. A total of 1,501 valid questionnaires were returned. As a starting point, 403 respondents (27%) indicated that they use the Internet. The analysis which follows is limited to the characteristics of those who indicated that they use the Internet.

Internet use among the community types ranges from a high of 37% in CG1 to a low of 17% in CG5 (Table 2). This is hardly surprising given that CG1 communities serve as important regional centres and enjoy healthy economies. CG5 communities are much more isolated from larger centres and may also face barriers such as lack of appropriate infrastructure or long-distance dialing costs to reach an Internet service provider (ISP). The implication of this observation is that residents of more isolated communities are not participating in the information economy to the extent of those in larger centres near urban core areas.

Across individual communities within each of the community groupings there are some wide variations in the incidence of Internet use. In CG1 the incidence of use in Kinkora (PEI) is 10% below the group average of 37%. In CG2 the incidence of use in Clair (NB) is 7% below the group average of 34%. In CG3 the incidence of use in Anchor Point (NF) is 8% below the group average of 22%. In CG4 the incidence of use in Springhill (NS) is 19% and in Wellington (PEI) it is only 15%; these are well below the 40% of the people in Stephenville (NF) and 32% of the people in Campbellton (NB) who use the Internet. In CG5 the incidence of use ranges from a low of 13% in Tignish (PEI) to a high of 18% in Pacquetville (NB), so there is a high degree of consistency across the group.

Table 2: Incidence of Internet Use by Community

	CG1	CG2	CG3	CG4	CG5
NS	Kentville 41%	Port Hawkesbury 39%	Barrington 23%	Springhill 19%	Guysborough 16%
PEI	Kinkora 27%	Crapaud 31%	Souris 20%	Wellington 15%	Tignish 13%
NB	Woodstock 37%	Clair 25%	Blackville 23%	Campbellton 32%	Pacquetville 18%
NFLD	Gander 43%	Happy Valley-Goose Bay 39%	Anchor Point 14%	Stephenville 40%	Point Aux Gaul 16%
Group	37%	34%	22%	25%	17%

Importance of Community Access Centres

Public access terminals or community access centres play a more important role in some types of communities than others with respect to providing Internet access to community members. Table 3 shows the percent of Internet users in each community and community group who access the Internet at their local community access centre.

CG3 and CG5, the more isolated and less economically prosperous communities, have a higher percentage of their Internet users accessing the services of the community access centre. This confirms the important role that access centres play in bringing the Internet to rural communities. Internet users in CG2 are less likely to be using the community access centre for Internet access. For residents of Point aux Gaul, NF, Wellington, PEI, Kinkora, PEI, Pacquetville, NB, and Gander, NF, the community access centre is particularly important, where more than 30% of the Internet users are making use of the community access centre.

Gender

The incidence of Internet use among females (28%) is slightly higher than among males (26%) (Table 4). When looking at the gender of users by community

Table 3: Incidence of Internet Use in Community Access Centres by Community

	CG1	CG2	CG3	CG4	CG5
NS	Kentville 11%	Port Hawkesbury 6%	Barrington 27%	Springhill 18%	Guysborough 17%
PEI	Kinkora 35%	Crapaud 5%	Souris 20%	Wellington 40%	Tignish 20%
NB	Woodstock 13%	Clair 8%	Blackville 27%	Campbellton 0%	Pacquetville 33%
NFLD	Gander 30%	Happy Valley-Goose Bay 25%	Anchor Point 22%	Stephenville 29%	Point Aux Gaul 44%
Group	21%	11%	25%	23%	28%

Table 4: Incidence of Internet Use by Gender, Age, and Education Attainment

Gender	Region	CG1	CG2	CG3	CG4	CG5
Females	28%	39%	31%	21%	31%	16%
Males	26%	36%	37%	22%	21%	18%
Age						
18-29	38%	36%	38%	42%	26%	46%
30-44	32%	52%	35%	21%	37%	20%
45-64	29%	36%	40%	25%	26%	14%
65+	7%	10%	10%	7%	13%	2%
Education Attainment						
University	60%	64%	62%	53%	60%	57%
Some University	45%	40%	45%	48%	33%	69%
Community College	35%	43%	43%	31%	27%	30%
Some Community College	24%	27%	28%	5%	32%	23%
High School	18%	26%	26%	15%	18%	2%
Some High School	11%	15%	14%	9%	9%	11%
Grade School	8%	20%	4%	6%	14%	4%

group, we find that slightly more males than females use the Internet in CG2 and CG5, but more females do so in CG1 and CG4.

Age

At the regional level, 36% of individuals aged 29 and under indicated they use the Internet. The second highest Internet users are those aged 30-44 (32%). Internet use among those aged 65 or more is quite low (7%) (Table 4). This difference across age groups poses significant challenges for those wanting to reach broad audiences for e-commerce or information dissemination purposes.

The general pattern of Internet use among those in different age groups among the community types for the most part follows that of the region as a whole, but with some important differences. In CG3 and CG5 there are significantly fewer people aged 30-44 or 45-64 who are using the Internet compared with those in the same age categories in CG1, CG2, or CG4. This might reflect the relatively higher unemployment rates and the relatively higher concentration of employment in the primary resource sector in those communities (such as Guysborough, NS, Blackville, NB, Souris, PEI, or Anchor Point, NF), which reduces the likelihood of access and use through a place of work. However, these two community groups (CG3 and CG5) also have a higher incidence of young adults using the Internet, suggesting a positive sign that perhaps older students are making use of the Internet for education purposes. Use among those age 45-64 years is significantly higher in CG1 and CG2. The incidence of young adults using the Internet in CG4 is very low at only 26%.

Education Attainment

The percentage of the population in Atlantic Canada using the Internet increases steadily with rise in education levels (Table 4). Internet use is highest among those who have completed university (60%).

In a similar fashion, Internet use in the different communities based on the level of education completed follows the same pattern. Some anomalies do exist, however. Those with some community college education in CG3 have the lowest incidence of use in those communities, and far below those with the same education level in other community groups. Those who have completed high school in CG5 have the lowest incidence of use in those communities, and far below those with the same education level in other community groups. In both of these community groups, then, there are significant population segments which could be targeted for access and involvement strategies.

Those with some university education in CG5 have the highest overall incidence of Internet use. This suggests that those who are able to return to the community after completing university have the means to provide themselves with access to the Internet.

Household Income

Policy makers and those who implement programs have been concerned about increasing access to computers and the Internet for everyone, and particularly for those who cannot afford to buy their computer or who find the connectivity charges beyond their means. This has spurred, for example, the Community Access Project to put publicly accessible, Internet-connected computers into as many rural communities as possible.

Not surprisingly, the survey results (Table 5) show that Internet use increases as household income increases. For example, 54% of those with a household income of over $60,000 use the Internet, compared with only 10% of those with a household income of $10,000-$19,999 and 13% of those with a household incomes of less than $10,000.

The pattern of Internet use and income across the community groups varies somewhat. For example, in CG2 (Clair, NB, Crapaud, PEI, Happy Valley-Goose Bay, NF, and Port Hawkesbury, NS), use is much higher among those in the $40,000 to $49,999 income category than in the other community types. In CG4 (Wellington, PEI, Campbellton, NB, Springhill, NS, and Stephenville, NF), the incidence of use among those in the two lowest income categories is higher than it is among those in the next two income categories ($20,000 to $39,999). This suggests that in these larger communities, a concerted effort is being made to facilitate access to the Internet among all household income groups. In CG1 the incidence of use among those in the $30,000 to $39,999 category is much higher than it is among those in the same income bracket in other community types. This is reflective of the relatively prosperous economy in those communities (Kentville, NS, Kinkora, PEI, Woodstock, NB, and Gander, NF).

Table 5: Incidence of Internet Use by Household Income and Employment Status

Household Income	Region	CG1	CG2	CG3	CG4	CG5
$60,000+	54%	55%	59%	43%	56%	44%
$50,000-59,999	40%	56%	35%	34%	33%	25%
$40,000-49,999	34%	24%	50%	31%	29%	35%
$30,000-39,999	22%	38%	21%	21%	11%	17%
$20,000-29,999	16%	22%	20%	14%	10%	15%
$10,000-19,999	10%	13%	8%	4%	15%	10%
$9,999 or less	13%	17%	9%	17%	28%	8%
Employment Status						
Student	87%	100%	100%	100%	60%	83%
Working Full Time	42%	51%	42%	38%	42%	34%
Working Part Time	19%	33%	29%	13%	22%	9%
Unemployed	16%	28%	35%	8%	5%	16%
Homemaker	16%	21%	21%	28%	7%	0%
Retired	9%	12%	17%	8%	9%	4%

Employment Status

With respect to the employment status of the respondents, it is not surprising that those who are students report a very high incidence (87%) of Internet use (Table 5). Approximately two of every five full-time workers, but only one of every five part-time workers, use the Internet. Relatively fewer of the unemployed, homemakers, and retired persons use the Internet.

An examination of the incidence of Internet use based on employment status among the five different community groups shows that there are significantly fewer students using the Internet in CG4 and CG5 relative to those in the other community groups. Internet use among the unemployed is much higher in CG1 and CG2 and much lower in CG4 and CG5. Internet use among both full-time and part-time workers is lower in CG3 and CG5 compared with those in the other community groups.

Improving Skills with IT Courses

One strategy employed by many to improve their opportunities in the workplace or simply to improve their overall skills is to take formal or informal courses or training related to the Internet and information technology. These might be taken in the workplace, in community access centres, at private institutions offering training programs, or through self-directed learning materials.

Across the Atlantic region, 14% of the respondents indicated they have taken some type of information technology-related course in the five years prior to this survey. While this seems to be a low percentage, it does represent 49% of all the people who have taken some type of training, skill development courses, or other activities to improve themselves. Slightly more than half of these (52%) have taken an information technology-related course at a community college, 30% at night school, 20% at university, and 12% at other locations (including private training institutions, community access centres, or the workplace).

Younger respondents were much more likely to have taken an information technology course in the five years prior to this survey. One in five (19%) of those aged 30-44 years and 17% of those 29 years and younger had taken a course in the previous five years. This compares with 11% of those aged 45-64 years and 2% of those aged 65 years and over.

Table 6 shows the percentage of respondents who have taken an information technology course in the last five years, and the percentage of those who have taken any type of course in the last five years and who have also taken an information technology course, by community type. More respondents from CG1 (19%) and fewer respondents from CG5 (12%) and CG3 (11%) have taken an information technology course in the last five years. However, it is interesting to note that for at least half of those people taking any type of course in the last five years in CG1, CG4, and CG5, an information technology course was chosen.

Community colleges are the most likely place in all community types for taking an information technology course. However, in CG3 and CG5, close to half have identified that they took an information technology course at night school. About 26% of those in CG2 have taken an information technology course at a university, and 20% in CG1 identified that they have taken an information technology course in some other place.

About 42% in both provinces took these courses at a community college, while another one-third in both provinces took these at night school. In PEI and Newfoundland the percentage of those taking an information technology course at a community college is higher (about 70%) and at night school is lower (less than 25%) than in the other provinces. In all four provinces about 12% of those who have taken an information technology course identified "other" as the location for that course—a community access centre, at work, or at a private institution.

Discussion

While the incidence of Internet use does not vary widely by gender, it does vary quite significantly by community type and by most other demographic characteristics.

In CG1 and CG2, which are characterized as being regional service centres with a relatively higher population, with locally based, stable market econo-

Table 6: Incidence of Information Technology Courses by Province and Community Type

	Of the Respondents	Of Those Taking Any Type of Course
Region	14%	49%
CG1	19%	55%
CG2	15%	43%
CG3	11%	44%
CG4	14%	51%
CG5	12%	51%

mies, respondents have a much higher Internet use than other community types. Those in CG5, which are relatively smaller and more isolated, with globally focused, fluctuating market economies, and a relatively lower level of human resource capacity, have much lower rates of Internet use. It is clear, then, that more proactive public participation strategies are needed in more isolated communities.

Internet use is more prevalent among those in younger age groups. People with higher education and income levels are more likely to use the Internet than those with lower income and education levels. Use among students is much higher than those who are working, unemployed, or retired. Creative strategies for encouraged those in "niche" markets of potential Internet users are required, where introductory sessions, demonstrations of Internet applications, and subsequent training courses must be tailored to demonstrate the everyday utility of this tool. Those who have not used this medium will require clear evidence of its use in their daily lives, be it for recreation, leisure, communication, job search, or other purposes.

Students have the highest incidence of Internet use. This is not surprising given their relatively greater access to computer technology through the education system. In contrast, those who are unemployed or working part time are less likely to be using the Internet. This raises an opportunity to more aggressively seek ways for these people to access the Internet in their communities, especially through community access centres.

The fact that almost half of all respondents who indicated they have taken any form of skills upgrading have taken an IT course illustrates that Atlantic Canadians view information technology skills in general and the Internet in particular as valuable resources for competing in the job market and with other economies. Of particular note is the fact that in CG4 and CG5, with "lagging" economies, half of the people taking any type of course or upgrading choose to take an information technology-related course.

Opportunities and Challenges

The profile of Internet users suggests that there are both significant opportunities and challenges for individuals, communities, community organizations, and governments to further increase the incidence of Internet use among the general population.

For communities where Internet use is low and there is a community access centre already present, there are opportunities for more aggressive marketing and promotion of its services. This can be done through creating meaningful and relevant opportunities for people and organizations to experience how the Internet might be used in their daily lives.

For governments there continues to be an enabling opportunity to support increased Internet use, providing a range of support services to community access centres (e.g., through the development of generic courses or programs which can be tailored to meet local need; the arrangement of global partnerships with private sector organizations to deliver courses, hardware, and software through access centres; and by providing funds for human resource support in the centres).

Furthermore, governments should continue to work with the telecommunications industry to facilitate the deployment of appropriate infrastructure and reasonable connectivity costs. Governments also have a role to play in providing relevant and easily accessible service information which can encourage people to use the Internet for many transactions. Pilot projects which examine the potential for driver's licence renewal over the Internet, for example, provide one such opportunity.

For individuals there are opportunities to become more entrepreneurial and responsible for their personal growth and development. One means of doing so is to take advantage of the community resources available through the presence of a community access centre.

There are, however, significant barriers to be overcome. Members of generations who have not grown up with the computer experience or who have not been exposed to this experience through their work or education are less likely to be users and must be eased into the use of the Internet with a more basic intervention about computers in general. Despite the declining price of computers, those with lower incomes are less likely to be using the Internet because of the inability to afford such costs. Community organizations have an opportunity to encourage these people to use the community access centre and to arrange for the resale of old computers, which can be connected to the Internet, to people with lower incomes should they wish to own one.

From a policy perspective, government must continue to provide information and services to the general population through both conventional and Internet mediums, since there is still a wide disparity within the population in terms of use of the Internet. Government must also ensure the sustainability of community access centres in the smallest and more isolated rural communities. While the ideal situation is for these centres to be self-sustaining through revenue generation activities, the reality is that in the smaller centres this poses more of a challenge. At the same time the findings identified in this chapter suggest that it is in these very communities where the access centres are most important for developing a more Internet-literate community. Finally, individual communities, led not only by their municipal government but by others as well, must make it their policy to make the most use of the Internet for its activities, ranging from promotion to discussion groups to much more, as a visible signal to its local residents that the Internet can and should play an important part of community functioning.

We are still in a state of great transition in our rural communities and small towns. Change from resource-based activities to other forms of economic activity, coupled with the pervasiveness of the Internet and technology in our general society, allows for both a great deal of uncertainty and a great deal of optimism and potential to coexist. How individuals and communities choose to manage the transition and make use of the Internet and other information technology tools will determine their relative success.

References

Atlantic Progress. (1995). Information Technology: Measuring its Importance. *Atlantic Progress.* 2 (4): 1-3.

Atlantic Provinces Economic Council. (1994). Major Opportunities in Atlantic Canada: the 1994 Inventory. *Atlantic Report,* 29 (1): 3-16.

Beesley, K., D. Bruce, D. Ramsey, and S. Suffron. (1998a). *Business Use of Information Technology in Rural Maritime Canada.* Research Paper No. 29. Truro, NS: Rural Research Centre, Nova Scotia Agricultural College.

Beesley, K., D. Bruce, D. Ramsey, and S. Suffron. (1998b). *Household Use of Information Technology in Rural Maritime Canada.* Research Paper No. 29. Truro, NS: Rural Research Centre, Nova Scotia Agricultural College.

BICON Consulting Associates. (1998). *Economic Linkages in Canadian Rural Communities.* Ottawa, ON: Adaptation Division, Policy Branch, Agriculture and Agri-Food Canada.

Bruce, D. and P. Gadsden, (1999). *Internet Access and Use in Rural and Small Town Atlantic Canada.* Sackville, NB: Rural and Small Town Programme.

Bruce, D. (1998). Building Social Capital and Community Learning Networks in Community Internet Access Centres. *Learning Communities, Regional Sustainability and the Learning Society.* Conference Proceedings, Volume 1. Edited by Ian Falk. Launceston, Australia: Centre for Research and Learning in Regional Australia, 45-54.

Bruce, D. (1997). Chaos, Chasms, and Champions: The Internet and Community Economic Development. In *Rural Research in the Humanities and Social Sciences IV.* K.B. Beesley (Ed.). Truro, NS: Rural Research Centre, Nova Scotia Agricultural College.

Caragata, W. (1998). The Net Grows Up: More Than a Third of Canadians are Now On-line. *Maclean's.* 111 (47): 104.

Computing Canada. (1996). Atlantic Canada Leads in Internet Use (Phase 5 Survey). *Computing Canada.* 22 (22): 44.

Dickinson, P and G. Sciadas. (1997). *Access to the Information Highway, the Sequel.* Statistics Canada Analytical Paper Series Number 13. Ottawa, ON: Minister of Industry.

Dillman, D. A. (1978). *Mail and Telephone Surveys: The Total Design Method.* New York: Wiley-Interscience.

Gurstein, M. (1999). Flexible Networking, Information and Communications Technology and Local Economic Development. *First Monday.* 4 (2). Available http://www.firstmonday.dk/issues/issues_4_2/gurstein/index.html.

Jordan, P. (1998). *Living in Atlantic Canada, 1998: Perceptions About Life in Rural and Small Towns.* Sackville, NB: Rural and Small Town Programme, Mount Allison University.

Mandale, M. (1998). Silicon Shores. *Atlantic Progress.* 5 (5): 148-153.

Norris, D. and K. Johal. (1992). Social Indicators from the General Social Survey: Some Urban-Rural Differences. *Rural and Small Town Canada.* Ray Bollman (Ed.). Toronto, ON: Thompson Educational Publishing Inc., 357-368.

Reimer, B., (1997a). *A Sampling Frame for Non-Metropolitan Communities in*

Canada. Canadian Rural Restructuring Foundation. Available http://artsci-ccwin.concordia.ca/SocAnth/CRRF/crrf_hm.html.

Reimer, B., (1997b). *The NRE Rural Sample Site Database—Logic and Approach*. Canadian Rural Restructuring Foundation. Available http://artsci-ccwin.concordia.ca/SocAnth/CRRF/crrf_hm.html.

Rural Secretariat. (1998). *National Rural Workshop Report*. Available http://www.rural.gc.ca/nrw/english/report_e.html.

Chapter VI

Building the Information Society from the Bottom Up? EU Public Policy and Community Informatics in North West England

John Cawood and Seamus Simpson
Manchester Metropolitan University, United Kingdom

If information and communications technologies are to be used to support community efforts to achieve social and economic development, then policies which link emerging technological opportunities with the social and economic context must be devised. In the European regions, the many organizations which promote community informatics have increasingly taken up the concept of the Information Society as both an objective and a rationale for such policy making. The key player in proselytizing this concept has been the Commission of the European Union and monies from EU Structural Funds have been deployed in a series of community, regional and interregional informatics initiatives which started in the early 1990s. In these initiatives, increasing emphasis has been laid on the devolution of decision making to the community, user access and the empowerment of those who risk exclusion from the potential benefits of ICTs. In brief, the Commission has sought to create a policy environment which encourages local and regional actors to build the "Information Society" from the bottom up.

Whilst European policy on ICTs in the 1980s was characterized by programs such as Esprit which sought to close the technological gap with the USA and Japan, policy in the 1990s has been characterized by a growing awareness of the social aspects of information and communications technologies. Not only have the economic and social policies of the EU been affected, but the strategies for research and development have been influenced too. The present Fifth Framework program pays special attention to the "creation of a user-friendly information society" which is both coherent and socially inclusive. Greater emphasis on the social context of ICTs has allowed community groups, local authorities and small businesses to take advantage of EU funding to develop or establish a wide range of informatics projects. Their objectives cover not only improving the performance of SMEs, but education and training, the reduction of social exclusion, the improvement of health services and the promotion of tourism. While the central aim of most EU programs remains regional and local economic regeneration, a much deeper appreciation of the role that informatics can play in this process has been developed. Explicitly or otherwise, at an EU, national and regional level, policy makers in the public, private and voluntary sectors are picking up items from the agenda of those theorists whose vision of a post-industrial society fuelled the Information Society debate in the 1970s and '80s.

In this discussion, our concern centers on the interaction between the policy makers in the European Commission and those who act at regional and local level. We shall briefly examine the origins of the Information Society concept and then investigate the different ways in which it has appeared in European policy making. At the European level, we shall pay particular attention to those agencies of the European Commission responsible for regional development, employment, social affairs and informatics which are working together in the Regional Information Society Initiative and the Information Society Project Office (ISPO). Moving to the local level, we shall argue that European regional and metropolitan authorities have entered into a cross-national, multilevel nexus of policymaking with national governments and the European Commission: a network which provides mutual support for informatics initiatives.

At the regional level, we shall focus on informatics projects and their promoters in North West England in order to explore the role of this network in the implementation of Information Society policies and their consequences for the growth of community telematics. The watchwords of such EU programs have, in recent years, been partnership, participation and integration. In fostering attempts to create the Information Society, the European Commission through "top-down" policies has promoted the idea of a "bottom-up" approach and has often been seen to bypass national governments in order to deal directly with regional and local actors. Regional bodies have been encouraged to build links with other European areas and a methodology advanced which is built on the idea of shared experience. We argue that such public policy intervention has been crucial in the realization of community informatics initiatives.

In conclusion, we shall draw on the North West experience to assess the relative importance of the different levels of policy making and to identify the nature of the interaction between local actors, regional authorities and European agencies. In this way we hope to show the extent to which European Commission efforts made from the top down have been effective in stimulating or strengthening initiatives coming from the bottom up.

The Roots of the Information Society Idea

The idea of the Information Society has its roots in "post-industrialism," a body of theory developed in the 1960s and '70s which identified the end of industrial capitalism and the advent of a service and leisure economy. Part sociology, part futurology, its proponents drew attention to science and technology as key social components, the emergence of new social groupings defined by noneconomic criteria, the distinctive importance of information in technology and the decline of traditional social conflicts. In *The Coming of Post Industrial Society* (1974), Daniel Bell argued that knowledge and information would replace labor and capital as central economic factors. Alain Touraine (*The Post Industrial Society*, 1974) foresaw new social divisions emerging as high-technology production developed and with it new groups of workers with high knowledge and skill levels. There was common agreement that the emergence of new political groupings such as environmentalists, feminists and consumer rights organizations, none of which were primarily economic in nature, were important indications of a trend toward post-industrial society.

Overlapping, and in part emerging from the debate on post-industrialism, is the Information Society thesis itself. Alvin Toffler's concept of a *Third Wave* (1980) in which industrial society is transformed into information society is the most well know of a series of futurologies which included Naisbitt's *Megatrends* (1982) and Masuda's *The Information Society as Post Industrial Society* (1983). Popularization of these accounts were often expressions of cheerful technological determinism in which the inevitability of (beneficial) information and communications technologies was allied to warnings of the dire social and economic consequences which would surely follow failure to embrace them. However, not all Information Society theorists are technological determinists. Recently, Manuel Castells has taken up the concerns of the post-industrial theorists in *The Rise of the Network Society* (1996) in which he proposes a new mode of development, informationalism, in which different relationships of production, experience and power are being forged. Significantly Castells and other prominent researchers into the social and economic dimensions of ICTs were part of the High-Level Expert Group which made the final policy report, *Building the European Information Society for Us All*, to the Employment & Social Affairs directorate of the European Commission in April 1997.

While the elaboration of these theories fuel a debate which this sketch can only hint at, the effects of the debate have been felt in the more practical world of commercial policy and economic strategy. As early as 1982, the Science Council of Canada had used the idea of the information economy to underpin its national strategy for microelectronics. By 1986, British Telecom was using Toffler's term, *The Third Wave*, as a title for its pamphlet on the "telecommunications revolution" and Masuda's ideas have been taken up by Japanese planners. A decade later, the idea of the transformation from an industrial to an information society is taken as given by the proponents of local informatics initiatives and strategic planners in the European Commission. Quoting Toffler, the Secretary General of the EU Regional Information Society Initiative framed his 1999 report in the context of a move from an "Industrial Age in which capital has been the main factor of production" to "the dawn of the Information Age" (Hughes, 1998). This is not an isolated reference; the concept of the Information Society has permeated the strategic thinking of important sections of the European Union, and in the next section, we will explore how it has helped to shape policy.

The Information Society in European Policy Making

In the early 1990s, the development of the Internet rekindled the Information Society debate and stimulated national actions. The U.S. National Information Infrastructure program was matched by Japanese, Korean and Australian plans for technical and social developments based on innovation in ICTs. Europe's response to these moves is of special interest to community informatics since the concept of the Information Society came to be linked with the principle of 'subsidiarity' in EU strategic thinking. Subsidiarity, a somewhat vague Eurospeak term, implies that, where possible, decisions should be taken at levels close to the citizen. The notion has been the subject of heated political debate but in the context of community informatics, it has contributed to the idea of 'partnership' which informs both regional and Information Society policies.

The European response began in 1993 when growing concerns about employment trends led to a White Paper on Growth, Competitiveness and Employment, prepared by Jacques Delors, then president of the European Commission. Delors dealt with a wide range of strategic problems facing Europe but, significantly, highlighted the threats and opportunities presented by the growth of information and communications technologies. In his opinion, "The move towards an Information Society is irreversible, and affects all aspects of society and interrelations between economic partners" (Delors, 1993). In December of 1993, the European Council Summit asked for a more detailed response to the challenge of the Information Society and a "High Level Group of Experts," led by Martin Bangemann, commissioner responsible for industry and ICTs, was established. Six months later it produced its report "Europe and the Global Information Society" (Bangemann, 1994) which urged the EU member states to coordinate efforts across a range of areas such as teleworking, public administration, health care and air traffic control. In relation to the regions of the EU, it was argued that there would be "new opportunities to express their cultural traditions and identities and, for those standing on the geographic periphery of the Union, a minimizing of distance and remoteness" (p. 5).

The Bangemann Report was underpinned by the belief that market forces would be the driver of the emerging information society. It contended that all remaining regulatory impediments to their unfettered operation should be removed and, in particular, the telecommunications sector was set out as a target for immediate liberalization. The thrust of the Report was economically and technologically determinist in nature, advocating that "the creation of the Information Society should be entrusted to the private sector and to market forces" (p. 34). It argued that, "There will be no need for public subsidies, because sufficient confidence will have been established to attract the required investment from private sources" (p. 35).

Soon after the publication of the Bangemann Report, further developments aimed towards building EU Information Society policy emerged in the form of an Action Plan, entitled "Europe's Way to the Information Society" (European Commission, 1994). Over the course of the following two years, a number of measures in fulfillment of this plan were undertaken. In 1996, having taken stock of progress made and future desirable activities, the Commission suggested that future EU IS policy should concentrate on the four main policy lines of developing the market and regulatory conditions to favor businesses, investing in ICT-based research, integration of relevant parts of the structural funds with IS policies on public services and social issues, and examination of global ICT policy issues such as trade (European Commission, 1996a). These were soon after given official endorsement by the EU, as the basis for a revised and updated set of initiatives (European Council of Ministers 1996).

As a consequence, the Commission devised and put forward a formal proposal for a multiannual Information Society program (European Commission, 1996b), underpinned by the belief that "the multifaceted nature of the Information Society implies a need for greater policy coordination between the various Community policies and between the Community's different instruments and funding mechanisms" (p. 5). The proposal was unanimously agreed upon by the Council of Ministers in December 1997 and the program has been allocated a budget of EU 25m between 1998-2002. Its broad objectives are to increase public awareness and understanding of the Information Society, to maximize the socioeconomic benefits of the Information Society, and to increase the EU's role in the global Information Society (European Council of Ministers, 1997a). More specifically, the program held some important implications for regional Information Society initiatives. Activities would span sectors and would be complementary to EU activities in other policy domains. Further, the program "should also provide a common framework for complementary and synergetic interaction at European level of the various national/regional/local initiatives for the development of the Information Society" (European Council of Ministers, 1997b, p. 3). In terms of the objective of raising the awareness of the ICTs in Europe, the EU resolved to address the needs of local initiatives, gather information on the needs of individual people, heighten the profile of the Information Society to the general public, create interest in the ICTs among small and medium-size enterprises (SMEs), and demonstrate their impact at the regional level. Under the second objective, the EU has resolved to evaluate "the opportunities and barriers which disadvantaged social groups and peripheral and less-favored regions may face in accessing and using Information Society products and services" (European Council of Ministers, 1997b, p. 4). The Commission has made calls for IS initiatives at the local, regional and national levels to develop ICT infrastructures and services (European Commission, 1996c).

Through the structural funds, principally the European Regional Development Fund (ERDF) and the European Social Fund (ESF), the EU has been developing a policy of economic aid to its disadvantaged regions since the late 1960s. Two landmark reforms of regional policy occurred in 1988 and 1992, resulting in successive doubling of funding levels to reach ECU 141 billion. Attention is focused on a number of priority areas, *inter alia* the development of lagging regions (Objective One), regeneration of areas suffering serious industrial decline (Objective Two) and the creation of work for the long-term unemployed (Objective Three). The structural funds are underpinned by the principles of cohesion, partnership, subsidiarity and additionality (Hooghe and Keating, 1994, pp. 374-376). The idea of partnership is particularly important since it brings together actors from the regional, national and EU levels in the form of monitoring committees whose role is management of the technical and bureaucratic elements of the programs (Hooghe, 1995, p. 182), though the idea has also come to be associated with the generation of initiatives.

For any region, administrative, economic and political resources are crucial to secure government funding (Keating, 1992) and there is evidence that sub-national representatives, particularly in the UK, have become increasingly aware of the advantages to be gained from having a lobbying presence close to the EU (Rhodes, 1996, p. 168). Apart from around 10% of the structural funds budget allocated to Community Initiatives, pilot projects and networking tasks, over which the Commission has control (Hooghe and Keating, 1994, p. 383), funding is determined as a result of a tripartite negotiation process involving primarily the national member state and the Commission, and to a lesser extent the sub-national representatives (Ansell, Parsons and Darden, 1997). This has led to the suggestion that a system of multilevel governance (see Marks, Hooghe and Blank, 1996) has recently emerged in EU regional policy.

A significant feature of Information Society policies which have emerged at the EU level is the way in which they have moved beyond the technologies themselves to the economic and social context. Whilst the Bangemann Report presented a familiar conjunction of economic and technological determinism, stressing the economic potential of a liberalized technological regime, later policy has concerned itself with how ICTs will be used and what role they will play in the lives of individuals and communities. Whilst a sense of the inevitable suffuses EU pronouncements on the Information Society, the opportunities and risks associated with the diffusion of ICTs have also been recognized. One example of how the Commission's Action Plan for the Information Society has been translated into initiatives reflecting the regional and local emphasis of EU policy is the "Telecities" project involving more than 60 European cities. Another is the promotion of initiatives for multimedia services in rural areas and digital towns as part of the Telematics Application Program (March, 1995). We go on to focus on a part of the EU Information Society initiative in which regional issues were central and in which the Commission has aimed to build partnership and participation in socio-technical change—the Inter-Regional Information Society Initiative.

The Inter-Regional Information Society Initiative

In late 1994, the European regions of North West England, Central Macedonia (Greece), Nord-Pas de Calais (France), Piemonte (Italy), Saxony (Germany) and Valencia (Spain) took a lead in mounting regional initiatives which could gain access to the structural funding linked to EU Information Society policies. They shared a common experience of economic decline, unemployment and marginality. Their objectives were economic regeneration, reskilling and the integration of rural and peripheral areas into core economic activity. The Information Society was to be their means of economic and social transformation and in November 1994, they announced a Memorandum of Understanding in which they committed themselves to a long-term, common effort to build the Information Society (IRISI, 1996a). This effort would be based on shared experience, the exchange of information and common implementation of new applications and services. The interregional cooperation would be based on a partnership within the regions of public, private and voluntary interests. Target areas for regional action and interregional cooperation were to include teleworking, distance learning, university and research center networks, telematics services for SMEs and city information networks (Kleinwaechter, 1996, Section 3).

In January 1995, the Commission launched the Inter-Regional Information Society Initiative (IRISI) as a pilot scheme which was resourced from the European Regional Development Fund. The participating six regions were to develop a regional strategy and action program for identifying priority applications and for assessing their technical and financial viability, commercial viability, social acceptance and economic contribution to regional development (IRISI, 1996b). These activities were to inform the best allocation of existing financial resources including support from the EU structural funds. The work plan for each region specified a common methodology based on regional partnership. Significantly, the work was focused on awareness raising, strategy development, partnership and consensus building, networking in the region and the stimulation and assessment of new applications. Thus IRISI was to be an essentially socioeconomic program within Information Society policy in which local and regional concerns informed technological development. IRISI was conceived as an explicitly bottom-up organisation.

At the regional level, IRISI is overseen by Regional Information Society Steering Groups with high level representation from all key players in the region and Regional Information Society Units (RISUs) were established whose task was to provide administrative support and project management. The RISUs are resourced by DG XVI. At the interregional level, representatives from each region and from DG XIII and DG XVI sit on a Network Management Committee which was established to coordinate regional activity. DG XIII provided the funding for the establishment and operation of a Network Bureau in Brussels which offers managerial and technical assistance to the regions and support for interregional collaboration (Cornford and Gillespie, 1998). It should be emphasized here that IRISI and its successor organizations do not provide funds but offer contacts, advice, exchange of experience and the possibility of networking with public and private partners. Initiators of projects must apply for funding from the relevant EU programs such as the ERDF, the ESF and the Framework programs.

In the pilot phase of IRISI (18 months), the apparatus of the initiative described above was set up and regional strategy and action programs elaborated and published. A series of interregional workshops and seminars were organized and, in October 1995, the first IRISI conference took place in Turin. At Turin, in spite of regional differences, it was found that common problems had been identified in local and regional informatics activity. The most important of these were the development of regional infrastructure, low public awareness of the potential of informatics, uneven levels of education and training for the "information revolution," the problems of funding the development of telematics applications for SMEs and a lack of clarity concerning the long-term social consequences of the information age (IRISI 1996c).

Other regional and interregional activities took place in Brussels, Thessaloniki and Manchester. At the Manchester regional conference whose focus was thematic work groups, the participant regions adopted the so-called IRISI Manchester Statement which summarized the IRISI experience, reconfirmed the long-term commitment of the six regions and invited other European regions to join the initiative (IRISI, 1996d). The European Commission was recommended to continue its support for a regionally orientated, "bottom up" approach. In 1996, the Commission selected a further 22 European regions for funding under the Regional Information Society Initiative (RISI) Pilot Scheme and in 1998, the six original IRISI partners and the new RISI regions set up a European Regional Information Society Association to further the social and economic development of the regions in the context of the Information Society (Sullivan, 1998).

Information Society Developments in North West England

Background

The North West of England, a region for planning purposes but not a political entity itself, consists of the metropolitan districts of Greater Manchester and Merseyside and the counties of Cheshire, Lancashire and Cumbria. The region has been characterized by the diversity of its economy and the fragmentation of its political institutions. The long-standing rivalry between the cities of Manchester and Liverpool and indecision about its territorial boundaries are just two factors which reveal its heterogeneity. In the immediate postwar period, there had been very few attempts to create a regional response to the decline of its traditional industries and compared to other English regions, the initiatives that had been taken were relatively unsuccessful.

In the late 1980s, local politicians, business leaders and other leaders from the public and private sectors in the North West began to work for coordinated action to counter regional economic and social decay. When it became evident that EU structural funds for regional development would depend on evidence of grassroots coherence, the North West Regional Association (NWRA) was set up with all the local authorities and the North West business lobby on board (Burch and Holliday, 1995). One of the first tasks of the NWRA was the production of a regional economic strategy which was intended to become the independent regional input to the 1994-99 programming period of EU structural funds. Preparation of the strategy was supported by £100,000 from the Commission.

In 1995, the NWRA and the North West Business Leadership Team formed a new organization called the North West Partnership, which brought together local authorities, business and higher education and sought to become a permanent lobby on behalf of the region. In 1998, the partnership published a Regional Technology Action Plan after a consultation process with both large and small firms, regional universities, public sector innovation support organizations and European funding program administrations. The Plan takes up the issue of the Information Society and declares:

> Promoting the Information Society is a priority for both the UK Government and the European Commission. The ideas underlying the Information Society are fundamental to the kind of innovative thinking which the RITAP (Regional Technology Action Plan) aims to encourage. The RITAP has been infused by the Information Society throughout its development and this will continue in its implementation. This has already been recognized and the North West Partnership's IRISI initiative continues to research and campaign for greater awareness and take up of Information Society opportunities. RITAP and IRISI should be read side by side and taken as mutually supportive" (North West Partnership, 1998, p. 10).

North West IRISI was a creation of the interests represented in the NWRA and was an original member of the EU Inter-Regional Information Society Initiative. It is a primary example of cooperation between the European Union and the NW region and is a testament to the success of the North West Regional Association in demonstrating regional coherence in order to secure structural funding in the changing policy environment. Most of the local and regional informatics projects which are running in North West England are represented directly or indirectly on the Steering Committee of the NW IRISI which acts as a forum, advice center, awareness raiser and initiator for Information Society projects. The largest group on the Steering Committee is made up of 12 local authority officers. There are seven representatives of large companies (of which four were telecommunications service providers) and representatives from higher education, the NW Government Office, development agencies, labor and voluntary organizations. In the period 1994-96, NW IRISI received a total of £114,000 for phase three of the RISI project from the European Regional Development Fund. In the next section, we move to a closer examination of the network of players which NW IRISI represents and the projects which they have initiated and manage.

The Structure of Local and Regional Telematics Initiatives in North West England

Players

Community telematics projects in the North West reflect the political and economic diversity of the region and are, at the same time, an indication of the efforts which have been made to build cohesion and participation. The picture is complex as there are several key bodies which have interests in Information Society activity. Some operate at both the regional and local level and there is a high degree of overlap and interconnection. One step from direct Information Society activity and working at the regional level is the North West Partnership (and its constituent associations) and the Government Office for the North West which has offices in Manchester and Merseyside. (A single Government Office for the North West was created in 1998 in a merger which reflects a growing regional perspective.) The main interest of these bodies lies in regional economic and social development. However, as we have seen, Information Society activity has become part of that objective . The Government Office, as the representative of Whitehall, has a significant role to play in the relationship between the region and the European Union funding bodies, between the region and UK Government departments which have interests in informatics (e.g., the Department of Trade and Industry's Information Society Initiative) and as a source of stimulus advice and information.

In the context of community informatics, the Government Office has the regional responsibility for the contracts of the Training and Enterprise Councils (TECs) and Business Links (Department for Education and Employment, 1998). These organizations, initiatives of the Departments for Education and Employment and Trade and Industry respectively, have been important players in local and regional informatics projects. The TECs, funded by government, work as partnerships of local interests. They aim to raise skill levels and employability, and to develop a culture of lifelong learning. Similarly, Business Links are private-sector partnerships of local enterprise agencies, chambers of commerce, and local authorities which provide advice and guidance to SMEs. They provide a base for a series of technology support centers which are part of the Department of Trade and Industry's Information Society Initiative. Business Links are funded by the Government Office through the Training and Enterprise Councils. The network of TECs and Business Links in the North West has a significant role in the implementation of the local economic regeneration components of UK and EU Information Society policy. In the domain of economic regeneration are also to be found the economic development agencies created by the local authorities. A prime example of these is Enterprise plc (formerly Lancashire Enterprise), originally created by Lancashire County Council and responsible for such telematics programs as Competitiveness Through Telematics in Lancashire (CTTL) which is part of the DTI's Information Society Initiative.

At the local government level are the Metropolitan, County and Borough Councils and in particular their economic development units and telematics advisory groups where they exist. It is important to remember that regional political authority resides at this level since there is no regional tier of government in England. Our analysis of North West informatics projects shows that local government, usually in partnership with other North West authorities, business, education and local interest groups, is the key player in most, if not all, of the community informatics activity which takes place in the region. Within local government, it is often officers of the authorities rather than elected representatives who play the leading role in the initiation and implementation of projects. Local authority officers are the largest group of representatives on the steering group of the North West Regional Information Society Initiative and some of them have developed a high level of expertise in securing European funding.

Strong links are to be found between some local authorities and local universities and further education institutions. Thus in Salford, the University plays a leading role in GEMISIS, a telematics partnership. In Liverpool, the Knowlsley Telematics Center, a project established by Knowlsley Borough Council and Business Link, is operated and staffed by the John Moores University. In Manchester, the City Council's Economic Initiatives Group has worked closely with Manchester Metropolitan University (MMU) to develop informatics projects for local business and the community. In some instances, community informatics projects have direct links to academic departments. For instance, the Liverpool Connect project is based in the Liverpool University Department of Computing. Others, for example, at MMU, where the Manchester Institute for Telematics and Employment Research (MITER) is located, specific informatics units have been created which work in partnership with local authorities, business and community interests.

While representatives of business in the North West have played a significant role in the North West partnership and the development of its economic and technology strategies, the involvement of business in informatics at the local level is less evident and somewhat uneven. Some large telecommunications service providers such as Nynex and British Telecom and other companies with a direct interest in telematics such as ICL (a North West computer manufacturer now part of Fujitsu), IBM and Newbridge Networks (an international network products and systems supplier) are involved in informatics projects but, with some exceptions, the involvement of small and medium sized firms is not widespread. (We are speaking here of partnerships with business rather than the uptake of telematics by SMEs.) In many instances, the Business Links and the TECs act as surrogates for small businesses. (See Sullivan, 1998, for a view on problems of representing the interest of SMEs in Information Society projects.)

Outside the metropolitan centers of Manchester and Liverpool, there are few grassroots informatics initiatives in the North West which have started without local authority or other institutional support. Perhaps in areas designated as economically and socially disadvantaged by the European Union, it would be surprising if community groups which had the resources necessary for significant informatics activity were to spring up. This situation reinforces the importance of public funding to the development of community informatics in the region. Notable exceptions to this picture are to be found in Manchester where several pioneering telematics projects began in the late 1980s. The establishment of electronic village halls and other schemes which are associated with the Manchester Host project have become well known and serve as models for other projects. It is worth noting that the arts and cultural industries community in the city has been the location of a series of innovative telematics projects. Indeed, this example gives support to those who argue that telematics is more likely to flourish in creative areas such as these than in the revitalization of traditional activities. Significantly, the grassroots developments in Manchester have not been repeated in many other parts of the region.

Projects

Information Society, community informatics and telematics projects in the North West region can be categorized according to at least three sets of criteria:
- geographical location,
- economic and social purpose and
- control and funding.

Not surprisingly, given the connection of their initiators, there is a significant degree of overlap and linkage. In addition, we have found that the common source of funding for almost all North West informatics initiatives is a partnership of local government and one or more of the European structural funds. Given these common factors, we can use the first two criteria alone to classify informatics programs. Thus North West informatics projects aim:
- to support SMEs and provide telematics skills and training for employment at a regional or a local level,
- to build informatics infrastructure for particular communities,
- to foster the social and economic development of specific communities or groups,
- to raise the awareness of the Information Society and its consequences at the local or regional level and
- to do most or all of these things.

Just one structural fund, the ERDF, lists 35 North West Objective 2 projects in the period 1994-96 (NW Government Office). We have selected four of these and one Objective 1 project to illustrate the five categories set out above. In this selection we have tried as far as possible to give examples from different parts of the region. In the space available, it is impossible to do full justice to each project. What follows are very brief summaries in which the emphasis is on those features which best reveal the interaction of the initiators and implementers of policy

Representative North West Informatics Projects

GEMISIS (Salford and Part of Manchester)
Gemisis is a wide-ranging, collaborative initiative set up by Salford University, Cable and Wireless Communications, the Cities of Salford and Manchester and Manchester TEC. The project's objective is "to develop user-driven applications that exploit the sociological, economic and technological benefits of the Information Superhighway to assist the regeneration" of North West England. This is a big project by North West standards and has secured funding from the ERDF, the EU Single Regenerative Budget and the Advanced Communications Technologies and Services program of the EU Framework IV initiative. Taking into account its partners' contributions, the value of the project is more than £12 million. In the later

phase of its development, the Gemisis partnership has been enlarged to include Newbridge Networks and ICL.

Gemisis covers both the technological and socioeconomic aspects of the Information Society. The technical side of the project draws on the resources of its University and business partners and since 1995, it has been working to develop applications that exploit the broadband fiber optic cable which provides the platform for the various facets of the project.

These include business, community, health and education service areas. In the business area, Gemisis has helped develop a Virtual Chamber of Commerce which delivers services to over 450 businesses. In the health area, it is working with local hospitals and public health agencies to develop applications and information services which meet user needs in primary and secondary care. Gemesis is active in the community service area developing a community campus, providing community information and facilitating access to and training in the use of new technologies. Finally, in the education area, it is developing multimedia support materials, fast access educational information, services for schools and an electronic forum for colleges to share information.

CONNECT (Merseyside)

While Salford is an Objective 2 region, Liverpool, regarded by the EU as an area whose economy was significantly lagging behind the European norm, was given Objective 1 status in the early 90s. Monies granted under this heading were used to establish Connect, an Internet center for Merseyside business which is operated by the Liverpool University Department of Computing. Set up in 1994 to help Merseyside SMEs make better use of the Internet, Connect offers services to the community and to unemployed people as well as business. For small firms, there are Internet Awareness Days which provide hands-on experience and information on Web-based commerce. In addition, there are Internet-related short courses and research and development services. Connect offers Web exposure as part of an integrated Merseyside presence called MerseyWorld. Businesses are added to the site through a Quick Start program, a process for getting organizations on to the Web rapidly.

In an area of relatively high unemployment, Connect has developed a "stepping stone" approach for a return to work. The modular course is aimed at people with previous computer experience, leads to a postgraduate certificate and claims a success rate of 60% in finding employment for its students. There is also a work experience program for unemployed people who wish to consolidate existing skills, and a system of student contracts which follows it and includes a job search element.

A novel feature of Connect is the Internet Express, a roving Internet access facility which takes up to 20 machines to libraries within Merseyside for three weeks at a time. Internet Express offers free browsing facilities and access to the skills and work experience courses outlined above.

GENESIS (Cumbria)

Compared with the metropolitan areas in the south of the region, Cumbria is a relatively isolated part of the North West where both rural and urban economies have been in decline. Its difficulties include vanishing traditional industries; poor infrastructure; dependence on agriculture, tourism and the nuclear industry; lack of educational opportunity; and the exodus of young people. The local authority has attempted to overcome economic, educational and social problems by addressing the questions of communication and the transport of information. The Genesis project (or Cumbrian Lifelong Learning) is one response to these difficulties. It has three main strands:

- lifelong learning supported by on-line technology and linked to research,
- electronic links to all levels of local government for all local communities and
- an electronic information highway which links together higher and further education, schools, public libraries, village halls and community centers.

Phase one of the Genesis project received £324,200 from the ERDF and provides 20 public access sites across Cumbria which have personal computers with touch screens or keyboards. The scheme provides tourism, business and educational information and includes a series of learning projects based in schools and colleges across the county. The geography of Cumbria provides a particular challenge to informatics. Urban centers are located along the coast and around the edge of the central upland region of lakes and fells. Genesis has the task of linking these widely separated communities.

Genesis is very much a bottom up informatics initiative in its aim to provide access to information, build participation in local politics, provide information and address specific social problems. To provide a basis for planning Genesis, a local opinion survey was carried out which included focus groups, a household survey and a questionnaire for the individuals who had seen the demonstrator equipment (Cumbria Local Government Briefings, 97/49 & 99/10).

G-MING (Central and Southern Manchester and Salford)

G-MING is included as an example since, although it has expanded into a wide ranging initiative which includes applications programs, multimedia and telemedicine, it began as an essentially technical project. It thus provides a contrast with the more socioeconomic initiatives. The project began after a feasibility study (1993) had examined the scope and potential for a high speed telecommunications infrastructure for the educational community in Greater Manchester. Representatives from the six local universities, Salford University College and the Royal Northern College of Music made up a steering group to oversee the project and a development team was appointed to manage its implementation and operation. The core objective of the scheme was to provide "a pervasive multimedia service which could enable new distributed multimedia teaching and remote learning facilities to be set up and shared between institutions" (Mills and Strom, 1995).

Funding for the scheme was obtained from the higher education partners and JISC (the Joint Information Services Committee of the UK higher education funding councils). Significantly for this study, the initiators of G-MING realized the importance of the local authority in securing further resources and a relationship was formed with Manchester and Salford City Councils. By linking to the economic regeneration policies of the two cities, G-MING won a major contribution to its funding. In 1995, the ERDF gave financial support to the connection of strategic sites which would allow the inclusion of the Town Hall and Manchester Science Park in the ATM network. The core network became operational in December of that year and since then student halls of residence, academic libraries, research centers and business parks have been connected.

Two of the G-MING partners, Manchester City Council and Manchester Metropolitan University, are also the cosponsors of MTTP, the Manchester Telematics and Telework Partnership, which includes Manchester Multimedia Center, the Manchester Technology Management Center, the Electronic Village Hall network mentioned above and Networking for Micro and Small Enterprises. The expansion of G-MING's horizons and activities is evidence of the potential which partnerships of local authorities, higher educational institutions and enterprising individuals can realize in community informatics when public funding allows innovative projects to be launched. Key individuals in MTTP were involved in the pioneering projects mentioned in the last section.

Informatics Projects Managed by Enterprise plc

The last example has been chosen to illustrate the role that economic development agencies play in community telematics projects in the North West and to show another way in which local government is linked with EU funding through informatics initiatives.

Enterprise plc was set up in 1982 as Lancashire Enterprise, the economic development and job creation arm of Lancashire County Council. It became a public limited company and in 1990, it opened one of the first economic development offices in Brussels, providing a link with the European Union. In 1995, it was admitted to the Stock Exchange via the Alternative Investment Market. It has since changed its name to Enterprise plc and now has a staff of more than 300 people working from Preston and a network of offices in Manchester, Liverpool, London and Brussels. Its clients include local authorities and regional bodies, the Government and other public and private sector organizations.

Enterprise plc coordinates the North West network of Technology Management Centers which aims to provide a range of innovation and technology services to improve the competitiveness of business and industry in the North West. Tech Web is a partnership and involves over 30 organizations in the region from the public and private sectors. These include large companies such as IBM and British Aerospace, local authorities in the region, the TECs and the Business Links. It is supported with funds from the ERDF and the ESF. The TMCs are an important resource for two informatics projects which are managed by Enterprise plc., the Lancashire Telematics Application Programme (LTAP) and Competitiveness Through Telematics in Lancashire (CTTL).

Lancashire Telematics Application Programme (LTAP)

Established in 1995, LTAP was developed in association with NW IRISI and is supported with funds from the ERDF and KONVER (an EU fund designed to assist diversification of defense dependent areas). LTAP provides a management and consultancy service for SMEs and, in this role, it seeks to facilitate the take up of a variety of telematics programs. It runs seminars and demonstrations which are directed at specific groups (for instance, solicitors and aerospace). LTAP's approach to the diffusion of telematics is focussed on the provision of seminars hosted by professionals who are selected for the high quality of their delivery style. Once an interest is established, seminars are followed up with demonstrations for those companies which want to go further. Demonstrations are held at Technology Management Centers located throughout the LCC region, for example at Blackburn, Preston or Skelmersdale.

At Blackburn, there is a multimedia unit which is part of the North West Tech Web. This program aims to provide a range of innovation and technology management services offered by the Technology Management Centers in the Lancashire area. Each center has a specialization. Blackburn's specialism is multimedia and it concentrates on the development of a range of innovation and technology CD-ROM products.

Competitiveness Through Telematics in Lancashire (CTTL)

The focus of CTTL is awareness raising amongst SMEs. CTTL provides a first point of contact for all telematics and electronic commerce-related inquiries. It seeks to inform SMEs on issues such as the benefits of the integration of new technologies into business activities. CTTL offers seminars, demonstrations and technical support to encourage awareness of and participation in the Information Society amongst SMEs in the Lancashire area, free access to Business Telematics Centers at the TMCs and in-company information audits to identify how a company might benefit by telematics use. It also gives support and partial funding for company-driven telematics implementation projects which address business needs. Enterprise plc has been able to offer some matched funding to SMEs wishing to develop a Web site. This has been on a fifty-fifty basis, with the company putting up half the funds.

One of the major objectives of CTTL has been to help get SMEs onto the Internet in order to enhance their access to information on competitors. Competitiveness is central to the work of CTTL but it believes connection to the Net has three other benefits for SMEs:

- enhanced communications with customers, suppliers and partners through e-mail, electronic data interchange, video conferencing and other technologies;
- improved information gathering and research e.g., on-line databases, catalogues, bulletin boards and WWW; and
- information dissemination, marketing and customer support, e.g., on-line CD-ROM based catalogues, electronic mailing lists, WWW sites.

Stage one of the CTTL project was completed in December 1997 and Stage two is being run under the auspices of the North West Technology Web project. The project is funded with support from the ERDF and the ESF.

Key Features of Informatics Projects in North West England

Our brief summary of North West Informatics projects which is based on a more extensive empirical survey, reveals a number of key features which inform our analysis. Firstly, European funding is central to informatics in the small business and community arenas. There is little or no community informatics activity without it and the EU structural funds are the major resource driving public initiatives in the North West Region. All our projects show partnership of one kind or another and in these, local government is the key player. It is the expertise in NW local authorities or in the agencies established by them which secures EU support and promotes informatics initiatives.

Secondly, all our examples relate to specific local issues—the problems of declining industries, rural isolation, lack of relevant skills and so on. The exacerbation of local problems is the obverse of the process of globalization. Economic and social regeneration are key objectives of North West informatics projects and thus the concept of the Information Society is for most policy makers a means to an end. There are no projects in the NW whose objective is simply to build the Information Society. Many of those who implement projects at the local level may be Information Society "enthusiasts" but they work toward the socioeconomic aims of the organizations which employ them. A significant danger which emerges from these first two points is that activities can become funding driven and projects designed to address local needs reworked to meet Information Society funding criteria. Of course, change to meet funding criteria is one way in which policy objectives are achieved but the danger remains of either distorting existing projects or creating projects which may not tackle real local problems.

Thirdly, a great deal of effort has been expended in promoting informatics to SMEs in North West England. Many projects are devoted solely to that aim. However there is very little evidence of SME initiatives without support from either local government or regional business development agencies. This is not a criticism but rather an indication of the obstacles which SMEs face in either adopting ICTs or in finding benefit from their use.

Finally, large firms do not seem to play a significant role in our examples except in a small number of big projects such as Gemisis in Salford or G-MING in Greater Manchester. In both these instances, we also see the presence of other major resource holders such as university systems developers or JISC (a UK national information systems agency). This absence underlines the obvious fact that corporate Europe has a rather different agenda to that of public authorities and may only participate in community initiatives when there is a clear benefit in doing so.

Interest Groups and Interactions

A large number of organizations and interests interact in the shaping of Information Society policy measures and their implementation in NW England. The European Commission, the NW government office, metropolitan boroughs, county councils, local authorities, big business, and to a lesser extent SMEs and community groups, have all contributed in some form or other to the series of measures which makes up a loosely bound public informatics policy. Certain groups and interests carry more weight than others and there is an uneven spread of resources. The power and interests of each of the actors and their interrelationships might usefully be explained by policy network analysis (Marsh, 1998). This could contribute to an understanding of the different policy tools chosen (Schneider and Ingram, 1990), how they are implemented and their likely effects (Bressers and Honigh, 1986). While policy network analysis is part of our future research agenda, what we hope to do here is to delineate the relative importance of North West actors and the nature of the interactions between them.

The European Commission (and in particular its relevant Directorates) is the key player, since as shown, EU finances support the majority of policy initiatives, as well as the IRISI forum. However, the Commission itself gains a number of advantages from its participation in regional informatics initiatives such as information and feedback on projects, as well as an enhanced profile and justification for its actions in the region and at the EU level. Another very important actor is the North West Government Office. While its role as funder in the TECs and Business Links has been significant, it also serves as an important agent for the UK central administration. This is crucial, since most bids for support under the structural funds must be channeled through Whitehall. As already indicated, local government plays a key role in regional Information Society activities and it has a pivotal position in mediating European policy for local consumption. Its officers are experienced negotiators at the EU level and have learned to "play the Brussels game".

A significant feature of the development of an informatics policy in NW England has been the creation and functioning of IRISI. In tandem with its sponsor organization, the NW Partnership, our case study highlights its ability to act as a locus for the exchange of ideas, the provision of advice and strategy formulation. It is particularly useful in maximizing regional capacities to obtain funds. Its eclectic membership (as evidenced through its Steering Committee) allows inputs to be made from all relevant actors. However, it is important to note that this organization is only a forum for exchange of what might be described as non-concrete resources and it has no political decision-making power. Outside the immediate confines of the region, the European IRISI network is significant for its role as a forum for exchange at the EU level as a whole. It shows the existence of a mechanism for policy learning and policy change in the NW informatics initiatives as a result of European-level activities.

The role of large business interests in NW informatics policy is selective. Their technical, human and financial resources make them powerful actors in the higher levels of regional policy debate such as in the North West Partnership. However, as we have seen they are mainly absent from community projects although they do participate when they can take advantage of financial assistance and the chance to build links with institutes of further and higher education.

Further and higher education institutions have played an enthusiastic and important role in certain projects, and also stand to benefit from access to funds and partners, as well as being seen to play a worthwhile part in their locality. In a rather weaker position lie SMEs and community groups. They are very much dependent on financial and other resources and are thus reliant on the actions of the more powerful actors, in particular local authorities and the European Commission. Given their tight operational margins and "limited degrees of freedom," it is often a challenge to convince them of the value to be derived from involvement in Information Society initiatives.

Conclusion

In a relatively short period of time, ideas propounding the importance of the Information Society in economic and social development have become widely accepted and in many respects common currency among a range of leading actors in the North West region. Burch and Holliday (1993, p. 42) have argued that "what is remarkable about the developments in the NW is that they constitute a 'bottom-up' attempt to occupy the regional tier through the development of a public/private sector partnership approach." However, as we have shown, the development of regional Information Society initiatives in the NW has been steered by the "top-down" hand of EU public policy. Within the framework of this policy, local authorities in NW England have developed strategies appropriate to their local communities which have been coordinated at a regional level. The degree of coordination is a witness to the direct influence of EU policy on the role of informatics in regional cohesion, although it could be argued that cooperation has been engineered to meet funding criteria.

Although there are many individuals and agencies ready to take action, without EU resources and local authority support, few initiatives would either get off the ground or develop to the extent to which they have. The message that the Information Society has a significant role in community regeneration has been clearly sent by the European Commission to the regional and local level in recent years. In considerable part, present strategies are a consequence of the debates on informatics and social change which informed policy making in Brussels in the early 1990s. However, the same debates influenced certain individuals and organizations in the regions and the policies coming from the Commission were met with similar ideas from below. Whilst there may well have been a meeting of minds in certain quarters, the power of any public policy tool is, in considerable part, determined by the budget made available. If, as Hooghe (1998) suggests they might, EU regional aid funding levels are reduced, then the extent to which IRISI and community informatics initiatives have taken root in the region will be truly tested.

References

Ansell, C., C. Parsons and K. Darden (1997). "Dual Networks in European Regional Development Policy." *Journal of Common Market Studies*, 35(3), 347-375.

Bressers, H. and M. Honigh (1986). "A Comparative Approach to the Explanation of Policy Effects." *International Social Science Journal*, 38(2), 267-87.

Bangemann Report (1994). *Europe and the Global Information Society. Recommendations to the European Council*. Brussels, European Commission, May 26, 1994.

Bell, D. (1973). *The Coming of Post Industrial Society: A Venture in Social Forecasting*. New York: Basic Books.

Burch, M. and I. Holliday (1993). "Institutional Emergence: the Case of the North West Region of England." *Regional Politics and Policy*, 3(2), 29-50.

Castells, M. (1996). *The Rise of the Network Society*. Oxford: Blackwell Books.

Corford, J. and A. Gillespie. (1997). "The Inter-Regional Networking Activities of the IRISI Initiative, Report of the Evaluators" Hughes, G. (Ed.). (1998) *Shaping the Information Society in the Regions: The Experience of the IRISI Initiative*. Brussels: European Commission, p. 209.

Department for Education and Employment (1998). *TECs. Meeting the Challenge of the Millennium*. Available at www.dfee.gov.uk/tec/contents.htm

European Commission (1993). *Growth, Competitiveness, Employment - the Challenges and Ways Forward into the Twenty First Century*, Brussels: Com(93)700.

European Commission (1994). *Europe's Way to the Information Society. An Action Plan*. Luxembourg: Com(94)347, July 19, 1994.

European Commission (1996a). "Communication from the Commission to the European Parliament, the Council, The Committee of Regions and the Economic and Social Committee on the Information Society. From Corfu to Dublin—the New Emerging Priorities." Available at http://ww~v.ispo.cec.be/infosoc/legreg/docs/corfudub.html.

European Commission (1996b). *Proposal for a Council of Ministers Decision Adopting a Multiannual Programme to Stimulate the Establishment of the Information Society in Europe*. Brussels: Com(96) 592.

European Commission (1996c). "Communication from the Commission to the European Parliament, the Council, The Committee of Regions and the Economic and Social Committee on the Implications of the Information Society for European Union Policies. Preparing the Next Steps." Available at http://www.ispo.cec.be/infosoc/legreg/docs/nxtstpen.html.

European Commission (1997a). "Information Society, Final Report of the High Level Group of Experts." Press release, available at http://ww~v.ispo.cec.be/hleg/buildpress.html.

European Commission (1997b). "Building the Information Society for Us All—Final Report of the High-Level Expert Group." Available at http://www.ispo.cec.be/hleg/Building.html#Introduction, April 1997.

European Council of Ministers (1996). *Council Resolution of 21st November 1996 on new Policy Priorities Regarding the Information Society*. Brussels: OJ NO C 376, 12.12.96.

European Council of Ministers (1997a). Minutes of the 2054th Telecommunications Council Meeting. Brussels, Dec.1, 1997. Available at http://www.ispo.cec.be/promise/telcouncil.html.

European Council of Ministers (1997b). Council Decision of 12th December 1997 Adopting a Multiannual Programme to Stimulate the Establishment of the Information Society in Europe. Available at http://www.ispo.ce.be/promise/12988en.html.

European Parliament and European Council of Ministers (1998). Decision of the European Parliament and European Council of Ministers Relating to the Fifth Framework Programme of the European Community for Research, Technology and Development and Demonstration Activities (1998-2002)—Joint Text Approved by the Conciliation Committee. Brussels PECONS3626/98, Nov. 25, 1998.

Hooghe, L. (1995). "Subnational Representation in the European Union." *West European Politics*, 18, 175-98.

Hooghe, L. (1998). "EU Cohesion Policy and Competing Models of European Capitalism", *Journal of Common Market Studies*, 36(4), 457-77.

Hooghe, L. and M. Keating (1994). "The Politics of EU Regional Policy." *Journal of European Public Policy*, 1, 53-97.

Hughes, G. (Ed.) (1998). *Shaping the Information Society in the Regions. The Experience of the lRlSI Initiative*. Brussels: European Commission.

IRISI (1996a). Discussion Paper 1, "Some Guidelines for Developing Regional Information Society Initiatives." Brussels, June.

IRISI (1996b). Discussion Paper 2, "Guidelines for Developing Feasibility Studies." Brussels, June.

IRISI (1996c). Discussion Paper 3, "Results of Preliminary Research into Regional Labour Markets and Human Resource Development in the Context of the Global Information Society," (First Report to DG V on the results and conclusions of the preliminary phase of research undertaken by the IRISI regions) Brussels, June.

IRISI (1996d). Discussion Paper 4, "Progress Report on the Achievements of the Inter-Regional Information Society Initiative." Brussels, July.

Keating, M. (1992). "Regional Autonomy in the Changing State Order: A Framework of Analysis." *Regional Policy and Politics*, 2(3), 45-61.

Kleinwaechter, W. (1996). "Regional Development and Information Society—The IRISI Initiative as a Pilot Action of the European Union". Paper presented to National and International Initiatives for Information Infrastructure Conference, Cambridge Massachusetts. Available at www.harvard.edu/iip/GIIconf/klein2.html

Marks, G., L. Hooghe and K. Blank (1996). "European Integration from the 1980s: State Centric v. Multi-level Governance." *Journal of Common Market Studies*, 34(3), 341-78.

Marsh, D. (Ed.) *Comparing Policy Networks*. Buckingham: Open University Press, pp. 3- 17.

Masuda, Y. (1983). *The Information Society as Post Industrial Society* Bethesda, Maryland.

Mills, P. & J. Strom. (1995). "G-MING: A High Performance Multiservice Telecommunications Infrastructure for the Greater Manchester Education Community."

Proceedings of the 6th Joint European Networking Conference, May 1995, and "A MAN for All Reasons" available at www..g-ming.net.uk/ together with other relevant material.

Naisbitt, J. (1982) *Mega Trends*. London: Sidgwick & Jackson.

NW Government Office (1999) *ERDF User Summary Pack*. Our thanks to Paul Devaney of the Manchester Government Office for supplying this information.

NW Government Office (1998). *Corporate Objectives 1998-2001*. Manchester: NWGO.

North West Partnership (1998). *Regional Technology Action Plan*. No place of publication given.

Rhodes, M. (1996). "Globalization, the State and the Restructuring of Regional Economies." P. Gummett (Ed.) *Globalization and Public Policy*. Aldershot: Edward Elgar, pp. 161-80.

Schneider, A. and H. Ingram, (1990). "Behavioral Assumptions of Policy Tools." *Journal of Politics*, 52(2), 510-29.

Sullivan, P. (1998). "Strategic Planning at Local and Regional Level for the Information Society. Developing a Local Information Society." Speech to European Information Society Conference, October 1998.

Toffler, A. (1980). *The Third Wave*. New York: Collins,

Touraine, A. (1969). *La Société post-industrielle*. Paris: Editions Denoël S.A.R.L.

CI and Community Networking

Chapter VII

New Communities and New Community Networks

Doug Schuler
Evergreen State College, USA

A Call for a New Community

> I can't predict what kind of community it will be, but the new community
> will be in reaction to the crushing bigness of systems.
> — *Theodore Roszak (Krasny, 1994)*

Global forces—economic, political and technological — threaten communities in many ways. On the one hand, citizens may feel like they're part of an undifferentiated crowd with no personal identity. On the other hand, they may feel isolated and alone, disconnected from the human community. In either case, people—especially those with fewer economic resources—feel that they have little control over their future. The consequences of powerlessness, real or perceived, transcend the individual; society as a whole suffers, for it is deprived of social intelligence and energy which could be tapped for the amelioration of social and other problems. As a matter of fact, many of this century's most pressing issues— the environment, women's issues, sexual identity, and others—have been brought to the fore through the efforts of *citizens* (Habermas, 1996).

Disempowering the individual and the community was probably not part of a master plan any more than degrading the environment was. Yet in many ways this is what has happened. Rebuilding the community—like cleaning up toxic dumps or reclaiming buried streams—will be a long process that will require diligence and patience. Rebuilding—and *redefining*—the community, therefore, is not optional, nor is it a luxury. It is at the core of our humanity; rebuilding it is our most pressing concern.

Geographically based communities are a natural focus for addressing many of today's problems. For one thing, many current problems—poverty, crime, unemployment, drug use, and many others—are concentrated in geographic communities. These problems are manifest in the community and are best examined and addressed by the community. Communities are also a familiar and natural unit. Smaller units can be clannish, unrepresentative, and powerless, while larger units are often too anonymous and unwieldy.

The old concept of community, however, is obsolete in many ways and needs to be updated to meet today's challenges. The old or "traditional" community was often exclusive, inflexible, isolated, immutable, monolithic, and homogeneous. Moreover, increased mobility coupled with widespread use of communication systems is de-emphasizing geography as the sole orienting factor in a "community." And, although problems may be manifested in specific geographic communities, the contributing factors of the problem may exist in New York, London, Tokyo, or other nodes in today's "Network Society" (Castells, 1996). A new community—one that is inclusive, fundamentally devoted to democratic problemsolving, outer-directed as well as inner-directed—needs to be fashioned from the remnants of the old community.

Defining a New Community

A new community is marked by several features that distinguish it from the old community. The most important one is that it is *conscious*. In other words, more than ever before, a community will need a high degree of awareness—both of itself (notably its capacities and needs) and of the milieu in which it exists (including the physical, political, economic, social, intellectual and other environmental factors). Further, the consciousness of the new community must be both intelligent and creative. The intelligence of a new community comes from its store of information, ideas, and hypotheses; its facility with negotiation, deliberation, and discussion; its knowledge of opportunities and circumstances; its ability to function collectively; as well as its application of technology and other useful tools. The creativity of a new community comes from its ability to reassess situations and devise new, elegant, and sometimes unexpected methods for meeting challenges to the local community and to the broader world to whom it belongs.

In addition to consciousness, the new community has both *principles* and *purpose*. Its principles are based on equity and sustainability, because a lifestyle based on overconsumption is illusory and ultimately self-defeating. Using these principles as a foundation, a new community also has goals and objectives that it strives to attain. Having purpose, the new community is oriented around *action*. This action must be consonant with its principles and it must be flexible. Projects and processes need continual reevaluation and adjustment, and projects and processes based solely on faith, tradition, or conventional wisdom will often be inequitable and ineffective.

As an inevitable consequence of its consciousness, principles, and purpose, the new community will have increased *power*. This power will be manifested in its ability to resist unwanted outside influences and to ensure desired outcomes. This new power could establish communities as rivals of government and business, or at least serve to moderate some of their vast power. This power is also a power that — like all power — could be abused. Hopefully, the power would be wielded according to the principles of the new community to the advantage of people everywhere.

It is clear that communities need to be responsible to a large degree for addressing their own problems and this is being done in many different ways by individuals and groups all over the world. Besides looking in—at their problems and at their resources — communities also need to be looking out. Sometimes the problem is caused by forces outside of the community, sometimes the problem must be shared by forces outside the community, sometimes it is necessary to communicate with others outside the community, and sometimes it is necessary to reach out because resources to deal with the problem aren't available locally.

The new community needs to contain elements of the old community. At the same time, many elements of the old society have outlived their usefulness. Modern circumstances have made change constant and new communities must learn to adapt. Modern circumstances have also made conflict likely so the new community must learn to discuss issues effectively. Finally, modern circumstances have created huge, inequitable chasms between economic classes so the new community must be built upon justice and compassion (see, for example, Castells, 1996).

Architect Christopher Alexander and his colleagues have developed an intriguing "pattern language" (Alexander et al., 1994) for designing rooms, buildings, and towns. This "language" embodies a powerful vocabulary of over 250 architectural patterns that are life-affirming and convivial. Although we are just beginning a similar discussion, it is probably not too early to begin thinking about an analogous "pattern language" that knits together a collection of civic, social, political, economic, and environmental patterns into a coherent and compelling vocabulary or language through which people can conceive, discuss, and build new communities.

Who Will Create the New Community?

Government has often been called in to help *solve* problems in communities. As we shall see, it is unlikely, as well as ultimately undesirable, for government to attempt to solve — *by itself* — the problems of deteriorating communities. Government at the national level is often too big, paralyzed, and partisan to be able to respond effectively. Also government — at least in the United States — is manipulated to an uncomfortably large degree by powerful corporate interests. For one thing, corporations and the very rich contribute a large percentage of political campaign dollars, thereby becoming the constituency to whom politicians are beholden. For another thing, corporate lobbyists pore over every piece of possible legislation, looking for sections to support, change, or remove that might give them an advantage in some way. The situation is often repeated at more local levels of government as well. Professional lobbyists, generally better versed, skilled, and recompensed than their citizen counterparts, are a major presence at state capitals. At a municipal level, business elites often enjoy a tight relationship with elected officials and the media. Needless to say, unaffiliated citizens play a minimal role in many aspects of the political process. The best reason not to expect (or want) a government "bailout," however, is that real solutions to community problems need strong community participation, and the government (at least in the U.S.) has rarely shown itself capable of being an equal participant with citizens in community projects. The truth is that government resources will be essential in any substantial community-building program. This, of course, includes the development of information and communication infrastructures. Any government program along these lines that is truly interested in improving community life will need strong citizen participation.

Is business then likely to play the leading role in rebuilding communities? Taking business' current practices and philosophy into account, the answer is no; in fact, business to a large degree has been one of the major forces behind community deterioration. Since business places profit-making as its highest priority, it necessarily favors "profits over people," as the slogan goes. If relocating factory work out of a community means greater profit for the corporation, the factory will likely close. Perhaps more significant is the fact that business is increasingly *not* a part of the community even when it's physically located there. Business, with its singular preoccupation with profit, is rarely accountable to the community beyond making merchandise and services available for a price. Indeed, the actions of business often suggest independence from and indifference to the broader needs of the human community. Business and government will both likely have important roles to play in this endeavor, but neither business nor government should be allowed to dominate.

The world is looking for new approaches to community problemsolving, as many of the old institutions (including the church, government, business, academia, and the science and technology establishment) and their traditional methods are being stalemated by new — and old — problems that are manifested locally but transcend the purely local. At the same time, it is becoming clear that the specialist or expert model is obsolete, and new approaches must be inclusive, discursive, participatory, and community oriented. Increasingly, these new approaches may be idiosyncratic and vary from place to place. Interestingly, many signs are pointing to *democracy* as the viable public problem-solving approach it was originally intended to be.

Core Values of the New Community

A human body has a nervous system, a digestive system, and a skeletal system, among others, that work together to sustain life. A community, likewise, has *systems* or *core values* that maintain its "web of unity" (MacIver, 1970). These six core values—conviviality and culture; education; strong democracy; health and well-being; economic equity, opportunity, and sustainability; and information and communication (Figure 1) are all strongly interrelated: Each system strongly influences each of the others, and any deficiency in one results in a deficiency of the whole (Schuler, 1996). It has long been known to illustrate the interconnectedness of the core values with just one example, that an individual's education and economic levels strongly determine the amount of his or her political participation (Greider,

Figure 1: Community Core Values

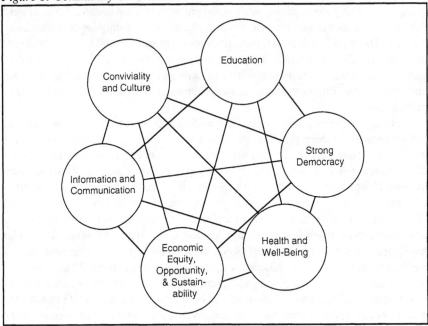

1993; Goel, 1980). Strengthening these community core values, particularly along the lines suggested by the attributes listed in Figure 2, will result in stronger, more coherent communities.

It is important to realize that addressing these core values is a dynamic and interdependent process that will necessarily change over time. Furthermore, there is no static final state, a utopia, that we are striving for, that we would recognize if and when we ever arrived there. Instead, community members need to develop a range of small and large projects that strengthen these community core values. Some projects will help individuals, some will help larger groups. Some will be short term, while others will be longer term. If all of these projects, however, follow from the principles of the new community, they are likely to positively reinforce each other, forming a civic movement that spreads throughout society.

Community Networks Can Support New Communities

The associations in community are interdependent. To weaken one is to weaken all. If the local newspaper closes, the garden club and the township meeting will each diminish as they lose a voice. If the American Legion disbands, several community fundraising events and the mainte-nance of the ballpark will stop. If the Baptist Church closes, several self-

Figure 2: Attributes of the Core Values

- **Conviviality and Culture**: belonging, being supportive, inclusive, active, conversational, affirming

- **Education**: equitable, empowering, effective, lifelong, inquiring, flexible, providing individual attention, creating communities of learning

- **Strong Democracy**: deliberative, equitable, proactive, functioning every day, voluntary, pluralistic

- **Health and Well-Being**: equitable, holistic, preventive, humane, community-oriented

- **Economic Equity, Opportunity, and Sustainability**: responsive, responsible, fair, cooperative, people-oriented

- **Information and Communication**: participative, trustworthy, affordable, universal, civic, pluralistic

---

help groups that meet in the basement will be without a home and folks in the old peoples' home will lose their weekly visitors. The interdependence of associations and the dependence of community upon their work is the vital center of an effective society.
- John McKnight (1987)

Networks of civic engagement embody past success at collaboration which can serve as a cultural template for future collaboration.
- Robert Putnam (1993)

Before computers took center stage, the term "community network" was a sociological concept that described the pattern of communications and relationships in a community. This was the web of community that helped us better understand how news traveled and how social problems were addressed in the community. New computer-based "community networks" are a recent innovation that are intended to help revitalize, strengthen, and expand existing *people-based* community networks much in the same way that previous civic innovations (such as the print-media inspired public library) have helped communities historically.

Currently, community members and activists all over the world are developing these new community-oriented computer services, often in conjunction with other local institutions including colleges and universities, K–12 schools, local governmental agencies, libraries, or nonprofit organizations. There were nearly 300 operational systems as of the mid-1990s (with nearly 200 more in development) (Doctor and Ankem, 1995) and the number of registered users exceeded 500,000 people worldwide (NPTN, 1995). These community networks (sometimes called civic networks, *Free-Nets* [a service mark of the National Public Telecomputing

Figure 3: Example Services for a Community Network

Conviviality and Culture
- Forums for ethnic, religious, neighborhood interest groups
- Recreation and parks information
- Arts, crafts, and music classes, events, and festivals
- Community calendar

Education
- On-line homework help
- Forums for educators, students
- Q&A on major topics
- Distributed experiments
- Pen pals

Strong Democracy
- Contact information for elected officials—"Ask the Mayor"
- E-mail to elected officials
- E-mail to government agencies
- Forums on major issues
- On-line versions of legislation, judicial decisions, regulations, and other government information

Health and Well-Being
- Q&A on medical and dental information
- Alternative and traditional healthcare information
- Community clinics information
- Self-help forums
- Public safety bulletins
- Where to find help for substance abuse, etc.
- Resources for the homeless; shelter information and forums

Economic Equity, Opportunity, and Sustainability
- Want ads and job listings
- Labor news
- Ethical investing information
- Job training and community-development projects
- Unemployed, laid-off, and striking worker discussion forums

Information and Communication
- Access to alternative news and opinion
- E-mail to all Internet addresses
- Cooperation with community radio, etc.
- Access to library information and services
- Access to on-line databases
- On-line "Quick Information"
- Access to on-line periodicals, wire services

Network, or NPTN], community computing-centers, or public access networks), some with user populations in the tens of thousands, are intended to advance social goals, such as building community awareness, encouraging involvement in local decision making, or developing economic opportunities in disadvantaged communities. In other words, the community network's services should support the core values of the community (Figure 3) (Schuler, 1996).

A community network addresses these goals by supporting smaller communities within the larger community and by facilitating the exchange of information between individuals and these smaller communities. Another community-network objective is to aggregate digital community information and communication thus focusing attention on civic matters. This is done in a variety of ways: by using discussion forums; question and answer forums; electronic access for government employees; information and access to social services; electronic mail; and in many cases, Internet services, including access to the World Wide Web. These networks are also integrating services and information found on existing electronic bulletin board systems (BBSs) and on other computer systems. The most important aspect of community networks, however, is probably their potential for increasing participation in community affairs. The Internet's current design promotes symmetry between *consumers* and *producers* of information and thus provides a far greater potential than that offered by traditional media such as newspapers, radio, or television.

Community members interact with community networks in various ways. Community-network terminals can be set up at public places like libraries, bus stations, schools, laundromats, community and senior centers, social service agencies, public markets, and shopping malls. Community networks can also be accessible from home via dial-up computers and from the Internet. In recent years, activists have also been establishing community computing-centers where people, often those in low-income neighborhoods, can become comfortable and adept with computer applications and network services (see, for example, http://www.ctcnet.org).

Community networks are currently local and independent. Many were affiliated with the National Public Telecomputing Network (NPTN), a now-defunct organization that helped establish a large number of community networks—or *Free-Nets* in NPTN's terminology. New organizations, such as the Association For Community Networks (AFCN) in the U.S., and the European Association of Community Networks (EACN), have recently been launched but, in general, community-network developers have not explored, in theory or in practice, the idea of stronger and closer relationships between them. Historically, community-network systems have had a difficult time financially, but increased public interest and some financial infusions from the government, businesses, and foundations have relieved—at least temporarily—some of the problems with some of the systems. Even so, very few of these systems are adequately staffed or have adequate resources for office space, hardware, software, and telecommunications. Whether or not community networks "implode," as Mario Morino of the Morino Institute has warned (1994), in the near or intermediate future, is an important concern that hinges on the question of whether or not community resources can coalesce around the idea of community computer-networks as a permanent institution worthy of financial and other support.

Community networks offer a new type of "public space" with similarities as well as major differences between other public spaces that our society currently offers. Steve Cisler, former senior scientist at Apple, forecasted (1993) that "just as electrical systems began to transform urban and small-town America a century ago, community computer networks will do so in the 1990s." Regardless of whether that forecast turns out belatedly to be true, community networks offer an important and rare opportunity for communities to democratically develop and manage their own communication technology.

The Seattle Community Network—
A Whirlwind Tour

Community networks take different forms in different communities, depending on who develops them and how people use them. Historical and demographic factors in the community and what types of services and institutions — computer based or not — already exist in the community, are also important factors. Changes in computer technology (new databases, graphical interfaces, plug ins, and distributed applications) are also influencing the design of future systems, currently moving beyond the earlier text-based systems to the World Wide Web. Although the Seattle Community Network (SCN) was launched with the text-based Free-Net software pioneered with the Cleveland Free-Net, it is now largely Web-based. Over 13,000 people are now registered users of SCN and the Web site gets hundreds of thousands of visits every year.

When a user encounters the Seattle Community Network (Figure 4) the first thing he or she sees is the SCN logo which blends communication metaphors and Seattle imagery: Hermes, the Greek messenger of the Gods, reclining on snowcapped Mount Rainier, beckons to future users, the Seattle Space Needle, an icon from the 1963 World's Fair, now retooled as a communications beacon held tightly in his hands. (The uncropped version of the logo features a postal stamp perforated border referring to another relative of digital communication's — the traditional letter.) Clicking the "About SCN" link will provide information about SCN's policy and principles. In addition to providing free Web space and e-mail, SCN also provides free support for electronic distribution via the *Major Cool* program. Basic informa-

Figure 4

tion about contacting SCN, getting an account and publishing information on the SCN Web site is found on the left side of the page, below the logo and the welcome message. Under that, the "Seattle Site of the Week" is featured; the last time I looked, it linked to "Community Powered Radio" projects in Seattle.

The SCN developers decided in the early design phase not to employ the building metaphor which was often used to organize information in Free-Net Systems ("Post Office," "Public Square," "Arts Building," "School House," "Sciences and Technology Center," "Library," etc.). Instead they devised less concrete descriptors such as "Activism," "Arts," etc. These main SCN categories ("Activism," "Arts," "Civil," "Earth," "Education," "Health," "Marketplace," "Neighborhoods," "News," "People," "Recreation," "Sci-Tech," and "Spiritual") are lined up along the right edge of the page. Since the categories are arranged alphabetically, the "activism" category heads the list. While the placement is accidental, its prominent location does help ensure prominence of "activism" in terms of the SCN Web site and of the idea in general, a major part of the SCN project philosophy. Commercial search engines and other major portals on the Web are, of course, unlikely to highlight this category at all: selling things is the primary objective of those systems and social activism is generally neutral or even hostile to the economic concerns and objectives of corporations.

If we click on the "activism" link on SCN, we will discover a wide range of information including links to "Environmental," "Human Rights," "Hunger and Homelessness," "Women," and "Miscellaneous" (Figure 5). All of this information relates to activism generally in the Seattle area and generally on SCN, but not always. The activism page, like all the other category pages on SCN, is coordinated and

Figure 5

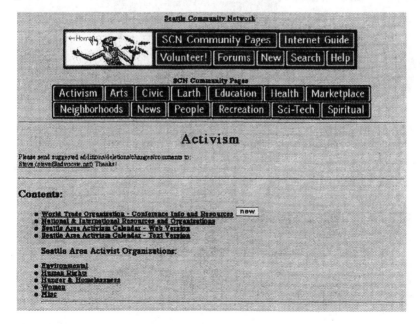

managed by a "Section Area Editor," one of the many volunteer roles at SCN. SCN, as of this writing, is run entirely by volunteers with few top-down directives. Subject Area Editors, then, are basically free to organize their Web page in the manner they prefer as long as they include the basic SCN header (which contains links to the other SCN subject pages) and are responsive to the information providers (IPs) who are adding information in that subject area.

The "Civic" section (Figure 6) has links to social services, politics, legal, nonprofit, philanthropy, public agencies, and international. The *Sustainable Seattle* project deserves particular attention because of its potent model which integrates community research, activism, and civic engagement. Sustainable Seattle has been developing a set of sustainability "indicators" which—taken as a whole—provide a meaningful snapshot of the Seattle region's ability to provide long-range social and environmental health for all of its inhabitants—human and otherwise. Given Seattle's natural surroundings and environmental ethos, it's not surprising that the "Earth" section (Figure 7) is fairly rich. Links here include "University District Farmer's Market," "Wannabe Farmers," and "Save Lake Sammamish."

The fact that the SCN "Neighborhood" (Figure 8) section has been growing steadily over the years is important to the SCN organizers because supporting geographical communities of various sizes has been a primary motivation from the project's onset. Although, ideally, a community network would exist for every community on the planet, it's clear that this is unlikely to happen in the near future. Although SCN places its main focus on the Seattle area, it is not intended to be its

Figure 6

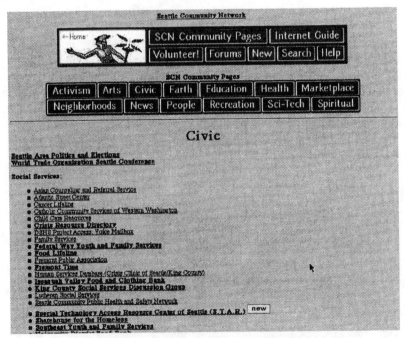

Figure 7

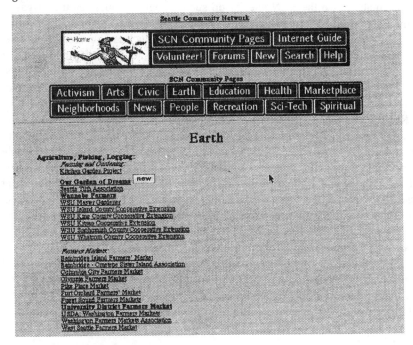

Figure 8

exclusive focus. Therefore non-Seattle neighborhoods such as Kenmore, Lakewood, and Bellingham use the SCN site, as do such global neighbors as the "Uganda Community Management Program" and the "USTAWI: Promoting Self-Sufficiency in Africa" sites, which can be found on the "Civic" section area. Neighborhood coalition groups such as the "Seattle Neighborhood Coalition" and the "Washington State Neighborhood Networks Consortium" also have links on SCN.

The "Science and Technology" (Figure 9) section points to a large selection of important resources both on SCN and other locations. This page also lets you post a URL to science and technology resources or post information about upcoming meetings. There are links to several innovative projects such as the Community Technology Institute (which offers free voice mail to homeless and phoneless people around the U.S.); *The Network*, a cooperative effort of community network activists and researchers worldwide; and the volunteer-run Vintage Telephone Equipment Museum in Seattle. The "Ask Mr. Science" service, out of service at this writing, allowed people to submit science questions to "Mr. Science" who would then post the answers online. This feature, based on Cleveland Free-Net examples, was one of SCN's oldest features and has been used by many Seattle area classrooms.

SCN, as of this writing, is a thriving computer system that has served as a model for many others throughout the world. On the other hand, its future is far from secure; after many years it still relies on volunteer labor and on donations. The search for sustainability is not an issue for Seattle alone, however. Most, if not all, community

Figure 9

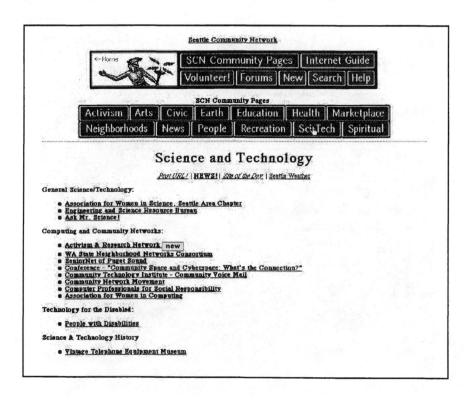

networks in the world are finding it difficult to find the necessary support. For that reason many people suspect that community networks will need to rely on the government in the future for support, although this view is not universally shared. There is also a strong fear that as billions of dollars are invested in commercial Internet ventures community networks and other new civic institutions that employ digital technology may simply become more and more marginalized as time goes on.

Technology and the New Community

> People need not only to obtain things, they need above all the freedom to make things among which they can live, to give shape to them according to their own tastes, and to put them to use in caring for and about others.
> —*Ivan Illich (1973)*

Both community and technology are inseparable aspects of the human condition. A community is a *web*, a web that is real yet intangible, a web of social relations and potentialities. Ideally, the web of community is a unity, a cohesive force that is supportive, builds relationships, and encourages tolerance. Sadly, the web of community is growing weaker in many ways (Putnam, 1995). Technology, too, is a web of sorts, for it also connects people in tangible and intangible ways. Technology mediates communication between people, changes social space, and alters roles and relationships in society. Humankind has fashioned and used technology for over 70,000 years—to multiply force or shrink distance—and technology, like language, is a natural and inseparable extension to our world and our world view. Yet especially in recent years, technology has become out of balance and out of control in many ways. Increasingly, communities are at the mercy of a seemingly autonomous technological imperative.

There is an apparent tension between the concept of "community" and the concept of "technology" that needs to be addressed. The stereotypes persist that communities are warm and fuzzy, whereas technology is steely cold, unyielding, mysterious, and dangerous. Part of the reason for those generalizations can be found in history—the grim and merciless toil in the factories of the industrial revolution—and part can be found in our collective imagination of the idyllic and convivial communities that theoretically existed "once upon a time," in the "good old days"—*before* technology.

Technology is often viewed as complex and incomprehensible. It is seen as larger in scope than the more familiar and comfortable spheres of the individual or community. Technology can be complex and inhumanly vast. But if people don't demystify the technology, it will forever be daunting, and people will continue to be victimized. The truth is that the culture of humankind can't be separated from its tools or from its technology. And although technological systems may seem complex, incomprehensible, and overwhelming in size, they need not be. Existing systems can be tamed and new community-oriented systems can be devised. By reasserting our control of our technological systems, some of the tension between "community" and "technology" can be removed and technology can be made to better serve human needs.

188 Schuler

The Vision of the New Community

There are a multitude of ideas and projects — both computer-oriented and not — that could help support the core values and aims of the new community. In general, those ideas and projects embody a set of value preferences (Figure 10) that indicate the general perspective of the new community. They are not binary choices; the rights of individuals are not to be abrogated, for example, just because there is a focus on community. Nor, for example, does the figure imply that commercial interests are not important, only that the focus here is primarily noncommercial.

Computer technology — in concert with other efforts — can play a positive role in rebuilding community by strengthening its core values. Whether these aims are realized will depend on citizens from all walks of life. Truly democratic systems can only be developed through broad participation. This endeavor must not be a charitable good-works project of elites nor a rebellion of the underclasses. It should be open to citizens of all races, economic classes, ethnic origins, religions, genders, ages, and sexual preferences. It must be global in nature, because a confluence of perspectives, experiences, and skills is needed in order to succeed.

Saul Alinsky, the premier American community activist, says (Boyte, 1989) that "the radical is that unique person who actually believes what he says. He wants a world in which the worth of the individual is recognized. He wants the creation of a society where all of man's potentialities could be realized." The vision of a new community is a *radical* one. Building it will require care and diligence, patience, and intelligence. The broader the effort is and the more tightly the efforts are interwoven, the stronger the force it will become. The momentum for positive change will be irresistible.

Figure 10: New Community Preferences

- Geographical over "virtual"
- Community over individual
- Public over private
- Community culture over mass culture
- Capacity building over needs-orientation
- Home-grown over specialist
- Empowering over disempowering
- Multiway conversation over broadcast
- Discussion over propaganda (e.g., talk radio)
- Inclusive over exclusive
- Process over goal
- Fundamental over superficial
- Democracy over autocracy
- Civic over commercial
- Voluntary over coerced
- Real needs over artificial needs
- Networks over hierarchies
- Sustainable over depletive

References

Alexander, C., Ishikawa, S., and Silverstein, M. (1977). *A Pattern Language: Towns, Buildings, Construction*. New York, NY: Oxford University Press.

Boyte, H. (1989). *CommonWealth: A Return to Citizen Politics*. New York, NY: Free Press.

Castells, M. (1996). *The Network Society*. London: Blackwell.

Cisler, S. (1993). *Community computer networks: building electronic greenbelts*. In Bishop (1994).

Doctor, R. and Ankem, K. (1995). A Directory of Computerized Community Information Systems. Unpublished report. Tuscaloosa.

Goel, M. (1980). *Conventional political participation*. In Smith et al. (1980).

Greider, W. (1993). *Who Will Tell the People?* New York, NY: Simon and Schuster.

Habermas, J. (1996). *Between Facts and Norms: Contributions to a Discourse Theory of Law and Democracy*. Cambridge: Polity.

Illich, I. (1973). *Tools for Conviviality*. New York, NY: Harper and Row.

Krasny, M. (1994, May/June). What is community? *Mother Jones*.

MacIver, R. (1970). *On Community, Society and Power*. Chicago, IL: University of Chicago Press.

McKnight, J. (1987, Winter). Regenerating community. *Social Policy*.

Morino, M. (1994). *Assessment and Evolution of Community Networking*. Washington, DC: The Morino Institute.

NPTN. (1995). Unpublished e-mail from Tom Grundner.

Putnam, R. (1993, Spring). The prosperous community — social capital and public life. *The American Prospect*.

Putnam, R. (1995, January). Bowling alone: America's declining social capital. *Journal of Democracy*.

Schuler, D. (1999). Community Networking Movement: http://www.scn.org/ip/commnet/

Schuler, D. (1996). *New Community Networks: Wired for Change.* Reading, MA: Addison-Wesley.

Smith, D., Macaulay, J., et al. (Eds.) (1980). *Participation in Social and Political Activities*. San Francisco, CA: Jossey-Bass.

Chapter VIII

CTCNet, the Community Technology Movement, and the Prospects for Democracy in America

Peter Miller
Community Technology Centers' Network, USA

In a large, airy room there is a crowd of young people and adults all working at computers. In one group students are having their first experience using a spreadsheet on an IBM PS/1. At the same time, in another corner, a senior adult is teaching herself to use a database on an IBM PC. A young man is updating the church's membership files and printing mailing labels. A young woman is at the Macintosh working on a desktop publishing project, and two teenagers are in another corner debating how best to make the Logo Turtle do what they want it to do. Others are casually 'messing about' with simulations. They are all using these technologies to achieve their own personal goals and objectives.[1]

This description of an open public access session at the Playing to Win (PTW) Harlem Community Computing Center in the late 1980s contains a vision that extends far beyond technology access and education in a single storefront setting— for PTW was one of the country's earliest examples of a technology education and access program established in a non-school community center, specifically for people of low-income and low literacy. It gave birth to what has become the Community Technology Centers' Network (CTCNet), a nationwide support project and membership association of more than 300 community organizations establishing similar technology education and access programs for disenfranchised communities or some special subset of its members. And PTW did this by embodying and making explicit a vision of democratic education around technology that is transformative and liberating for individuals, organizations, communities, and society at its best, one that is applicable to the community technology movement in general and to the wider and broader struggle for democracy and social change.

An Overview of Community Technology in the United States at the Turn of the Millennium

A brief sketch of the community technology movement in the United States today will help serve as a vantage point from which to view its origins historically and philosophically, and uncover some of its democratic educational and political foundations and orientations.

The community technology movement in America is a diverse and multidimensional family of efforts, practical and intellectual, each of which originated under particular technological, political, and historical circumstances, but are more and more frequently brought closer and closer together through a convergence of technologies and public policies that shape their development and distribution. Thanks to the deregulatory policies of the 1996 Telecommunications Act and the trend in digital technologies to merge previously distinct services such as cable and telephone, corporate developers are busy forming huge new corporations across the now outmoded regulatory, technology, and sectoral divisions. Along with these mergers, a convergence of public interest power is also taking place among those who have begun to build new linkages across these no-longer-distinct lines.

As the executive directors of CTCNet, the Association for Community Networking (AFCN), and the Alliance for Community Media (ACM) wrote in their recent introduction to the jointly produced *Community Technology Review* focusing on telecommunications policy:

> Our distinct organizations and individual constituents may focus on different technologies, but ...[t]he astounding emergence of new technologies in recent years and the complex policy environment which has resulted represent both a threat to and an opportunity for our efforts to ensure that the benefits of the digital society are available to all. ... [W]e represent centers and institutions dedicated with real resources and skills to help people defend their basic rights to economic opportunity, educational parity, health and safety, and democratic participation through technology and media. And this rapidly changing and complex environment represents an opportunity for us to collaborate in new ways, and has created new points of entry into policy processes for new kinds of constituencies. While corporate entities seem to increasingly control the process, there is also increasing interest in supporting and promoting telecommunications in the public interest. ...And the technology itself affords an unprecedented opportunity to level the playing field and allow previously unheard voices to be heard.[2]

Thus here is one place where three different organizational strands have been united by the convergence in technology and public policy—community technology centers and community networking projects, which grew independently from the mid-1980s, and community cable television, which originated two decades earlier. As a community technology center-oriented project, CTCNet is joined by a number of related specialized programs, including those of SeniorNet, the Alliance for Technology Access and its network of centers and partnerships targeted specifically to people with disabilities, Libraries for the Future and other libraries adopting technology programs, the Neighborhood Network and Campus of Learners initiatives for subsidized and public housing developments sponsored by the federal Department of Housing and Urban Development (HUD), the new federal Community Technology Center (CTC) program in the Department of Education (along with those in the Department of Education's 21st Century Learning Center program which have technology components), and a number of other associations of programs.

The community technology center movement includes, too, those physical facilities that grow out of community networking projects, a somewhat different tradition devoted specifically to establishing on-line community resources. The movement, which began in the mid-eighties with the Cleveland FreeNet, saw the growth and spread of FreeNets across the country, the birth and death of the National Public Telecommunications Network,[3] and the birth of a new generation of community networks, nourished in part by the federal Telecommunication Information Infrastructure Assistance Program (TIIAP, recently renamed Technology Opportunities Program, TOP) in the Department of Commerce and by the establishment of the Association for Community Networking.[4]

Finally, there are those CTCs that are primarily community cable television access centers, many of which have broadened their program and mission to provide public access to communications technologies to include the Internet, computers, and digital media. These public, educational, and governmental (PEG) access cable centers are one of the few forms of CTCs with a source of sustainable funding, a result of the franchise regulations establishing in the early 1970s. These PEG access centers are represented and organized by what was originally the National Federation of Local Cable Programmers, founded almost a quarter century ago, and transformed in the 1990s into the Alliance for Community Media.

This movement is widespread and growing — and no wonder. Computers and on-line resources are powerful tools for helping everyone, and individuals from disadvantaged groups are no exception: adult literacy students gain confidence and facility in reading and writing English through use of the word processor; unemployed workers prepare résumés and cover letters and learn and improve keyboarding, business applications and systems skills for reentering the job market; after-school and day care children learn how useful and fun computer applications can be; participants of all ages improve their communications, writing, keyboarding and literacy skills and gain knowledge of the world and others through growing telecommunications options—on-line chats, e-mail and pen pals, contributing, posting and commenting on essays and stories, producing their own as well as exploring others' Web pages, and working on joint projects frequently involving graphics, video, audio, and desktop publishing. All told, along with a host of independent local projects, many not even calling themselves by any particular name like community technology center, there are something in the neighborhood of 3,000+ community facilities across the country.

The sections below will focus on the development of Playing to Win and the networks which grew from it—the Playing to Win Network and then the Community Technology Centers' Network or CTCNet. We'll then look back at some of the early history in the development of community cable access television before looking at present and future prospects. Before turning to these things, however, it will be helpful to see this core of active center-based community programs as one aspect of an even broader movement to generate technology support across the whole nonprofit spectrum. This is a movement that is reflected in the growing trend among nonprofit and community-based organizations, social service agencies, churches, and community centers for acquiring and integrating computers and computer-based technologies into their day-to-day organizational life as well as into their programs, a movement that is occurring in all forms of organizational life across the spectrum.

The broader community and nonprofit technology effort is led by a series of special efforts to provide technology assistance to the nonprofit community which has coalesced around such projects as the National Strategy for Nonprofit Technology, Circuit Riders, and OMB Watch's Technology and Nonprofits Program, building upon the work begun in the late 1980s through the coming together of the Technology Resource Consortium. It includes those involved in the struggle for access and alternatives to other media and technology, from newspapers to radio; it is informed by the wide range of media and community issues that groups like the Benton Foundation, the Civil Rights Forum on Communications Policy, the Center for Media Education, the Media Action Project, and Computer Professionals for Social Responsibility are involved with; it is related to a number of electronic democracy projects and has roots with those projects and theorists concerned with democracy and technology on a deeper political level, and the individuals and the emerging programs in higher education that seek to guide and educate people involved with these issues.[5]

As computers become more and more ubiquitous, their appearance in programs and agencies which serve primarily poor people and people of low literacy is part of their "natural" development and distribution. Yet this growth and movement which, in many instances, has been the product of an emphatic desire to be current and up to date or to provide the opportunity for jobs or recreation or to increase organizational or programmatic efficiency, does have very social and political transformative possibilities beyond these immediate goals. These possibilities rest upon the conviction that basic tools of daily life need to be accessible to everyone. As a pronounced radical and self-conscious philosophy, it is found most fully articulated among those organizations which have established community technology facilities in a deliberate fashion rather than adding them on piecemeal. Among these, Playing to Win has been an important influence.

Playing to Win, 1980-1991

Try to remember or imagine the early, mid, and late '80s and even early '90s, when the whole idea of a community computing center was somewhat foreign and exotic. Try to remember or imagine the world before "Windows," before the Apple Macintosh, where "booting up" a computer resulted in technical prompts appearing on the screen, requiring esoteric languages, when almost all applications were written rather than bought, when the whole idea of computers for poor people or people of low literacy not only sounded strange but impossible—because what would such people possibly be interested in doing with computers in the first place? And, beyond this, didn't computers require such a high degree of literacy to begin with that it would be counterproductive, if possible at all, to introduce them to a low literacy audience?

In these early years it took foresight and a vision of possibilities, well beyond seeing the small pool of technical jobs that might be open to an impoverished ghetto dweller who might by happenstance receive some computer technology training. Anticipating these possibilities, Antonia "Toni" Stone founded Playing to Win in 1980 as a program for prisoners, three years before the opening of the community computing center in a public housing basement which bore its name, to promote the educational use of computers and offer technical assistance to prisoners and rehabilitation agencies. PTW was established on the principles that technology is a tool to help people achieve their own goals; students work collaboratively as much as individually and learn as much from play as from work; teachers are facilitators, resources and participants in the learning process; curriculum is project-based.

With the help of interested researchers and students, Toni and other PTW staff evolved special techniques for helping literacy students acquire computer and literacy skills concurrently and in a mutually reinforcing manner through a series of exercises and guides which, supported through a succession of corporate and foundation grants, evolved into *Keystrokes to Literacy*.

Keystrokes details a wealth of activities that can be created using basic productivity tools—word processors, databases, spreadsheets, and graphics—that are adapted to individual needs in learning and teaching literacy and computer skills in a complementary and reinforcing manner. The key to *Keystrokes* is its design as a tool to be used in service of the learner's needs and interests—whether the learner be the teacher or student. There are no literacy prerequisites. *Keystrokes* activities can be designed around learning the alphabet or the spelling of a student's name.

As Toni Stone explained in an overview of *Keystrokes* in the Fall-Winter 94-95 issue of the *Community Technology Center News and Notes*:

> Each activity focuses on a small-scale but clearly defined computer skill. For example, learning to erase is an important first step. Most people using a computer for the first time are likely to put something on the screen that they don't want there. Getting rid of it pronto inspires confidence that mistakes are easily corrected. So some of our first activities involve simply using the backspace (or delete) key to erase. For example, a word processing screen might be prepared as follows:
>
> cornx xx xbeans xxx xxbaconxx xpizzax xxx xxcake

The objective is to erase all the letters that don't spell out words. A learner has only to position the cursor (a skill covered in a prior activity) and use the backspace key. What results when all the x's are removed is a list of (in this case) foods, but could be any set of words that the learner is becoming familiar with. For more advanced students, the x's might be interspersed with the letters in the words, or a variety of letters could be used as candidates for erasure. The activity could be further extended to introduce the return and tab keys to line up the words in a column as a list, but that would be up to the individual instructor and student.[6]

Two related examples elucidate some of the cultural dimensions of *Keystrokes'* pedagogy. First, in illustrating how teacher and learner can frame their work around subjects of common interest, the teacher uses the "erase" activity below to reinforce the student's concern with blotting out drugs.

```
******************************************************************
Say NO to Drugs. Erase them completely!
Drugs mess with your head. They change your thoughts. They can take your life.
Say NO to drugs. Erase them from this screen and from your mind.
Cocaine      Ecstasy        Marijuana        Heroin
Alcohol      Crack          Angel Dust       LSD
Weed         Uppers         Downers          Speed
Barbiturates Ice            PCP
Name:                          Date:

******************************************************************
```

Although the task is similar to that in the preceding exercise—positioning the cursor after the last letter in the word to be erased and pressing the delete or backspace key as many times as needed, this same template can also be used for blocking/ highlighting entire words, lines, or sections, and then deleting, cutting or pasting, or other exercises. Entering name and date, saving to the student's data disk, printing a copy to take home, are all related activities that can be built upon.

Along with deleting and cutting, the following screen was developed for a related exercise—copying and pasting:

```
******************************************************************
We shall overcome,
someday.
Deep in my heart
I do believe

Name:                          Date:
******************************************************************
```

The template can be used to construct the entire refrain of the well-known civil rights song—any song that is familiar that has repeated lines—something common to many folk and religious songs—can be used.[7]

All *Keystrokes* activities are kept simple. Other features of the *Keystrokes* approach include:

- **Providing each learner with a printed copy of his or her work at every session.** This can be shown to family and friends as proof positive of increasing computer skill. And it doesn't hurt to slip a second copy into the learner's portfolio as a record of accomplishment.
- **Putting the learner in charge**—at the keyboard from the very first moment. Computer skills must be acquired through hands-on experience. Lectures and demonstrations are time wasters.

- **Encouraging experimentation.** The temptation to run through the menu bar of a word processing program, explaining every option, is to be avoided. It's too much to remember anyway. Instead, learners are encouraged to try things out, to see what happens. If they can't extricate themselves from something, they'll ask because they need to know.
- **Answering questions with questions.** "What do you think happened?" "What have you tried?" "What has worked before?" — anything rather than a direct answer.[8]

The *Keystroke* work was picked up in a diversity of settings, including Baltimore Reads, one of the first municipal literacy programs in the country, and spun off in a number of directions. In collaboration with the Union Settlement House and United Way, PTW produced *Key Strokes for Keystrokes*, a companion volume for ES(O)L students. A *Keystrokes to Mathematics* was also produced. It was during this period, too—beginning even earlier— that Toni Stone, reflecting the multidimensional aspect of technology equity, also coauthored with Jo Sanders *The Neuter Computer*, designed to help educators, parents, students, teachers, trainers and policy-makers overcome the computer gender gap. Like *Keystrokes*, *The Neuter Computer* offered a series of templates and activities, customizable for specific situations, 56 computer activities for kids, 96 strategies for increasing girls' computer use, along with an overview of the computer gender gap, guidelines for planning and evaluating a computer equity program in a school (and elsewhere), and forms, questionnaires, and resources.[9]

Growing out of this work, as well as the experience at the Harlem center and in training others, an explicit "Playing to Win Credo" was developed and later adopted by the Network project:

****

PLAYING TO WIN CREDO[10]

Purpose is:
    universal technological enfranchisement
    to broaden the scope of personal capability and interest
    to enable learning and functioning through technology

Technology is:
    tool
    information resource
    vehicle for communication

Students can:
    learn to operate machines and programs
    learn how to create programs
    learn how to use programs as tools
    learn from programs
    learn with programs

Students are:
    participants in learning process
    working collaboratively
    in control

    tinkerers
    actively engaged
    playing to win

Results are:
    empowerment: skill in tool use, success in learning
    increased self-esteem
    ability to use resources
    ability to articulate process and need
    recognition of personal contribution
    respect for contributions of others
    habits of self-assessment

Teachers are:
    facilitators, guides, coaches, gardeners
    resources
    participants in the learning process
    role models

Activities are:
    project based
    reference real-world activity
    respect and use background, culture, skills of participants
    provide for team work

Assessment is:
    the joint task of participant and teacher
    based on personal accomplishment
    substantiated by personal portfolio

****

The Credo incorporates an amalgam of democratic political education and pedagogy. Its description of students and teachers and emphasis on their collaborative partnership as well as the entire description of their activities reflect much of Paulo Freire. Technology as appropriate tools shares much with the ethos of the *Whole Earth Catalogue* and expresses an especially close kinship with Ivan Illich's *Tools for Conviviality*.[11] "Playing to win" and "conviviality," in fact, are key orientations that stand in useful contrast to what is frequently taken to be the best approach to computer training and education—work—as in the expressions "Workforce Training" and "Welfare to Work."

As John Dewey reminds us in *Democracy and Education*, play and work have more in common with one another than they are in opposition and both are required aspects of democratic education.[12] And lest we overemphasize the drudgery, the alienating and unfulfilling nature of much low-income community preparation for "work," it is preferable to note its playful aspects.

Recreation, as the word indicates, is recuperation of energy. No demand of human nature is more urgent or less to be escaped...If education does not afford opportunity for wholesome recreation and train capacity for seeking and finding it, the suppressed instincts find all sorts of illicit outlets, sometimes overt, sometimes confined to indulgence of the imagination. Education has no more serious responsibility than making adequate provision for enjoyment of recreative leisure (p.205).

Work and play need to be brought together, to fulfill the promise of "a truly democratic society, a society in which all share in useful service and all enjoy a worthy leisure" (p. 256). The idea of "playing to win" captures the duality, especially in providing resources for the working poor.

As the title of Dewey's *Democracy and Education* suggests, education and politics intersect and inform one another at their highest reaches. The political implications of PTW pedagogy were made explicit when, in 1988, Playing to Win adopted its mission statement.

Playing to Win Mission Statement

Playing To Win was established in 1980 to confront the prospect that, in an increasingly technologically dominated society, people who are socially and/or economically disadvantaged will become further disadvantaged if they lack access to computers and computer-related technologies.

Playing to Win promotes and provides opportunities whereby people of all ages who typically lack access to computers and related technologies can learn to use these technologies in an environment that encourages exploration and discovery and, through this experience, develop personal skills and self-confidence.

Playing to Win will operate one or more community-based technology learning centers as arenas for creating, testing, and evaluating effective models of education in and with technology, and it will initiate efforts to share the results of its practical experience with others seeking to provide similar opportunities in their communities.

Playing to Win will be a leading advocate of equitable access to computers and related technologies. It will invite and actively encourage other organizations throughout the United States to join in this mission, and will attempt, with them, to build a national network of neighborhood technology learning centers so as to promote the development of a technologically literate society.[13]

The Playing To Win Network (1991-1994) and the Community Technology Centers' Network (1994-Present)

While the seeds of the network were present from the beginning, it took well into the next decade to become explicit and then established. Throughout the 1980s, as Playing to Win was building its programs, hundreds of visitors came by to witness the new experiment underway. Many Manhattan-based community agencies came by to consider or follow-up a partnership/collaborative program being developed. From the beginning, three kinds of programs characterized PTW: 1) structured classes, seminars, and presentations, covering specific areas in technology development and how to use it; 2) open public access hours, specifically designed for community members to "drop in," check out the resources, get some hands-on experience and support, undertake a project—much like the opening scene describes; and 3) partnerships/collaborations with, say, local literacy, day care, library, youth, or seniors programs.

And just as many local New Yorkers came by to see the new experiment in operation, there were a stream of people—academicians as well as practitioners—from across the country, too. *MacWorld, PC Magazine, Business Week,* and other publications representing and covering the emerging computer industry frequently ran stories about the unique goings-on at PTW as an example of what could be done with and for the technology "have nots."

As a result of this experience, the idea of a "network" of such centers began to grow. It arose as a way to meet the needs of the growing number of visitors, in order to devote attention to their requests for information and help in setting up similar programs with their own organization, and as a way of spreading the ideas found in the PTW mission statement. Negotiations began with the National Science Foundation in the late '80s to support this idea, bolstered by the growing interest and development of a national advisory board that included a broad array of influential members of the technology education and research community, especially as that bore upon community-based learning in nontraditional settings. Board members included: Stephanie Robinson, a former vice president of the National Urban League and then with the Commission on Chapter 1; Shirley Malcolm, with the American Association for the Advancement of Science; Carleton Reese Pennypacker from Lawrence Berkeley Labs; Ann Lewin from the National Learning Center; Don Holznagel, Northwest Regional Education Laboratory; Pedro Pedraza, Center for Puerto Rican Studies at Hunter College; and Myles Gordon, Vice President of Education Development Center (EDC) in Newton, Massachusetts, among others.

By the late 1980s, in fact, a small cadre of centers had developed around PTW. First, in Somerville, Massachusetts, "Playing to Win at SCALE (the Somerville Center for Adult Learning Experiences)" was being transformed into the independent Somerville Community Computing Center (SCCC), a low-cost collaborative model—in contrast to PTW's growing stand-alone storefront which had moved out of its public housing basement origins to comparatively spacious street-level facilities on 111th at 5th Avenue. A collaborative with the city's adult education program and community action agency, the SCCC offered the same broad range of program options as PTW, bolstered by the educational technology resources of TERC, a nearby Research and Development (R&D) organization just over the border in Cambridge. The major partnership with SCALE provided the space, utilities, and building amenities. In return, the SCCC provided the hardware, software, management and technical assistance for an adult ed program class to begin to integrate technology into their curriculum, and used other hours for partnerships with other programs and agencies, and for open public access.

In Pittsburgh, the New Beginnings learning center for young children, and in Washington, DC, the Future Center at the Capital Children's Museum were developing similar programs with PTW assistance and financial support from the Echoing Green and Reynolds Foundations. Finally, in 1991, NSF awarded PTW a three year, $1 million demonstration project grant, "to develop a network of 30 centers similar to PTW, primarily in the Northeast U.S.," and the Playing to Win Network

(PTWNet) was officially launched. By June 1992, almost two dozen people gathered for the second annual affiliates meeting at PTW where the original four members were joined by representatives from El Barrio Popular education program in Spanish Harlem, the Fox Hill housing services agency on Staten Island where PTW Network staff had spent a great deal of time trying to develop a center, and the United South End Settlements/Harriet Tubman House in Boston, along with visitors from three other agencies. The following year, 11 centers attended the meeting, with new representatives from the Fortune Society, an NYC organization that provided a range of support services to former prisoners that had ties with PTW from its pre-center days; and two from the Boston area, including the Roxbury YMCA and a "Clubhouse" program that had been initiated out of the Computer Museum in downtown Boston. In addition, network staff had made contact with and were developing a program with The Bridge/Family Health Services in Jacksonville, Florida.

The "slow" growth of the network during its early years can been attributed to several sources. First, with the "retirement" of Antonia Stone, coincident with the arrival of the NSF grant and the change in project management, PTW underwent an organizational crisis, a not-uncommon occurrence that takes place in many organizations when its charismatic founder steps down. One needs also to take note of the nature of the community computing center enterprise in the early 90s and the resultant approach that was used. As the following report of the second annual meeting suggests, the growth was hardly slow at all from the perspective of the times.

Besides a radical commitment to equalizing computer access across economic, literacy-levels and gender lines, PTW has a unique approach to technological access and education, based on viewing the computer as a tool which anyone can learn to use for his or her own purposes. PTW helps people learn to use tool software (word processors, databases, spreadsheets, graphics, telecommunications, etc.) through project-based activities relevant to learners' lives, usable with learners at all levels or skill and experience, and resulting in an appreciation for collaborative work and empowerment.

At this year's gathering were *a core of activists who had come together for the first time only a year ago* (emphasis added), led by network coordinator Laura Jeffers and, with the retirement of Toni Stone, new PTW Executive Director Ramón Morales. The meeting featured a full schedule of workshops and community-building activities and a second day of hands-on math and science activities for adults and kids, geared toward the PTW approach.

PTW had just received its Echoing Green and National Science Foundation demonstration grants, the meeting report went on to discuss, and noted that affiliates would be chosen on the basis of:

- Location in areas populated by the socioeconomically disadvantaged, where technology access is limited.
- Commitment to expanding the technology resources of their communities and to work with the PTW approach to using technology as a tool of empowerment.
- Willingness to give time and attention to addressing the issues of the network

and the development of its resources and to participation in its ongoing evaluation.

- Ability, or potential ability, to support participation with adequate and appropriate space and personnel, and to underwrite staff development time and attendance at the annual all-affiliates meeting.[14]

A network of community computing centers—what's that? Up through the early '90s, the approach was, almost of necessity, a franchise model, primarily based upon the success of Playing to Win. Clear models to learn from and follow, and a definitive commitment to financial and staff dedicating resources to what was still a new and experimental idea, were the approaches of the day.

By February 1994, two major changes had taken place that indicated the network was ready to enter a new phase. First, network reorganization took place and leadership moved to the Boston area. Secondly, the Internet, what was then commonly called the National Information Infrastructure, was starting to make an impact on the nation's consciousness as something more than just another technological specialty development. In part as a result of the Clinton-Gore campaign and election, what had formerly been the province of a small, technological elite was starting to receive more and more public notice. In the language of the technology industry, telecommunications was becoming the new "killer app." The growing interest in computer technology among community organizations across the board was given an additional boost by telecommunications.

Much of this was reflected in the national expansion grant written for the NSF in early 1995, in partnership with Education Development Center, a large, innovative R&D organization in Newton, Massachusetts, which had come to house the project shortly after its move to Boston, thanks largely to the work of Myles Gordon, PTWNet Advisory Board member and EDC Vice President and one of EDC's founders. By this time PTWNet had grown to almost 50, with more than half its membership split between New York City and metropolitan Boston, five in Washington, DC, two in Pittsburgh, and six elsewhere across the United States, plus six abroad. The growth reflected a number of changes in the conception of the network—the movement from franchise to technical assistance approach, a change which took place in an environment where more and more community agencies were starting to develop community computing center-type programs on their own.

The process of a community agency developing a community computing center program frequently came about in the following way: computers were becoming more and more important, some members of the agency would argue. Yes, others would say, but how can we be expected to devote resources to a new program when we can hardly maintain our current programs and are so uncertain about what to do. Back and forth the debate would go. Then, in growing numbers, community organizations would take tentative steps to move in the new direction. And here a host of questions would frequently arise. What kinds of computers do we get and what software? What do staff using them have to know? How do we use this stuff in our programs? What else is it good for? How do we fund such programs? More and more, what was renamed the Community Technology Centers' Network when the NSF grant arrived at EDC came to be a technical assistance resource for such community organizations, responding to requests for assistance rather than having to go out and recruit, and doing so by developing a growing mutual self-help approach as well as central office and regional field staff expertise.

Among other major changes, telecommunications—originally used solely as a means for the directors of affiliates to communicate with each other, and little used at that—was starting to become a program tool.

Like PTW, one of the marks of some of the early affiliates was the development of their own philosophical and theoretical justification and foundation. Two of these in particular bear mention—Future Center at the Capital Children's Museum in Washington, DC, and the Clubhouse at the Boston Computer Museum.

Future Center's philosophical approach to computers can be found right in their manual for staff and volunteers. The Capital Children's Museum was itself a project of the National Learning Center, founded in 1984, to develop and showcase a range of progressive programs which included informal environments for lifelong learning, individualized instruction in a formal school setting, and, in both environments, the use of computers as personal tools for learning. The Future Center manual was explicit about "Our Learning Philosophy" and acknowledged that its educational practices were "guided by a robust collection of learning theories put forth by some of today's leaders in cognitive psychology. Among these noted psychologists are Howard Gardner, Harvard University; Reuven Feurstein, Hadassah-Wizo Institute, Jerusalem; Mihayli Csikszentmihayli, University of Chicago; and David Perkins of Harvard University." Out of this came a credo of their own, focused in a series of beliefs about their program participant students who:

- possess a personal intelligence profile that is unique to them (much like a thumbprint) that cannot be measured with a number or I.Q. and that is made up of a combination of these several known "multiple intelligences": linguistic (verbal ability), logical-mathematical, spatial, bodily-kinesthetic, interpersonal, intrapersonal, and musical;
- should be given ample opportunity to explore and grow within their strongest area of intelligence;
- should be assessed through performance, portfolios, learning journals, and/or discussions with the teacher;
- can learn regardless of their age, present condition, or reasons for that condition;
- learn best through direct interaction with the tools and materials in an environment;
- must be given the time to so enjoy their learning that they enter a state of mind called FLOW;
- cannot be expected to perform and learn well in a time-driven environment with clocks, bells and externally imposed schedules;
- can, however, expect to learn and retain what has been learned in a learner-driven environment where the student decides with constructive mediation when, what, and how to learn;
- have the best learning experiences on the computer through project-based activities that incorporate their interests and experiences;
- should use the computer as a tool to explore phenomena and solve problems and through this type of use they will gain the most meaningful and powerful

use of computer technology; and

- are to have total control of the computer and interaction with it at all times.[15]

With such a philosophy, their guide to "Being a Tutor" followed very much along the lines of the PTW's Credo. As it begins:

First and foremost, your role here is to empower our Members with the skills and confidence to make computer technology work for them. To do this, you must allow the student as much control over the operation of the computer as possible.

NEVER TOUCH THE COMPUTER, even to turn it on, unless it is absolutely necessary or the student is so totally confused that they are not getting anywhere...(23).

Treat the "student" as a partner in the learning and exploratory process: "if you happen not to be familiar with a piece of software, be honest with the student about it and be willing to work together with him/her to solve the problem." The guide even incorporates an explicit "playing to win" attitude: "This is a place to explore and discover; let students 'mess about.' ...Most of all, you and others should always have fun!" (23-24).

A similar self-consciousness and deliberateness characterized the development of the clubhouse program for kids. Its founders at the Boston Computer Museum, Mitch Resnick of the MIT Media Lab and Natalie Rusk, now with the Minneapolis Museum of Science, outlined their philosophy of technology access and education in the July-August '96 issue of *The American Prospect* in their essay on "Computer Clubhouses in the Inner City.

Initially, they presented the clubhouse as distinctive in terms of its "high end" approach to technology and the use of mentors: "At many other centers, the main goal is to teach young people basic computer techniques (such as keyboard and mouse skills) and basic computer applications (such as word processing). At the clubhouse, in contrast, the goal is for participants to learn to express themselves fluently with new technology" (6). But, as they go on to say, "The clubhouse is based not just on new technology, but on new ideas about learning and community." Explicitly, "young people become designers and creators—not just consumers—of computer-based products."

At its foundation, four core principles drive the development of the clubhouses: (1) support learning through design experiences; (2) help youth build on their own interests; (3) cultivate "emergent community"; and (4) create an environment of respect and trust. Of these, the most philosophically rich is the first, where the authors explain in greater detail:

Design activities engage youth as active participants, giving them a greater sense of control over and responsibility for the learning process, in contrast to traditional school activities in which teachers aim to "transmit" new information to the students. Design also encourages creative problem solving and fosters a search for multiple strategies and solutions, instead of the focus on getting one right answer that prevails in most school math and science activities. Design projects are often interdisciplinary, brining together concepts from the arts as well as math and sciences.

Design activities, moreover, can create personal connections to know, since designers often develop a special sense of ownership (and caring) for the products and ideas that they design. Yet design also promotes a sense of audience, encouraging youth to consider how other people will use and react to the products they create. And design projects provide a context for reflection and discussion, enabling youth to gain a deeper understanding of the ideas underlying hands-on activities.

This emphasis on design activities is part of the broader educational philosophy that MIT Professor Seymour Papert has termed "constructionism." Constructionism is based on two types of "construction." First, it asserts that learning is an active process, in which people actively construct knowledge from their experiences in the world. People don't get ideas; they make them. (This idea is based on the "constructivist" theories of Jean Piaget.) And, second, people construct new knowledge with particular effectiveness when what they make is personally meaningful.[16]

By the summer of 1999, with over 300 members, CTCNet was standing at the verge of its next major development. Under the leadership of new Executive Director Holly Carter, CTCNet held its eighth national conference in July in Chicago as it looked ahead to becoming an independent organization separate from its National Science Foundation and Education Development Center-supported origins. The week before saw the deadline for the first federal applications for the newly established Community Technology Centers (CTC) program, initiated with $10 million for the establishment or enhancement of 40-60 new centers across the country, scheduled to be announced in the early fall.

A Look Back at the Origins of Community Cable TV

To help appreciate possibilities for the future, it is useful to turn our attention to some of the history and lessons of community television, the community cable access movement, for there are a number of surprising parallels to the emergence of CTCNet out of Playing to Win. In fact, it was two decades earlier that another radical development in technology carried with it a very similar promise and movement for democracy, when another new kind of community technology center was set up in New York City that gave birth to another organization and helped lead a movement.

From its origins in Canada in the late 1960s, concurrent with that country's own "War on Poverty" the broad-based political movement of the times, the story of the access movement is rich with living legacies, with three particular moments or aspects of note here: its origins in the development of a new technology which provided the material foundation of the movement; a lesson involving its tactical alliances with the corporate community; and, finally, returning to the technology, its use as an integrated tool for democratic reconstruction.

There were, to begin with, two major technology developments that played a key role in the emergence of the community cable access movement: cable itself and the invention of portable video equipment. Cable began to be used in the United States coincident with the spread of television in the late Forties, but on a very small scale, primarily as a way to broadcast signals to small rural communities that lacked regular television stations in their area or suffered poor reception due to geographical barriers, primarily mountains. Local entrepreneurs, often hardware or electronic dealers, erected towers that received broadcast signals. Coaxial cable ran from these towers or community antennas to local households for a modest fee. Community Antenna Television (CATV) was born. As Ralph Engelman tells us in his excellent monograph, "By 1952 some 70 systems served about 14,000; a decade later, 800 systems had almost a million subscribers. Early in the next decade, cable spread to cities such as New York and San Francisco with special reception problems that made cable a feasible option. Considered to be a 'common carrier,' cable television had no concern or control over content and was welcomed initially by the major networks as a way to extend their markets."[17]

In 1968 Sony introduced the portable video camera and recorder unit, or portapak. It weighed only 20 pounds and as important as its mobility and lack of obtrusiveness was its ease of use. The contrast with the documentary filmmaker's 16-millimeter movie camera and related equipment was striking. Not only could one person carry the entire unit—a nonprofessional could use the equipment with comparatively simple training, much as one can do today, though the revolutionary change at the time was dramatic. No special treatment or lab was required for developing tape or synchronizing video and sound, and though editing equipment was required for final products, much of this could be done by erasing and rerecording.

The combination of cable and new video technology provided the technological foundation for the community cable access movement. Concurrent with the spread of cable, community and local production with portable equipment provided a new source of material. Cable operators could provide people with an avenue to originate their own programs. CATV evolved from an auxiliary service to a distinct medium—"cable television."

From initially favoring the spread of cable television because it meant new subscribers and markets, the major networks—CBS, NBC, and ABC—became increasingly wary and even fearful of cable television and used its influence to lobby for restrictions on its growth. In 1966, the Federal Communications Commission (FCC) extended its jurisdiction over cable TV and banned further distant signal importation into the nation's 100 largest markets. There was, in Engelman's words, "a virtual freeze in the expansion of cable television that would last until 1972."

It was during this period—from the late 1960s through the early 70s—that the cable access movement and the cable industry worked hand-in-hand. Located above a movie theater on Bleecker Street in Greenwich Village and funded by a major grant from the Markle Foundation, the Alternate Media Center at New York University became an assembly point for apostles of video and cable technology, public access pioneers, members of experimental video collectives, educators, city planners, and federal policymakers from around the nation. Here experiences and tapes were shared and strategies developed for a public stake in development of cable TV.

The Alternate Media Center took initiatives in both production and policymaking. Demonstration tapes were circulated nationally. Pilot projects—in Reading, Pennsylvania, for example—would demonstrate the potential of community television and its interactive, two-way capabilities. Alternate Media Center interns helped establish access centers throughout the nation. Out of the Center's internship program would develop the National Federation of Local Cable Programmers, which became...the most important membership and advocacy organization of the community television movement. The Alternate Media Center sought to set policy as well as precedents. In 1971, it immediately became involved in the struggle over the cable franchising process and the introduction of community channels in New York City—which many considered a bellwether of the future of cable television in the United States. In addition, George Stoney and Red Burns [cofounders of the Alternate Media Center], in collaboration with Nicholas Johnson, the maverick commissioner of the Federal Communications Commission, were instrumental in the creation of federal requirements for access channels in 1972.

According to George Stoney, generally regarded as the father of the access movement, Red Burns became co-director of the Alternate Media Center because of her ties to the entertainment and cable industry. "And wonder of wonders," Stoney added, "our strongest champion was Irving Kahn, head of Teleprompter, and at that time, the most powerful influence in the cable industry. He was eager to give us a welcome on an access channel as soon as his franchise was operational in Upper Manhattan" (21).

For the cable industry, public access programming and channels could make cable appear to be a socially responsible medium. Cable corporations could point to the access movement as a unique form of public service, giving people a voice in presenting themselves and producing their own media. Public access could make cable appear a viable alternative in contrast to the "vast wasteland" that was commercial broadcast television. Public access could usefully bolster the cable industry's quest for legitimacy. At a critical moment in history, the interests of the cable industry and the community television movement coincided.

Born in this environment and closely tied to congressional, executive, and FCC politics and currents, cable access centers and their organization were, from the beginning, politically savvy, in contrast to the centers which emerged in the '80s and '90s—witness the franchise fee setasides which have been mandated by law to provide one sustainable source of ongoing support. One constancy throughout, however, has been the key role foundations have played. The Markle Foundation's crucial role in helping establish the Alternate Media Center came after major foundation studies on the future of cable were undertaken in response to the FCC's request for information and research. Two years before seeding the AMC, Markle had teamed up with the Ford Foundation to fund a series of studies on the future of cable television by the Rand Corporation, and around the same time the Alfred P. Sloan Foundation established the Sloan Commission of Cable Communications. In sum, writing in 1990, Engelman tells us: "Major private foundations—Carnegie, Rockefeller, and Ford, among others—have played a critical role shaping public policy in general, and communications policy in particular, in the United States" (35).

There is one final lesson to note here about the democratic possibilities of the new media, from the early origins of the cable access movement in Canada. It took place around what has come to be known as "the Fogo Experiment" which was undertaken by Challenge for Change, founded by the National Film Board of Canada in 1966.

Challenge for Change/Société Nouvelle, in the words of the Commissioner of Film, was begun "to help eradicate the causes of poverty by provoking basic social change." It was established as a collaborative of both its English and French sections and with other federal agencies and departments—Agriculture, Communications, Health and Welfare, Indian Affairs and Northern Development, Labour, Manpower and Immigration, and Regional Economic Expansion—with representatives from each serving on a joint commission. In the words of George Stoney, Guest Executive Director of Challenge for Change from 1968-70, just before he went on to found the Alternate Media Center, the program was "a social contract between the people who were in charge of a government program—an agency or social service—and the people who were the recipients of that program or service, designed to find out how they felt about what was being done and what would they like to see changed" (5).

Challenge for Change, Engelman tells us, drew on, among other things, the tradition of the social documentary pioneered by Robert Flaherty and John Grierson. Flaherty's *Nanook of the North* (1922) had been revolutionary in its portrayal of social reality without a studio, a story, or professional actors. In the 1930s Grierson, who followed this tradition, was named the first government film commissioner. The National Film Board, he said, "will be the eyes of Canada. It will, through a national use of cinema, see Canada and see it whole...its people and its purpose." Challenge for Change was rooted in his vision. In his "Memo to Michelle about Decentralizing the Means of Production," published in the Challenge newsletter, Grierson wrote:

> The basic tendency of the Challenge for Change program is to follow decently in the original cinema-verité tradition which the English documentary people associate with *Housing Problems* (c. 1936). With that film there was talk of "breaking the goldfish bowl" and of making films "not about people but *with* them" (6).

The point about subject participation and involvement was underlined by problems that developed with Challenge for Change's maiden film, *The Things I Cannot Change* (1966). As Engelman summarizes them,

> The well-meaning filmmaker, Tanya Ballantyne, lived with a poor Montreal family with 10 children for three weeks. Her film sympathetically portrayed the hardships of the family's everyday life and was broadcast on Canadian television. ...[T]he broadcast had a devastating impact on the family: the children were ridiculed at school and the parents felt humiliated in their neighborhood. The family had neither an opportunity to see the film prior to its release nor advance warning of its date of broadcast (7).

The contrast between high intentions and social commitment, on one hand, and outcome, one the other, was stark. Later that same year, Danserau, working in a small town in Quebec, took the next step by inviting participants to screen rushes, censor anything they found objectionable, and control the final screening process. This procedure minimized the danger that subjects would be surprised, offended, or suffer any of the indignities that befell the Montreal families earlier.

Moreover, through such a process, the film's subjects not only had greater control to prevent the problems that had arisen from *The Things I Cannot Change*—their whole consciousness was changed through the objectification of their lives and their involvement in a supportive though critical process of examining them. They "gained a heightened awareness of their own situation through their involvement in its portrayal on film" (8).

This approach was "crystallized" during the Challenge for Change Fogo Island project. Fogo Island, off the eastern coast of Newfoundland in northeast Canada, not only had half its 5,000 residents on welfare at the time due to the declining fortunes of the fishing industry—the government was in the process of relocating the entire population, which was divided into 10 separate settlements. Here was the site for the new experiment in using film as catalyst for social change.

With the involvement of the islanders directly in the filming process as well as members of Newfoundland's Memorial University and community organizers, original plans to produce a traditional documentary-length portrait gave way to numerous shorter vignettes that elicited greater response and interaction from the islanders themselves, "linear chunks of reality," in the words of one regional project director. Fogo Islanders were filmed only with permission, helped choose the sites and subjects, were integrally involved as first viewers and editors. Twenty-eight short subjects, six hours in all—*Fishermen's Meeting, The Songs of Chris Cobb, Billy Crane Moves Away* as examples—provided an overview through vignettes and individual stories. Participant consent was required for showings outside the village. As Engelman summarizes the process, "Group viewings were organized all over the island to foster dialogue within an isolated, divided population. The films and discussions heightened the awareness of the people that they shared common problems and strengthened their collective identity as Fogo Islanders" (9).

The impact was dramatic. Not only was the relocation plan abandoned, but—with fishermen now talking directly with cabinet ministers—the case for a coopera-tive fish-processing plant was pressed, and ultimately successful, along with the establishment of a boat-building collective and a consolidated high school. A follow-up film, *The Specialists at Memorial [University] Discuss the Fogo Films* (1969), documented the impact politically as well as on the academic community involved.

As Engelman characterized them, "The passive subjects of the film *The Things I Cannot Change* were transformed into active participants on Fogo Island... documentary films were now being made virtually by the people depicted instead of merely with them" (11-12). Anthony Marcus, a Canadian psychiatrist, summarized his analysis of the liberating potential in *Citizen Communications*:

> The simple device of reflecting an image magnifies the individual's self-image. The emotional dilemma induced by the gap between the image on the screen and the subjective feeling of the viewers, produces a crisis in which the person attempts to bring the two aspects into harmony, thus increasing his self-knowledge (12).

And on a community level, the transformative possibilities were even greater. In using technology as a tool for portraying oneself and one's world, the possibilities for social change as well as individual transformation are limitless.

Looking Ahead — Democratic Prospects

The origins of the community cable access television movement shed useful light on the prospects for CTCNet and the current community technology movement. The corporate alliances of the past provide guidance for the present as cable, long distance, baby Bell, and other utility corporations jockey for position in the burgeoning communications marketplace and policy arena.

And just as change in video technology provided one of the key material bases for the movement for democracy in the '70s, so too has the emergence of current technology provided the dramatic basis for social change today. In fact, the development of computerized graphics, video and audio, and their distribution over the Internet is arguably even more dramatic than the invention of the portapak and spread of cable in its democratic potential—in terms of cost, accessibility, and training requirements, to the degree the lessons of the Fogo Experiment stand to be learned and experienced on a scale of virtually limitless possibilities.[18]

Developed in 1999, Frank Odasz's *Cross-Cultural K12 Internet Guide*[19] is as applicable as much in informal learning environments as it is in kindergarten through high school settings and can be considered a technologically current, Internet-oriented update of the *Keystrokes* manual. Rooted in rural Montana and developed as a distance education guide for Alaska out of a community networking tradition, it nonetheless, like the urban-informed *Keystrokes*, was put together for teachers and students alike, geared to individual customization, need and use, with a similar emphasis and orientation towards self help, individual and community empowerment.

The *Internet Guide* starts with an introduction to search engines as a beginning step in becoming a self-directed independent learner, the first of four progressive levels which go to make up the "Internet style of learning." Prefaced with thoughts about "Students as producers, teachers as facilitators (shift from being the 'sage on the stage' to a 'guide on the side')" and tips on how to get a computer and Internet accessibility as cheaply and easily as possible, the Guide rests upon three "historical firsts" that the Internet brings with it: 1) fingertip access to the world's knowledge base; 2) inexpensive global self-publishing for teachers and students; and 3) students, who often know more about computers and the Internet than teachers.

The succeeding levels of mastery are:

1) *Browsing and searching*—With extensive lists of Web site descriptions that can be used as customized "templates" and an activity checklist for Web browsing, searching, and, yes, copy and paste, so that the material can be taken directly into documents and presentations of one's own (with useful copyright, fair use, and appropriate material guides).

2) *Students as producers*—Beginning with the easiest self-authoring feature on the Web: animated musical greeting cards (http://www.bluemountain.com), which can be used much in the way that "Print Shop" was one of the basic program for beginners in the early days of Playing to Win and is still a valuable introductory software tool for computer novices to use for an easy and useful product result. Moving on to creating Web pages with Netscape Communicator, easily modifiable to any basic Web-authoring software. The resources in

this section include tours of student-created Web sites and portfolios, worksheets and tutorials for progressively more complex levels of multimedia production.

3) *Project-based collaboration*—The work model of the future, Odasz tells us, is "multiple short term projects requiring highly developed groupwork skills" (64), and the third level provides a series of activities that can be consulted by students and teachers interested in learning how to plan, implement, and evaluate projects that involve others.

4) Real world problem solving — Finally, there are guides here involving building learning communities, learning to earn, and electronic democracy, all done through project-based approaches to real community issues and problems—primarily an invitation to explore and connect with the myriad Web sites and projects which are engaged in this work (and play).

In conclusion, from *Keystrokes to Literacy* to the "Internet Style of Learning," the community technology movement has resources, leadership, impulses, and philosophies that are guided by a rich vision of democratic possibilities and regeneration. As recent coverage of these trends were summarized: "A growing gap threatens to separate the techno-haves from the have-nots—perhaps permanently. To some, closing this divide is about providing computer access. To others, it's about teaching computer skills. But some change agents are playing for much higher stakes: To them, it's about fundamental social change."[20]

The profiles of Joseph Loeb and Break Away Technologies in Los Angeles, emerging out of the church and civil rights activism of the 1960s; Tony Streit and the work of Street Level Youth Media in the north, south, and westside neighborhoods of Chicago; and Bart Decrem, Magda Escobar, and Plugged In in East Palo Alto, California, as recently as 1992 the per capita murder capital of the United States, are all illustrative of community technology development being guided by and rooted in a vision of democratic regeneration. And there are many other individuals and centers, too, that can usefully be noted for their specifically democratic political origins and agendas: Lauren-Glenn Davitian at Chittenden Community TV and the Old North End Community Technology Center in Burlington, Vermont; Fred Johnson and the Media Working Group in Covington, Kentucky; Carl Davidson and the Coalition for Information Access and *cy.rev* in Chicago, and Pierre Clark and NeighborTech there as well; Dirk Koning and the Community Media Center in Grand Rapids; Steve Snow and Charlotte's Web; and Amy Borgstrom and ACENet in southeast Appalachian Ohio. These experimenters are not only leading the community technology movement—they bring leadership to the broader movement for democracy.

Endnotes

1   http://www.ctcnet.org.
2   Holly Carter, CTCNet; Amy Borgstrom, AFCN; Bunnie Riedel, ACM, "Convergence and Collaboration at the Cusp of the New Millenium," in "Communications Policy on the Front Lines: Ideas for Change," *Community*

Technology Review, Summer-Fall 1999, p. 3, on-line version available at http://www.ctcnet.org/review99.htm.

3   See the author's "Requiem for the Boston Computer Society and the National Public Telecommunications Network," September 1996, http://www.ctcnet.org/requiem.html.

4   See http://www.afcn.org. For the other organizations in this section, see: http://www.alliancecm.org, http://www.civicnet.org/comtechreview/communications_policy_resource.htm and the database at http://www.civilrightsorg.linkalpha.htm.

5   The relevant Web sites are http://www.igc.org/trc, www.nsnt.org, www.civilrightsforum.org, www.cme.org, www.cpsr.org. Among those interested in the wider reaches of democratic theory and practice are Ed Schwartz, founder of the Center for the Study of Civic Values (see *NetActivism: How Citizens Use the Internet*, Sebastopol, CA: Songline Studios, Inc., and O'Reilly and Associates, 1996, and http://www.libertynet.org/community/phila/natl.html); Benjamin Barber, *Strong Democracy: Participatory Politics for a New Age*, Berkeley, CA: U of Cal Press, 1984, 1990, esp. pp. 273-78; and *The American Prospect*'s Electronic Policy Network at http://epn.org. While there are a large number of individuals involved with both the practical and educational components of the community technology movement, there is a scattered but growing interest in schools of higher education in the United States that are approaching this programmatically. The University of Michigan's School of Information, the LBJ School of Public Affairs at the University of Texas/Austin, and MIT, home of the Media Lab and the Center for Reflective Community Practice, are some key examples. See especially http://www.si.umich.edu/community/.

6   Antonia Stone, "Keystrokes to Literacy—An Overview," *Community Technology Center News and Notes*, Fall-Winter 1994-95, p. 23.

7   Antonia Stone, *Keystrokes to Literacy*, NY: National Textbook Company, 1991, ISBN #0-8442-0679-2, pp. 57, 94.

8   Antonia Stone, "Keystrokes to Literacy—An Overview," *op. cit.*, p. 24.

9   *The Neuter Computer: Computers for Girls and Boys*, Jo Schuchat Sanders and Antonia Stone for the Women's Action Alliance, NY: Neal-Schuman Publishers, Inc., 1986.

10   Ideas on which the Credo was based came out of research by Jan Hawkins in the mid-80s. For many years, Jan was the Director of the Center for Children and Technology, which became the New York office of Education Development Center (EDC), housed in Newton, Massachusetts. Her own individual work was importantly influenced by the Coalition of Essential Schools, whose principles are also reflected in this credo. The Credo has appeared in most issues of the *Community Technology Centers Review*, e.g., Summer-Fall 1999, page 48.

11   Howard Reingold's *Virtual Community* (Reading, Massachusetts: Addison-Wesley, Harper Perennial, 1993) is a paean to the experience of one of the first community networks, the Well in the Bay Area. See Paulo Freire, *Pedagogy of the Oppressed*, NY: Continuum Publishing Co., 1993; *The Whole Earth*

Catalogue, 30th Anniversary Celebration Edition, Winter 1998, San Rafael, CA: the Point Foundation; and Ivan Illich, *Tools for Conviviality*, NY: Harper Colophon, 1980.

12 John Dewey, "Play and Work in the Curriculum," Chapter 15, *Democracy and Education*, NY: MacMillan Co., 1916, 1944.

13 Found in all the *Community Technology Centers News & Notes* and *Community Technology Center Reviews*, 1994-present. Adopted with some slight revisions when the Playing to Win Network was transformed into the Community Technology Centers Network in 1996.

14 "The Playing to Win Community Computing Center Holds 2nd Annual Affiliates Meeting in Harlem." *Impact*, publication of the Boston Computer Society Social Impact Group and Public Service Committee, August 1992, p. 15.

15 "Handbook for Staff and Volunteers at Future Center," *Impact*, publication of the Boston Computer Society Social Impact Group and Public Service Committee, November 1993, pp. 22-23.

16 Mitch Resnick and Natalie Rusk, "Computer Clubhouses in the Inner City," *The American Prospect*, July-August 1996, p. 62.

17 Ralph Engelman, "The Origins of Public Access Cable Television, 1966-1972," *Journalism Monographs*, the Association for Education in Journalism and Mass Communication, Department of Journalism, University of Texas at Austin, March 1990, p. 21.

18 One place to see a direct line of development from the Frogo Experiment to a community development/revitalization approach based on the democratic implications and uses of Internet technology is at the Don Snowden Program for Development Communication at the University of Guelph in Ontario, Canada. In 1967 Snowden was director of the Extension Department at Memorial University in Newfoundland and on the original project team. By the mid-1970s Snowden and his colleagues were experimenting with the "Fogo process" in the Arctic and Alaska, Africa and Asia. He died in 1984, working with the Canadian International Development Agency in Bangladesh while using the Fogo process approach to build communities through small-scale water control structure development. The Snowden Center was established originally at Memorial University later in 1984 and relocated to the Rural Extension Studies Program at Guelph in 1994. See http://www.snowden.org.

19 http://lone-eagles.com/guide.htm

20 Sara Terry, "Across the Great Divide," *Fast Company*, July-August 1999, p. 192.

Chapter IX

Community Networks for Reinventing Citizenship and Democracy

Fiorella de Cindio
University of Milan, Italy

When the Milan and Bologna community networks were designed and launched in 1994, the goal was not simply to give access to all, but, by doing so, to take a step toward reinventing citizenship and democracy. This chapter first presents the "vision" behind that goal and then summarizes the outcomes of these early experiences in terms of the original goal. Citizens' communities emerge as the true engine of innovation and therefore the chapter outlines how to translate the "community network vision" into design principles and implementation guidelines.

Introduction

The first Italian community networks were conceived and designed in 1993-94 during a unique period in our country's history. In 1992 and 1993 a group of public prosecutors from the Milan court — called the "Clean Hands" Pool — revealed the corruption that had characterized Italian democracy since the end of the Second World War. Italy had been frozen for 45 years by a set of political parties that consisted of the two major parties (the Christian Democrats, who had led the national government since 1948 without interruption, and the Communist Party, which was constantly in opposition during the period) that were surrounded by a few satellite parties. Among these, the Socialist Party acquired a major role during the eighties. In the 1993-1994 period, most of the outstanding statesmen were indicted and, in many cases, imprisoned. The political class was undergoing radical change, while civil society—known in those days as the "fax people" because its voice was heard through the faxes people sent to the newspapers—asked to play a greater role. These events transported the country from the First to the Second Italian Republic.

The promoters and designers of the early community networks (Milan, launched in September 1994; Bologna, active in January 1995), because of their personal history, not only technicians, but also were well aware of this socio-political context. They conceived of the Net as a way for giving voice — more than faxes can — to those who never had a chance to play a role in the civic and political arena. The Net was seen as a way for contrasting a *closed* Italian society, dominated by the "membership" culture (I trust only those in my own "church": a party, a group within it; my identity is rooted in the group I belong to). This "membership culture" and its consequences are well described in Colombo (1996) with a virtual world, *open* according to the Internet tradition. The Net is therefore conceived as a new *tool*, which may open *new domains of possibility* to everybody, and, even more, as a kind of *new world* to be colonized possibly better (more open) than the real one.

In other words, the goal of opening the Milan and Bologna community networks was not simply to give access to all, but, by doing so, to take a step toward reinventing citizenship and democracy. Probably, there was also an intuition of the role network communities can play. These nets were imitated by several other towns, where other people shared the same goals: using the Net to enhance citizen participation in the *res publica*. In order to make this participation effective, it is of course necessary to involve public institutions and force them to use the Net to experiment with new forms of communication between administrators and citizens, electors and the elected.

Five years later, it is now possible to sum up a very first attempt: do the several community networks confirm the original "vision"? Do they suggest general guidelines for pursuing that vision? The goal of this chapter is to discuss these points. But it is worth remarking that, even though five years seems long, tiring, and stressing to those who have carried on day-to-day management of these initiatives, it is also a very short time for the design and implementation of such complex socio-technical systems, and for driving ultimate conclusions over such an ambitious dream. This is why we call them *early* experiences.

After a presentation of a few experiences — those the author has been directly involved in or knows best — the chapter discusses the root of the vision, so to speak, i.e., why one can assume that the Net may support reinventing citizenship and democracy. This is followed by a summary of these early experiences in terms of the original goal. On this basis, a section is presented that outlines some design principles to (re)orient further developments toward approaching the original goal.

A Short Introduction to Community Networking in Italy (and Europe)

This section gives a short presentation of community networking, starting from Italy, with a few references to the other European experiences. The aim is by no means a complete survey, but rather to focus on certain characteristics that are well worth presenting to aid in subsequent discussion of citizenship and democracy issues. We will use "local computer network," as a generic term denoting any telematic initiative rooted in a geographical area, typically around a town.

The Community Network Approach

RCM, The Milan Community Network

Rete Civica di Milano (RCM) is the Milan community network that was launched in September 1994 as an initiative of the Civic Informatics Laboratory at the Department of Computer and Information Sciences (DSI) of the State University in Milan, Italy. RCM was set up with a small investment by the university and the donations of sponsors, mainly from the information and communication technology (ICT) industry. Since then, it has survived thanks to several research contracts, mainly, but not exclusively, with the local government.

RCM is strongly inspired by Internet values, such as being free and open. Its major goal is to introduce each component of the Milan community — citizens, associations, public institutions, and businesses — to a free-of-charge, effective, user-friendly electronic environment where everybody can personally try out the potential of electronic mail (both local and worldwide through an Internet address), conferencing, and chatting. In this way, RCM offers the various participants in everyday Milanese life, both as individuals and as groups, the chance to gain a hands-on experience (not just talking or listening about it) of the actual potential of networking applications. We particularly target citizens in their private and professional roles, as well as associations and schools, local government institutions, small- and medium-size enterprises, and, in general, everybody who is not yet online.

In other words, RCM's ambition is to integrate the principles and goals of community networks as outlined in Schuler (1994) with services provided by local government institutions and by business enterprises too. In that way RCM aims to extend the meaning of "community network" to encompass and support *all* the components of the town — the whole local community. We often use the picture in Figure 1 to illustrate this concept.

Another design principle, very clear in the minds of RCM's promoters from the very beginning, was the idea that computer professionals are nothing more than

Figure 1

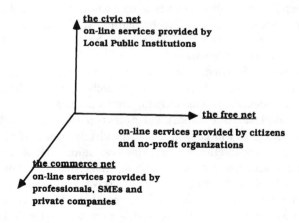

the civic net
on-line services provided by
Local Public Institutions

the free net
on-line services provided by citizens
and no-profit organizations

the commerce net
on-line services provided by
professionals, SMEs and
private companies

enablers of the social actors, responsible for providing them with a reliable environment. The staff focused on "keeping the network running," both at the technological level (managing and updating the computers and the network) and at the "social" level, guaranteeing adherence to specific netiquette rules, which we call "Galateo" and require that each subscriber accept in order to become a member of RCM. Moreover, the staff is engaged in involving more and more subjects: citizens, schools and associations, the local institutions and local businesses, helping them develop their own content and projects. Only in few cases have RCM administrators taken the initiative of suggesting areas of discussion or services. Usually, they listen to citizens' proposals for opening a new forum, note overlaps and similarities, suggest the use of aliases to solve them, and rearrange areas together to facilitate navigation. To stress this concept, we adopted the slogan *la rete siete voi* ("you are the network") to inspire everyone who logs onto the network to be an active member and promoter. And indeed, RCM has mainly grown from the bottom up.

RCM architecture consists of two tightly coupled servers:

- The local community server runs the FirstClass Intranet Server, by SoftArc Inc. (1998), which is accessible free, via the FirstClass client, through 32 telecom lines. Users who have an Internet account connect to it through the Net (reducing the need for more and more telecom entries) using either the FirstClass client or a standard Web browser.
- The associated Web server (http://www.retecivica.milano.it) enriches the FirstClass Intranet Server with programmable facilities, e.g., for handling statistics, for developing specific applications, etc.

Five years on, the state-of-the-art at RCM can be presented through the achievements in each of the three above dimensions. In fact, we believe that the community network's (CN's) development and sustainability is a *process* that aims at including and balancing such dimensions.

The Development of the RCM Free Net

- About 7,000 registered members present themselves to the virtual community through a (short) "résumé" that each person is required to prepare when s/he subscribes. Every member connects using a unique User ID and a password, but in messages sent is identified by her/his "real" full name, which also constitutes the left-hand side of her/his Internet address (firstname_lastname@rcm.inet.it).
- Some 80 nonprofit associations have their own areas in RCM, including hobbyists (like motorcyclists, kite lovers, archers, etc.), volunteers (SOS, Caritas, etc.), thematic groups (including the local delegations of all the major political parties and trade unions), and professional associations. These areas are directly supplied with content and independently managed by a member of the association who informs citizens of the association's initiatives and carries on on-line discussion with those who are interested. In a few cases, the association also has a private conference that serves as a kind of intranet among association members. Increasingly often, these areas, developed and managed inside RCM, are made available through a homepage on the Web. A similar pattern holds for some 60 schools connected to and through RCM.

- About 400 moderated public forums are promoted and managed by the members of the on-line community. These cover a large variety of interests.
- Some 300 volunteer moderators are in charge of approving the messages sent to the forums and animating discussion. They are not censors, but are asked to check that the Galateo is not violated (pertinent messages, no insults, no swearwords, and so forth) to preserve RCM as a friendly, virtual place where everybody, including children, can feel at home. (This is not a restriction of freedom: to express ideas without obeying these rules of fairness, one can use the Internet newsgroups available through RCM.)
- Since 1997, in order to solve the conflicts that naturally can and do arise (typically concerning the approval of messages) and, more generally, to support the staff in the decisions about network management (for instance, improvements to the Galateo, reorganizing areas to keep navigation in RCM easy, and the like), three moderators are elected "Coordinators" through an on-line election every year. They are called "ASSI" (Intermediate Structure among Members and Staff).
- Traffic can be estimated as follows: 50,000 daily hits on the FCIS server (through Web browsers and FC clients) plus 15,000 daily hits on the Web server.

The Development of the RCM Civic Net

When RCM was started in September 1994, none of the local public institutions were involved. That is, RCM was born as a "pure" free net. But the staff immediately worked to involve city hall and the other local public institutions. The task was not easy because most of them were seriously late in the computerization process. Now, RCM cooperates with city hall, including the public libraries, the province, the regional authority, the chamber of commerce, several schools, and some health services. All of these provide citizens with on-line services that can be grouped as follows:

- Distribution of information useful to the public includes a daily listing of events of all kinds (theater and culture in general, politics, sports, children's events, etc.) that take place in and near Milan.
- Interactive services enable citizens to make reservations to visit the old city hall building, allow students to preregister in the city schools, etc.
- "Direct lines" are public forums in which citizens can ask whatever question and get a public answer, either from public officials or administrators (e.g., Milan's Deputy Mayor). Note that, in this case, questions and answers are public, so all citizens can read previous exchanges, learn from them, and ask something more specific. This is not an old service provided in a new way; it is a radically new service where the new opportunities offered by telematics are used to advantage.

It is also worth noting that these "direct lines" make it possible to bypass delays in computerization. For instance, the Milan public libraries still do not have their catalog on line, but if one needs to know if and where a certain book is available, s/he posts a question through RCM and the librarian searches the library's databases (through queries and manual searches) and then forwards the results, again through RCM. It was recently observed in Pacifici et al. (1999) that preserving this kind of personal interaction between citizens and public officials will enable them to find solutions for less trivial questions and problems. Within these exchanges, new services can be envisaged and designed in a participatory way. We discussed the advantages of this participatory approach to the design of on-line public services in Casapulla et al. (1998).

The Development of the RCM Commerce Net

In this dimension, our experiences are few but significant. The most successful are forums in which groups of professionals provide consultancy to other citizens: there is the lawyer-on-line, the architect-on-line, the tax-expert-on-line, the trade-unionist-on-line, etc.

Moreover there are areas managed by small private businesses according to the same pattern followed by nonprofit associations. And there are larger service companies (e.g., a local electric power company and the Milan trash-collection company), which have their own Web sites but manage an area within RCM to increase direct dialog with citizens.

Because this topic is beyond the scope of this paper, we omit further discussion of the successes, failures, and prospects of this dimension, which were presented in De Cindio (1999).

The Role of the University

As a final remark on the development of RCM, it is worth stressing the role played, through the years, by the Civic Informatics Laboratory (LIC) at the University of Milan. We believe that the strict link between RCM and the LIC is one of the key factors in RCM's success for two basic reasons.

First, the community network is managed by a public but nonpolitical institution. In the context described in the introduction, this has meant being perceived by citizens as impartial, free, and open to their opinions and proposals. Even more, some of them consider the CN a kind of watchdog authority that can defend them from the abuses of public and private organizations. Moreover, the university "umbrella" makes RCM independent from any specific local institution.

Second, the university context continuously supplies competent students (primarily, but not only, from the computer-science school) willing to develop their theses around RCM, because the LIC gives them the opportunity to develop their professional (technological) skills while acquiring familiarity with the public and social issues, a mix that is increasingly required by the market but seldom provided by Italian academia. On the other hand, student activity provides RCM with a constant willingness to innovate and with the enthusiasm of youth.

From RCM to A.I.Re.C.

In designing RCM, we were careful to make choices that would guarantee the replicability of the initiative in smaller towns. This goal had obvious consequences at the implementation level. In fact, although the Department of Computer Science of the State University of Milan is an environment where the ability to manage the most advanced ICT exists, we decided to pay the price of adopting proprietary software and running the community network on BBS technology, namely FirstClass (1998). At the same time, we implemented the largest possible integration with standard Internet environments, typically with the Web. This is not the place to explain and discuss this choice (which we have done elsewhere, see, e.g., Sonnante et al. (1998)), but it must be stressed that it is partly thanks to this choice that community networks have then been launched in several small and medium-size towns in Lombardy, the region surrounding Milan.

Figure 2 shows an overview of the community networks that have somehow drawn inspiration from the RCM experience. Nevertheless, these nets vary greatly because of several factors, including:

- *The Promoters:* While RCM started inside the university, in most of the other cases the initiative came from a local government, city hall (in the case of Desenzano, Treviglio, Mantua, and Biella), the province (Sondrio), or local-government consortium (Cremona and Polirone). There are cases in which the community network was founded by citizens themselves (Bergamo, Varese, San Donato, and Novara), but only the last two have been successful. Finally, in one case (Como), the promoter is a small ISP, which indeed acts as such to support the willingness of a group of people to launch the local CN.
- *The Technology:* Some of these CNs use a BBS (FirstClass or WorldGroup) as community server, more and more tightly coupled with a Web server to make (part of) the content available via standard Web browsers; others adopt a pure Web server (Cremona and Mantua).
- *Local Community Identity:* of course, coupled with the nature of the promoters, influences CN evolution, from a more institutional approach (e.g., Sondrio, Cremona, and Mantua) to a more community-oriented approach (Milan), to a mix of the two (Desenzano and Treviglio).

Figure 2

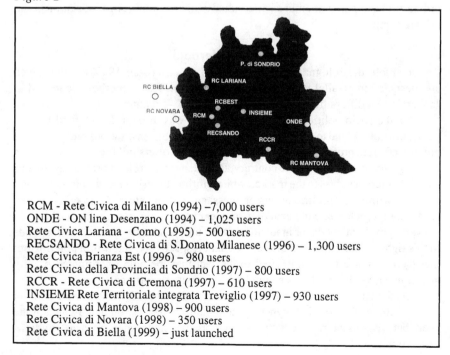

RCM - Rete Civica di Milano (1994) –7,000 users
ONDE - ON line Desenzano (1994) – 1,025 users
Rete Civica Lariana - Como (1995) – 500 users
RECSANDO - Rete Civica di S.Donato Milanese (1996) – 1,300 users
Rete Civica Brianza Est (1996) – 980 users
Rete Civica della Provincia di Sondrio (1997) – 800 users
RCCR - Rete Civica di Cremona (1997) – 610 users
INSIEME Rete Territoriale integrata Treviglio (1997) – 930 users
Rete Civica di Mantova (1998) – 900 users
Rete Civica di Novara (1998) – 350 users
Rete Civica di Biella (1999) – just launched

With two exceptions, all the community networks listed in Figure 2 are now members of A.I.Re.C., the Association of Informatics and Community Networking promoted by the Lombardy Regional Authority (AIREC). They can be considered successful experiences because they have been able to involve and connect hundreds of citizens, local government institutions, and some private companies. They have been recognized by the national government as promising initiatives for empowering local communities in the global information society.

EACN

Since 1996, RCM and its "sister" Italian CNs, have met with the promoters of similar initiatives elsewhere in Europe. In the summer of 1997, RCM, A.I.Re.C, and the Barcelona Community Network (BCNET) organized the First European Conference on Community Networking (ECN97) in Milan. From that experience, through ECN98, held in Barcelona in the summer of 1998 (ECN98), a European Association for Community Networking (EACN) came to light, promoted by activists from Belgium, France, Italy, Spain, and the United Kingdom.

It is worth pointing out that both in Milan and in Barcelona the biggest "local computer network" in Europe, i.e., De Digitale Stad (Amsterdam) (DDS) is also invited and present. While the DDS board of directors has decided to stay out of the EACN promoter group, DDS inspired RCM to set up a foundation to put its development on a sound basis. The RCM Foundation was indeed set up in December 1998, adopting the new form of "participatory foundation," presented in Bellezza and Florian (1998), i.e., a foundation that allows individual membership. The RCM Foundation includes as founding members, local public institutions (city administration, the province, the region, the chamber of commerce, and the university) and, as ordinary or supporter members, individual citizens, nonprofit associations, and private companies.

Bologna, Iperbole

Iperbole, the Bologna community network (Pacifici et al., 1999) was conceived and designed in parallel with, and independently of, RCM. Iperbole, launched in January 1995, differs from RCM in two — related — features.

On the socio-political level, the difference was in the role assigned both to public institutions and to the market. Believing that public institutions must make the greatest effort to promote the use of the Net, Iperbole offers full Internet access, free of charge, to everybody. On the contrary, RCM leaves the role of providing citizens with full Internet access to the market, while entitling them only to an Internet e-mail account (within certain limitations on message size and storage). In other words, RCM and Iperbole share the goal of guaranteeing the *right to electronic citizenship to everybody*, but they differ in identifying the universal service which corresponds to this right — full Internet access and e-mail, respectively. It is worth pointing out that the difference is not unrelated to the fact that Iperbole is promoted by city government, while RCM was begun by a university department.

On the technological level, Iperbole is therefore based on a standard Internet server, with the rich graphic, hypertext, and multimedia environment typical of the Web, but weak in interactive features and more difficult to manage, as is any first-generation site.

Iperbole and its evolution is fully described in Pacifici et al. (1999). Let us take a few numbers from there to picture its current level of development (March 1999) quantitatively: about 15,000 registered citizens; 108 schools; 810 nonprofit associations and similar institutions; 60,000 daily hits; and 35 local forums.

In more qualitative terms, the evolution of Iperbole has been influenced by changes that have occurred in the city administration and by the role of "model" attributed to it. For the sake of conciseness, we can summarize these changes as follows:

a) Because of city elections one year after Iperbole's launch, its political management changed. "Institutional" leadership by a member of the city government replaced the original "visionary" leadership of the founder — a professor of philosophy from the university, who had been given over to politics.

b) To meet the increasing costs required to provide full Internet access to all the citizens who asked, in 1998 the city administration decided to abandon the fully free approach and to ask everybody for a small, onetime fee (about $40).

c) Because Iperbole is increasingly considered and presented as a "jewel" of the left-governed town, in terms of innovation, when the national government shifts to the left, Iperbole becomes a test bed for the most relevant innovations taking place in Italian rules of government (the so-called "Bassanini laws" and, in particular, the digital signature project).

The overall consequence of these factors is that the focus of Iperbole is more and more on providing effective on-line services from city hall to citizens (i.e., according to a top-down model) rather than encouraging the citizens' own initiatives and community development according to a bottom-up approach. This is shown, so to speak, by at least two facts:

- Pacifici (1999) presents "Iperbole - la rete civica" as just one of several, more institutional projects, of Bologna Digital City.
- As seen in the "numbers" above, the 15,000 registered users have developed only 35 local discussion forums.

The Telecities Approach

Rewarded in 1997 by the European Union as the best project in the "Public Administration" section of the "Bangemann Challenge," Iperbole became a standard of reference for several towns in Europe.

While some of them, such as Tarragona, a town in Catalonia 100 kilometers south of Barcelona, "copied" Iperbole by opening "Tarragona InterNET" (TINET), which preserves the whole vision, most of the imitations simply developed a Web site presenting the city in a "shop-window" style (which is not difficult in Italy where almost everybody has some old building to exhibit) combined with a few services, including:

- supported access for citizens and/or special categories of public or nonprofit institutions, such as schools and libraries;
- interactive public services, including, e.g., certification, access to public acts, and payments; and

• the ability to send (private) e-mail to certain local ministries and/or public offices or officials.

Over time, the "telecities-like" approach increasingly loses its community orientation. This is, for instance, the picture that comes out of Mambrey (1998) regarding the German experience.

Such an approach can lead to "frozen" digital cities, but also to conflict, as in the case of Rome, the Italian capital and currently president of the Telecities Consortium. Here, the control over what is still called the "Rete Civica di Roma" is firmly in the hands of the assistant director of City Hall, who sees it as a way to exhibit the town to the world on the occasion of the 2000 Jubilee. As a result, a number of conflicts, concerning what people publish on the "Rete Civica di Roma," have arisen between city leadership and local network associations. In one case, the director decided to bar access to all nonprofit associations because of a (questionable) mistake by one of them. Despite the shared goal, declared by everybody, this prevented Rome from having an actual, active community network.

Democracy, Participation, and Communication

Almost all experiences of local computer networks list, among their aims, the goal of giving everyone access to the Net. RCM and Iperbole stressed, from the outset, that this is not only a matter of physical access. Their ultimate goal was indeed *to reinvent citizenship and democracy*, exploiting the opportunities that the net — seen as a tool, a new domain of possibilities, and a new world — offers.

Here, we first discuss the mutual interplay among democracy, participation, and communication and then present the "vision" that ICT, by supporting, enhancing, and extending communication among human beings, can empower democracy. We then consider one more issue that contributes to setting up a democratic context and discussing the experience presented previously in terms of these goals. It is worthwhile to point out that the aim of this paper is not a comprehensive discussion of democracy issues at this end of millennium, even though the author is well aware of the political debate surrounding these issues, in particular in Italy and Europe. Its more modest goal is to contribute to this debate with considerations firmly based on the concrete experience gained in these years, rather than on ideological opinions weakly rooted in the network reality, as is often the case.

The Interplay Between Democracy, Participation, and Communication

It is obvious that forms of democracy have been evolving, through the centuries, in close connection to the forms of communication available at each historical turn. Communication, after all, is the means that leads to concrete opportunities for citizens to participate in the *res publica*, and it is such participation that underlies the idea of democracy itself. In Athens, communication took place through meetings and discussions in the *agora*. Even as late as the end of the seventeenth century, communication technologies had not substantially improved.

As a result, in order to support increasingly large cities and states, the political institutions, like all other social organizations, adopted a hierarchical structure as the basis for handling complexity. Representative democracy is therefore based on delegated authority and guarantees efficiency by creating channels for communication from the top down and vice versa, as representatives are elected to make decisions and transmit the will of the people. The structure of political parties basically reproduces the state structure, from the smallest units rooted in the society, upward to local "federations," to boards (at city and regional levels), and on to national boards.

Therefore a modern democratic system essentially allows citizens to choose from a set of options drawn up by others (whom we may term professional policy makers). Voting in an election is essentially like putting an "x" on a predefined set of options: names of candidates, of coalitions, and the like.

However, at any time of rapid social change, the principle of delegated authority lapses into crisis. Everybody wants to take part directly, to join in the discussion of plans and projects, and to have a say in decisions. We can see a continuous push and pull between representative democracy and at-large democracy, from the French Revolution onwards. In many constitutions, the institution of referenda was included to meet this demand. But the Italian experience has clearly shown that this institution, which had a progressive role in innovating Italian society in the seventies, can be abused, and without adequate support from the technology, is soon rejected by the citizens themselves. Even today's most widespread communication technologies — the telephone, fax, and television — are unable to guarantee mass participation for any sustained interval. As noted above, during the Clean Hands investigation we saw the rise of the 'fax people' who made themselves heard. But fax technology (from the periphery to a center, which, in this case, was the newspapers that could then report on the faxes received) does not support horizontal communication and the coordination of individual voices.

Even in "normal" times, the limits of this level of participation are increasingly evident. Again without any attempt at completeness, let us mention that delegation is one basis for bureaucracy. Over time, delegation implies a decreased participation in elections. Polls are then used to 'take the pulse' of the electorate. But polls, too, go no further than recording reaction to questions made up by other people. Even participation in political parties and movements, which allows citizens to play a more active role in determining political choices, is becoming increasingly burdensome in the hectic lifestyle of today. When they indeed are willing to contribute, people prefer a more limited, but tangible, activity within volunteer organizations.

The "Vision"

After years of direct experience on the Internet, typically gained through professional positions in the research and university environment, many people envisaged ICT and the Net leading to a leap in the means of communication that also invites radical change in the very idea of participation and democracy. Let us summarize the roots of this vision:

- The Net is, by its very nature, an *open* environment, while society is partitioned by membership in different groups and, thus, a typical closed environment.

- The Net grows in accordance with a *bottom-up* process: there is no a top-down plan designed and driven by someone. Discussion lists and newsgroups are put together through the volunteer efforts of people who share a particular interest.
- In a society broadly dominated by mass media that simply broadcast to a passive audience (particularly, television networks), network communication, as an intrinsically interactive medium, pushes for *active involvement*.
- Computer-mediated communication also increases cooperation, open exchange, knowledge sharing, and frank dialogue. Even in a competition-based context, it opens up middle ground to share resources, leaving the following selection to market rules.
- In a culture governed by images, electronic mail (on which all newsgroups and mailing lists rely) restores importance to the written word, in contrast with the many "shop windows" opened on the Web.
- In a world still mainly ruled by hierarchical organizations — quite in bad shape, to be honest: i.e., corporate holding companies, political parties, centralized governments — telematics makes room for flatter organizations and their network-oriented structure, in part thanks to a technology that enables effective control of communication overload.

Let us delve further into this last issue through the words of Terry Winograd (1981) who wrote:

"In the decentralization of work and the distribution of expertise, we see a movement toward reducing our dependence on centralized structures and expanding the importance of individual "nodes" in a network of individuals and small groups. The use of computer communication in coordination makes it possible for a large heterogeneous organization to function effectively without a rigid structure of upward and downward communication. This shift does more than reorganize the workplace. It puts forth a challenge to the very idea of hierarchical organization that pervades our society. It may open a new space of possibilities for the kinds of decentralized communal social structures that have been put forward as solutions to many of our global problems."

The Net, or, to put it better, networking technology, affords such opportunities for two fundamental reasons:

1. As Winograd emphasizes, ICT makes even communication not based on a strictly hierarchical system efficient, and also effective, sustainable in time, etc. Proof of this can be seen in the paradigm shift from the company based on a Ford model to a post-Ford structure that makes liberal use of groupware technologies. That shift can be exported to other environments, typically from corporations to social organizations.
2. Second, 'virtual' activity frees individuals from *time* and *space* constraints (and, to some extent, from economic constraints, too) that belong to the 'physical' world. This interplay between the 'virtual' and the 'physical' (or the 'actual' as some believe it better to say) extends human opportunities as it has been shown by changes within the research community, the first community

significantly exposed to this innovation.

Therefore, the "vision" held that, at least in principle, the Net *may* open the society and increase participation and democracy. Let us try to be clear. The "vision" is not that ICT "erases" the need for hierarchical structures to govern complex organizations. But it reduces the need, thus supporting the "flattening" of organizations and the simplification of procedures. It increases accountability, leaving open the possibility of short circuits between nonadjacent levels of the hierarchy, improving the chance for distributed control over delegated functions.

In the socio-political context, this is not a claim in favor of replacing representative democracy with direct democracy. We are claiming that a lot of people saw the opportunity for ICT to make possible a kind of *participatory democracy*.

One More Issue of Democracy

Although this chapter makes no claim of completeness, it is not possible to sidestep one further issue. Democracy implies that the fundamental rights of the individual be guaranteed. But this not only requires identifying the various the rights to be defended. Because conflicts among different rights arise, it is necessary to fix priorities and achieve a sound balance among them. According to Alexis de Toqueville (1835), press freedom is the utmost guarantee of democracy, since it allows for democratic control. Unfortunately, in the Western world, press freedom and, more generally, freedom in the media system are by no means guaranteed. This is particularly true in Italy, where the press is almost entirely owned by economic powerhouses and the owner of the largest media company is also the leader of one political coalition.

The Net, which gives everyone the chance to publish worldwide without any filter, appears to us to be the only chance to overcome this undemocratic situation.

Implementing the "Vision"

The experiences of "local computer networks" presented earlier contain a variety of examples concerning the above vision. In the following remarks we abstract some general considerations from them.

As one can imagine, there are examples that validate the vision, but also difficulties and problems. On the one hand, when allowed to fill the Net with their own content and projects, citizens use the new opportunity in a deep, creative, and innovative way. They exploit their "virtual" activities to enrich their "real" life and to overcome the constraints that the physical world often brings with it. The vast majority (there are, of course, exceptions, but these are confined to a very limited number of people) shows a willingness to build the virtual world as a cooperative, open, and peaceful shared space. They also manifest a clear interest in participating in the political life. In RCM the forum dedicated to political debate, *Polis*, has had — steadily over the years— some of the highest traffic on the Net.

However, they have to face the contradictory behavior of politicians—the "democracy" professionals and experts. Even those of the new generation that has come upon the scene in the Second Republic, despite the number of words they spend to appear innovative, modern, and "online," are still far from being aware of the modifications and opportunities the Net may offer.

A question therefore arises: are politicians simply ignorant or are they apparently in favor of the use of the ICT but, in reality, afraid of the accountability that ICT introduces? Probably the truth is somewhere between these two extremes. On the one hand, they are unprepared to sustain the peer-to-peer discussions the Net imposes, and, on the other, they are afraid of losing their power, which can be questioned in a peer-to-peer (public) discussion.

In any case there is a need to connect more and more citizens and to make them the actual engine of democracy. Furthermore, the "Net engine" does not consist of individuals but of communities of individuals. The "fax people," through the Net, become a community and gain strength and power, much as the "programming ants" of the freeware/open-source community, all together, are able to develop software (as the Apache Web server) that becomes more popular than competing Microsoft platforms.

Connecting more and more citizens is therefore a prerequisite for achieving this goal. It requires a huge labor of education. In Western countries (and we apologize for omitting the "rest of the world" here) the problem is less and less a matter of infrastructures. In Milan, since the spring of 1999, private entrepreneurs have been providing full Internet accounts free to everyone. But having an account is by no means sufficient.

We believe that community networks can play a fundamental role as 'learning communities' to introduce the use of the Net not only to less-advantaged people but also to the vast majority of ordinary people. The key factor is the "sense of community" that gives rise to a kind of mutual support, to chains of self-help. Each person can act as "Net-mentor" to others, in the real and in the virtual worlds. From these "Net-mentors," people learn the functions of networking technologies (e-mail, newsgroups, mailing lists, the Web, search engines, now the portal, etc.) and, at the same time, what is surely more important and difficult: how to use them for their own purposes. In this way, people understand the Net as an active technology and as a cooperative framework, in contrast to the dominant mass media that simply broadcast to a passive audience. Such communities are the appropriate context for new relationships between citizens and public administrators, elected representatives and electors, and, why not, customers and salesmen.

Design Principles

If it is true that the citizens' communities are the actual engine that can exploit the Net to return full citizenship rights to people, offsetting the power of the many "empires" based on economic, institutional, media, and political power, we must bear very clearly in mind that a pure city Web site or a digital city that does not make any relevant effort to promote citizen participation cannot contribute to positive innovation. A city Web site with no real participation is like a city with no inhabitants. As Jean-Jacques Rousseau put it in *Du Contract Social*: "Les maisons font la ville, mais les citoyens font la cité." That might be roughly rendered as: "Buildings may create an urban environment, but only citizens can make it civilized." Phrased in Latin terms, *civitas* is rooted in the *urbe* but it means much more than that. A city hall that provides information and/or on-line services can be a good starting point, if it evolves toward a platform for empowering the whole local community. Unfortunately, a lot of local computer networks are currently nothing more than frozen virtual towns.

The question is therefore how to exploit the best practices and principles of community networking to make them alive.

To this purpose we need to translate the "vision" into design principles and guidelines, both at the methodological and technological levels. Below we sketch four design principles abstracted from and shared by the A.I.Re.C. community networks.

(i) The community is a resource for itself, and the community network is a resource for the community. Together they sustain, enhance, and support the redesign of the *res publica* and the renewing of civic identity, through the involvement of the entire local community in a new kind of citizenship and democracy.

(ii) The community network is a *virtual town* interwoven with the *real* town, that supports it for communication and cooperation, extends the community into new spaces for discussion, strengthens local community identity, and opens the community to the outside world and to the future.

(iii) The community network is a free-of-charge and easy-to-use environment, based on existing infrastructures for hands-on and facilitated education toward the Net. Moreover, it is a "learning community" that supports citizens in an incremental understanding of the Net.

The principles outlined above are what distinguishes community networks from the large majority of local computer networks that take on various names: digital cities, 'telecities,' civic networks, city nets, and city Web sites. All these initiatives also claim that they are interested in improving relations with the citizenry. But, in fact, they contrast with community networks in the *role* they assign to citizens. This defines a fourth principle.

(iv) Citizens need user-friendly interfaces to use ICT applications but they cannot be conceived of simply as *users*, as computer professionals tend to do. They need good services, provided by public and private institutions, but they cannot be conceived of simply as *customers* for such on-line services, as often found in public-sector striving to provide efficient services by applying business principles. People are *sovereign citizens* with the right to participate in the transformation of the very idea of the city and society that globalization brings with it.

Implementation Issues

Putting the above principles (or requirements) into practice requires the ability to deal with both the socio-political issues and the technological issues. In fact, as pointed out in Stolterman (1998), "Even very small technological changes can radically change the space of possible actions for the users." Therefore, we must pay very close attention to the methods and technologies developed in the ICT environment.

Socio-Political Issues

Let us discuss a very concrete point that well demonstrates the kind of concrete decisions one must take when managing a community network.

The notion of *citizenship* traditionally has geographic implications. I'm citizen of Milan, Italy. Therefore I vote for the representative of Northwest Italy in the European Parliament. And I vote for the Milan representative in the Lombardy Council. And so on. As citizen of Milan, I am entitled to ask the Milan mayor and his council (and analogously at the other levels) about the way in which they govern the town. What happens if I move to Paris to work for one year? What about students who live in Milan for five or more years but are still legally resident in their hometowns, in the South of Italy or elsewhere? RCM enabled my friend Alessandra, who has been working in Grenoble for four years, to continue to feel rooted in the Milanese community. Not only can she read Italian newspapers on the Web, she can also discuss the candidates in an election and, should they be willing, she could discuss issues directly with the candidates.

Therefore, a question arises: what notion of citizenship is adequate for a world in which the Net erases the geographic borders? This question is not a question for the future. We have to answer it right now.

RCM admits as members everyone who asks. With roots in a university department, it would not have been acceptable to reject students not resident in Milan! Because we offer only free e-mail, this is not a problem. On the contrary, Iperbole, managed by city government, is conceived of as a service for those who are Bolognese citizens: they pay the taxes and they get the (full Internet access) service. Nonresidents (students, workers, etc.) are excluded.

Similar questions concerning if and how to import the rules that govern real life into the virtual town often arise. The CNs directly managed by city governments tend to be much more conservative. RCM, born in a university framework and now managed by an independent participatory foundation, has been free to envision the net as a domain for carrying out innovation and experimentation. We always make reference to real life in managing new situations, but instead of reproducing the real into the virtual, we try to use the virtual to extend, enhance, and improve the domain of possibility.

ICT Issues

At the methodological level, *participatory design* (Schuler and Namioka, 1993) is the natural approach. This means that ICT must be designed not only *for* those who will ultimately use it, not only *with* them, but also *by* them (Briefs et al., 1983). In Henderson, Kuhn and Muller (1998), there are several examples, including the RCM experience (Casapulla et al., 1998).

At the technological level, after the first generation of CNs running on BBS technology, the second generation has been based on the Web, but this encourages information broadcasting and predefined interactive services, rather than active and lively participation. While the former effectively supports *discussions*, the second broadcasts *information* through the rich graphic, hypertext, and multimedia environment typical of the Web. Moreover, good *interactive services* require groupware platforms.

It is therefore necessary to realize that the technology is not yet mature for fully supporting communities. A meaningful discussion (e.g., of a political issue) should be based on a precise awareness of the issue at hand (e.g., a proposed new law). This is to say that information qualifies the discussion, but the reverse is also true: the discussion enriches, qualifies, and certifies information and services.

The need for a third-generation environment that integrates the three functionalities above, on the server side and on the client side, is apparent. The prototype developed in Aletti (1999) is an effort in this direction, to be compared and complemented with several other proposals (see, among others, Bentley et al. [1997] and Navarro et al. [1997], and those presented and quoted in Cisler [1998]). We believe that building the skills accumulated in creating and managing community networks into a piece of software can yield something of interest in running any network community, including business-oriented communities (see, e.g., Hagel and Armstrong, 1997).

(Excessively Optimistic?) Conclusions

In this paper, we have argued that the early experiences of the community networks launched in Italy and Europe since 1994 support the vision that the Net may open the possibility to *reinvent citizenship and democracy*. Citizens and, more importantly, citizen communities, seem to be the fundamental engine of this process. Enabling them to connect and become active is not at all straightforward: (on-line) participation must be nurtured and designed. Previously we suggested four design principles in this direction. In our opinion, these principles are what distinguish community networks from the vast majority of local computer networks that take on various names: digital cities, telecities, city nets, city Web sites, and often look like cities with no inhabitants.

However, citizen participation in the design and development of the community networks has further positive consequences. Vise President Al Gore suggested (see, e.g., Gore, 1995) — and we have actually experimented (see Casapulla et al., 1998) — that citizens may contribute to *reinventing government*. And people's active involvement in lively communities is increasingly considered crucial for successful on-line business, in North America (Hagel and Amstrong, 1997) as well as in Italy (Micelli, 1997).

Since this covers the major dimensions of a society, we might say, by way of a slogan, that community networks may indeed constitute the basis for reinventing — in the era of the *globalization* — citizenship, government, and business, too, starting from the local level, i.e. from cities.

This is indeed a challenge for Italy, and for Europe, as well, — envisaged, e.g., in Schiavone (1998) — because the cities have, in the past, been the basis for social life, social relationships, and business. It is useful in this regard to consider the idea historians use in characterizing the Italian maritime republics as *global cities*. Venice had a strong identity of its own but was at the same time a port (should we say: a "portal"?) open to the world — at that time, the Mediterranean. The challenge facing cities at the turn of the twenty-first century is to exploit community identity to build better communities in the information society.

References

Aletti, M. (1999). "Una piattaforma Web-based per reti civiche di terza generazione." Master thesis in Information Science. University of Milano, A.A. 98-99, June (in Italian).

Airec. (1999).Workshop on "Community Networks for the Local Economy Development." Milano, February. Available at: http://www.airec.net section "Eventi."

Briefs, U., Ciborra, C. & Schneider, L. (Eds.) (1983). *Systems Design For, With and By the Users*. North-Holland.

Bellezza, E. & Florian, A. (1983). *Le Fondazioni del Terzo Millennio*. F. Angeli, 1998 (in Italian).

Bentley, R., Horstmann, T. & Trevor, J. (1997). The World Wide Web as Enabling Technology for CSCW: The Case of BSCW. In Computer Supported Cooperative Work. *The Journal of Collaborative Computing*, 6(2-3) , 111-134.

Casapulla, G., De Cindio F., Gentile O. & Sonnante L. (1998). A Citizen-driven Civic Network as Stimulating Context for Designing On-line Public Services, in Henderson R.C., Kuhn S., Muller M., (Eds.), *Proc. of PDC98, the 5th Biannal Participatory Design Conference: "Broadening Participation."* Seattle, WA, USA, November.

Cisler, S. (ed.) (1998). *Technology Issue, Community Networking*, quarterly of Assoc. For Community Networking, December.

Colombo, G. (1996). *Il vizio della memoria*, Feltrinelli. (in italian).

De Cindio, F. (1999). Community Networks: a learning community for networking and groupware, in *SIGGROUP Bulletin*, 20(2).

de Toqueville, A. (1835). *La Démocratie en Amerique*.

FirstClass Intranet Server. (1998). *Administrator's Reference Manual*. Softarc Inc.

Gore, A. (1995). Common Sense Government: Works Better & Cost Less. *3rd Report of the National Performance Review*, September.

Hagel, J. & Amstrong, G. A. (1997). *Net Gain Expanding the Market through Virtual Communities*. Harvard Business School Press.

Henderson, R.C., Kuhn S., & Muller M. (Eds.) (1998). *Proc. of PDC98, the 5th biennal Participatory Design Conference: "Broadening Participation."* Seattle, WA, USA, November.

Mambrey, P. (1998). Do Cultural Difference Matter? Local Community Networks in Germany. Presented at the workshop "Designing Across Border: The Community Design of Community Networks," CSCW98/PDC98 Workshops, Seattle, WA, USA, November, 1998.

Micelli, S. (1997). Comuniti virtuali di consumatori, in *Economia e Management*, (2).

Navarro, L., Rodriguez, G., Rozo, M.C., Serra, A., & Turri, J. (1997). Final Prototype Design Report, EPITELIO Tech. Rep. Available at http:// www.pangea.org/epitelio/tools/doc/D4_2.html.

Pacifici G., Pozzi P. & Rovinetti A. (1999). *Bologna Citta Digitale*. Franco Angeli (in Italian).

Schiavone, A. (1998). Italiani senza Italia. Einuaidi (in Italian).

Schuler, D. (1994) Community Networks: Building a New Participatory Medium. *Communications of the ACM*, 37(1).

Schuler, D. & Namioka, A. (Eds.) (1993). *Participatory Design: Principles and Practices*. Hillsdale, NJ: Lawrence Erlbaum Assoc.

Sonnante, L., Casapulla, G., De Cindio, F., Gentile, O. & Grew, P. (1998). Managing Community Networks Using FirstClass: the Milano Experience. In Cisler, S. (Ed.), *Technology Issue, Community Networking*, quarterly of Assoc. For Community Networking, December.

Stolterman, E. (1998). Technology Matters inVirtual Communities, presented at "Designing Across Borders: The Community Design of Community Networks." CSCW98 Workshops, Seattle, WA, USA, November 1998.

Winograd, T. (1981). "Computers and the New Communication," unpublished note, August 12.

Web Sites

[AIREC] Association for Informatics and Community Networking—Lombardia. http:// www.airec.net

[BCNET] Barcelona Community NETwork. http://www.bcnet.upc.es

[DDS] De Digitale Stad. http://www.dds.nl

[EACN] European Association for Community Networking. http://www.eacn.org

[ECN97] 1st European Community Networking Conference: "Put People First," Milano (I), July 1997. http://www.retecivica.milano.it/ecn97/

[ECN98] 2nd European Community Networking Conference: "Models for Digital Cities: New Roles for Community Networks," Barcelona (E), July 1998. http:// www.bcnet.upc.es/ecn98/

[IPERBOLE] Bologna Community Newtork. http://www.iperbole.bologna.it

[RCM] Milano Community Network. http://www.retecivica.milano.it

[TELECITIES] Telecities Network. http://www.edc.eu.int/telecities/

[TINET] Tarragona InterNET. http://www.fut.es

Acknowledgment

This chapter relies on five years of existence of the Rete Civica di Milano. Therefore it is impossible to thank everybody who deserves it. However, I would like to acknowledge the RCM administrators and staff, the RCM members and moderators—without them this would not have been possible. Particular thanks to Philip Grew who helped us in unfolding the typical twisting of the Italian style into an English text.

Chapter X

ICT and Local Governance: A View from the South

Susana Finquelievich
University of Buenos Aires, Argentina

*Argentina is slowly walking along the path of ICT uses for social and civic purposes. Local governments and community organisations are understanding the potential advantages of Community Informatics, and facing a myriad of prejudices and material obstacles to implement it. This chapter shows the first results of a three-year research on the subject of information technology, local governance, and community networks in the City of Buenos Aires. It deals with two intimately interrelated issues: a) **Local government's use of ICT for local management and communication with citizens: results and obstacles**. The government is opening slowly to the use of ICT to decentralise urban functions, increase the flow of horizontal institutional information, update urban management, inform the citizens, and increase public participation in urban affairs. However, prejudices, fear of technology, and above all a resilient institutional culture are still considerable obstacles for informatization. The paper surveys the technological changes implemented by the Government of Buenos Aires City and studies the social actors who were responsible for them, as well as the social processes that made them possible, as a necessary framework to understand the slow development of electronic community networks. b) **Emerging Electronic Community Networks**. From 1997 onwards, they have multiplied in various sectors: education, culture, community health and wellness, citizens' rights, participation in urban affairs. The chapter studies the local particularities of community, focusing on the differences between large and small community organisations, and their conceptions of time and space, linked to the use of on-line resources. It finishes by analysing the link between the local governmental context referring the use of ICT and the slow emergence of electronic community networks.*

Introduction

Citizens, politicians and researchers have been searching for answers to the defies and new social processes which emerge at the end of the millennium, characterised by three main trends: informatization, urbanisation and globalisation (Castells, 1998). In Latin American cities, these global processes have had many adverse effects: urban fragmentation, increasing unemployment, poverty, socio-spatial dualisation, severe cuts in social services, higher costs in urban infrastructures and services generated by their privatisation and difficulties of local governments to manage the increasingly complex cities and to satisfy the populations demands. One of the possible answers that local governments are beginning to implement is precisely the use of informatics to increase the efficacy of institutional management, to collect the demands of the population, and to obtain political consensus.

On the other hand, there are social movements, community organisations, non-governmental organisations (NGOs), citizens, briefly, a set of social agents who constitute civil society, which are implementing alternatives to create alternative action spaces, and searching for solutions to the local problems triggered by global processes. These social agents are also beginning to use informatics to create local and international networks, get strength through the dissemination of their actions, access to international funding sources and exert pressures over national and local governments.

These processes do not evolve without severe difficulties. Organisational inertia, resistance to the use of information and communication technologies (ICTs), scarce access by the population to ICTs, inadequate policies carried on by telecommunication enterprises, translated into high Internet and telephone costs, contribute to make slow and painful progress along this path. However, some of the existing social agents assume new roles: progressive sectors of local governments and the more active members of the largest community organisations find a common ground in claiming a broader citizens' participation using electronic communication.

A set of questions emerges when working on the subject of Community Informatics in Latin America: How are information and communication technologies transforming urban life quality in a developing country? How are technological decisions related to the local political context? How is information technology used in communication between urban managers and citizens? Which are the factors that facilitate or inhibit the implementation of ICTs in the cities? How do community networks emerge in this context? This chapter tries to answer these questions, by contributing to the understanding of the potential advantages of Community Informatics in Argentina, and the prejudices and material obstacles it faces. It shows the first results of a three-year research on the subject of information technology, local governance, and community networks, coordinated by the author.

This research, "New Paradigms in Citizens' Participation Through Information Technologies" (1997-1999), carried on at the Gino Germani Research Institute, University of Buenos Aires, was based mainly on field work. It included documental study, close direct observation of the work of civil servants and community organisations, interviews with national and international organisation consultants, and two series of in-depth interviews: one with 30 Buenos Aires Government civil servants, and the other with 30 leaders and members of electronic community networks. The case study is the City of Buenos Aires (13.5 million inhabitants, including Greater Buenos Aires), where its local government system has changed in 1995, from a mayor chosen directly by the Republic's president, to a citizen-elected mayor, with the consequent changes in attitudes towards the use of ICTs in local management and communication with the citizens. The chapter deals with three intimately interrelated issues:

- **Local government's use of ICT for local management and communication with citizens: results and obstacles.** In 1995, the Government of Buenos Aires City underwent a structural transformation, from a system in which the mayor was chosen directly by the Republic's president, to a citizen-elected mayor. That year is also a landmark, with respect to the local government's attitudes towards the use of ICTs in local management and communication with the citizens, changes which are related to the transformation of the political system. The current government is opening to the use of ICT to decentralise urban functions, increase the flow of horizontal institutional information, update urban management, inform the citizens, and increase public participation in urban affairs. However, prejudices, fear of technology, and above all a resilient institutional culture are still considerable obstacles for informatization. The paper surveys the technological changes implemented by the Government of Buenos Aires City and studies the social actors who were responsible for them, as well as the social processes that made them possible.
- **Emerging Electronic Community Networks.** Computer-sustained community networks are slowly emerging in Argentina. From 1997 onwards, they have multiplied in various sectors: education, culture, community health and wellness, citizens' rights, participation in urban affairs. The chapter studies the particularities of community informatics in this developing country, focusing on the differences between small and large community organisations, on their diverse use of time, space and on-line resources.
- **The relationship between the local governmental context referring the use of ICT and the slow emergence of electronic community networks.** This is a necessary framework to understand the slow development of electronic community networks, unassisted by the State. The fact that the local government has been reluctant for years both to the use of ICT and to citizens' participation in urban affairs, explains more than budgetary limitations, the reason why the use of on-line resources for community organisations has not been encouraged. These limitations are due not to technological drawbacks, but to political and cultural reasons.

Community Informatics: A View from the South

Although Community Informatics, as a subject included in a wider one, the relationships between technological innovation and urban society, has been studied as well as implemented in industrialised countries for more than a decade, it is still a novelty in South America. Since the early eighties, authors from diverse nationalities and disciplines have contributed significant works to the subject of Community Informatics. Some of the most known are the research works of Manuel Castells. In the first volume of his trilogy "The Information Age: Economy, Society and Culture" (1998), he tackles the subject of how local democracy is being enhanced through experiments in electronic citizen participation through which citizens debate public issues and make their feelings known to the city government. He also analyses, albeit briefly, the issues related to virtual communities. This particular subject seems to be extremely attractive: several authors have written about virtual communities from different point of views. Rheingold (1994, 1993) describes the relationships that individuals from different countries establish around common interests in the WELL community. Harasim (1993) believes that networlds will change almost every area of social life. Schuler (1996) studies those communities in Cyberspace, which are devoted to democratic problem solving and the renewing of the very concept of community. Mele (1997) deepens into the subject of how organised electronic networks can shift the position of community organisations in the local power structure. Wellman (1988, 1996) studies electronic communities as significant social networks. Serra (1998) faces the challenge of fighting social exclusion through the Civic Networks in Europe. These authors share some characteristics in approaching virtual communities. Most often social equity, democracy and solidarity is the discourse of the decade. Agren (1998), who emphasizes theoretical understanding of participation in virtual communities, states that "a kind of utopianism is not hard to recognise" among the former authors. Other scholars, such as Jones (1998), Wellman (1996) and Baym (1998), work on the social structures of virtual communities, or, as Jones states, "on the social construction of reality that exists online, which is (...) not constituted by the networks CMC users utilize, it is constituted in the networks" (1998, p. 5).

Research on cybersociety is still scarce in Argentina, although authors such as Piscitelli (1996, 1998) and Cafassi (1998) have explored the complexities of cyberculture. Others, such as Finquelievich, Vidal, and Karol (1992); Finquelievich, Karol and Kisilevsky (1996); and Finquelievich and Schiavo (1998) have worked on community informatics, approaching the subject from the process of informatization of local governments and community organisations. Work on virtual communities in Argentina is still in its embryonic state. However, this chapter tries to summarise the results of the research developed to the present, based on the research work developed by Herzer and Kisilevsky (1998), Kisilevsky (1998), Baumann (1999), and the author herself (1998, 1999), within the research team "City, Society and Cyberspace," at the University of Buenos Aires.

Passive Resistance: The Inertia of Organisational Culture

How do local governments incorporate informatics as a new tool? How do civil servants perceive these technologies? Do these perceptions change with governments from different political colours? We have surveyed some of the most significant organisational and technological changes which have been implemented in the management process of the Buenos Aires Legislative Body. In order to provide comprehensive background information regarding this topic to understand the political processes related to the implementation of information technology, it is necessary to travel briefly though the history of the recent political decisions.

The City of Buenos Aires acquired its legal state of Autonomous City with the reforms introduced in the National Constitution in 1994. The citizens, or "porteños" (people from the Port), as they call themselves, won the right to elect their own mayor—until then chosen by the president of the nation—and to implement their own legislation. On August 6, 1996, Dr. Fernando De La Rúa assumed his position of Elected City Government Chief. The very same year, the Constitution of the Autonomous City was sanctioned, and a new Legislative Body was elected in 1997, to replace the old City Council. The New Legislative Body changed some of the procedures in order to increase public participation in urban affairs. It introduced mechanisms of semi-direct democracy: public audiences, referendums, and popular initiatives. However, it postponed most of the decisions concerning the use of information and communication technologies (ICTs) to a near future, in which a new "intelligent" building would be equipped for the Body.

By 1998, the Legislative Body had acquired some informatics equipments that were placed in the building (actually the San Martin City Theatre), occupied transitorily while waiting for the new one to be finished. It included computer terminals, an internal information network and a server for the official Web page. However, no attempt was made to train the legislators and the civil servants in the use of informatics or to hear their needs about this issue. As a matter of fact, many legislators claimed that they had no information whatsoever about official plans to improve the technological conditions of the institution. Some of the Commissions in the Legislative Body benefited from the new equipment and from Internet connections, but most of the legislators, when interviewed, said that the informatics equipment they had in their offices was their own, and that they lacked Internet connection. The different capacities and qualities of the working computers, and the fact that many legislators' offices were not in the main building but in its neighbourhood, were strong physical and financial obstacles to incorporate them into the internal network. Besides, this network, planned as a technological support for the legislative work, was not introduced to the civil servants, which ignored all the dependencies connected to the Internet and how to use it. The presentation of new projects, the delivery of the daily bulletin about the work done by the different Commissions and the results of the sessions, and the documentation and information generated by this work, were still done according to the old ways: by hand, in the corresponding counters.

Many remnant members of the old City Council did not know how to use a computer, least of all how to connect to the Internet, and they did not believe that their work would be benefited by the use of ICTs. When interviewed, they stated their antitechnological prejudices: "Computers have no heart," or "Those who are skilled in computers and Internet are still too young to partake in public decisions." Those who usually used the computers—although simply as text processors—were the legislative advisors. All these facts show that the use of ICTs was quite limited. The restrictions were attributed to the precarious situation of the temporary building, and to the accepted fact that "changes come slowly." Hopes for progress, technological as well as organisational, were postponed until after moving to the new "intelligent building."

Public Participation and On-Line Activities

Different attitudes and behaviours towards on-line activities are related to diverse styles of municipal governance. Understanding the participation paradigm the legislators had is important to think in which ways ICTs can be used in community participation plans. Interviews with the legislators included a series of questions about this subject, which detected two apparently opposite attitudes:

a) The first paradigm was *restricted participation*. Legislators who adhered to it stated that voting was the only participatory action citizens should have. Eventually, they admitted that citizens' participation could go as far as expressing their opinions within the spaces created by the legislators themselves, such as a committee meeting or neighbours' visits to the premises of the political party where the legislator developed his political activities. Legislators who adhered to this scheme firmly believed that elaborating the city's norms and laws to regulate the city's management was solely and exclusively the function of the legislators. They were also the most antagonistic towards the use of ICTs, since their political practices are supported by clientelistic, face-to-face relationships, in which the legislator is the exclusive owner of all pertinent information, and held the power of management and resource control.

b) The second paradigm was *extended participation*. Legislators who supported it maintained that social organisations and neighbours' associations should have access to decision making in urban issues, and that spatial physical and social institutionalised participation spaces should be implemented for that objective. Such spaces were conceived as places in which the interests from the different social agents were discussed, negotiated and eventually conciliated. Within this scheme, the legislator became a mediator between the different interests of neighbours and the city, negotiating between the parts, and between the citizens and the executive. This scheme allowed small interstices for extended participation, even if citizens' demands reached the legislators though traditional means, such as phone calls, letters, and face-to-face meetings. Not surprisingly, legislators who supported extended participation were those that perceived ICTs as useful tools to disseminate information to the citizens, collect their suggestions and complaints, and to improve civic participation in the resolution of urban problems.

An analogy can be traced between participation schemes and the perception legislators have about their own public role. For legislators who supported limited participation, an opinion survey among citizens was "useful, so we won't take unilateral decisions, and won't make mistakes all by ourselves." Legislators who adhered to extended participation were in favour of creating participation spaces in order to conciliate different interests and allow community organisations and individual neighbours to intervene in the different phases of conception, decision and implementation of urban policies (Kisilevsky, 1998).

What has Changed Between Past and Present?

By mid-1998, the Legislative Body proposed to redesign the Legislative Palace. It was recycled into an "intelligent building," with 600 computer terminals with voice and data, Internet and intranet connections, laptops and e-mail to receive messages from neighbours, a cable TV channel, and direct transmission of the Legislative sessions through a giant screen. The goal was to adapt it to the exigencies of a modern Legislative Body; to design and implement management technologies, emphasising the informatics aspects. These changes were aimed at guaranteeing the values of transparency and efficacy, and to train the staff in updated management skills, including informatics.

Moreover, each legislator and each parliamentary commission would have their equipment connected to the network, as well as access to the Internet. The Legislative body would have its own Web site, and both the administrative and parliamentary circuits would be managed from the network. Also, the network would have public access. The project included the creation of a participatory citizens' network to allow a fluid interaction between citizens, community organisations, and the Legislative Body, based on "free access to all the information in the network." The first phase consisted of a pilot experience: connecting five to seven neighbourhood centres, conceived as physical spaces for computer terminals; training courses; and other related activities.

Some legislators were openly sceptical about the modernisation of their work habitat. One of them said, "I'll be completely frank. I don't know if this new stuff will be really useful. I can't even image it. What will happen if a neighbour or a community organisation calls me in the middle of a session? If they give me new information, a new complaint, will it be fair to change my position and my vote right then and there?" Another confessed, "We used to laugh about the informatization process. It seems it's going to be a reality after all, but to us, it is still like a dream. It was all so precarious in our temporary offices at the San Martin Theatre... we didn't even have a miserable computer. I had to buy my own computer, because the Legislative Body wouldn't provide it. This change to the intelligent building seems too wonderful to be true. I still don't believe it."

There is a gigantic gap between the financial resources employed in techno-logical upgrading and the scarce efforts made to inform the staff about it or to train them in the use of the new equipment. Is this technological modernisation project perceived as a priority issue in the Legislative Body's everyday life? If we take the place it occupies in the internal circulation of information as a parameter, it decidedly is not. The internal newsletters barely mention the subject. Many legislators learned the news about the "intelligent building" through the newspapers. Even if they are more open-minded than the members of the old City Council, when interviewed, they talk about their lack of skills to use computers and the fears that "computers require abilities that are well above our capacities." Other legislators expressed that they were "too old to acquire new, complex skills." Some legislators felt that the process of working with no technological equipment at all, to the use of Internet and intranet, is too fast: "It's like if you take a guy who lives in a shanty town, rip off his rags, and dress him in a tuxedo. He'll be bewildered and confused, and will have no idea about how to wear his new clothes."

When asked about their feelings towards the information that should be delivered to the general public, the scene became blurry. Most legislators declared that information plays a main role in the process of encouraging citizens' participation, and that fluid circulation of information is one of their top priorities. However, they did not discuss which type of information should be disseminated, or the appropriate channels that should be used to increase the efficacy of this dissemination. Nor did they debate the key issue of how to articulate massive communication means, such as newspapers, radio and TV, with other, more flexible levels of communication, such as Internet or local FM radios. Moreover, the legislators expressed doubts regarding the democratic potential of informatics. They argued that ICTs can reinforce social segregation and widen the gap between information-rich and information-poor. However, when the researchers who interviewed them suggested the implementation of public-access computer terminals in the City's Decentralisation Centres as a way to bridge the gap, the legislators showed little or no interest in such a solution. Instead, they chose to continue to have face-to-face meetings with the neighbours, as a continuation of the most traditional way to negotiate political issues (Herzer and Kisilevsky, 1998). One of the legislators stated that "direct over-the-counter contact solves problems that are not even tackled by phone, letters, least of all by flimsy, immaterial e-mails."

If present legislators have similar prejudices regarding the use of ICT as the old City Council members, what has changed between the past City's administration and the present one? In spite of the appearances, a deep transformation has started in late 1998. One of the most significant changes in institutional culture is that at present, the implementation of ICT in everyday work has become a top priority of the Legislative Body. The Grupo Gestor (Managing Group) was created with the purpose of leading a structural institutional transformation agreed to by all the political parties. Training campaigns were started in October 1998, in order to educate the staff in the use of informatics, particularly the Internet and intranet. The incorporation of ICTs is considered as the instrumental support of the Modernisation program. The new legislators, particularly the younger ones, are much more favourable regarding the use of informatics. However, they accept to use it without giving much thought to its potential, the real needs they have for information technology, how they could benefit from its further uses, where and how should it be implemented and which priorities should be defined (Herzer and Kisilevsky, 1998).

Two initiatives of the Buenos Aires Government regarding interaction with citizens via electronic means should be mentioned: the city's Web site, and the Legislative Body's Web site. The first one, (http://www.buenosaires.gov.ar), provides complete information about Buenos Aires' government. The site includes the Government's organization, and the names and addresses of the current civil servants (although not their e-mails, phone or fax numbers), the city's budget, information about tourist places and a site dedicated to the Centros de Gestión y Participación (CGPs, or Centres for Management and Participation). It also gives a complete description of the UNDP-managed Decentralisation Program. When compared to the previous information void, this Web site is encouraging. However, it is little more than an informative guide. It does not offer the possibility of interactive communication, except by a single electronic address to which citizens can send comments and proposals. Visitors cannot send e-mails to the Government's offices, nor use the Web page to make any consultation nor procedures.

The Legislative Body's web site, http://www.legislatura.gov.ar, goes further and offers genuine possibilities for interaction. It was implemented the very same day as the opening of the "intelligent building" that is at present the official Legislative Palace, on February 22, 1999. The site provides several links, from "Getting Acquainted with the Legislative Body" to "Internal Structure." The site provides information about each legislator, including his or her photo, résumé and e-mail address. Through "Live Transmission" it is possible to "assist" to the Legislative Body's sessions, while "The Legislative Network" offers opportunities for real interaction among civil servants and citizens: chats with the legislators, virtual forums, surveys, etc. Although it is still too early to evaluate the degree of success of these electronic services, they promise to become valuable channels for active citizens' participation.

This site is a considerable qualitative progress in the field of local government's use of informatics, not only because it offers detailed information about the legislators, but also because it provides spaces for citizens' expression and participation through the Internet. However, technologies by themselves do not cause changes if they are appropriated and adequately used by the potential users. The existence of the Legislative Body Web site is undoubtedly a progress in the path of facilitating access to information to the citizens and expands the possibilities of popular participation in urban matters. But the final result depends on the involvement assumed by the civil servants with these tasks, and this involvement requires constant training, but above all, a will to transmit transparent information and to encourage information flow in all directions. What is the use of a Web site if the available information is not displayed in it? Which are the real possibilities of popular participation if the really important, decisive information is kept hidden? One of the most serious obstacles met by the present modernisation process is the lack of engagement of many legislators and civil servants with these new communication and information channels.

Joys and Obstacles of Electronic Urban Decentralisation

The project of incorporating the use of ICTs in the Centres of Management and Participation (CGPs) work is one of the most important promises offered to local participation through electronic communication in Buenos Aires... and it is also the one that faces the greatest obstacles for its implementation. The process of decentralisation and deconcentration of urban management carried on by the Government of the City of Buenos Aires, through the creation of CGPs, is included in the Modernisation and Decentralisation Program: **"The purpose of technology is to allow the citizens to have a wider access and participation in the Government's decisions through the use of computer and telephone networks."** (Programa de Modernización y Descentralización, 1997). The Program is carried on by the **Subsecretariat of Decentralisation and Modernisation**, which depends directly on the City's Executive Power and is assisted by UNDP. Its goal is to implement ICT as a management tool, based on the thought that a multimedia digital network gives the government both the possibility to inform the citizens about their work and to provide a communication channel to encourage the participation of the City's inhabitants. According to UNDP, a geographically decentralised organisational structure responds faster to citizens' requirements than a centralised structure. Three key factors support decentralisation: a set of common values for the whole organisation, an efficient communication channel with the official core, and a clear strategy to achieve the established goals.

Based on interviews carried on from November 1997 to November 1998 (Baumann, 1999), with the Directors of five CGPs in northern neighbourhoods of Buenos Aires, as well as with the UNDP consultants responsible for the project, we have identified the following obstacles met by the decentralisation project:

a) *Delays in the budget assignation:* The project starts to be implemented in early 1996, but the budget was effectively assigned in April 1997, delaying the whole process.

b) *The underutilisation of available computer equipment:* The CGPs started to acquire computer equipment in early 1997, at an average of three computers per Centre, equipped with modems, laser printers and dial-up connections to the Internet. However, until October 1998, the computers were used merely as word processors, and only then some of the CGPs' Directors started using e-mail and the Internet. By then, an intranet system with a centralised service, the Claim Central Unity (CCU), was implemented. Through the CCU, the CGPs can transmit the neighbours' claims or complaints to the different city departments, to be solved. However, once the claim is sent, the CGPs cannot control its parcourse, nor follow the procedures for its response.

c) *Lack of trained staff:* CGPs have a reduced staff, mostly employees from the previous government, without any formal training in informatics or telecom-munications. The three UNDP consultants charged with the informatization process are too busy to assist 16 Centres. Besides designing and implementing the new networks, they are charged with their maintenance, as well as hardware and software repairs. No technical team has been foreseen to assume these tasks. In many cases, the CGPs' Directors ignore the potentials and multiple uses of the Program. This is explained by the absence of clear transmission, not only of the Program's goals and strategies, but also of the information about the informatization process.

d) *Lack of understanding, at various levels of the city's administration, of ICTs' potentials and possibilities as participation instruments, about the availability of infrastructure, and above all, about the need to adapt the institutional culture to their use:* Civil servants are preoccupied with how to distribute information to citizens and to other governmental levels. They usually use alternatives as bulletin boards, local media, newsletters, etc., but they do not think about information technology as a useful tool. Even if computers are available at CGPs, ICTs are seen as a remote instrument to be used in an indefinite future, not as a concrete, present possibility.

In synthesis, the three City government levels have implemented quite differ-ent strategies regarding ICTs uses, which are not necessarily complementary. The executive's Web site is conceived as an information place, or virtual bulletin board. The Legislative Body's Web site is designed as an interactive space, an open door for citizens' participation. Meanwhile, the CGPs Informatization Project seems to be reduced to using the intranet to transmit neighbours' claims. To the obstacles pointed earlier—which are common to the three Government levels—are added three questions which must be solved in order for the Government's initiatives to become effective tools for administrative decentralisation as well as for citizens' participation.

The first issue is *organisational:* it is necessary to promote deep transformations in the organisational culture of the public administration. It is not merely a technological change, directed to increase or change the availability of hardware and software. Neither is it an issue of training the staff in the use of information tools. It is not even a problem solved by hiring technologically trained specialised personnel. On the contrary, it is what in Organisational Sociology is called "orgware." Its goal is to incorporate genuine network logic in all the levels of the Administration. This requires not only changes in technical equipment, skills, and abilities (the "hard" organisational aspects) but also deep transformations in institutional habits, hierarchies, organisational mobility, information flows, and communication strategies. Above all, it needs to tap the maximum of existing human resources, which implies constant training and consultation about their needs and perceptions within the organisation. *Adaptation to network communication implies the reengineering of the "soft" aspects of public institutions, as well as the creation of a true citizen-centred culture of service, as opposed to the present institution-centred organisational culture.* These changes will not be effective if there is not a political will to change the procedures, rules, standards, and statutes that structure and paralyse the institution. It becomes necessary to include a new rationality, goal-oriented instead of rule-oriented (Finquelievich, Vidal, and Karol, 1992; Finquelievich, Karol, and Kisilevsky, 1996; Baumann, 1999).

Emerging Electronic Community Networks: From BBSs to Citizens Rights

In such a background, how did electronic community networks develop? Mostly, they developed randomly. The Internet was launched in Argentina in 1995. Although it was awaited with expectation, it is still an expensive tool for the local elites, although it is disseminating among the middle-income groups. It is estimated that less than 400.000 persons (0.7% of the population) have Internet connections in a country of 34 million inhabitants. High costs of Internet connections and telephone tariffs make Internet prohibitive for the majority, given the fact that 27.1% of the Argentine population earn less than $148 a month (ECLAC, 1999). However, since 1985 some groups were already creating virtual communities through BBSs. Later, universities implemented discussion forums though e-mail threads, but it was not until 1995 that the electronic community network started to play a social role. These emerging networks are for the time being the realm of the middle classes, which is explained because they possess both the financial capital to acquire computer equipment and the cultural capital to use them.

The on-line research of virtual communities related to citizens' participation, during July and August 1998, was deceiving in the first phase (Baumann, 1999).

Virtual communities as defined by Howard Rheingold (1994): *"... Social entities that emerge from the Web when a sufficient number of people carry on public discussions during enough time, and with enough human feelings to establish personal relationships in cyberspace,"* could not be identified at first. What were found were discussion forums and newsgroups, but none of them discussed urban or local policy issues. The search was thus focused on NGOs that worked on issues related to urban everyday life, such as human rights, environmental problems, or survival strategies for the impoverished middle class. Among them were found Conciencia (Conscience) and Poder Ciudadano (Citizens' Power)—both of them working on the defense of citizens rights, Greenpeace, the Asociación Voluntarios de Parque Centenario (Volunteers for Centenario Park), and the Paraguas Club (Umbrella Club, a network which links together unemployed middle-class professionals and micro-entrepreneurs). All of them displayed recent, modest and rather rudimentary Web sites, where the NGOs provide information about their goals, work and achievements. In all the sites there is e-mail address where the visitors can contact them, but only as a bidirectional communication. No efforts to establish a network are made.

The research was reformulated to focalise on NGOs as possible embryos of virtual communities. A sample of 12 NGOs were grouped around a series of variables: *organization* (goals, action areas, participation in NGOs forums, human resources), *economic structure* (funding, budget) and *use of technology* (use of ICTs, perception of community informatics, usual means of communication and information). **Two categories of NGOs were identified: information-rich and information-poor.** The first and smaller one (two organisations out of 12) has computers and Internet connections. They have e-mail, Web sites, and they constantly explore new ICT uses and consider information technology as indispensable for their work. These organisations have considerable high budgets (from $500,000 to $700,000 per year). They do not have a local territorial belonging, but work at national levels. They also have highly qualified paid staff (between 15 and 25 employees), besides a relatively high number of volunteer workers (around 40 persons), who use ICTs daily, particularly the Internet. Other characteristics are memberships to national and international networks of NGOs, the inclusion of community informatics as a permanent issue in their agenda, and the development of massive communication strategies. Last but not least, these organisations receive financial support from different national and international institutions and foundations, enterprises or individuals, and have efficient fundraising systems.

The NGOs within the second category (information-poor) do not have informatics equipment, or they have some computers, but they are underutilised. None of these NGOs has access to the Internet. They have strong geographic links with a given neighbourhood or urban area, and they have extremely low budgets, which are often informally managed. They have no paid staff, and they work with part-time volunteers, generally professionals in Law and Urbanism who do not use ICTs, but approve of its use. (The staff members explain that they do not use ICTs themselves, pointing to the insufficiency of financial resources and the lack of training in computer use). These organisations do not belong to NGOs, national or international, and do not include ICTs as a priority issue in their agendas. Their massive communication strategies are defined according to the moment's requirements, and they do not have sustained relationships with the local media. Occasionally they have spaces in neighbourhood newspapers or newsletters, or in FM radios, but they have serious financial limitations when they must pay for publicity. They are strongly dependent on external financial support, through discontinuous small funds from different State institutions, local enterprises, and/or individuals, and they have serious difficulties keeping autonomous funding systems, since they do not have permanent, self-reliant, and organised fundraising systems.

In short, *there is a direct correlation between the financial situation of a community organisation, its territorial scope, and the use of ICTs*. Information-rich community organisations are those which are also financially solvent, have national and/or international scopes, manage massive communication campaigns with the media and have highly specialised paid staff. On the contrary, information-poor community organisations are financially unprotected, have no self-reliant funding strategies, have strictly local roots and goals, and depend on volunteer work. However, these are not the only variables: the dualisation of community organisations concerning ITC use is closely linked to their relationships with space and time.

Time, Space, and ICTs

Our field work pointed to the fact that the difficulties experimented by smaller NGOs to incorporate ICT use are directly related to their geographical roots and scopes. The very essence and objectives of these organisations are rooted in specific spaces, neighbourhoods or limited urban areas within a neighbourhood, where they implement their actions (Baumann, 1999). Even their names denote the geographic and social areas where they work: "Friends of Palermo Lake," "Volunteers for Centenario Park," "Creative Neighbours of Saavedra and Núñez," etc. Small NGOs operate at the neighbourhood level, in face-to-face networks, which constitute their very identity. This is one of the causes that until recent times made their members think that they did not need ICTs, since they have direct contact with other members and with the population they address to. Their target population is the nearby neighbours, and their actions are focused in the defense and conservation of concrete public physical and social spaces. Most of their activities are centred on impeaching the transformation of local places through the global spatial logic, through monumental public works and/or private development projects (large closed condominiums, mega malls, etc.). However, some changes have been detected in the last year: these NGOs have realised that their strength resides in their capacity to integrate local urban networks. When they do, they add demands, get further training, are able to influence in decision-making levels and to impose a local logic to the civic society, thus contradicting the global logic of international enterprises, without losing their local power. Some of these networks have achieved remarkable success, becoming important social agents in the process of urban planning, through their participation in the City Government areas where the Buenos Aires 2000 Strategic Plan is being debated. These NGOs are admitting the advantages of ICT's use as valuable tools for their everyday work and the dissemination of their goals and achievements, but as mentioned above, they still have serious financial and training difficulties to implement an efficient use of community informatics.

On the contrary, the largest NGOs are not rooted in physical places (Baumannn, 1999). Their goals reach wider, more general areas, and are focused on environmental or ethical issues. Their names do not refer to any geographical area, but to social principles: Citizens Power, Greenpeace, Conscience, etc. Larger NGOs participate in wider national or international networks. This is a significant comparative advantage, since they have access to financial resources and strategic alliances, which strengthen their actions, and provide them information and training. Since their target population refers to the whole country, or to many other countries, the use of ICTs is fundamental to participate in these networks, mainly through e-mails, chats and Web pages. Their space is the globalised world.

The ways in which community organisations conceive, use and manage their relationships with time is also a significant variable when evaluating their relationships with ICT. If the space of Information Society is the space of flows (Castells, 1998), its *time* is atemporal. This is the emergent dominant form of social time in the network society. Atemporal time belongs to the space of flows, to the global space, while biological time, the sequences determined by each society, characterises physical local places. This becomes evident in our fieldwork. A member of a neighbours' organisation says: "In our organisation, we can't make promises. We can't even promise our free time. We can't plan what we'll do next week. All the time we have to offer activities that are attractive enough to get the neighbours to participate in them. We all invest our time, our efforts, in this organisation, but we have to work in our jobs to earn our livings. It often becomes very difficult. Our organisation works with socio-biological time rhythms. People come to work here, but there is always somebody who gets married, has a child, gets divorced, loses his or her job, has exams at the University... There are times in which no one comes, and we have to wait until our members have solved their personal problems to reassume community work."

The relationship NGOs establish with time is correspondent with the relationship they establish with space. Large NGOs have a number of specialised, paid permanent staff, trained in ICT uses, who use time in flexible ways, and are in constant contact with other organisations in real time, disseminating and receiving information around the world, and around the clock. ICTs allow them to participate in global or national campaigns, and to take actions in real time. They can plan the most effective strategies, since their networked actions affect power where it is: the space of flows. On the contrary, small NGOs are strongly rooted in local spaces and times, away from the space of flows. Their relative isolation, as well as their reduced budgets, keeps them far away from the space of flows. They seldom have access to funding possibilities and to acquire computers as well as skills in ICT use. Hence, they do not have a rapid answering capacity towards the advances of the space of flows over the local spaces they try to preserve. Their time use is more biological and social than the atemporal time managed by the enterprises.

These difficulties and limitations are related to their self-perception. An NGO volunteer sadly stated: "We are the falling middle class." Those who actually fight to preserve urban green areas and neighbourhoods' life quality from the unyielding development enterprises are those who do not want to give up the benefits of social welfare, those who believe in social redistribution of the national product. The deterioration of public spaces in Buenos Aires is a consequence of the privatisation of life quality in the space of flows. As we stated in a previous work (Finquelievich, Vidal, and Karol, 1992), those who live in the space of flows—ironically, limited to restricted urban areas, the wealthier neighbourhoods—have more urban services, cared-for green areas, recreation facilities, safety, because they have an income that allows them to afford these commodities. On the contrary, those who live in local space fear the State will abandon the increasing absence of public physical and social spaces. For these last social sectors, ICTs, particularly the Internet, may be especially effective, in order to strengthen the links between community organisations and the whole society, gather together NGOs that work on compatible issues, and thus exert the necessary pressure on the State when necessary.

The Relationships Between Local Governments, and Electronic Community Networks

The development of electronic community networks, in most countries, is linked to national and/or local public policies to encourage, make feasible, and help finance civic networks. This is related to an institutional culture which is centered around the citizen, goal-oriented, which values a large horizontal information flow, and which encourages citizens' participation in public issues. In Argentina, the panorama is completely different. How would a local government whose civil servants ignore most of the ICT potentials, who privilege vertical, hierarchical information systems, and who don't believe in full public participation, strive for the creation and dissemination of efficient electronic community networks?

At present, political leaders and civil servants alike have serious preventions towards ICT dissemination in their organisations. Politicians are accustomed to using an "information strainer," which filters the information that arrives to the public through traditional media. The most valuable information is kept hidden in restricted circuits, within different governmental levels, political parties, enterprises and unions. The holes in this strainer will become larger as citizens learn to claim their rights to be informed. An interesting example is the frustrated TIC (Transparency in Information to the Citizen) Web page that should have been implemented by an agreement between the City's Executive Power and Citizens Power Foundation. The page was designed to publicise all the large buyings done by different City Government's levels and Departments. After a year of insistence, Citizens Power could not get any information from directors or secretaries. The Web page never saw Internet lights.

This panorama changed in the light of October 1999 Presidential Elections. For the first time in Argentina's history, access to the Internet and informatics literacy campaigns became valuable stakes at the politics games. Both candidates from the two major political parties updated their political platforms, promising cheaper and easier access to the Internet, and measures to reduce the digital divide. Dr. De la Rúa, Mayor of Buenos Aires, and presidential candidate from the Radical party—he was elected President on October 24—spoke for the first time in favour of encouraging the creation of citizen's networks, and opened three free-access Internet centres in low-income neighborhoods.

At the regional level, political groups are admitting the importance of access to the Internet a political stake. Some of them are actually working on the implementation of "cybercities." The most advanced projects are La Carlota, in the Province of Cordoba, and the Benito Juarez-Laprida Cyberpole, Province of Buenos Aires. Both urban areas have been provided with computer terminals, located in public places. However, while La Carlota has not advanced much into the Information Society, the Benito Juarez-Laprida Cyberpole has interested not only the inhabitants in both towns, but also the nearby municipal governments. The difference resides in the fact that the local organisations have been invited by the organising institutions and the local governments to discuss intensively about their perceptions and needs regarding public access ICTs, that a small but active group of neighbours has been organised by a local leader—a lady pharmacist interested in ICTs—and that a number of teachers are being trained in ICT uses to train later the towns' population. Most of all, both mayors of Benito Juárez and Laprida have become involved with the project from the beginning, and have negotiated favourable conditions for telephone and Internet access with Telefónica de Argentina, the phone company that provides that geographical area.

The general trends detected in our research work suggest that in Argentina the role of the State, both at national and local levels, is essential to the creation and survival of electronic community networks. Generally, community organizations are created and thrive *without* any assistance from the State. Moreover, many of them are born with the goal to oppose State decisions involving urban life quality, identity or environmental issues. But to bridge the gap between the space of places and the space of flows (according to Castell's definition), to broaden their territory to include cyberspace, and to conform local and international electronic networks, most organizations need the push of State-implemented technological policies. These can adopt many forms: regulation of telephone and Internet access costs, education and training in ICT use, reach out for public consensus through the use of ICT, or facilities to have public access to ICT. Only when the local State will be mature for the use and dissemination of ICT will Argentina local organizations take a real leap into Information Society. We may not like the concept, but in most of the studied cases, the adoption of technology by local communities has been a top-down affair.

It is important (and there is never too much insistence on this point) to guarantee citizens access to the use of ICTs. As far as present socio-economic conditions, mainly those that concern the distribution of the national income and access to socio-cultural goods and services, continue to polarise the current Argentine social structure, and to generate an increasing social exclusion, there is no possible democratic use of informatics. It is a necessity, in the first place, for a State promotion of initiatives regarding the implementation of free computer terminals in public places, such as libraries, community centres, CGPs, NGOs, etc. In the second place, it becomes essential to include ICT uses in education, particularly in the first levels. It is not enough to teach students on the use of informatics, but also to the constant use of ICTs in the classrooms and at home, as a means to obtain and produce information, and to integrate all areas of knowledge.

Bridging the Informational Gap

When participating in discussions and debates about the Informational Society it is common to hear theoretical thoughts about the "black holes" in global cities, "dualisation of the informational society" or "third digital worlds." These concepts refer to the large masses of population excluded from the benefits of the globalised worlds, lacking goods and services to consume or sell (not even their devaluated labour force), and socially irrelevant for the system. The borders separate these social groups from those who do benefit from the informational society area drawn by their participation in such a society, their management of informatics tools and above all, by their participation in the processes of creation and dissemination of knowledge, information and technological production. The city of Buenos Aires is a typical example of these concepts. This is why it has become fundamental— a generation of national and local State active policies to disseminate ICTs use, to promote an active, massive and free access to them for those who cannot afford them, and to train the population in the use of ICTs having taken into account their social uses. We suggest a series of items to be considered when designing public policies:

- The inclusion of training in new technologies from the earliest stages in education, in order that children may be able to internalise the use of ICTs as tools for information and participation. This should include massive informatic education campaigns, articulating the private, public and voluntary sectors.
- The creation and dissemination of public, free electronic networks. This is feasible through the implementation of a structure of computer terminals connected to Internet. The terminals could be accessed at CGPs, community centres, schools, public libraries, public phone services, etc, or encourage the creation of Free Nets, Civic Networks or Public Access Networks.
- State strategies to provide or help provide computer equipment and Internet connections to NGOs, through "soft" loans from State Banks, agreements between the State, electronic firms, and communication enterprises, subsidies, etc.
- Training and funding of technical teams for permanent training and research, with the goal to train and assist NGOs and community organisations, orienting them towards the different uses of ICTs and adapting informatics tools to the particular needs of each social sector.

ICTs by themselves do not generate an increase in citizens' participation, nor encourage the surpassing of social or economic barriers. ICTs are not intrinsically democratic. They are (only?) tools to communicate, establish links and relationships, and support the huge amount of information in which the dominant economic system is based. Increasing the use of information technologies does not imply the disappearance of social or the emergence of more democratic societies: these ideals depend on the policies adopted by governments and the civil society. But access to information technologies is a *sine qua non* condition, an indispensable step for any social project that searches to promote those values.

Buenos Aires still has a long way to go before its complete incorporation to the

Information Society. The creation and consolidation of a virtual space for public participation, democratic reinforcement, and the strengthening of solidarity community networks requires constant efforts from both the local government and the civil society. We are still far from the generation of a critical mass of ICTs users and community organisation members who could stimulate a synergetic movement for the constitution of electronic citizens' networks. Community organisations are key agents in this process, and the trends detected in our research suggest that they are heading towards this direction. However, progress in this area needs the generation and extension of a double process. On one hand, it implies the local Governments efforts to change its institutional culture, disseminate and reinforce its own electronic networks and collaborate in the generation of electronic community networks. On the other, it requires the community organise efforts to create and disseminate their own networks, and claim free access to information. Only when these two movements converge on the Web, will Argentine communities be able to plan a better civil society and a fuller democracy.

References

Agren, P. (1999). "Virtual Community Life: A Disappearance for Third Places for Social Capital." http://www.informatik.umu.se/poagren.

Baumann, P. (1999). Usos sociales de TICs y Participación Ciudadana. El caso de la Ciudad de Buenos Aires. Internal Report, Research Project. "New Paradigms in Public Participation Through ICT Use. The case of Buenos Aires." 1997-1999. Instituto de Investigaciones Gino Germani, University of Buenos Aires.

Bassi, R. (1998). Informe de Internet en Argentina, Número 1.4 al 5 de Marzo de 1998, May 1995 to March 1998, ftp://planeta.gaiasur.com.ar/pub/reporte.exe.

Baym, N. (1995). "The Emergence of Community in Computer-Mediated Communication." Jones, S. (Ed.) (1995), Cybersociety: Computer-Mediated Communication and Community. Thousand Oaks: SAGE Publications.

Castells, M. (1998). The Information Age: Economy, Society and Culture," Vol. I: "The Rise of the Network Society. Malde, MA: Blackwell Publishers.

Castells, M. (1998). The Information Age: Economy, Society and Culture," Vol. II: "The Power of Identity. Malde, MA: Blackwell Publishers.

Castells, M. (1998). The Information Age: Economy, Society and Culture," Vol. III: "End of Millennium. Malde, MA: Blackwell Publishers.

Corvalán, P. (1997). Informe de Avance Proyecto PNUD ARG/97/007, Sistemas de información, October.

Corvalán, P. (1998). Informe de Avance Proyecto PNUD ARG/97/007, Sistemas de información, January.

Finquelievich, S., J. Karol & A. Vidal. (1992). "Nuevas tecnologías en la ciudad. Información y comunicación en la cotidianeidad," Centro Editor de América Latina, Buenos Aires.

Finquelievich, S., J. Karol & G. Kisilevsky (1996). "¿Ciberciudades? Informática y gestión local", Ediciones del CBC, Universidad de Buenos Aires.

Finquelievich, S. & E. Schiavo, (Eds.) (1998). "La ciudad y sus TICs", Editorial de la Universidad Nacional de Quilmes, Argentina.

Finquelievich, S. (1998). Comunidades virtuales: ¿Nuevos actores en el escenario local? EN.RED.ANDO, http://www.enredando.com.

Finquelievich, S. (1999). "¿Lo que mata es la velocidad?" La Puerta, http://cyn.delmercosur.com/lapuerta.

Harasim, L. (1993). "Networlds: Networks as Social Space." Harasim, L. (Ed.) *Global Networks, Computers and International Communication.* Cambridge, MA: MIT Press.

Herzer, H. and G. Kisilevsky. (1998). "Realidad y Ficción de las TICs: Su aplicación en la Legislatura de la Ciudad de Buenos Aires." Paper for the International conference "La ciudad en.red. ada," Instituto Gino Germani, University of Buenos Aires and University of Quilmes, Quilmes, 9-11 December.

Jones, S. (Ed.) (1995). *Cybersociety. Computer-Mediated Communication and Community.* Thousand Oaks: SAGE Publications.

Kisilevsky, G. (1998). "La informatización del Gobierno de la Ciudad de Buenos Aires. Del Concejo Deliberante a la Legislatura." Internal Report, Research Project "New Paradigms in Public Participation Through ICT Use. The Case of Buenos Aires," 1997-1999. Instituto de Investigaciones Gino Germani, University of Buenos Aires.

Mele, C. (1997). "Cyberspace and Disadvantaged Communities: The Internet as a Tool for Collective Action." P. Kollock and M. Smith (Eds.), *Communities in Cyberspace.* Berkeley: University of California Press.

Piscitelli, A. (1995). Ciberculturas, Paidós, Buenos Aires.

Rheingold, H. (1994). *The Virtual Community. Finding Connection in a Computerized World.* London: Secker & Warburg.

Rheingold, H. (1994). *The Virtual Community: Homesteading on the Electronic Frontier.* Reading, MA: Harper Perennial.

Serra, A. (1998). Training and organizing community networks in EPITELIO. Training Development Pack and Network Preparation Report. Epitelio, WP 6.1. cANet-UPC. December 31.

Schuler, D. (1996). *New Community Networks. Wired for Change.* New York: Addison-Wesley.

Turkle, S. (1995). *Life on the Screen. Identity in the Age of Internet.* New York: Simon & Schuster.

Turkle, S. (1996). "Virtuality and its discontents: Searching for community in cyberspace," *The American Prospect*, No. 24, 50-57.

Wellman, B. (Ed.) (1988). *Social Structure. A Network Approach.* Cambridge: Cambridge University Press.

Wellman, B., J. Salaff, D. Dimitrova, L. Garton, M. Gulia, C. Haythonthwaite (1996). "Computer networks as social networks: collaborative work, telework, and virtual community." *Annual Review of Sociology*, Vol. 22, 213.

Wellman, B. & M. Gulia (1996). "Net surfers don't ride alone: virtual communities as communities." *Communities in Cyberspace.* P. Kollock, M. Smith, (Ed.) Berkeley: Univ. California Press.

Chapter XI

Community Informatics for Electronic Democracy: Social Shaping of the Digital City in Antwerp (DMA)

Jo Pierson[1]
Vrije Universiteit Brussel

Introduction[2]

People living together in harmonious communities is the primary goal of most modern societies. The way these communities are built depends on the ideas, values and ideals of the society in which it is carried out. Campfens discerns two perspectives:

"From a humanitarian perspective, it may be seen as a search for community, mutual aid, social support, and human liberation in an alienating, oppressive, competitive, and individualistic society. In its more pragmatic institutional sense, it may be viewed as a means for mobilizing communities to join state or institutional initiatives that are aimed at alleviating poverty, solving social problems, strengthening families, *fostering democracy*, and achieving modernization and socio-economic development" (1997: 25).

Yet any community is only viable when all members can communicate with each other. Nowadays, the possible ways of communication have expanded enormously, especially since the convergence of informatics and telecommunication into information and communication technologies (ICT) offers a powerful tool.

The expansion of new means for communication like the Internet can thus be used for supporting communities in their efforts for social, economic or political development. This refers to the concept of 'Community Informatics' (CI). CI is defined as:

"an approach that starts from the perspective that ICT can provide a set of resources and tools so that communities and individuals living in communities can use ICT to pursue their goals in such areas as local economic development, cultural affairs, civic activism, community-based health and environment initiatives and so on." (Gurstein & Dienes, 1999)

Furthering the well being and welfare of a community through the development and use of ICT can thus be seen as a basic aim of CI. In this chapter we examine the political viewpoint, which entails the feasibility of promoting the democratic processes in a (municipal) community by means of ICT.

ICT and Democracy

It is generally accepted that ICT have the capacity of organizing virtual democratic processes like information exchange, interaction with local officials, public debate or even voting. ICT applications that enable these kinds of activities contribute to a so-called 'electronic democracy.' Related concepts are 'digital city,' 'cyberdemocracy,' 'teledemocracy,' 'push-button democracy' or 'point-and-click democracy,' dependent on the (utopian or dystopian) labels attached to it (Brants, Huizinga & van Meerten, 1996, 233). These kind of technological services, inspired by socio-political motives, rely on a community-based information network. The purpose and the democratic potential of such a network depends on the initiating actors and their motives and values (Arterton, 1987).

Based on an analysis of a broad range of citizen networks or community-based information networks, Friedland (1996) discerns four broad network models that can overlap. Besides 'advocacy networks' and 'electronic public journalism' he also distinguishes networks that are driven by technology in local or regional communities, the so-called 'community networks.' The latter refers to CI initiatives like Freenets (free networks), civic networks and community bulletin boards. They generally pursue goals in areas of civic activism, on-line public debate and advocacy. This concept has also found its way to Europe, with the 'De Digitale Stad' (DDS) in Amsterdam as a prominent example.

However the model most relevant for our analysis refers to the networks which have grown out of broadly defined governmental or local planning activity. Because the local governments often hold many powers, as the Belgians do, some relevant democratic innovations have come from local governments in new forms of economic planning and community development. This model of 'government-community economic development' applies to most current digital cities initiatives in Europe. In general they take the initiative of providing public information and services by means of ICT in order to optimize the communication and information exchange with citizens and the local representative democracy. However despite the rhetoric, one can question to what degree these kind of top-down initiatives really succeed in deepening democratic processes in a community, as envisaged by CI.

Digital Cities in Belgium

Belgium and Flanders have no tradition regarding public funding of community access. In comparison to other countries there is a lack of national and regional public support for societal and community projects regarding ICT (see, e.g., Pierson, 1997). Consequently the support depends largely on the willingness and the efforts at the local level of the city or community. This explains the diversity of initiatives, ranging from an ordinary Web site with a picture of the mayor to a fully interactive information and communication forum via Internet technology. A typical case striving for the latter and supported by local ICT funding is the digital city of Antwerp (Belgium).

Digital cities are a rather recent phenomenon in Belgium. Antwerp was the first city in 1995. Since then the number of cities that deployed a virtual counterpart on the Internet has increased exponentially. In March 1997 a study counted 44 initiatives in the whole of Belgium (Gerard et al., 1997). Combining findings of recent studies, this number has risen to 357 in the first half of 1999, which corresponds to more than 60% of all Belgian cities and municipalities (Steyaert, 1999; van Bastelaer, 1999).

Digital City of Antwerp

The initial idea for the digital city in Antwerp came from staff members at Telepolis, the local EDP-centre[3] responsible for the information and communication technologies of the departments in the Antwerp administration. This institution is directly dependent on the city council. In that capacity it is also responsible for the implementation and exploitation of the optical fibre backbone 'Metropolitan Area Network Antwerp' (MANAP), owned by the city of Antwerp. In an agreement with the local authorities, it is stated that MANAP and all the ICT applications using it as a backbone (like the Antwerp digital city) should benefit the political and economic development of the Antwerp region. These developments show that MANAP and thus also the digital city application are entrenched in the local political and economic development planning.

Furthermore the case of Antwerp is very instructive because it is generally acknowledged that their virtual city was and is among the 'forerunners' in Belgium not only in time, but also with regards size, graphic styling, content and creativity. It has been awarded a number of times,[4] while it also ranked in numerous top five lists on the Internet. Also featured in special newspaper editions,[5] and in the Internet press, the Antwerp case has been presented as the pioneer and the exemplary digital city.[6]

Approach

We will discuss how the development of the digital city application, i.e., the Web site 'DMA—Digitale Metropool Antwerpen' (Digital Metropolis Antwerp)[7] interacts with the broader framework of the local technology policy. Based on Hughes' concept of an 'open (technological) system' (1987, 52-53), we discern three interrelating levels (network, platform and application). We introduce these levels in our study for analytical purposes. This subdivision was not explicitly posed as the framework for the digital city development. Yet they are implicitly present in the actual development process, because the application of the digital city and the platforms for consulting grew out of the implementation of the local broadband network MANAP. All three levels are also interwoven in the more general project on open communication and efficient service provision, with the same name as the Web site, 'DMA—Digital Metropolis Antwerp.'[8]

In order to explain the current situation, we will give a historical account of the context and the interests that influenced the genesis and coming about of the digital city initiative on those three levels. Yet in contrast to standard historical accounts of technological development, we regard the design, development and appropriation of technologies like ICT as a social process. We refer to similar analyses in the field of innovation and technology studies (Marvin, 1988; Flichy, 1995b; Winston, 1998). More particularly for our case we draw useful insights from the 'social shaping of technology' tradition (MacKenzie & Wajcman, 1985). The basic assumption is that the technological features are situated within specific social dynamics that enable and constrain the development and design (and thus also the appropriation) of an ICT application, like a digital city. In our case the development process in each of the three levels is determined by social shaping which incorporates the political, economic, organizational and technological dynamics. In that way our findings contrast with the simplistic utopian visions of the future, which predicts the electronic democracy as an inevitable outcome of the proliferation of ICT in society. On the contrary we will demonstrate how the social shaping process on each of these three levels configures the democratic character of the digital city, as is envisaged by CI. This relates to three major points of concern regarding access: technological access, social access and policy access.

Social Shaping of the Antwerp Digital City on Three Levels

CI can be deployed to empower communities in the political deliberation and decision process. The opportunities for political choice and participation should increase through possibilities of two-way, interactive electronic networks (Lyon, 1988, 12). However we should be aware of extreme techno-optimistic views that claim that new forms of participatory democracy will emerge automatically because wired-up citizens will engage directly in voting and will contribute substantially to the political decision process (Lyon, 1988, 86). This leads to uncritical statements like:

> "Information technology will change democracy. Public and political administrations are beginning to use computers and communication technology to inform and engage the citizens." (http://www.challenge.stockholm.se)

This quote is taken from the presentation of the Global Bangemann Challenge 1997-1999. It is an international competition between cities with an IT Award for best practice of communication technology for citizens and communities. The extended Antwerp digital city project mentioned earlier, also called DMA, has won the award in the category of 'Public access and democracy.'

Statements like these give a good impression of the unrestrained faith in the possibilities of information and communication technologies for changing and improving democracy. This fits in with the more general techno-deterministic rhetoric that attributes all kinds of positive social, political and economic progress to the mere introduction of ICT into society.

It is hard to obtain useful insights for our study with these kind of boom visions, as well as the opposite doom visions of a surveillance society. In our view a deeper and more realistic understanding of the development of an ICT initiative like the digital city of Antwerp is only possible if we place the technology within a broader societal framework. In that way we avoid viewing the digital city on the Internet solely within a technological logic. This follows the idea of Mansell and Silverstone.

"When the veil of technological inevitability is challenged, as Raymond Williams suggests, we begin to see that information and communication technologies are being employed by producers and users in ways that depend on and alter highly culturally specific understandings about how communication relationships and the production and exchange of information are integrated within social, political, and economic life. Simplistic utopian or dystopian visions of the future provide us neither with an understanding of how these changes come about nor with an understanding of the longer-term implications." (Mansell & Silverstone, 1996, 3)

In the same way Arterton dismissed these two extreme visions regarding the democratizing effectiveness of communication technology. He found that:

"Essentially, what I had taken to be an examination of the capabilities of different technologies proved to be an exercise in evaluating a number of institutional arrangements or contexts in which citizens participate politically." (1987, 26)

In connection to this observation the theoretical tradition of social shaping of technology (SST) offers an adequate way for analyzing the development process. The SST approach in the study of technological evolution was originally elaborated by MacKenzie & Wajcman (1985). This theoretical viewpoint fits within the observation that the design and selection of technical systems are complex social processes emerging from continuous negotiations among supplier, user, and policy-making communities. These kind of observations are appearing more frequently in the literature on technical change and innovation (Mansell, 1996, 17). It contrasts with traditional approaches based on technological determinism, which concentrate on the 'effects' of technology and on the 'impact' of technological change on society. As a continuation of post-Enlightenment traditions, the latter does not problematize technology and technological change, but took it for granted. The scope of enquiry was limited to monitoring the social adaptations that were required for the technological progress. SST moved the focus on the content of technology and the particular processes involved in innovation. Technological change is in that way a social process of interplay between on the one hand the technical and on the other hand the social in a 'seamless web' (Williams & Edge, 1996). 'Social' includes political, economic, organizational, cultural, as well as technological factors.

So in order to assess the social shaping process of our case, it is essential to involve the context of the system in the reconstruction of events. Hughes states that technologies do not come in the form of separate, isolated devices but as part of a whole, as part of a 'technological system' (MacKenzie & Wajcman, 1985, 12). These systems contain 'messy,' complex, problem-solving components that are both socially constructed and society shaping. Among these components Hughes discerns physical artefacts, organizations, scientific components, legislative artefacts and natural resources (Hughes, 1987, 51). It entails that technological factors can also play a role on their own in the shaping process, which runs against social reductionism. The latter excludes every possible influence by technological artefacts. It is this way of *"intégrer dans une même analyse technique et société, sans choisir un terme au détriment de l'autre [...]"* (Flichy, 1995, 121), which will lead to the most valuable insights on technological innovations and ICT applications like digital cities.

For our analysis we can label the 'Digital Metropolis Antwerp' Web site as an open (technological) system that relates to two environments on which it is dependent (Hughes, 1987, 52-53): the ICT level (delivery platforms) and the network level (MANAP). So to discuss the development of the digital city initiative in Antwerp we should distinguish three interrelating levels: network level, ICT platform level and application level.[9] These levels lead to a threefold approach of the Antwerp digital city development: the 'information highway' metaphor that stimulated the idea of a digital city, DMA as a tool that can be consulted via different platforms and the internal structure and content. However these levels should not be seen as strictly separated entities, but more as three interrelating lines of approach for analysing the social shaping of the Antwerp digital city.

First of all the MANAP network constitutes the backbone channels making connections physically possible. In the so-called 'information society,' this is indicated as the 'information highway.' In general the latter refers to the infrastructure or 'broad band network' (BBN) that conveys mainly information, communication and entertainment services (Burgelman, Punie & Verhoest, 1995, 121-122). It is the widespread usage of these kind of multimedia services that determines the viability and the future development of the channels. One of the interactive services offered by the local authorities in Antwerp is the DMA site on the Internet. In contrast with most of the other services on MANAP—which are for internal municipal use— this application is also aimed at external use by and between citizens. However services or applications like these can only be transmitted if the (potential) users have adopted and appropriated the necessary ICT or multimedia delivery platforms for consulting the services. Besides the World Wide Web on the Internet this refers to computers and modems in the home or in a (municipal) cybercafé. The digital city can also be consulted on one of the information kiosks in different public places ('cyberbooths').

We will discuss each level separately. First the discourse is treated, which gives an account of the official viewpoints and statements for the initiatives taken, related to the digital city. Next we try to get a deeper understanding of the background and the motives by reconstructing the social shaping process. This can be supplemented by some indications of the actual performance in reaching the citizens.

Network Technology Level

Metropolitan Area Network Antwerp (MANAP)

It was in fact Telepolis Antwerp that initiated the digital city. This institution also set up, implemented and manages the municipal MANAP infrastructure which is the 70-km long optical fibre backbone for all ICT applications in the Antwerp region. Telepolis is a nonprofit association that depends on and is financed by the city of Antwerp. Telepolis, formerly known as 'Informatica Centrum Antwerpen' (ICA, Informatics Centre Antwerp), is responsible for information and communication technologies in the city of Antwerp, the OCMW (Public Centre for Social Welfare) and the harbour of Antwerp. The board of management of Telepolis includes representatives of the three aforementioned authorities. ICA/Telepolis was founded in 1988 as a merger of the EDP-centres of the Municipality of Antwerp and the Public Centre for Social Welfare. At the time the main tasks of these computing centres were the development and management of programmes on the central mainframe. Meanwhile the workload was extended with the introduction of office computer services, telephone services, telecommunications and the installation of PCs and LANs. It employs 237 persons and has a yearly turnover of 37 million Euro.

The decision to set up MANAP fit with the idea that access to the 'information highway' for local councils should be simplified. On the other hand, it was motivated by the idea that it would considerably improve the provision of services by the city administration and public welfare services. Only after the network was installed policy documents stated that this digital channel could also help bridge the gap between citizens and policy makers. Top-down it should bring all kinds of information to the people (administration, regulations, employment, education, social and medical care, transport, environment, etc.). Bottom-up it should serve as a communication channel for all citizens and give them the opportunity to participate in the 'information society,' for example by means of a digital city application. As a consequence, it should also involve those people who do not have the necessary skills, (financial) possibilities or education. For this reason information kiosks or 'cyberbooths' were placed in public places. This social goal is complemented by a more political goal. The municipal services can be decentralised and spread around the whole city without endangering the homogeneity and consistency of service by creating isolated cells. The network enables 'local' contact with and for people. The city authorities consider this as an important contribution to the process 'deconcentration' of political power, which should reach its full implementation in 2000. The first step of this political process was the establishment of district centres.

Social shaping process on network level

With the installation of MANAP in 1994, the city council aimed in the first instance at cost reduction and quality improvement of the public administration in Antwerp, mainly aimed at intra-municipal telephone communication (Van der Cruyssen, 1997). In that way the development of a 'digital city' was originally only a kind of spin-off of the main MANAP project. However, not long after its introduction, it was picked up by the local authorities and put forward as one of their major contributions to 'bridging the gap' between politics and citizens in the information society.

For this we can refer to the new political situation after the elections. The ultra-right nationalist party Vlaams Blok became the biggest group on the city council. In order to keep a democratic majority, all the main parties[10] were obliged to form a coalition. In their government agreement 1995-2000 the coalition firmly set out the intention to diminish the so-called 'cleavage' between politics and the citizens. This political intention was echoed through all the municipal departments. As for the informatics department ICA (later Telepolis), they also put forward their possible contribution in the 'dienstverleningsovereenkomst 1995' (service provision agreement 1995). Within the paragraph on 'participation' in Chapter 1 on 'Administrative organization,' is stated:

"Via the Digital Metropolis Antwerp (DMA) the inhabitant should have the possibility to establish contact with municipal departments or even directly with the mayor by means of a PC at home." [Own translation] (ICA, 1995)

This policy priority was also in line with the restructuring of political responsibilities in the new municipal government in 1994. In the latter alderman Bruno Peeters combines the powers and means for general information and communication with those for telecommunication. This led to a better integration of both areas in comparison with the former government in which the responsibility for telecommunication resided in the department of public works. It is therefore no surprise that the municipal government agreement for 1995-2000 stated that ICT services on MANAP should support the inner-municipal decentralisation, the refinement of local democracy and the improvement of the communication. In this way information and communication technologies like these should lead to an improved quality of life, following the technological deterministic and techno-optimistic view found in many European policy documents, like the Bangemann report (High-level group on the information society, 1994).

It is clear that MANAP, with its fibre optic and ATM-switches,[11] presents a real step forward in technological terms. In order to make adequate use of this high performing network, there has constantly been a quest for appropriate services for the citizens, like DMA. This search for appropriate content has already been around in public discourse since the construction of the network (Marain, 1995). Yet this kind of technology-oriented approach has some advantages. Firstly the Antwerp public administration as initiator seems to have some vision on the future development of the local 'information society.' The use of 'information highways' for CI should for example incorporate aspects of equal access (Internet lessons) and low financial thresholds (free use of 'cyberbooths'). In order to go from vision to action, a 'push' approach is sometimes needed to implement elements of change. Secondly once the appropriate infrastructure has been installed, all possible services and content have then a means for distribution. In that sense one avoids the vicious circle where content isn't developed because of the lack of infrastructure, while (private) infrastructure investments will only be made if there is enough content. Moreover the introduction of this kind of broadband network adds an important asset when tendering for innovative projects or programmes with official bodies like the EU.

Platform Level

WWW site via computers and information kiosks

This level refers to the combination of on the one hand to the physical delivery platform (computer, public kiosk etc.), and on the other hand the Internet protocol

(TCP-IP) and World Wide Web coding (HTML), which enables different kind of information and communication activities. The initiators proposed from the start a graphical WWW interface for the first version of the 'Digital Metropolis Antwerp,' in contrast with the first textual digital city in Amsterdam. They argued that a text-based interface would create a higher threshold for people, because it would be less user-friendly. This textual interface, however, seemed not to hinder the success of the first version of the digital city of Amsterdam in January 1994, which got an enormous response. In 10 weeks time some 13,000 virtual inhabitants were registered (Schalken & Moorman, 1995, 237-238).

Also several efforts have been made to make the municipal services on the fibre optic network accessible to as many people as possible by establishing the free dial-in, the cybercafés and the information kiosks. In addition several measures are taken for support and training. Besides the standard FAQ (Frequently Asked Questions) list on the DMA site and the possibility of asking questions by e-mail, Telepolis, by order of the city council, organizes free Internet lessons to lower the threshold and to help people using the worldwide network. Another initiative is the 'DMA Internet Service' (DSI). Several DMA inhabitants and users have joined in order to support and help each other online. Finally within the framework of a larger economic support programme, a bus has been equipped with computers with Internet access. This so-called 'cyberbus' traveled around in the province of Antwerp and could be reserved by sociocultural associations for events, markets and other social occasions. We will now discuss some of these access possibilities in more detail.

Every citizen can receive the DMA site at home when he/she has a PC, a modem and a telephone connection. The user only needs to call a specific dial-in telephone number that gives free access. One only has to pay the cost of the (local) telephone connection. It is of course also possible to consult the digital city via a standard (commercial) Internet link. To get access to all the services of the digital city one needs to subscribe as an 'inhabitant.' Everybody (not only citizens in Antwerp) can sign up as an inhabitant of DMA by filling in an electronic form with their personal data. Advantages of subscription are free Web browser software, a free e-mail address and space on the server (maximum 750 Kb) for one's own homepage. In addition inhabitants can participate in the chat channel.

People can also consult the digital information and communication services of DMA for free at the cybercafés in public libraries.[12] A survey in February 1997 on the use of the cybercafé at the central public library showed that almost 68% of the visitors took their first step on the Internet via this cybercafé (K. Meulemans, pers.comm., 11th February, 1998).[13] In addition in the Telepolis building there is a 'cyberlab' with computers connected to DMA and the Internet. This is a kind of multimedia classroom, which is put at the disposal of schools and associations.

It is also possible to use one of the 23 indoor information kiosks ('cyberbooths') in public places. Most of these public terminals not only give access to DMA but to the whole Internet, with the possibility for printing and electronic mail.[14] They are located at several locations like the city hall and the seven districts centres[15] and are directly connected with the MANAP network. Another 13 kiosks are placed in museums,[16] as part of the European INFOCITIES project.[17] Finally additional information kiosks are located in two 'wijkkantoren' (quarter offices).

Social shaping process on platform level

Several platforms for accessing the digital city were provided. However figures show a large difference regarding the use of the various access possibilities. When we look at the data for January 1998,[18] we see that 70.3% consult the digital city via a commercial Internet link, while 15.7% use the free dial-in connection. The internal use by the city administration amounts to 8.6%. The access via the public cybercafés corresponds with 3.3%, which is still three times the use via the information kiosks (1%). Finally the 'cyberlab' generates 1.1% of the consultations.

Most striking in these figures is the limited use of the information kiosk for consulting DMA, despite the fact that these appliances were in fact especially developed for accessing the digital city. Indications of this limited use also appeared in a small study in cooperation with Telepolis (Verfaellie, 1998). Only 8% of the respondents who used these booths declared to use them for accessing the digital city. Most users (78%) were only interested in Internet access. This supports the thesis that the cyberbooths seem to overshoot the mark of being an efficient digital link between government and citizens. However they could lower the threshold for general Internet use by offering this free access. But then the platform should be known more widely. The same study found that based on a telephonic survey of 321 respondents in the Antwerp region only 8% had heard about the information kiosks. Only two persons in the survey had also used this appliance. This gives some indication of the limited significance of these platforms for the digital city, despite their cost of almost 6,200 Euro each.

When focussing on the access possibilities we should also consider other characteristics. For example design features influence the way technology will be used in everyday life. The design element refers to the issue of social shaping as control through technology.

"Designers visualise a script of preferred reactions to the artefact, and they try to shape the technology in order to make these reactions as mandatory as possible. [However] in the end, [...] the predictability of the outcome of employment of technologies remains uncertain because of the flexibility of interpretation of technology."[19]

The kiosks have for example some features that limit the actual use in a sense. For example all the 'cyberbooths' are placed inside municipal buildings, which makes accessibility dependent on open hours and days. To a large extent this overlaps with working hours of many citizens. Yet the cost would be much higher if the 'cyberbooths' were to be water-resistant and withstand vandalism. As a consequence this economic factor in a sense limits the actual use. Another aspect of design is the keyboard that is rather uncomfortable to use for many people. This is attributed to the fact that it complies with the regulatory provisions and technological standards for designing an interface suitable for everybody, including children and wheelchair users (P. Scheyltjens, pers.comm., 27th January, 1998, & F. Steenhoudt, pers.comm, 27th January, 1998). Still most people won't use the kiosk for too long, because it is not possible to sit in front of the screen.

Application Level

Digital Metropolis Antwerp (DMA)

DMA is part of the larger local technology policy aimed at implementing several municipal public services applications via the MANAP. On this horizontal 'wide area' network, the applications are in fact offered to mainly two categories of users, first of all the city staff and second the city population. When the newly elected administration took office in 1994, it organized the implementation of the first network applications, besides telephone services for the city administration. These services can in fact be classified in several telematics areas.[20] DMA belongs to the area of services for citizens and visitors.

The idea of developing a 'digital city' was introduced by a group of four staff members in the public informatics organization Telepolis at the beginning of 1995. There was already an interactive information service in Antwerp: the Regional Interactive Teletext (RITT) (P. Cremers, pers.comm., 27th January, 1998). But according to the initiators of DMA, this system had no influence on their plan for a digital city.

They were in fact inspired by the success of the digital city in Amsterdam. In contrast with the latter, however, there is no tradition of Freenets in Antwerp, not even in the whole of Belgium, in which representatives of the societal field build a virtual community. The persons who developed the idea made an official proposal, which they submitted to the Telepolis' board of management. This board consisted exclusively of representatives of local public institutions (city, OCMW and CIPAL). The proposal was immediately accepted by the board. A possible explanation for this easy approval could be that the political representatives saw DMA as an additional legitimisation to citizens for the investments made in MANAP. It was included in the general 'service provision agreement 1995' between Telepolis and the city council.

First the licence of the Amsterdam concept was bought, which covers the use of pages and its maintenance. For the practical and organizational setup of DMA, the initiators were advised by the developers of DDS Amsterdam. This formal collaboration between the two cities resulted in visits to Amsterdam as well as visits from Amsterdam people to Antwerp. This kind of support in social shaping of DMA has led to the many resemblances with DDS. For the first version of DMA, which was opened on 14th June 1995, Telepolis worked together with the Internet service provider Riverland because they had the original licence of DDS. This version consisted mainly of static information sites. On 18th December 1995, Version 2 was launched which allowed more interactivity. The third version was introduced on the Net in February 1996. Since June 1998 the fourth update (Version 3.5) is on the Web.

Figures show that DMA now has between 20,000 and 30,000 virtual inhabitants, i.e., persons who have registered as virtual inhabitants. However earlier figures showed that only 40% of the virtual inhabitants actually live in the Antwerp region. The e-mail facilities generate over 500,000 messages per month (Telepolis, 1999). Since the latest update there seems to be a strong increase of visitors. While for January 1998 some 66,000 hits were counted, this number has risen to 78,000 in July, 100,000 in December and 190,000 in April 1999 (D. Beukeleirs, pers. comm., 27th may, 1999). No (qualitative) in-depth user studies regarding community building aspects were carried out yet.

Social shaping on application level
Involvement of municipal departments and information suppliers
The main focus of attention in the development of the Antwerp digital city lies with the internal municipal employees and information suppliers. In the past regularly meetings were organized with the municipal departments in order to exchange ideas and to show new developments in DMA. This operational level then fed the so-called 'editor's board' within Telepolis which sets out long-term strategies. This systematic approach has led to several improvements of the information supply and document transactions.

With regards to electronic services and transactions there were some catches in the organization of the response process until 1996 (Steenhoudt, 1997a). Not all city employees had the necessary computer equipment at their disposal. If a citizen did send a demand for an official document or a complaint concerning the public administration via DMA, this would be transferred most of time to the Telepolis team. The latter would then make a printout of the electronic message and send it via the internal mail system to the services or persons in question. The consequence of this procedure was that an electronic DMA message did arrive immediately at the public administration, but not at the person who could take care of it. At that time sending a letter by 'snail mail' directly to the department or person in question would generate quicker response than using the digital way.

The situation today has changed however (D. Beukeleirs, pers.comm., 11th June, 1998). All electronic requests are being directly received by the department concerned. For example the electronic 'meldingkaart' (reporting card)[21] is transmitted to the Department for Information, while an order for a city guide arrives at the Department for Tourism. When a new electronic form is implemented, Telepolis first makes sure that the department in question disposes of sufficient equipment enabling them to deal with electronic requests. Telepolis only supports the logistics, not the procedures themselves. This demonstrates that the initiators are constantly trying to optimize the digital city on the basis of former experiences. It could be described as (social) learning by trial-and-error with the municipal departments as the main source of information and knowledge.

Citizens' involvement in development
The efforts for gathering information on needs and usage of municipal departments that (want to) participate in the supply of services and information for DMA contrast with the attention for the use of DMA by citizens. With regards to citizen involvement, the changes in the updates of the digital city interface were in fact mainly based on intuition and on complaints and reactions by citizens. So no systematic surveys or social experiments among the population were carried out yet. Evaluations are in fact based on the main avenues of action by users: exit, voice or loyalty (Hirschman, 1974). They reject the application or service, they give remarks or they just keep using it. Yet the initiators notice that in general the citizen is not

extremely involved with the digital city, despite the 20,000 to 30,000 virtual inhabitants (D. Beukeleirs, pers. comm., 27th May, 1999).

There is a significant imbalance between the search for input from municipal departments and the search for input from the citizenry. This can be attributed to the initial social purpose of the underlying MANAP network. The network was originally installed to improve the information and communication exchange between the municipal departments and between the city employees. So from the beginning the initiative was oriented internally, more or less conceived as an 'integrated application' (Williams & Edge, 1996, 886-887) involving databases and computer networks within the city administration. Therefore it is difficult to redirect and adapt this 'integrated application' for appropriation by the outside world.

Citizens' usage

The initial objective in the first three versions of the digital city of Antwerp (DMA Version 1, 2 and 3) was to actually mirror the real city. The interface should embody a virtual place for living where local authorities and the virtual inhabitants had merged. Central to this goal was the fact that the potential inhabitant had to choose between 14 virtual quarters or neighbourhoods as a place to stay, dependent on his or her interests.[22] This can be compared with the 25 squares in the DDS 3.0 of Amsterdam. In each quarter the related municipal departments and the virtual community were integrated, symbolized by a bridge. This reflected in a sense an ideal situation, where authorities and citizens live side by side on an equal level enabling more mutual communication. With the general objective of 'refining democracy' it should help to diminish the cleavage between politics and citizens. However in the newest DMA interface (Version 2.5), we see that the original notion of virtual quarters has faded away and more or less has become a menu for navigation. Most of the quarter names do still exist but they now serve as 'themes and functions,' as they call it themselves.[23] The link with the former quarters is only kept when registering as a virtual inhabitant. The citizen indicates one or more topics (i.e., former quarters) with which he or she feels most affiliated.[24]

In the first three versions of DMA there was also a strong impetus for not only simplifying and improving information and communication exchange between the local authorities and the community, but also within the community itself. To stimulate the latter 'wijkanimatoren' (quarter instigators) were elected by the virtual inhabitants. Besides the enlivening of their own quarter, they should bring the inhabitants closer to each other by all kinds of initiatives like organizing games, moderating discussions, bringing the latest news and links, urging inhabitants to follow the rules, etc. In June 1997 the inhabitants of each quarter separately could vote for a candidate, but only 5 to 10% really made use of this possibility. This could be explained by the fact that there was little choice in opponents. In only three of the 13 quarters citizens could vote for more than one candidate. This initiative has been cancelled, because the response was too low and several quarter instigators do not meet the Telepolis expectations on participation and taking initiatives. In addition Telepolis has organized 'bewonersdagen' (inhabitants days), which gives virtual inhabitants, visitors and municipal departments a chance to meet in real life.[25]

Access to the Electronic Democracy

Community informatics is aimed at the development and use of ICT for the benefit of a community and its members. In an ideal situation of CI the ICT should break down existing barriers for communication and cooperation, as well as optimize existing information and communication exchange in a physical community. This enables the community to use the ICTs as effective and efficient tools for pursuing their goals in several areas (Gurstein & Dienes, 1999). In this chapter we discuss political goals with CI in relation to electronic democracy.

Dutton (1999) indicates that the most important feature of ICT in the society is their capacity of shaping 'tele-access' to information, people, services and technologies. Within the realm of electronic democracy the concept of optimizing 'access' is a crucial prerequisite for adequately linking the local authorities and the citizens. Barber (1984, 272) already argued that 'equal access' to (new) public information resources is one of the paths to institutional reform for a strong democracy. When democratic processes are mediated through ICT, the degree of access to those ICTs then co-determines the democratic calibre of the local political setting. However tele-access is itself the result of a social shaping process involving technological as well as social factors (Dutton, 1999, 4-9). This refers to the discussion on the development process of our case. We see several indications how access is enabled or constrained within and by the social shaping process on the three levels of the digital city, which we will discuss further on. Within this approach three related dimensions of access appear to be pivotal: technology access, social access and policy access. These dimensions are to be seen as an adequate way of approaching a digital city initiative, because they enable an assessment of the democratic potential. Technological and social access refers to the major points of concerns in any ICT initiative when we accept that development and appropriation of ICT is in fact a social process (see part 2). Yet specifically for a CI initiative in the sphere of electronic democracy it is necessary to supplement these dimensions of access with the concept of policy access. The latter then refers to optimizing access to public information, but also to the way in which citizens are able to participate in the political process. We will now discuss how social shaping on each level configures the development and the respective access dimensions of the digital city. This can be helpful for assessing future CI initiatives regarding electronic democracy.

Network Level

The electronic democracy initiative in Antwerp is a direct consequence of the implementation of the Municipal Area Network Antwerp. This large-scale infrastructure project was set up for reasons of savings on municipal telecommunication costs and service improvement of the local public administration. However the political developments, with the successful election result of an ultra-right nationalist party and the restructuring of political responsibilities, have fostered the acceptance and implementation of the digital city idea, suggested by technical staff members. This demonstrates how technological and economic motives are complemented afterwards by political contingency. As a consequence of the latter the initiators brought in the network infrastructure in the local democratic processes by way of implementing a digital city.

The teledemocracy initiative also has symbolic surplus value for the city, because it creates an image of democratic innovativeness. The anonymous MANAP infrastructure can be presented as a modern technological means for giving more and better access to democratic processes. Providing the necessary technological infrastructure for entering the electronic democracy could be called enabling 'technological access.' But as our case study shows, this technological access is embedded within a larger societal context.

Platform Level

The way people can participate in the electronic democracy is also determined by the interface or platform that enables access. This is another form of optimizing technological access. Several of these access possibilities to the digital city of Antwerp were provided. First there is the standard in-home access by a commercial Internet subscription which accounts for more than 70% of all visits to the digital city. Yet one should keep in mind that more than two Belgian families out of three (67.5%) have no computer, while 94% have no Internet connection at home or use it less than once a week (Lowette, 1998). This indicates that the digital city is mainly consulted by those people who are already technologically involved in the so-called 'information society' (Internet users).

The latter relates to 'social access,' which refers to social or cultural barriers that limit the use of the digital city application (Gurstein & Dienes, 1999). In order to avoid these barriers other platforms with a lower threshold are provided. This refers to the local dial-in number, for which the citizen does not need an Internet subscription, and the free public cyber cafés. But the most prominent initiative with regards to promoting social access is the installation of information kiosks in several public places in Antwerp, as was also the case in DDS Amsterdam. However this means of access appears to be used only to a minor degree, while only a very small share of this use is related to the digital city itself.

In order to explain the latter one should look at the platform itself. The design and implementation features heavily influence the actual use. Designing the appliance without the possibility of sitting leads to more uncomfortable use. In addition the information kiosks are all placed inside municipal buildings, mostly under supervision of city personnel. The cost would be much higher if the 'cyberbooths' were to be water-resistant and withstand vandalism. Yet besides the restricted opening hours (office hours), it also creates hesitation for those citizens that are afraid that city employees could observe their incompetence with regards to ICT use. So the design and economic features in the social shaping process here determine the restricted use of the digital city services.

Application Level

Our case study shows a significant imbalance between the search for application input from the municipal administration and information suppliers on the one hand and the citizenry on the other hand. This is attributed to the original social purpose of the underlying local broadband network and the subsequent social shaping process of the digital municipal environment. However with regards to this CI application aimed at furthering the electronic democracy, it is important that the citizens can and will participate by means of ICT applications. This not only refers to the capacity of being able to use the applications (like DMA Web sites, e-mail or newsgroups), but also the way these applications in themselves enable and constrain citizens' involvement in the local democratic processes. The first dimension refers to *social access* of the digital city application DMA (see earlier), while the second dimension can be denoted as '*policy access.*'

As to social access in our case, we see that several initiatives have been taken for capability building regarding use of the Internet via different platforms. This includes helpdesk, free Internet lessons, support teams, assistance in the public cybercafés and the 'cyberbus.' However besides the mere use of these new technologies and platforms for electronic public information and services, citizens should also be capable and stimulated to access the public sphere. This 'electronic citizen access' or policy access refers to optimizing access to public information. It could even be extended to participation in decision making by way of dialogues or by way of plebiscites (Dutton, 1999, 174; Arterton, 1987, 63-68). However assessing the latter participatory level is no easy task, as Tsagarousianou already stated:

"The extent to which the particular applications of information and communications technologies enlarged the participatory process, their success in introducing other than top-down ways of political participation and in delivering their promise to 'promote community-oriented participatory democracy' through two-way communication between citizens and public officials are by no means easy to determine." (1998,175)

Yet, despite the difficult assessment, it appears that on this form in the 'policy access,' the case of DMA falls short. Or in the words of van Diemen (1997), the local citizens are more consumers of governmental services than co-producers of information and policy in the discursive public sphere and the decision-making sphere. We will discuss this more in detail.

For start-up the initiators took over the design and interface of DDS Amsterdam. This application had proven to be successful, also with regards to citizen participation. However the digital city of Amsterdam was initially organized by a group of citizens, as a joint initiative of De Balie, a centre for politics and culture, and the Hack-Tic Network foundation (later XS4all) which had its roots in the movement of hackers and computer activists (Schalken & Moorman, 1995). This grassroots genesis formed a fertile breeding ground for an active civil participation. Telepolis did copy the framework to the Antwerp context, but lacked the initial citizens' involvement. We now find that DMA can generate only a low degree of participation (in comparison to DDS Amsterdam) and therefore lacks the same kind of political community building through ICT.

The latter is also linked with the outlook and the interaction possibilities of the digital city interface. The outlook of the first three versions was based on the virtual metaphor of a city with its own virtual quarters. In its most recent version, it is now evolved more into a sophisticated means for conveying local government information. The original 'quarters' have been transformed to 'themes and functions.'

As to interaction, the democratizing potential of the DMA application can be assessed by looking at the way in which the electronic public sphere is widened and opened up (Tsagarousianou, 199,: 175; Brants, Huizinga & van Meerten, 1996, 243). This public sphere can be created via forums like chat-boxes and free e-mail.[26] These two means for interaction are indeed used intensively, but without special features for inducing a public sphere. The latter would require a forum for as many people as possible, 'rational' discussion where arguments and views can be confronted and the primary aim of systematically and critically checking on government policies (Verstraeten, 1996, 348). The forum is available but there are no special precautions for the last two requirements of genuine public sphere. In the former DMA versions there were some tentative attempts on checking policies, like chatting with the mayoress in the chat-box of the former 'Government' quarter. Yet this happened only occasionally. In addition the elected quarter instigators could moderate discussions in the quarters, which to some extent supports 'rational' discussions. However since the latest DMA update the chat-boxes have been centralized without the distinction of quarter or theme. The initiative of quarter instigators has also been abolished. The free e-mail can of course still be used to discuss local democratic issues, but no particular framework stimulates to do so. Only when the ultra-right party Vlaams Blok tried to use this mail system for attracting new members, did it rouse some emotions and lead to heated political discussions.

Conclusion

Broadening our perspective from the Web site DMA to the larger framework of the local technology policy, divided in three interrelating levels (network, platform and application), helped us to better understand this case within the context of CI. It is the development and set-up on each of these levels within their specific social shaping context that affects the access dimensions and with that the feasibility of ICT use for promoting democratic processes. This becomes even more clear when we apply the same analysis framework for the digital city of Amsterdam, to which DMA is closely related (see above).[27] DDS was developed within the spirit of a community network, where the function of the application came first. Introduced just before local elections, it should enable civil participation in local policy. In that way the underlying network technology (with 20 modem lines) was of less importance, as long as it could enable the conveyance of the necessary information and services. The choice of platforms was important to the extent that they should enable every citizen to participate. Hence the focus on free software from in the beginning (Freeport, Lynx, Mosaic) and the installation of public terminals in the city. DMA on the other hand emerged out of the search for useful functions of the new municipal broadband network. By copy-pasting the DDS framework the initiators also aspired to copy the success and the citizen involvement. Yet they overlooked the social shaping background of the Amsterdam initiative. We will discuss now more in detail the outcome of DMA and assess how it enables and constrains access on the three dimensions. Discussing the strengths and weaknesses of our case study regarding this access issue will also highlight some generic concerns for initiators and application designers of an electronic democracy initiative.

With no doubt we can say that for the digital city of Antwerp a lot of effort and resources are put in creating an interactive and creative electronic public information and communication channel on high performing network technology and platforms. As a result the Antwerp virtual city was and is among the 'forerunners' in Belgium not only in time, but also regarding technological access. However this is no surprise, given the technological background of the initiators, i.e., staff members of the EDP-centre Telepolis.

Besides optimizing technological access it is also important to optimize social access, aimed at the potential users of the implemented platforms and applications. Therefore special attention should be paid to the possible social or cultural barriers in design and implementation that could limit the use, like cost, required knowledge and skills or practical arrangements. We see in our case study that on the one hand there are the efforts of lowering the financial thresholds for access by offering free access possibilities (like free dial-in and information kiosks). On the other hand there were some initiatives of knowledge building (like free Internet lessons). Yet despite these efforts of lifting barriers of knowledge and means, figures show that users are still atypical to the actual city population. More specifically the virtual city mainly consists of those citizens who are already involved in the so-called 'information society.' This refers to the technologically engaged, to people who are already familiar with ICT and Internet.

The idea of developing DMA as a CI application for the Antwerp community was in fact only one of the many initiatives in the search for useful content on the new municipal network. The virtual community idea in DMA has led to facilities like free e-mail, free Web space, on-line discussion and chatting. Especially at the start-up there were efforts for optimizing democratic processes via this (virtual) community idea, with initiatives like chatting with the mayor, newsgroups and notice boards on regulatory and political issues or voting of quarter instigators. But as new interfaces were developed, this motive of participation in policy faded away. So despite the attempt to incorporate the community network idea of bottom-up citizen involvement, with regards to policy access, DMA has in practice always been more of a tool for the local government enabling adequate information provision between the city council and the citizens. The participation dimension in the policy access has been downgraded even more in the latest developments of the application.

The latter development follows in fact from the value choices made by the initiating actor. As Arterton remarked on citizen participation via communication technologies:

"The nature of participation evoked was a product of the values brought to the endeavour by the project organizers. [...] Thus I feel justified in concluding that the future of our political institutions will be determined by the value choices we make about our political process." (1987, 197)

In order to develop CI application for electronic democracy it is therefore essential to anticipate as much as possible on the motivation, objectives and value choices of the different social and political actors involved on every level (network, platform and application), embedded within their specific socio-political culture (Tsagarousianou, 1998, 173). Only in this way will it be feasible to enable communities using ICT for creating a stronger democracy.

References

Arterton, F.C. (1987). *Teledemocracy: Can technology protect democracy?* Newbury Park: Sage Publications.

Barber, B.R. (1984). *Strong democracy – Participatory politics for a new age.* Berkeley: University of California Press.

Brants, K., Huizinga, M., & van Meerten, R. (1996). The new canals of Amsterdam: an exercise in local electronic democracy. *Media, Culture & Society,* 18, 233-247.

Burgelman, J.C., Punie Y., & Verhoest P. (1995). *Van telegraaf tot telenet - Naar een nieuw communicatiebestel in België en Vlaanderen?* Brussel: VUBPress.

Campfens, H. (1997). International review of community development. In H. Campfens (Ed.), *Community development around the world: Practice, theory, research, training* (pp. 13-46). Toronto: University of Toronto Press.

Chambat, P. (1994). NTIC et représentations des usagers. In A. Vitalis (Ed.), *Médias et nouvelles technologies - Pour une socio-politique des usages* (pp. 45-59). Rennes: Editions Apogée.

Collinson, S. (1996). *Forecasting and assessment of multimedia in Europe 2010+— Final report to European Commission.* Edinburgh: University of Edinburgh.

Digitale metropolen—Antwerpen wil een intelligente stad worden (1997, March). *Inside Internet,* p. 37-39.

Dutton, W.H., (1999). *Society on the line: Information politics in the digital age.* New York: Oxford University Press.

Flichy, P. (1995a). *L'Innovation technique—Récents développements en sciences sociales vers une nouvelle théorie de l'innovation.* Paris: Editions La Découverte.

Flichy, P. (1995b). *Dynamics of modern communication—The shaping and impact of new communication technologies.* London: SAGE.

Francissen, L. & Brants, K. (1998). Virtually going places—Square-hopping in Amsterdam's Digital City. In R. Tsagarousianou, D. Tambini, & C. Bryan (Eds.), *Cyberdemocracy—Technology, cities and civic networks* (pp.18-40). London: Routledge.

Friedland, L.A. (1996). Electronic democracy and the new citizenship. *Media, Culture & Society,* 18, 185-212.

Gerard, P., Gobert, D., Rasmont, B., & Vandendriessche, G. (1997). *Relations Citoyens Administration - Etude des villes virtuelles belges.* Namur: FUNDP.

Gurstein, M., & Dienes, B. (1999). "A 'Community Informatics' Approach to Health Care for Rural Africa." Paper presented to The Africa Telemedicine Project: CONFERENCE '99 "The Role of Low-Cost Technology for Improved Access to Public Health Care Programs Throughout Africa," Nairobi, Kenya, February 19-21, 1999.

High-level group on the information society. (1994). Europe and the global information society. *Recommendations to the European council* ('Bangemann report'). Brussels: EU.

Hirschman, A.O. (1974). *Exit, voice and loyalty—Responses to decline in firms, organizations and states.* Cambridge (Mass.): Harvard University Press.

Hughes, T.P. (1987). The evolution of large technological systems. In W.E. Bijker, T.P. Hughes, & T.J. Pinch (Eds.), *The social construction of technological systems—New directions in the sociology and history of technology* (pp. 51-82). Cambridge: MIT Press.

ICA. (1995). *Dienstverleningsovereenkomst 1995.* Antwerpen: CIPAL-ICA Informatica Centrum Antwerpen.

Lowette, T. (1998). Internetmonitor—Electronic media in Belgium: how many users? (Report November 1998). Brussels: GRID.

Lyon, D. (1988). *The information society: Issues and Illusions.* Cambridge: Polity Press.

MacKenzie, D., & Wajcman, J. (1985). Introductory essay—The social shaping of technology. In D. MacKenzie & J. Wajcman (Eds.), *The social shaping of technology—How the refrigerator got its hum* (pp. 2-25). Milton Keynes: Open University Press.

Mansell, R. (1996). Communication by design? In R. Mansell & R. Silverstone (Eds.), *Communication by design—The politics of information and communication technologies* (pp. 15-43). Oxford: Oxford University Press.

Mansell, R., & Silverstone, R. (1996). Introduction. In R. Mansell & R. Silverstone (Eds.), *Communication by design—The politics of information and communication technologies* (pp. 1-14). Oxford: Oxford University Press.

Marain, F. (1995, March 13). MANAP maakt Antwerpen intelligent. *De Standaard,* p.14.

Marvin, C. (1988). *When old technologies were new—Thinking about electric communication in the late nineteenth century.* New York: Oxford University Press.

Pierson, J. (1999). Metropolitan Area Network Antwerp (MANAP)—Digital Metropolis Antwerp (DMA). B. van Bastelaer & C. Lobet-Maris (Eds.), *Social Learning regarding Multimedia Developments at a Local Level—The Case of Digital Cities* (pp. 151-172). Namur: CITA-FUNDP.

Pierson, J. (1998). Case study: Metropolitan Area Network Antwerp (MANAP)—Digital Metropolis Antwerp (DMA) (Not published). Brussels: VUB.

Pierson, J. (1997). Als een VIS in het water—De Vlaamse overheid en de Vlaamse informatiesnelweg (VIS): Overzicht en analyse. *Informatie en Informatiebeleid,* 15(4), 25-32.

Schalken, C.A.T., & Moorman M.A.H. (1995). Overheid en Digitale Steden. In Rathenau Instituut (Ed.), *Toeval of noodzaak? Geschiedenis van de overheidsbemoeienis met de informatievoorziening* (pp. 225-275). Amsterdam: Cramwinckel.

Steenhoudt, F. (1997a, March 3). Al is de digitale burger nog zo snel, een gewone brief achterhaalt hem wel (Techno Antwerpen Deel 2). *De Morgen - Metro,* p.9.

Steenhoudt, F. (1997b, March 3). De nieuwe Antwerpse snelweg (Techno Antwerpen Deel 2). *De Morgen - Metro,* p.8-10.

Steyaert, J. (1999). De gemeentelijke Web site in Vlaanderen. Paper presented at the 'Deuxième rencontre réelle de villes virtuelles,' Namur, Belgium, June 28, 1999.

Telepolis. (1999). DMA voor iedereen! *Telepolis Krant,* 5(1), 4.

Tsagarousianou, R. (1998). Electronic democracy and the public sphere—Opportunities and challenges. In R. Tsagarousianou, D. Tambini, & C. Bryan (Eds.), *Cyberdemocracy—Technology, cities and civic networks* (pp.167-178). London: Routledge.

van Bastelaer, B. (1999). "Les villes virtuelles en Belgique francophone et germanophone." Paper presented at the 'Deuxième rencontre réelle de villes virtuelles,' Namur, Belgium, June 28, 1999.

Van der Cruyssen, P. (1997). Antwerp, an intelligent city. *Global Communications Interactive 97*. (Hanson Cooke), 81-83.

van Diemen, D. (1997). "The public Internet terminal in Amsterdam: Embodying the ideal of citizenship in technological design." Paper presented to the international research network EMTEL on "Technology, Cultural Identity and Citizenship," Dublin, Ireland, May 9-10.

van Lieshout, M. (1999). The digital city of Amsterdam: between public domain and private entreprise. In B. van Bastelaer & C. Lobet-Maris (Eds.), *Social Learning regarding Multimedia Developments at a Local Level—The Case of Digital Cities* (pp. 101-149). Namur: CITA-FUNDP.

Verfaellie, N. (1998). *Project 'Telepolis.'* Antwerpen: Karel de Grote-Hogeschool— Katholieke Hogeschool Antwerpen.

Verstraeten, H. (1996). The media and the transformation of the public sphere—A contribution for a critical political economy of the public sphere. *European Journal of Communication*, 11(3), 347-370.

Williams, R. & Edge, D. (1996). The social shaping of technology. *Research Policy*, 25, 865-899.

Winston, B. (1998). *Media, technology and society—A history: From the telegraph to the Internet*. London: Routledge.

Interviews

Mr. Paul Van der Cruyssen, co-ordinator European projects (Telepolis):
  18th November 1996
  25th January 1998
Mr. Dirk Beukeleirs, project manager DMA (Telepolis):
  22nd September 1997
  11th June 1998
  27th May 1999
Mr. Tom Bosmans, helpdesk DMA (Telepolis):
  24th December 1997
Mr. Paul Scheyltjens, 'information kiosk' in district centres (Telepolis):
  27th January 1998
Mr. Peter Van Avermaet, 'information kiosks' in museums (Telepolis):
  25th February 1998
Mr. Frans Steenhoudt, journalist De Morgen:
  27th January 1998
Mr. Paul Cremers, Cabinet alderman Peeters:
  27th January 1998
Mr. Kurt Meulemans, Antwerp city staff member in communication
department:
  11th February 1998

Appendix

To get a better overview the evolution of the Antwerp digital city we give a short historical overview of the relevant events.

MANAP-DMA in its historical, technological, infrastructural and organizational context

1948	Scientific research points out that an improvement of the work process in the city administration is necessary.
1951	Formation of a 'mecanografic department' with Hollerith or cardpunch.
1967	Acquisition of the first computer (mainframe 16k memory).
1970	Introduction of teleprocessing in order to guarantee better services for the harbour.
1971	Reorganization for the establishment of a 'computing centre' with its own software department in order to develop customized applications.
1985	Consulting by Coopers&Lybrand leads to an improvement of the work process and the organization structure.
1988	The city council, the OCMW (Public Centre for Social Welfare) and CIPAL (Centre for the Informatics of the Provinces Antwerpen & Limburg) establish ICA (Informatics Centre Antwerp), a merger of the computing centres of the city and the OCMW.
1990	ICA is transformed from a department within the 'intercommunale' CIPAL to a non-profit organization, 'CIPAL-ICA,' as a more independent subdivision of CIPAL.
1992	The city council decides, in agreement with the Chamber of Commerce, to set up a telematics programme which becomes part of the 'Strategisch Plan Regio Antwerpen' (SPRA, Antwerp Region Stategic Plan). The first cables for a fibre optical network are installed.
1993	The city of Antwerp is Europe's cultural capital for one year, which became a kind of stimulus for the renewal of the city infrastructures and public resources in 1992.
1994	In June MANAP (Metropolitan Area Network for AntwerP) is the first optical fibre network based on ATM (Asynchronous Transfer Mode) in the world. In November the central computers of ICA and CIPAL merge.
1995	ICA changes its name to 'Telepolis.' On 14th June the first version of the digital city of Antwerp (DMA) on the World Wide Web is open for the public. The second version is already implemented on 18th December. The first municipal cybercafé is opened in the central library in the centre of Antwerp.
1996	The third version of DMA is opened in February. In June Telepolis moves into the new building in Antwerp.
1997	The first 23 information kiosks are placed in several public locations. The second municipal cybercafé is opened in the public library in Deurne. Antwerp wins three awards in the Bangemann-challenge.
1998	The fourth update of DMA, Version 3.5, is introduced in July. A third public cybercafé is opened in the library of Ekeren.
1999	The fourth municipal cybercafé is opened in Hoboken.

Endnotes

1 Research Assistant for the Fund for Scientific Research—Flanders (Belgium) (F.W.O.).

2 This chapter is based on findings from the European research program (DGXII): Targeted Social Economic Research (TSER)—Social Learning in Multimedia (SLIM) (Phase 3: Integrated Studies). The first results were discussed in the report on the Antwerp case study (Pierson, 1998). This report has been incorporated in the final report on *Social learning regarding multimedia developments at a local level – The case of digital cities* for the TSER-SLIM programme (Pierson, 1999).

This research is also part of an ongoing inter-university project, Research & interdisciplinary evaluation of the information society: Networks, usage and the role of the state, *funded by the Belgian federal institute for science policy (DWTC) with the partners SMIT (VUB), LENTIC (Liège) and CRID-CITA (Namur).*

In addition the study receives input from the Flemish three-year research project Medialab on Socio-economic and regulatory conditions for the innovation of multimedia services in Flanders.

3 EDP = Electronic Data Processing.

4 This refers to the DMA site as a whole (e.g., Silver Clickx-award, 1997), as well as several individual sites developed by virtual inhabitants.

5 See for example: Steenhoudt, F. (1997b, March 3). De nieuwe Antwerpse snelweg (Techno Antwerpen Deel 2). *De Morgen - Metro*, p.8-10.

6 See for example: Digitale metropolen - Antwerpen wil een intelligente stad worden. (1997, March). *Inside Internet*, p. 37-39.

7 <http://www.dma.be> or <http://www.antwerpen.be>.

8 More information on this project can be found on <http://www.challenge.stockholm.se/Winners/info_digi.htm>.

9 Williams discerns three similar levels (components; delivery systems/platforms; applications) in a revised version (http://www.ed.ac.uk/~rcss/SLIM/public/phase1/SSICT.html) of Williams & Edge (1996) as adapted from Collinson (1996). Also Chambat (1994) denotes these levels as the elements of incertitude with regard to the 'smart home' concept.

10 Socialists, liberals, Christian-democrats, democratic nationalists, green party and independent representatives.

11 ATM = Asynchronous Transfer Mode.

12 'Public' cybercafés can be found in the central library in the centre of Antwerp (15 computers) and in the public library in Deurne (five computers). In 1998 the third cybercafé was opened in Ekeren (five computers) and in 1999 a fourth in Hoboken.

13 Referring to the study 'Onderzoek naar het gebruik van het cybercafé in de COB van Antwerpen—Stad Antwerpen Afdeling Communicatie—Februari 1997.'

14 However Internet access on the kiosk in the city hall was shut off on request.

15 These 'Cyberbooths' are located in the city hall (Antwerp centre), the administration centre (Borgerhout), the 'Orangerie' (Hoboken), and in the district centres

of Deurne, Berchem, Ekeren, Wilrijk and Merksem.

16 The museums with an information kiosk are: Museum Plantin-Moretus/ Prentenkabinet/Nationaal Scheepvaartmuseum/Etnografisch Museum/ Rubenshuis/Vleeshuis/Volkskundemuseum/Stadsarchief/Hessenhuis/Prospekta/ Middelheimmuseum/Koninklijk Museum voor Schone Kunsten/ VandeKerckhove & Devos.

17 http://www.infocities.eu.int.

18 http://www.dma.be/statistiek/9801.htm (accessible until June 1998).

19 Sørensen, K.H. (1996). *Learning technology, constructing culture—Socio-technical change as social learning* (http://www.ed.ac.uk/~rcss/SLIM/public/ phase1/knut.html).

20 In general the ICT services and applications on MANAP can be classified in six telematic areas: services for better administration, tele-medicine, services for companies at the port and other SMEs, services for environmental monitoring, services for culture and tourism and services to the citizens and visitors.

21 Standard form for reporting problems in the neighbourhood.

22 The 14 virtual quarters were: Culture/Media/Going out/Tourism/Environment/ Sports/High-tech/Well-being/Economy/Government/Education/Forever young/ Living/Open house. The 15th area was called 'Construction' and contained the tools for building a personal Web site.

23 See the description of DMA on <http://www.challenge.stockholm.se/Projects/ Categories/PublServDemocracy/info_digi.htm> versus <http://www.challenge. stockholm.se/Winners/info_digi.htm>.

24 The quarter names 'Government,' 'High-tech' and 'Open house' have disappeared, while only 'Hobby' was added. The former quarters have been trans formed into topics: Culture/Media/Going out/Tourism/Environment/Sports/ Hobby/Well-being/Economy/Education/Forever young/Living.

25 The first day was organized on 19th September 1997 and the second one in June 1998.

26 In the former version there was also a notice board for exchanging messages.

27 The case of DDS Amsterdam has already been the subject of many studies. For a detailed description see e.g., Schalken & Moorman (1995); Brants, Huizinga & van Meerten (1996); Francissen & Brants (1998); van Lieshout (1999).

Chapter XII

Internet-Based Neighborhood Information Systems: A Comparative Analysis

Danny Krouk, Bill Pitkin and Neal Richman
University of California, Los Angeles, USA

Introduction

Networks, networks everywhere
No place is undisturbed.
The revolution's underway.
Its pace cannot be curbed.
-Vinton G. Cerf, September 2, 1999

This verse comes from a poem read by one of the key figures in the development of the Internet at a recent symposium held to celebrate the 30th anniversary of the first successful transmission of digital bits from one computer to another, which ushered in the era of computer networks (Kaplan, September 6, 1999). Perhaps not unexpectedly, participants in this commemorative event reflected on the rapid development of networking and what we today call the Internet and predicted its ubiquity in everyday life, likening it to electricity. Obviously, however, we are not quite there yet. Recent data from the U.S. Department of Commerce suggest that, despite rapidly increasing rates of computer ownership and Internet access in the United States, there are still many people who have been left out of the information revolution. Researchers found that Internet access is highly correlated with income, education level and race, leading them to conclude:

The information 'haves' have dramatically outpaced the information 'have nots' in their access to electronic services. As a result, the gap between these groups — the digital divide — has grown over time. (McConnaughey et al., 1999, p. 88)

In the context of globalization, this so-called "digital divide" exists between developed and developing countries, between rural and urban areas, and even within urban areas. According to two leading thinkers on the impact of the information age on urban areas, the digital divide has exacerbated existing socioeconomic disparities, creating "opposite and equally dynamic poles of the information economy" and leading to a socially-polarized "dual city" (Borja and Castells, 1997, p. 42).

If there exist such disparities in access to this new information media, what is being done to ensure that this gap does not increase even more, especially as the Net becomes more commercialized and subject to the needs and interests of business? As evidenced in the compilation of chapters in this book, there are numerous examples from around the world of people working to address this "digital divide" by promoting alternative, community-based Information and Communications Technology (ICT) projects. This "Community Informatics" approach seeks to "design electronically-enabled 'services' or applications so that they are as widely available and usable as possible" (Gurstein, 2000).

A natural venue for the development of such applications is the field of urban planning and community development, especially in light of recent changes in planning theory. The traditional "expert analyst" model of policy development, in which planners (i.e., the "experts") serve as interpreters of data for policymakers, has increasingly been challenged by both practitioners and theorists who have long sought to reduce social and economic disparities as a part of "community building" in urban areas. According to John Friedmann (1987), this is accomplished when planners successfully link technical knowledge with action, thus leading to processes of social transformation. Reflecting on how Habermas's theory of communicative action informs the experience of planning practice, several theorists contend that planning is a complex, dialogic process in which community residents must play a role (Forester, 1989; Innes, 1998). Therefore, planners serve as a link between the local government and residents, analyzing data and information in order to understand community dynamics and develop neighborhood plans in a collective process. Would this, then, not be an ideal place to explore the intersection between ICT and community planning and development?

One of the first inquiries into this issue took place in the spring of 1996, during a colloquium at the Massachusetts Institute of Technology's Department of Urban Studies and Planning. The impetus for the colloquium came from faculty members in both the computer technology and community development concentrations of the department, and the presentations from the series were published in an edited volume (Schön et al., 1999). The academics who participated in the colloquium painted a rather grim picture of the impact of ICT on cities, outlining how it is—and likely will even more—increase social and economic disparities between the information "haves" and "have-nots" and debunking hype of "digital utopia." Community activists, who had traditionally been skeptical of technology's role in community planning and development, also participated in the colloquium and were surprisingly optimistic about the potential role of ICT in community building. The editors of the volume attribute this optimism to the activists' realization that "at a time of declining government funding for inner cities, communities lacking electronic access to resource announcements will be disadvantaged in competing for scarce resources" (Schön et al., 1999, p. 374). Moreover, the colloquium gave the activists an opportunity to hear about innovative ICT projects that are increasing governmental transparency, creating space for grassroots participation and collaboration in planning, and providing cutting-edge educational opportunities for inner-city youth.

In this chapter, we will more deeply analyze three such innovative projects in order to reflect on how ICTs are changing the dynamics of community development. We are interested in understanding the underlying dilemmas facing these projects, all of which seek to improve community participation and collaboration in the planning process. As Internet-based systems, how do they help or hinder the process? How are they addressing the disparity between information "haves" and "have-nots"? Our ultimate goal is to see if perhaps the unique experience of each project might signal a middle way between the two extreme views on the role of ICT in community development: the unabashed optimism represented by proponents of "electronic democracy" and "e-topia" on the one hand, and the intemperate pessimism of those who denounce the socially and economically polarizing impacts of the new technologies on the other hand.

Background

The Internet and Community Development
Much has been written about the rapid growth of the Internet by counting the sites on the World Wide Web, and the rising numbers of global users. However, what is often overlooked is that the Internet has not only grown rapidly; it has evolved rapidly through a number of distinct stages.

- Stage One: The origin of the Internet as a defense department application to ensure rerouting of information in case of global attack.
- Stage Two: A university-based system whereby faculty and researchers could share information and work on common research projects through a network.
- Stage Three: Expansion of Internet sites and users outside of the university where participation only requires some hardware, a modem and a phone line.
- Stage Four: Growth of Electronic Commerce, including Internet advertisement, economic transactions, pay-per-view Web sites, and Wall Street investment.

The next stage will likely involve some convergence among information and communication systems, cable television, telephone computers, and hardware and software providers. There is too much economic potential to limit market penetration to only "technical experts."

In light of these trends, do these changes hold any significance to planning and community development? Interestingly, while physical freeway development (and obstruction) has been the bread and butter of legions of planners and community activists, few seem to have given the Information Superhighway serious attention. Cities are increasingly taking advantage of resources on the Internet, but most tend to utilize the Web to produce electronic brochures and other forms of "Web presence" (Nunn and Rubleske, 1997; Huffman, 1998).

While many Web sites publish public data, few have clearly identified neighborhood-based constituencies with whom there is an ongoing communicative exchange. As new information systems evolve, the question is unlikely to be who owns and controls the Internet. The anarchic nature of the system makes it difficult to parcel, subdivide and sell to the highest bidder. Rather, the question will be, who will shape the way that the Internet is utilized and how will they accomplish this? Local experiments in making information meaningful and accessible at the community level are defining new possibilities and, only in this way, establishing precedents for public access to neighborhood information in cyberspace. Will public information be freely available and utilizable by the public, or will these sources of data be sold to the highest bidder for private uses?

It is perhaps useful to remember earlier changes in the field of telecommunications, especially the battles waged over public access to certain cable channels. Although some space was allocated to public access and community programming, very few resources were made available to produce competing content. The results all too often were vanity productions that reflected more the desires of the local star or producer than those of any clearly identified audience. As private investment now begins to pour into the Internet and shape the substance and presentation, it is very important that a competing vision for using this electronic network keeps pace and defines approaches that address needs likely to be unmet on sites aimed at serving very large groups of consumers. While businesses parry for their future in this new domain for national and global commerce, along the margin local experiments are emerging, which suggest alternative uses of these information tools to support neighborhood planning and development to achieve community-defined objectives.

Research Methodology

This chapter examines three cases of Internet-based systems that propose to support neighborhood planning. The presentation of the cases is followed by a discussion of the possible implications of these systems for community development. Do we see viable new ways of going about planning and community development? What are some of the opportunities and challenges that we see for these efforts?

In selecting cases, we identified three criteria for what we call "neighborhood information systems":

1. Incorporates an Internet-based system that provides public access to property and demographic related information.
2. Provides access to record and/or statistical data at the neighborhood level.
3. Provides an interpretive framework for its information, aimed at nonprofessionals.

Moreover, it became clear that each system was based within an "information intermediary." These intermediaries play a role in the spaces between government, private parties and community-based groups to assemble, reinterpret, and present relevant data. In this research, three types of information intermediaries will be examined: a) a community-based, nonprofit research organization; b) a university technical assistance and outreach center; and c) a municipal program for neighborhood-based planning.

The purpose of the selection criteria outlined above was not to identify Internet-based planning applications that are somehow representative of some larger population and thereby assist in generalizing from them. On the contrary, selection was based on what Michael Quinn Patton (1980, p. 105) has called "sampling extreme or deviant cases" which in this case might inform us about the future use of new technologies for community-based planning purposes. In other words, we seek to study the *status nascendi* rather than the status quo on the Internet. Whether or not such applications are rising phenomena may rely on how effective these local experiments, and others like them, are in promoting neighborhood improvement.

The primary data sources utilized for the comparative case research were:

1. interviews with project developers, organizational leaders, community activists, site-visitors, funders and representatives from public agencies;
2. archival information, such as project plans and grant proposals;
3. quantitative information on web site utilization and public counter information requests; and
4. the content of the Web sites:
   - NEWS (http://www.cnt.org/news/demo/html)
   - PAN (http://www.ci.seattle.wa.us/)
   - NKLA (http://nkla.sppsr.ucla.edu)

In each instance, the aim is to (1) examine the impact of these projects on public access and the use of public data in neighborhood improvement and city policy, and (2) reflect on the implications for such projects on the emerging field of ICT in community planning and development.

Case Studies

Case 1 – Center for Neighborhood Technology: The Neighborhood Early Warning System

History

In 1988, several Chicago community groups got together to deal with the problem of housing abandonment. Among the problems that were identified, access to information was one. The Center for Neighborhood Technology (CNT) agreed to take on that piece of the problem.

Initially, the access to information problem focused on several specific communities. Several different information-gathering strategies were tested, including an effort to manually collect the information and keep track of it in a physical card catalog. Next, CNT developed a "Meta-file" program in which they worked with the city of Chicago to acquire data for those neighborhoods and delivered it to community groups on floppy disks. This was obviously a very time-consuming process, especially with regard to updating the data in the various neighborhoods. Eventually, CNT developed a more efficient way of distributing the data.

They brought the data back to their offices, merged the different pieces together in an Informix database, and distributed it via dial-up communications software.

This system allowed for text based searches of the database in a kind of "command line" mode. While command lists assisted users in navigating the system, use of the system was not a casual undertaking. In 1994, CNT expanded coverage from the initial five communities to the entire city of Chicago. Data was updated approximately each month. In 1995, CNT developed the Web interface for their database, opening up to access to a wider audience.

From the earliest days of the project, the city of Chicago was an important supporter. Representatives from the city's department of housing were among the initial participants in the housing abandonment task force. They helped facilitate initial access to the city's mainframe, and provided funding at different points. Over time, these specific individuals in the city moved over to the Planning Department. Eventually, city funding dried up. That more or less coincided with additional difficulties in acquiring city data, resulting in less frequent updates.

Overview

The NEWS system is one of many features of the Center for Neighborhood Technology's site. They provide a wide variety of information relating to their programs and initiatives including urban sustainability, location efficient mortgages, and their *Neighborhood Works* newsletter. Upon entering the NEWS area, users receive a brief introduction, and are allowed to choose their method of searching the database: single address, area (address range or census tract), or area statistics.

Since the development of the Web interface in November of 1995, the system has registered over 27,000 hits as of July 1998. (This probably underestimates actual hits. Some hit counter information was lost due to a crash of the system sometime after the move to the Web interface.) A recent analysis of their server logs shows a consistent hit rate of 1,500 hits per month.

CNT has a close relationship with many of the organizations that use its system. Through this relationship, CNT has largely anecdotal information about how people use their system. They report several kinds of use:

1. *Finding or obtaining information about abandoned buildings.* Commonly this includes trying to determine the owner of a building.
2. *Identifying vacant lots for community uses.* This often includes identifying vacant parcels, finding their owners, and requesting permission for community uses such as a community garden.
3. *Community planning.* Larger and more sophisticated community groups, such as Beth El New Life, may request sections of the database to do more traditional planning activities.
4. *Tenant organizing.* Tenant groups, such as the Metropolitan Tenants Group, use the system to target buildings for tenant organizing.
5. *Lead abatement.* Groups that focused on lead abatement use the system to identify older housing stock, which is more likely to have lead paint.
6. *Homeownership retention.* A group on Chicago's South Side developed a program to help seniors whose homes are at risk due to tax delinquency.

Some of these uses access the data through a Web interface. Others rely on CNT providing customized access to unique chunks of the database. CNT tries to encourage use of the system by providing training and assistance on special projects, as allowed by funding constraints. For example, CNT trained one group in how to copy data off of the Web, geocode it in GIS software, and display it on a map.

CNT speculates that there are also "casual" or "curiosity" visitors to the site. For example, there were significant peaks in site traffic that followed articles about the system in major Chicago newspapers. They admit that there may also be use of the system by real estate and development professionals. But they do not have any specific information to support or deny this.

While CNT is generally satisfied with the amount of site traffic, there is some concern that they should try to increase the amount of "direct use," as opposed to "casual" or "curiosity" use. Their future development plans include efforts at systematically identifying the access to and the use of their system, and increasing the portion of "directed use".

CNT has a number of ideas for improvements to its system. It appears that most are simply waiting for appropriate funding.

1. *"Interactivity."* CNT hopes to get increased involvement from community groups and individuals in generating and verifying the data and their system. For example, they could provide a "community comment" section along with each record in their system. Site visitors could record their comments about property conditions, the veracity of the data, etc., and those comments would be viewable by other visitors to the site. Or, CNT could work with students from local schools, perhaps the social studies class, to verify data in the field.

2. *"Expansion."* CNT has considered expanding the geographic reach of its coverage as well as the types of data that it provides. For example, they have considered expanding into the suburbs (a six-county area). Also, CNT would like to include other data sets that community groups would find useful, such as building and demolition permits.

3. *"Map interface."* CNT would like to include mapping capabilities in their Web site. Appropriateness of the mapping software, and cost of the software, are major factors.

4. *"User interface."* CNT would like to spend time evaluating improving its user interface. Goals of this effort could include "harvesting information" and making the site "easier" to work with. User interface design efforts could be included in a map interface or with the site as it is.

Impact

The NEWS system has had a wide variety of impacts in the Chicago area and beyond.

1. *Access to information.* The NEWS System has been providing access to difficult to obtain information for roughly 10 years.

2. *Promotion of a concept.* CNT has used the NEWS system to promote and disseminate knowledge about the problem of housing abandonment, and tools to address it. Much of this promotion, and distribution of tools, has occurred through CNT's many outreach and training programs over the years.

3. *Promotion of innovations.* Many of the most avid users of the NEWS system do not use it to locate, analyze, or prevent housing abandonment. With free and easy access to property records, groups have devised many different uses for

the data. Some groups have designed programs that, by their own admission, would not have been designed, and would not survive, without the NEWS system.

4. *Distribution of analysis.* The broad accessibility of information and CNT's outreach efforts have allowed innumerable organizations and neighborhood groups to analyze their urban environment, and demonstrate their analysis to others such as city officials. The NEWS system has played significant roles in the analysis of property deterioration, crime patterns, transportation plans, city demolition programs, etc. by neighborhood groups that might not otherwise have the information and resources necessary to perform and present such analysis.

5. *Promotion of neighborhood early warning systems.* CNT has received requests for information from cities all over the United States that are interested in developing a system similar to the NEWS system. CNT was instrumental in assisting in the start-up of the NKLA system.

Challenges

The main challenges facing CNT in providing this service can be broadly classed within the concept of "maintenance." In order of importance, they are:

1. *Obtaining and updating data.* CNT does not report experiencing any active resistance from public agencies in trying to obtain data. Nevertheless, the task of getting data and updating it is enormously time-consuming. Primarily the barriers are bureaucratic in nature. Overtime, CNT has had luck with certain data sets by building relationships with particular city employees. CNT is also actively pursuing relationships with different departments and levels of government, in the hopes of improving their access to data.

2. *Money.* The project is demonstrated that it needs an ongoing inflow of funds. CNT does not believe that there is a market for charging "per hit." There may be possibilities for charging for some more advanced services, such as a map interface. CNT estimates of the cost of simply maintaining the system, exclusive of hardware and software upgrades, is about $20,000 per year. Recently, they secured private foundation support in this amount for the next five years. They are enthusiastic about returning to regularly updating the data.

Case 2 – Neighborhood Knowledge Los Angeles

History

Neighborhood Knowledge Los Angeles (NKLA) was founded in the spring of 1996 by a professor and graduate student in the UCLA School of Public Policy and Social Research. Neal Richman, a professor of urban planning at UCLA, had been working on the redevelopment of a notorious slum property in the City of Los Angeles. One of the remarkable characteristics of the property was that over half of the purchase price went to pay back taxes and unpaid water bills. From this experience, Richman hypothesized that slum properties could be identified through unpaid tax and water bills and that such public liens could be useful for the process of redevelopment.

At the same time, Danny Krouk, an Urban Planning graduate student at UCLA, ran across CNT's Neighborhood Early Warning System while researching the relationship between property tax delinquency and abandonment. Using CNT's site as a model, Krouk and Richman began to develop NKLA as a tool to identify actual and potential slum properties, and provide resources for redevelopment. Databases were rescued from the "informational fiefdoms" of city departments and placed on the NKLA Web site.

NKLA originated at the UCLA Community Outreach Partnership Center, and staff developed a preliminary plan to transfer control of this information project to a resident-based group. Control of NKLA was given over to the Community Building Institute (CBI), a new tenant organizing group, in mid-1996, and the project continued to expand with funding from the Los Angeles Housing Department and a planning grant from the National Telecommunications and Information Administration (NTIA). As the project developed, however, a new consensus emerged that recognized the value of returning the NKLA project to its original home at UCLA. The Advanced Policy Institute (API), the UCLA School of Public Policy and Social Research's center for outreach, training and technical assistance, is now the home for the NKLA project.

Overview

The goal of NKLA is to provide public access to information that helps predict neighborhood disinvestment so that community organizations and residents can more efficiently target areas for intervention. It seeks to meet this goal by placing otherwise difficult to access databases (e.g., LA City Building Code Complaints, LA County Property Tax Delinquency, LA City Building Permits) on the Internet in an easily searchable format. Moreover, NKLA provides content on, and links to, resources for community developers, such as funding opportunities, community development research, and neighborhood data.

On a fairly consistent basis, NKLA receives at least 1,500 requests to its databases each month. Most frequently, site visitors request the following types of information:

1. *Demographic and other statistics.* Once users define the geographic area they area interested in, they are provided with census statistics, environmental information, and housing statistics.
2. *Building code violations.* Site visitors may request information on building code complaints by address or address range.
3. *Property tax delinquency.* Site visitors may view property tax delinquency information by address, address range, parcel number, or parcel area.

The primary users of NKLA have been community-based organizations, students (who account for approximately one-third of the traffic on the site), and city employees (from such departments as Housing, Building and Safety, Planning, and Cultural Affairs) who may otherwise not have access to the data. Other users include neighborhood activists, people looking to buy single-family homes, politicians, and curiosity seekers. There is virtually no evidence that the system is being used for nefarious purposes.

The usage logs, however, do indicate that NKLA has much more occasional use than it imagined. The typical user, from any category, will access the site for a particular purpose, over a limited period of time. Most users will return a maximum of three or four times, and many do not return at all. The number of users that regularly and consistently access the system has admittedly been fairly limited.

NKLA has a number of projects that are in development or are planned for the near future.

1. *Mapping.* NKLA has already experimented with a map interface. Using ESRI's MapObjects, NKLA developed an interface for code violation data, allowing users to drill down through Los Angeles geography, and access records for particular properties. A common map interface to all of the data was recently developed and due to go up in September 1999.

2. *Automated updates.* NKLA is currently working with the city to develop automated mechanisms to update NKLA data. Initially the project would focus on a single data set. Data updates would be sent over the Internet to NKLA at regular intervals.

3. *Community training.* NKLA staff have conducted extensive training on the Internet in general and using NKLA in specific, but the majority of this outreach has been to students and professional staff from city departments and nonprofit organizations. NKLA is currently designing plans for an intensive training program for community residents, especially those living in neighborhoods most at-risk of disinvestment. The outreach program is planned to begin in August 1999.

4. *Community asset data.* In response to the fact that its system has been perceived as "deficits-focused," UCLA recently initiated a community asset mapping project. Youth will identify neighborhood strengths and enter them into an interactive mapping application, thus creating a community-based data set that gives a more even image of their neighborhoods.

Impact

Despite a relatively short history, NKLA has had a significant impact on community development practice and policy in Los Angeles:

1. *Access to information.* In keeping with its mission, NKLA has provided public access to data that was previously inaccessible to the public at large. By applying Internet technology to a practical planning issue, NKLA has opened up opportunities for dialogue, action, and accountability in Los Angeles.

2. *Coordination and innovation within city government.* NKLA staff have been surprised at the level of usage by personnel from city departments. Instead of simply contacting another department for information, city staff have often found it easier to access the city's own data through NKLA. This incoherent situation has prompted calls for more coordinated action within city government with regards to data and information. Slowly, NKLA has helped local government realize the value of their data and the need to develop strategic plans for its dissemination and use. For example, the LA City Information Technology Agency has modeled its mapping program on the Pilot NKLA Map Room.

3. *Policy change*. Within the first two years of NKLA's existence, the Web site helped push addressing slum housing to the forefront of the local political agenda. Community development researchers used NKLA to demonstrate the growing patterns of neighborhood disinvestment in Los Angeles, as well as the insufficiencies of governmental response to these problems. This prompted the Mayor of Los Angeles to appoint a Blue Ribbon Citizens Committee on Slum Housing to tackle this issue, and many of the committee's recommendations were approved by the City Council, signaling an historic shift in local housing policy. In a sense, then, community groups concerned about the conditions of rental housing in Los Angeles gained what Sawicki and Craig (1996) refer to as "credibility and thus entry into the discussion."

Challenges

Despite the success of NKLA in several arenas, there have been several challenges in the project:

1. *Updating data*. The task of obtaining and updating data from a variety of city and county sources is extremely time-consuming. In some cases, there is resistance to the idea of providing data for a public access system. In particular, there have been conflicts of interest where agencies sell public data to generate revenue. More commonly, it is simply bureaucratic obstacles and low priority that get in the way.
2. *Grassroots use*. Despite the fact that NKLA was designed primarily with low-income residents in mind, it has been extremely challenging to engage community members in using the site. This may be partly due to an unfamiliarity with computers, as well as to cultural and language issues.

Case 3 – Neighborhood Planning with Seattle's Public Access Network and "Data Viewer"

History

In July 1994, the City of Seattle adopted a comprehensive plan for the 20-year time period from 1994 to 2014. This municipal planning process was a response to state-mandated growth management policies and sought to identify community values, municipal priorities, strategies and the beginnings of an implementation plan. This new vision for Seattle sought to address a number of concerns by managing growth so that it would not contribute to sprawl or other forms of environmental degradation, enhancing the coordination and responsiveness of municipal agencies to neighborhood concerns, and giving neighborhoods a much larger role in establishing their own development priorities. The city's ambitious program in community-based information outreach and training grew out of a large-scale, municipally coordinated and supported neighborhood planning process to establish new local goals and policies.

The shift toward neighborhood-based planning began before the formal adoption of the city's new plan. One of the central concepts was to promote an urban village strategy whereby housing and jobs are clustered around a neighborhood-commercial core. In general, "urban villages" reflect existing use patterns that the planning process seeks to reinforce, in part by encouraging growth in these areas in ways that preserve local amenities. Rather than select a few neighborhoods each year as pilot project areas, the city council requested in 1995 that the Planning Office embark upon the development and adoption of 37 neighborhood plans simultaneously. Community groups expressed needs for a wide range of data (e.g., census information, traffic volume data, etc.) as soon as possible so that they could meet the terms of their contracts with the city for neighborhood planning services. Since the city could not afford to hire a slew of trained staff in GIS and data systems, the Neighborhood Planning Office (NPO) began to explore other alternatives.

In the summer of 1996, NPO proposed the use of a digital information system whereby neighborhood groups could directly access planning databases, graphics and an easy-to-use mapping interface through CD-ROM disks. The neighborhood groups would be trained to effectively utilize all of the information on the CD-ROMs, and thus diminish reliance on the Planning Office and other municipal departments for data.

Overview

Because the original premise of the planning initiative was to disseminate data on CD-ROM rather than through the Internet, it is best to begin with that portion of the information system. The Information Package "Data Viewer" utilizes ArcView, ESRI's desktop GIS program, which was modified to make it easy to use by nonprofessionals and focused on identified neighborhood planning areas. Data sets include:

1. Base maps (arterials, blocks, boundaries, shoreline etc.)
2. Land Use/Value/Zoning
3. Housing, Health, Education, and Civic Locations
4. Crime and Public Safety
5. Landscape and Environmental Features
6. Municipal and District Boundaries
7. Population and Demographic Data
8. Streets and Transportation
9. Parks, Opens Space and Recreation
10. Utilities.

As noted, the Neighborhood Planning Office (NPO) Web site — hosted on the City's Public Access Network (PAN) Web site — primarily contains support material for the planning process, and provides neighborhood planning documents that are in the process of adoption. There are nine main sections of the NPO pages on PAN:

1. Program Overview
2. Program Information
3. Planning Resources
4. Monthly Events Calendar
5. Neighborhood Focus Newsletter
6. NPO Staff Assignment Info
7. Neighborhood Plans
8. Meetings Notices and Web sites
9. Other Agencies and Resources

There is only one searchable database in this portion of Seattle's PAN and that is a compilation of community development projects.

Impact

The real measure of the tool will be in the quality of the adopted plans and their influence in shaping local development. It is still too early to assess these results. But other criteria might include: Has this public information program broadened participation in the planning process and put real tools directly in the hands of the local citizen? Have there been real efficiencies related to relying less on city staff for analyzing information and more on neighborhood-based planning processes? While more in-depth case studies remain to be conducted as part of this research, some of the initial findings from city staff may be telling.

From the issuance of the disks in July 1997 through October of that year, all but three neighborhood groups signed the licensing agreement, and picked up the CD-ROMs, a total of 172 copies, distributed to 34 groups. During those "kick-off" summer months, 42 individuals were trained by the city on the use of "Data Viewer." Yet in a report written at the end of that period, NPO staff concluded, "Consultants appear to be the predominant user of the tool so far, indicating the need for continuing outreach and education of the citizen planner." Still, these trained consultants did work directly for the neighborhood groups who were each able to pay for professional services with city grant funds. Although grassroots training and utilization of the system was limited, this approach reduced the demand local consultants placed on the city's planning staff for ongoing mapping support.

One impediment with regard to information dissemination seems to have been hardware. Data Viewer is described as a very complex program requiring sufficient computer resources to function well. Nonetheless, the report also notes a growing demand for this information outside of the identified neighborhood associations, from both community-based organizations and private sector-businesses. The information dissemination program may have led to uncovering unidentified demand by many user groups.

Particularly, as the demand for CD-ROMs grows, the real efficiencies of putting information in the hands of the community (or its consultants) are likely to be realized. Staff estimates that producing the five-CD set for each of the 37 neighborhood groups costs $600 each. Grants to each group means that neighborhoods have no out-of-pocket expense. Staff cost for training support average out at $876 per neighborhood. This brings the total cost to approximately $1,500 per neighborhood and suggests significant cost savings in comparison to sole reliance on city staff for these services. Yet, cost comparisons to an Internet-based information system would also be useful. As of December 1999, all but two of the neighborhood groups have adopted their new local plans. Access to technology may have hastened project completion; yet, to accurately evaluate this effect, a control group without GIS access would have had to be established.

Challenges

The Seattle Neighborhood Planning Office and their supporters should be commended for this attempt at expanding access to municipal data and analytic tools.

The breadth of information on the CD-ROM and its presentation makes it an exemplar for information outreach. But questions remain if it makes sense to rely on desktop applications, Internet approaches or a combination.

The city of Seattle does not integrate its Web site with the CD-ROMs as complementary tools for neighborhood planning. It is difficult to even find any mention of "Data Viewer" on the NPO pages of Seattle's PAN. Also, there is no real inventory on the NPO site of the many valuable information resources on PAN that could be used for neighborhood planning. There is a connection to the PAN homepage, but no direct links whereby site users can find information on construction permits or code violations with PAN.

This result is not inconsistent with the city's objectives. The city never proposed to use the Internet for mapping or providing data for neighborhood research. There is still no such plan. From the outset the aim was to disseminate desktop mapping and data tools to support neighborhood planning activities and that goal has been achieved.

Summary of Case Studies

The matrix on the following page provides a summary of the three community information innovations, highlighting both the similarities and differences based on six separate criteria. The Chicago NEWS and NKLA share some obvious parallels. This is unsurprising since NKLA was modeled on the Chicago project. However, Seattle's GIS-based "DataFinder" represents another important case because it has begun to explore the potential of government-initiated public information projects that aim to broaden the opportunities for neighborhood-based planning.

One clear difference is in how planning is conceptualized. In Seattle, the information system seeks to change the built environment through harnessing the power of neighborhoods to change planning and zoning regulations as part of its general plan. In Chicago and LA where the information projects are located outside of government, the emphasis has been shifted away from regulatory and normative standards for planning towards the real power of capital investment. From this perspective, plans are meaningless unless there is the capital behind them to turn them into some physical reality. Thus, the NKLA and NEWS projects are focused on supporting community-based planning processes that pinpoint and respond to disinvestment. Yet, taking a broader view, all three systems in their own way are seeking to encourage more investment (public and private) within the city's boundaries rather than on the urban periphery.

Conclusion

The best way to view each of these three community development ICT projects is as works-in-progress. With limited funds and highly committed staff, each has tried to buck the trend whereby public databases are increasingly available only through commercial means to paying consumers—real estate analysts, planning consultants and others. The true meaning and significance of these projects will be more apparent over time, especially through the precedents they establish in

Summary of Case Studies

Criteria	Chicago Neighborhood Early Warning System (Chicago NEWS)	Neighborhood Knowledge Los Angeles (NKLA)	Seattle Neighborhood Planning Office (NPO)
Website URL	http://www.cnt.org/news/	http://nkla.sppsr.ucla.edu/	http://www.ci.seattle.wa.us/
Purpose	To combat neighborhood disinvestment & deterioration, support community-based organizing & development interventions.	To combat neighborhood disinvestment & deterioration, support community-based organizing & development interventions. Policy change & improving municipal implementation (code enforcement).	To support neighborhood planning in Seattle. Components in an integrated comprehensive planning process. Aims to promote livable, sustainable Seattle communities, planning for growth through urban village strategy.
Region of Activity	City of Chicago. City-wide information but interest especially in low income areas.	City of Los Angeles, with secondary emphasis on County of Los Angeles. City-wide information but interest especially in low-income areas.	Neighborhoods in the City of Seattle. Focusing on urban villages with an integration of housing, jobs & services.
Time Period	Started in late 1980s went on the Web in 1995.	Started in 1996 & continues today.	Began development & adoption of neighborhood plans in 1995.
Institutional Trajectory	Community-based non-profit planning & development consulting firm with some initial university collaboration.	Began as part of university-based partnership program. Temporarily part of new community-based nonprofit. Now housed at UCLA.	One City Government & one department.
Funding	Fits & starts in funding. Significant support from City of Chicago & TIIAP initially. New foundation support for ongoing five-year operations.	Primarily City of Los Angeles & TIIAP Grants. Using surplus from contract work to support public information project. Fannie-Mae Foundation & Microsoft grants for neighborhood planning.	City municipal funding & user payments. Community planning groups must pay for CD-ROMs, but there are planning grants for local groups. Project directly funds neighborhood planning.
Technological Trajectory	From card catalog to databases, to searchable integrated Internet-provided databases. Early warning information integrated by property address with context information—demographic, HMDA, etc. No interactive mapping component.	Web-based from the start with original emphasis on links to community development resources & information. Early warning databases from local agencies are integrated by use of parcel numbers. Interactive mapping component. Moving into complex queries of multiple databases.	CD-ROM planning tool to save staff time in providing information support for neighborhood planning. Internet site for information on planning process & products & access to non-electronic resource material. Few targeted Internet links to electronic data.
Data Type & Integration	Highly integrated context & administrative databases. No mapping integration. "Deficits" emphasis. Non-traditional planning information—early warning data.	Early warning databases are highly integrated. Interactive mapping as well as other search tools. "Deficits" emphasis, but moving rapidly in "asset mapping" alternatives.	Highly integrated physical planning, zoning, & context information on ArcView-based CD-ROM. NPO & Internet site not linked directly to CD-ROM. "Assets" emphasis. Mostly traditional planning data.
User Feedback Mechanisms	E-mail address on site.	Feedback forms on site. Community training & testing sessions. Plans for community-produced data sets.	Completed neighborhood plans posted on city Web site.

providing full and complete disclosure of public administrative/planning information.

We have chosen to not conclude with a traditional evaluation of these nascent, emerging public information initiatives based on the achievement of their long-range goals. Yes, there has been some success in broadening planning participation, in expanding access to technology, in sharing knowledge, and in supporting new development policies, plans and projects. Nevertheless, these are still mostly minor victories, representing at best first steps on a much longer journey.

Rather, in this conclusion we examine some of the common dilemmas faced by each of the ICT projects and the different ways in which each was (or was not) addressed in building these new community information systems. None of the leaders of these projects are naïve e-topians, nor are they convinced of the inevitability of information apartheid. With their hands immersed in managing these complex and often confounding projects, they have been seeking new information pathways to expand opportunities for their constituent communities. Each project in its own way speaks to the potential of ICT for planning and community development purposes, as well as to not-so-obvious risks.

Finally, we look to these three cases for clues about the future of ICT in the field of community development—the challenges and the opportunities. Are these projects harbingers of a new community information order? Or, will they be discarded local experiments in a new information hierarchy built to serve the interests of the powerful? If, as the prefatory poem states, "the revolution's underway," the question remains: an information revolution, for whom?

Seven Dilemmas Facing Community Development ICT

Informing Political Choices or Defining Them?

Each project speaks to the goal of equitable access to planning information, thereby expanding knowledge and the range of planning choices. However, each project also carries implicit assumptions embedded in both the available data and the manner in which this information can be displayed. The Seattle system makes traditional planning data available and easier to map, but incorporates the traditional planning lexicon and tools, hence inadvertently shaping the planning outcomes.

Both the Los Angeles and Chicago projects are also built around a set of implicit assumptions that neighborhood decline can be monitored through the use of a set of public databases and, in essence, that this information is predictive. The systems "pop" these interpretations out at the user as "early warnings," whether in text-based or mapped formats. Without more interpretive background for users, each project risks dangerous misinterpretation (e.g., a property owner with poor skills or bad luck conflated with one who is purposefully exploitative). Yet, even with a very strong set of underlying assumptions, the projects can still reasonably claim to have expanded the range of local choices. These data sets are organized as a counter-narrative, a view into processes of neighborhood deterioration that were until recently inaccessible to local residents. Moreover, the emphasis of these systems on signs of physical and economic decline broadens the range of possible local intervention.

Supplementing Community Knowledge or Supplanting It?

Obermeyer (1995), in identifying the "hidden GIS technocracy," has raised important epistemological questions about "what is the nature of knowledge" and "who constructs it." Accordingly, while technocrats tend to control both the kinds of data collected as well as the underlying programming of the information systems, they make claims about the democratic opportunities inherent in using these inexpensive, powerful, new tools. One concern with the Seattle project is that by making GIS ubiquitous, the project unwittingly establishes a new threshold for participation, defining the terms of planning the debate and raising the bar on necessary technical proficiency. In this instance, the real test of the use of technology to democratize planning would be a comparative analysis of plans adopted in Seattle between those that were developed with these tools and those that were developed by traditional methods. Did these public information systems supplement and improve local understandings or did they drive the planning processes?

Participants in community-based training on the NKLA system frequently corroborate the validity of the on-line data with local perceptions of problem properties. Yet, in one instance where a notorious building could not be found on any of the city databases, important questions were raised about municipal inspection procedures and gaps in local code enforcement. Therefore, these information systems also can be used to check public perceptions as well as municipal action/inaction.

Expanding Public Dialogue or Data Delivery?

The goal of each project is to expand information exchange; however, the systems are generally designed to deliver public data in a more integrated, but still fundamentally unidirectional manner. In the Seattle project, data is distributed through CD-ROMs and the Internet, is used by consultants and their community clients, and then is displayed as proposed neighborhood plans. Neither the CNT nor NKLA projects have incorporated mechanisms for community verification, content collection or even descriptive case studies of "information-in-action" on their sites. Each has plans to rectify this, but most effort still remains directed at the prodigious tasks of securing, translating and disseminating public data. The growing literature on the use of information technology in community development consistently lists community-based training as a primary requirement for successful implementation of these projects (Anderson and Melchior, 1995; Servon and Horrigan, 1997). The cases examined in this study confirm the importance of this type of outreach, especially in the neediest communities, where computer and Internet access tends to be lowest.

Highlighting Neighborhood Potential or Neighborhood Problems?

One risk of the CNT and NKLA projects is an overemphasis on neighborhood deficits, with the risk of reinforcing the processes of redlining by making problems in low-income communities much more visible. One might argue that these information systems now provide hard data to support informal discriminatory lending practices. However, researchers on these projects maintain that the data shows that nuisance properties are widespread and not solely located in poor communities. The challenge for lenders is to support projects that upgrade or redevelop those properties and thereby expand their lending markets. These projects provide the vehicle for this to occur in a more targeted and purposive way. Moreover, the UCLA project has begun a new initiative on neighborhood asset mapping that will provide some countervailing images of low-income neighborhoods.

Facilitating Access to Information or to Power?

Establishing a commission and charging them with the task of research is a classic political diversion taken by government when immediate action is needed. One may ask: "When people demand power, are these projects delivering them new research opportunities?" Are these projects linking low-income people into new electronic networks of power, or, in fact, are these projects only offering people window seats on political/economic processes beyond their control (e.g., opportunities to witness economic disinvestment dollar by dollar in "real time")? Yet, some important policy, planning and program initiatives have emerged from each system. The Chicago system, for example, is linked to the development of the federal low-income housing tax credit; in Los Angeles, NKLA had a role in the development of the new code enforcement program. Though mere access to information does necessarily translate into political power, these types of projects can serve as valuable building blocks in the right group of hands.

Contributing to Urban Synthesis or Fragmentation?

Each project in its own way takes a neighborhood focus, seeking to harness the interest and energy of local residents in shaping their own communities. The Seattle project specifically focuses on the development of neighborhood plans as integrated components within the city's general plans. However, the ease with which information can be organized in neighborhood-specific ways raises the risk of a narrowing of vision. The Chicago project is embedded in the long-standing, integrated and comprehensive planning work of the Center for Neighborhood Technology, and the Neighborhood Early Warning System is an on-ramp to broader sets of community development information. NKLA is developing a Policy Room to analyze early warning signs citywide and will be linking this project with a Web site API is completing on regional housing issues. In this instance UCLA will be working with the Southern California Association of Governments on an interactive regional Web site seeking to inhibit further metropolitan sprawl.

Empowering the Excluded or the Exploitative?

While expanding access to public information is seen by each project as expanding participation, these same systems can be used to increase the domination of certain groups. The NKLA and CNT projects provide a direct guide for pinpointing vulnerable properties and may inadvertently help those seeking to prey on neighborhoods. A disreputable lender could use the system to target low-income homeowners at risk of tax default. Such uses have not become apparent in either case, but NKLA has from the beginning required users to register and state their purposes for the data. In all three cases, project leaders have spoken about information access leveling the political playing field, with an underlying assumption that economic/political elites already have access to these materials.

Many of these same inherent tensions are likely to emerge in other ICT projects aimed at bridging the digital divide. Yet, an important question to ask is: Do these seven dilemmas disappear if information is held tightly by those who have traditionally had access (e.g., government employees, researchers, real estate developers, mortgage lenders). Researchers must be careful not to assume that the status quo is preferable because these sticky issues are buried in traditional power relations. Additionally, dangers of elitism can easily slip into the discourse, and ethical dilemmas arise when others exercise the same privilege that researchers hold by virtue of their advanced training and credentials.

The Future

We end with some crystal ball gazing, using these cases to envision some of the opportunities arising in the future. "Networks, networks everywhere, No place is undisturbed" ...We agree and, as the titans battle over the spoils from emerging business and investment arrangements, we realize that a new level of turbulence has entered into old collusive economic and political formations. We do not see the new formations as being necessarily more or less democratic and inclusive than the old. Rather, we see in this period of rapid change new opportunities for information insurgency, such as increased access to information that belongs in the public domain.

For example, with regard to real estate information, Microsoft has claimed that the National Association of Realtors is involved in a restraint of trade, because the organization has prohibited its agents from listing on the Microsoft home sales Web site. It is in just such internecine battles that we see the opening for left flank, democratic insurgencies, raising questions about why market data is proprietary at all.

Another emerging trend is the public sector shift towards reliance on intranet technologies, away from mainframes, in order to improve internal coordination and efficiency in program finance and management. Most agencies are building thick firewall protections to prevent tampering with their internal data. Yet, even if such gate-keeping is justified, there are few reasons that data sets cannot be exported out of the system in ways that link to wider access. During periods of investment in new public information systems, information advocates must point to the synergies associated with broader public access to data as both equitable and prudent, even if government only agrees to the release of a subset of data fields.

The three ICT projects examined in this chapter are each important because they have jumped in fighting for turf in this new information order, using the legitimacy of the university, of community planning projects, or neighborhood-based constituents to lay claim to information and related interpretive tools. Seattle has produced an information package and made it available at low to no cost to participating neighborhood groups and their technical advisers. UCLA and CNT have assembled data to which only the most sophisticated real estate investor previously had access, making it freely available over the Web and conducting outreach and training with community-based organizations.

Much of the direction of ICT in the field of community development will be driven by how quickly new information systems can build an active and demanding constituency for their product. It will be important that they pull together coalitions that represent the needs of the poor but also incorporate other more powerful interests. For example, UCLA has conducted outreach to the insurance industry on the NKLA information system, identifying a potential ally in the struggle for improving housing conditions and reducing risks to property and people. There must be pressure on government to make its information boundaries porous, especially if government does not take on the role of public dissemination.

Each of the projects analyzed here are noteworthy for their "practical excellence." Whether in fact others will heed the lessons that they are learning remains a very open question. The technology itself does not lead inevitably to more democratic forms of information sharing; but nor does it ensure a narrowing of information networks where the speed transfer creates a world without political friction. Much depends on the step-by-step progress of "Community Informatics" projects that expand the information boundary in the public domain. These initiatives, in turn, must rely on an organized constituency of users, with their base at the community level, who can demand what is rightfully theirs.

Appendix—Other Neighborhood Information Systems

Municipalities
Many cities and counties are making public records, such as assessment records, freely available over the Internet. Here is a sampling of some of the different approaches to doing so.

City of Oakland, Community Economic Development Authority. http://www.oaklandnet.com/government/ceda/ceda.html
The city of Oakland provides access to assessment, zoning, natural and physical infrastructure down to the parcel level through an interactive mapping application. This site is designed to promote and facilitate economic development by providing easy access to property information and development restrictions.

SanGIS. http://www.sangis.org/index.html
The city of San Diego provides access to federal, state, city and county data from the county to parcel levels, through an interactive mapping application. This site seems to exist primarily to provide public access, without particular applications in mind.

City of Ontario, CA. http://www.ci.ontario.ca.us/gis/index.asp
The city provides access to municipal data at the parcel level through an interactive mapping application. This site is designed to promote development and property reuse in Ontario.

City of Milwaukee. http://www.ci.mil.wi.us/citygov/assessor/assessor.htm
This site provides access to assessor records. However, it enriches that information with sales statistics and other items.

Non-Governmental Organizations

Many non-governmental organizations are also embarking on Internet-based public access projects.

Cleveland Neighborhood Link. http://little.nhlink.net/nhlink/

A guide to community services in resources in Cleveland. It includes demographic and other statistical information. However, it consists primarily of program descriptions and links. This site appears to be designed to help nonprofessionals find services information.

Minneapolis Neighborhoods. http://www.freenet.msp.mn.us/nhoods/mpls/

Neighborhood descriptions, maps, HMDA data, and other planning-related information about Minneapolis neighborhoods. This site appears to provide information for planning professionals and academics primarily.

Cleveland's CANDO. http://povertycenter.cwru.edu/cando.htm

Case Western Reserve allows access to a custom application that allows statistical analysis of demographic and sociological data at the neighborhood level. The primary audience would appear to be academic. However, the site does describe a training program aimed at community groups.

EGRETS. http://eslarp2.landarch.uiuc.edu/egrets/

The East St. Louis Geographic Information Retrieval System provides access to static and interactive maps portraying planning and demographic information concerning East St. Louis neighborhoods. This is the product of a university community partnership. Its origins were first in sharing information within the University. It was subsequently "turned outward" to work with existing community outreach programs.

Neighborhoods Online. http://www.libertynet.org/nol/natl.html

This site, based in Philadelphia, acts as a gateway to various federal government statistics sites that may be of use to different community development projects. In essence, it is a links page; however, it does not appear that way when you visit it.

RTKNet. http://WWW.rtk.net

The Right to Know Network exists to provide public access to federally controlled public information. Currently, it provides detailed access to a variety of housing and environmental data sets.

The West Philadelphia Digital Database.
http://www.upenn.edu/wplp/wpdd/wpddhome.htm

This is a university-community partnership project. They do not have an on-line database. However, they have developed a GIS community development/urban planning database program that they distribute.

INFORain. http://www.inforain.org/olmap.htm

An interesting mix of environmental and social information about the Pacific

Northwest available through a variety of text and map interfaces, including an MPEG fly-by viewer.

Commercial

Many of the big commercial information providers do not provide Internet access to the data. They have proprietary dial-up systems or distribute CD-ROMs. This is because much of their market is in relatively sophisticated real estate professionals, such as real estate brokers, banks, etc. Here are a few that are aimed at average consumers.

DATAQUICK. http://products.dataquick.com/consumer/
This real estate information provider has recently begun to port some of its products to the Web. Currently, you can access information about recent sales, crime statistics, and demographics, for a fee.

PropertyKey. http://www.propertykey.com/
PropertyKey is a real estate information service provider. Using their Web client software, one can access real estate records about property assessment, ownership, liens, etc. through various text-based queries. You can visualize information in a map, and you can see photos of properties. This is a fee for service.

References

Anderson, T.E. and A. Melchior. (1995). Assessing Telecommunications Technology as a Tool for Urban Community Building. *Journal of Urban Technology*, 3(1), 29-44.
Atkinson, R. (1997). The Digital Technology Revolution and the Future of U.S. Cities. *Journal of Urban Technology, 4(1), 81-98*
Borja, J. and M. Castells. (1997). *Local and Global: Management of Cities in the Information Age.* London: Earthscan Publications Ltd.
Calabrese, A. and M. Borchert. (1996). Prospects for Electronic Democracy in the United States: Rethinking Communication and Social Policy. *Media, Culture and Society*, 18(2), 249-68.
Eichenbaum, J. and F. Pearl-Schloss. (1995). Digitized Data: Raw Resource in the Municipal Jungle. *Journal of Urban Technology,* 2(3), 81-98.
Forester, J. (1989). *Planning in the Face of Power.* Berkeley, CA: University of California Press.
Friedmann, J. (1987). *Planning in the Public Domain: From Knowledge to Action.* Princeton, NJ: Princeton University Press.
Graham, S. and S. Marvin. (1996). *Telecommunications and the City: Electronic Spaces, Urban Places.* London: Routledge.
Gurstein. M. (2000). *Community Informatics: Enabling Community Uses of Information Technology.* Hershey, PA: Idea Group Publishing.
Huffman, L.A. (June 1998). More Cities Utilizing Technology. *Government Technology,* 14-15, 56.

Innes, J.E. (1998). Information in Communicative Planning. *Journal of the American Planning Association*, 64(1), 52-63.

Kaplan, K. (1999, September 6). Internet Founders See Its Future in Every Aspect of Life. *Los Angeles Times*, C1, C4.

McConnaughey, J., D.W. Everette, and T. Reynolds (1999). *Falling Through the Net: Defining the Digital Divide*. Washington, DC: National Telecommunications and Information Administration, U.S. Department of Commerce.

McConnaughy, J. and W. Lader (1998). *Falling Through the Net II: New Data on the Digital Divide*. Washington, DC: National Telecommunications and Information Administration, U.S. Department of Commerce.

McGarigle, B. (July 1998). Democratizing GIS. *Government Technology*, 14-15, 48.

Nunn, S. and J.B. Rubleske (1997). Webbed Cities and Development of the National Information Highway: The Creation of World Wide Web Sites by U.S. City Governments. *Journal of Urban Technology, 4*(1), 53-79.

Obermeyer, N. (1995). The Hidden GIS Technocracy. *Cartography and Geographic Information Systems, 22*, 1.

Patton, M.Q. (1980). *Qualitative Evaluation Methods*. Beverly Hills: SAGE Publications.

Sawicki, D.S. and W.J. Craig (1996). The Democratization of Data: Bridging the Gap for Community Groups. *Journal of the American Planning Association,* 62(4), 512-23.

Schuler, D. (1996). *New Community Networks: Wired for Change*. New York: ACM Press.

Servon, L.J. and J.B. Horrigan (1997). Urban Poverty and Access to Information Technology: A Role for Local Government. *Journal of Urban Technology,* 4(3), 61-82.

Schön, D.A., B. Sanyal, and W.J. Mitchell (Eds.) (1999). *High Technology and Low-Income Communities: Prospects for the Positive Use of Advanced Information Technology*. Cambridge, MA: MIT Press.

Chapter XIII

Community Impact
of Telebased Information
Centers

Morten Falch
Technical University of Denmark

Introduction

Telebased information community centers or just telecenters have been seen as the killer application to empower local communities in developed and developing countries to meet the challenges of the information society.

The point of departure has been different in various parts of the world, and a number of quite diverse models for development of telebased information centers have been applied. While centers in developed countries, with an almost universal coverage of telephony services, have been focussed on enhancing IT capabilities and access to IT-based communication services, developing countries have also focussed on provision of basic telephony.

This chapter presents the approaches taken in Scandinavia, Hungary, Western Australia and Ghana in order to reach these objectives, and discusses the experiences with the different models and the national strategies used for setting up telebased information centers with special attention to their applicability in developing countries.

Perceptions of Telebased Information Centers

The first telebased information centers established in Scandinavia were termed telecottages. They had much focus on provision of IT facilities and dissemination of knowledge on technology to the public. Since then the concept has developed, and telecenters are now in operation in most parts of the world.

The UK has also a fairly long tradition for telecenters and the number of telecenters is still increasing. However, the distribution of centers among regions is very uneven (with the highest concentration in Wales) and reflects variations in opportunities for public funding. Most centers function as telework centers and provide facilities for teleworkers. Eighty percent of the companies provide facilities for teleworkers coming from more than one company (Cogburn, 1998).

Germany has also a long tradition for telecenters. Telecenters were established in East Germany in 1992, to improve the access to telecommunication facilities after unification. Telecenters are also operating in the Western part of the country (Cogburn, 1998).

In France teleworking plays an important role in the creation of telecenters, the most successful centers act as IT service companies with little or no emphasis on local development objectives. Telecenters are a recent phenomenon. The telecenters are generally very large, but the number of centers was by the end of 1997 less than 10.[1]

In Southern Europe only very few centers have been established, but most countries have announced plans to become more actively engaged in this area. Italy had in 1997 only two rural-based centers but has announced a plan to create 57 new centers. Spain has about six centers all established quite recently. Also there more centers are planned. Training and upgrading of IT-related working qualifications are important ingredients in the centers operating in this region.

The concept of telecenters has also been used to promote rural development in Eastern Europe, most notably in Hungary. Estonia has also quite an active telecottage movement, which has received support from Sweden. More than 50 centers are operating in each of these two countries (CTI, 1998,1, and CTI, 1998, 2).

Australia provides, as one of the few high-income countries with a very sparse population, a unique experience of operation of telecenters in remote areas. Both in Australia and in Hungary and Estonia, a very broad range of activities is included in the concept. The most important is that the activities contribute to development of the local community, and some of the services provided have very little relation to IT or to telecommunication.

In the U.S. many telecenters have teleworking as their primary activity, in particular California has established quite a number of teleworking centers. However, there is also a number of community-based technology centers employed in training of marginalized people.

Telecenters have also been established in many third-world countries. Many of these are essentially phone shops, sometimes also offering fax or other supplementary services. In the Indian state Punjabe more than 10,000 such centers have been established on a franchise basis. In Senegal, 12,000 such centers are established as private franchises initiated by the national public telecom operator Sonatel. Similar centres are planned in Thailand (ITU, 1999).

In addition to this, telecenters have also been established through a local initiative by small entrepreneurs as a result of a growing market demand for IT and telecommunication services. As described below, Ghana provides a prominent example of this. Although their primary services are related to basic telecommunication services, they may also offer other business services like photocopying and typing. So far, these types of centers are mainly located in urban neighbourhoods, where a large population of customers without residential access to basic telecom services exist.

More ambitious centers offering a multitude of services like IT training, distance learning, tele-medicine, informational services, etc., have also been established. These types of centers are usually established in cooperation with international agencies like the ITU (International Telecommunication Union), FAO (Food and Agricultural Organization) and UNCTAD (United Nations Conference on Trade and Development) in an increasing number of low-income countries (e.g., Benin, Mali, Tanzania and Surinam) or as part of national programs financed by telecom operators (e.g., South Africa and Tunisia).

It follows that the concept of telecenters embraces a wide spectrum of strategies for using information and communication technologies in development of local communities. This variation complicates a strict definition of the concept that embraces all relevant developments.

The British Telecentre Association (TCA) makes a distinction between "telecottages" and "telecenters" (Simmins, 1999). While telecottages are community based and emphasize social objectives, such as learning, access to technology, access to work, etc., telecenters are more commercially focussed and emphasize provision of a working environment for teleworkers. This definition corresponds to the concept of shared facility centers developed in California (Gillespie, 1995), but it does not fit very well with the situation in third world countries, where telework plays a very limited role. Here the term telecenter is often used for small shops providing telephony services to the general public.

The concepts of Community Tele-Service Centers (CTSCs) and Multipurpose Community Information Centres (MPCICs) correspond to the concept of telecottages (Engvall, 1998; Cogburn, 1998). These are multipurpose centers that ventilate a number of different activities for the local community, in specific, local community within a rural area or a deprived urban area, so that communal use can be made of the facilities available.

This paper will use the concept of telebased information centers or just telecenters, for centers that supply one or more telebased services to the local community, and thereby contributing to cultural or economic development. Such centers may receive different sorts of funding or generate their own income by selling their services on purely commercial terms. This broad definition includes telecottages, as well as community tele-service centers and multipurpose community information centers.

Regional Experiences

Scandinavia

The Scandinavian countries share a very high penetration of telephone lines, PC and Internet users and have traditionally been in the forefront of advanced usage of information and communication technologies. They are however quite different with respect to industrial structure and geography. Denmark is, in contrast to the other countries, fairly densely populated and distances are short. Finland and Sweden both have a strong presence in the telecommunications industry in particular in mobile technologies.

Sweden, Denmark, Finland and Norway established their first telecottages almost at the same time in the late 1980s. The concepts used in the four countries were quite similar. The aim was both to reverse a trend of outmigration from rural areas, and to increase IT awareness and capabilities. Public funds were provided for the initial investment and for operations during the first years. However, the experiences have been quite different. While Sweden and Finland both still host a large number of telecenters, the telecenters in Denmark and Norway have with a few exceptions all closed down. The development in Sweden, Denmark and Finland is described below.

Sweden[2]

Sweden introduced the concept of telecottages (telestugar) as early as 1985,[3] and Sweden is often claimed to be the first country to set up telecottages, although a few British telecottages have claimed to be in operation since 1980 (Gillespie, 1995). Even today, Sweden is one of the countries with the highest penetration of telecenters.

The first telecottage in Sweden opened in Vemdalen, a village in the north not far from the Norwegian border. The person responsible for the telecottage in Vemdalen was Henning Albrechtsen who had retired to the area after a colorful career as an academic and author. He, in turn, had been influenced by the Dane, Jan Michel, who spoke at a seminar on economic development in rural regions using information technology.

The aim of setting up this first telecottage was to make jobs, vocational training and service facilities available to people in this remote part of Sweden (where there is less than one inhabitant per square kilometer). To do this, access to a variety of computers and modern telecommunications equipment was provided. The incumbent operator Telia and the Swedish authorities provided funding for approximately one mill. Swedish kroner (= 120,000 ECU) worth of computer and communication facilities.

From the beginning, the initiators of the project wanted to create a number of jobs where people could sit at home, each working with their own computer connected to the main computer in the telecottage. Initially the villagers were invited to attend courses free of charge so they could become familiar with the new technology.

The unions were concerned that workers might be exploited in their low-paid isolation. A solution was adopted whereby these individuals became members of the telecottage staff, working at home if they preferred, but with the choice of working alongside their colleagues in the telecottage. As employees, their rights are thus assured, as are their contacts.

The following years more than 40 telecottages were established all over Sweden. The sources of funding were, however, extremely limited. Apart from a few telecottages that managed to attract funding from Telia, the only financial support offered were SEK 50,000 maximum received after establishment from the public authorities. Therefore each center had to generate its own income through commercially oriented activities. A large number of centers had to close down in the following years and in 1993 only 23 telecenters still persisted. The problem was obvious: Some centers were established without good entrepreneurial skill and a good business idea. When the money was gone, the center had to close down. The downturn seems however to have stopped, as it is estimated that about 25 telecottages are in operation today.

Swedish telecenters can be divided into three different categories:

- Community Projects
- Small and medium-sized enterprises (SMEs)
- Cooperatives

Both the needs of local communities and new market opportunities have led to establishment of these centers. The initiative has come from both public and private sectors. Income is mainly generated internally from selling of the services provided by the center. Some centers have received external funding to set up businesses.

Most telecenters are located in small villages, but there are also telecenters in some of the larger towns.

The telecenters provide the following services:

- Computer facilities
- On-line facilities
- Networking
- Education and training
- Information
- IT support and service
- Administrative services, e.g., accounting
- Consultancy work

During the 1980s when the first centers were established, there was an enormous need for information and training in basic computer skills, and there was a huge market for information. There still is, but at that time the centers were really in the forefront in providing this information.

The centers provide services to both small and medium-sized companies and private citizens. Engagement at a telecenter is just one way to do telework. A large number of people are teleworking from home drawing on the data network and other facilities provided through the local telecenter.

The telecenters also act as focal points and as consultants for many local initiatives, e.g., the village network "BygdeNet" (http://www.bygde.net/bygde/bygde.nsf?open).

The telecottages are generally organized as privately owned companies. One reason for their commercial success has been their ability to network and establish contacts with other telecenters, customers and other small businesses. The organization for the Swedish telecottages (TC-S) aims to network telecenters, so competence not available locally can be obtained from other members of the association. The organization establishes not only contacts between telecottages, but also with other small and medium-sized enterprises.

The telecottages are also a part of an active movement engaged in development of rural areas. The stated objective of TC-S is to promote economic development in rural areas by use of IT and new communication technologies.

In Sweden telecenters have operated on a strictly commercial basis for many years. Lack of funding has forced the centers to base their activities on a clear business idea. Centers unable to do this had to close down in the beginning of the 1990s. One of the implications is that telecenters have become more oriented towards business services and consultancy work. In this way they have become an important part of the business infrastructure in rural areas, improving the general environment for small and medium-sized enterprises operating in rural areas. This function is enhanced by the networking of telecenters partly mediated through TC-S.

On the other hand, the commercialization of telecenters limits their ability to engage in local activities which are not commercially viable. Telecenters can act as local kiosks, where public information is available. However, they have to be paid for this service. This is a new kind of cooperation between centers and local development. It is a form of community business, where different local actors collaborate, and where the centers play an active role.

Denmark

The development of telecenters started in Denmark in the late 1980s; 16 projects were initiated with public funding, and in 1990 there were 10 telecenters, six in rural areas and four in municipalities and provincial towns. During the next few years still more centers were created. The main purposes of these centers were education and consultancy.

Some of the centers faced technical problems in particular related to communication and video applications. Therefore the main applications were simple computer applications such as word processing, local databases and computer-games. In spite of this, an evaluation of some of the centers concluded that the centers have:

- Made new information technology accessible for a wider section on the population in Egvad. The major part of the users has obtained elementary knowledge of IT, and a smaller group has profited practically and professionally.
- Local businesses had been somewhat strengthened — in particular mink farmers have benefited from development of a mink program for controlling breeding provided by the telecenter in Egvad (Jæger a.o., 1995).

In spite of this relatively optimistic evaluation, all of the centers included in the evaluation have now been closed. This development is also seen in the rest of the country and so far none of the original more than 60 telecenters have survived, and less than a handful of newcomers exist. The most sustainable seem to be training institutions for unemployed which are based on public funding from various sources.

One of the problems was that none of the centers managed to attract work, which could create local employment. A telecottage on the Danish island Fejø tried to attract administrative work (for example typing) from the Ministries, but they never succeeded, and when the funds initially allocated for the experiments ran out, it was very difficult to continue the operations.

Despite the low survival rate the large number of projects has created considerable awareness and, while in operation, spread of skills. Many of the participants in the telecenter projects have at a later stage been involved in other IT-related dissemination activities.

Although most of the centers are closed some of the activities continue under other auspices. Computer training is done at training centers established as part of local employment programs, and IT access is provided at public libraries and Internet cafés. In addition more than half of the households have their own computer.

Telework is becoming more widespread but telework is done from home and the concept of shared facility centers have not been really used in Denmark. Only one company—a software company—has established a working environment at their own telecenter. One reason may be that the short distances allow teleworkers to come to their place of work a few times every week even from the most remote places.

On the other hand there might be a need for community centers as sites for creation of local content services. The current explosion of community Web sites in Denmark has up to now been driven by local administrations. There might be a chance that — particularly in the smaller communities — local associations may be formed in order to maintain such community services with the local administration services only being part of the total site.

Only a few local communities have taken this path yet but there seems to be a great potential in such a development — based on the elements of Danish culture, like association (a prime instrument for organizing) and the cooperative tradition, etc. A local community association for operating such services may also aid the psychological aspects of the dynamics between citizens and administrations by shifting the ownership of the (virtual) center. The relevance of such arrangements naturally depends on local culture.

Finland[4]

In Finland development of telecenters has from the beginning been closely connected to regional development. One of the first telecenters in Finland was established by Digital in 1988 in the outskirts of Helsinki with the objective of reducing costs of office space. But from 1989 most telecottages were started in cooperation with regional development projects. The objective was to use telecottages as information centers for local initiatives and to disseminate IT awareness into rural areas. A number of projects were funded by the Ministry of Domestic Affairs, local authorities and the national telecom operator. About 70 telecottages were established, but when the schemes for public funding ended in 1991, only the most well-functioning economically could continue their operations. In 1993 there were about 45 telecottages, and in 1997 the number had further been reduced to about 40. Most of these are to be found in rural areas.

Later on other telework initiatives were established. Finland's Advisory Committee for Rural Policy was responsible for drawing up the country's Telework Development Program in 1994. Among the program's suggestions were to encourage the public sector to adopt teleworking arrangements, to start provincial telework projects, to initiate a national 'Telework for the disabled' campaign, and to undertake a national marketing campaign to promote teleworking. These initiatives do not address telecenters directly but aim to encourage other types of telework, e.g., carried out from people's private homes.

A major share of the telecenters are organized as cooperatives. This has in certain instances proven to be a cost-efficient way to organize formerly unemployed freelancers who do not wish to start their own enterprise. An example of such a cooperative is Taitoverkko (Skill Net). This is a multisectoral enterprise, which acts as a node for local know-how, and has created cooperation between people from different professions.

Western Australia[5]

Western Australia provides a very interesting example of a viable model on how telecentres with limited financial support can give an important contribution to community development. Western Australia is an extremely sparsely populated area and the distances are enormous. Although it is the largest state in Australia, it has only a population of 1.7 million, with two-thirds living around the capital. More than half of the rest live in regional centres. Some 200 communities spread over a vast area service the rest of the population.

It is to these communities that the WA Telecentre network directs its services. Fifty-one centers have been established and are supported by the WA Telecentre Network; 800 people participate in the operation of the network. The centers can receive financial support not only for the initial investment but also for their operations.

The network today comprises 51 operational centers with an additional eight having recently received funding. A further 61 communities have indicated their desire to join the network. In total, it is planned to establish 100 telecenters. The centers are located in towns of 200-600 persons and usually at least 50 km from each other.

Telecenters can receive public funding from the State Department of Commerce and Trade. New centers can receive funding for equipment, including telephone connection costs, up to 30,000 AUD. Existing centers can be supported with up to 20,000 AUD per year for salary assistance.

It is necessary to have some support from the local Government or the local community to become eligible for support. For instance, it is expected that the telecenter building is provided rent-free/maintenance-free by the local community. Additional funds for specific projects can be applied for, if the projects can demonstrate regional development or employment opportunities created by the proposed project.

Although the centers may receive state funds, they are community owned and managed. It is also important that they generate their own income. The most common sources of income are:
- Membership fees
- Telebased training or fee-paying courses, e.g., in IT skills
- Secretarial services
- Hire of services and equipment
- Fee from establishing agencies
- Internet access and e-mail services
- Reception services, phone-answering and e-mail receival for SMEs
- Provision of professional rooms
- Labor market programs

One of the most profitable services has been the provision of e-mail services to backpackers, who want to send or receive e-mail.

The telecenters also engage in many activities, which do not always have a direct relationship to information or communication technologies. The most important is that the service has a community impact and is economically feasible. Fifty percent of the telecenters produce a local newspaper for their hometown and one of the telecenters has even established a bakery. This has been economically feasible because the center attracts people and because sharing of facilities lowers the costs.

The telecenters have given substantial impetus to cultural and economic life in remote areas. Educational opportunities is one of the most important benefits. This includes computer training as well as distance learning in a wide range of subjects. The telecenters also provide opportunities for gaining working experience locally. For example a six-month program has been developed for the unemployed. More than 40 people received working experience and new skills by working at a telecenter 12-15 hours a week.

Hungary

Hungary is the most advanced country within Eastern Europe with regard to telecenters. The first telecentre in Hungary opened in 1994, and in 1998 there were 53 in operation. In December 1997, a symposium on telecenters in Hungary was held, a National Telecottage Program was announced, and a plan to establish 100 more telecenters in Hungary was made. The Hungarian Telecottage Association is also prepared to help neighboring countries to start up national programs on telecenters and build similar associations.[6]

The National Telecottage Program has received funding 0.4 mill USD from the Government (CTI, 1997) and international funding has been received from USAID and the Soros foundation. However, the telecottages are established as local community initiatives and do not consider themselves as part of the public authorities.

Their objective is to reduce migration from rural areas through provision of access to information and telecommunications, job training and career counseling, etc., for the local population. The most common services provided by the telecottages are (U.S. Aid, 1998):

- Information dissemination
- Education
- Office and business services
- Communication services
- Consulting
- Community services
- Social care

It follows that a very broad definition of a telecottage is applied. Many of these services provided do not relate to use of IT or telecommunication services. Education and training in computer skills are provided in most centers, but training in other subjects is also provided. Consulting has included agricultural extension services and help to market and export local food products. Some centers also provide legal advisory services and organize local transport (e.g., car pooling) and bus services.

Thus the Hungarian model has extended the range of services provided from

IT-related services typically provided by a telecenter to new areas of major importance for local communities but with little relation to IT.

Another interesting aspect is that the Hungarian Post offices also have begun to extend their range of services to include some of the typical telecenter services. Thus, Hungary hosts both an active grassroots-based telecottage movement and a more centralized initiative building on a similar concept. Time will show which of these two initiatives will be the most successful in the long term.

Elek Straub, chairman-chief executive officer at the Hungarian telecom operator Mátáv, has summarized the community impact of Hungarian telecottages as follows:

"Telecottages will become part of the economy, due to their market value as they are capable to deliver the most varied services and products into areas where they otherwise would have been impossible or very difficult to obtain:
- Telecottages as business mediators: They are in direct contact with the customers and responsive to their needs.
- Telecottages are able to distribute information very quickly by use of their own network.
- The public infrastructure provided by the telecottages offers new possibilities for the cooperation of business, civil and public spheres."[7]

Ghana
Ghana has a penetration of telephone lines of less than one per 100 inhabitants. The telephone lines are highly concentrated around the capital Accra. But although more than 80% of the lines are located in the Greater Accra region, penetration is still below 5%. No other region has a penetration of more than 0.4%. Even in these regions the phone lines are concentrated around the regional capitals. Only 48 out of 110 district capitals are connected to the national grid (Anyimadu, 1999).

This leaves the majority of the population without access to telecommunications. However, the number of lines has almost tripled since 1993 and it is expected that the supply will continue to grow also in the rural areas where wireless local loop services are being introduced.

Although a few telecenters have been established through international grants the overwhelming majority of the centers are purely commercial in their orientation and established by private entrepreneurs as small private enterprises. This development has not been initiated through national or international development programs; the initiative has come from the entrepreneurs themselves. Some of the telecenters were established even before it became legal for private businesses to resell telecommunication services to the public. Already back in 1992 a number telecenters were established and a study from 1997 reports 50-60 centers in the Greater Accra region (Mansell and When, 1997), and the market is still growing.

A field study covering telecenters in three selected regions of Accra (Accra central, Madina and Lima) and two provincial towns (Akatsi and Sogakope) was carried out in May 1999, in order to obtain more information on the economic viability and the community impact of Ghanaian telecenters. At this stage a few provisional results from the study are available.

More than 20 telecenters were recorded in each of the two neighborhoods of Lima and Madina. Lima is a typical low-income residential area with small businesses as well. Madina is a high-income suburb of Accra situated close to the university campus. The number of telecenters in the business district of Accra central seems to be more limited. This may be due to a higher penetration of private phone lines in this area. Outside Accra telecenters are located in larger provincial towns and even in a few villages using wireless local loop.

In Accra the standard equipment for a telecenter is two telephone lines, two phones, fax, photocopier and one to two computers. The telecenters outside the capital are however generally less advanced than the telecenters in Accra. These centers usually only have one or two phone lines and no computers.

All centers have telephony services as their main source of revenue, and many centers provide both outgoing and incoming calls. In Accra local and international phone services are the most important while the calls made from Akatsi and Sogakope mainly are trunk calls. Many of the telecenters are either established in conjunction with other businesses or try to expand their revenue by offering complimentary services, such as typing and photocopying. One telecenter also offered collection of the daily revenue for small businesses too small to have their own banking account (the so-called su-su service).

E-mail services are offered from a few centers in Accra. An e-mail address can be obtained free from Africa online and e-mails can be sent for about 0.25c per message. This service is also available from post offices (Sangonet Database).

The major source of funding seems to be earnings made abroad either by the manager or by a close relative, but telecenters are also established as part of other small businesses financed by business income generated locally.

Most of the centers interviewed were very new and not very profitable. The daily revenue varied from 2 USD to 60 USD; most centers reported a revenue at around 25 USD per day. Many of the centers in Accra which have been in operation for some years, complained about increasing competition. Many have seen a telecenter as an easy way to establish a new business and it was the impression that in Accra there were too many centers to make the business viable.

The centers also felt the competition from the increasing number of phone booths, which offer lower rates. Use of phone booths demands that the customers have invested in a phone card. This is a major expense and an important barrier to usage of phone booths by the rural poor. People using phone booths are generally more educated than the customers using telecenters.

One telecenter in Madina was established by use of radio wave technology. The center was established in 1992 long before Ghana Telecom extended their service to the area. The owner of the center did not see the telecenter business to be profitable anymore, but he considered setting up a new business in a rural area not yet served by any of the telecom operators.

The busy hours in the Accra-based centers are early morning and evening, indicating that the services mainly are used for contacting family members and friends. However, particularly in Nima, a number of small enterprises rely on the services from the telecenters in their business. An indication of this is that some centers combine their tele-services with other business-related services.

In Akatsi market days are the most busy. Villagers use the market day to do the phone calls they need. A survey of users in the Akatsi region concluded that telecommunication services mainly are used for enhancing their business. The main benefit was information on market prices enabling farmers to decide when to travel to a specific market to sell or buy products (Anyimadu, 1999).

Models for Generation of Income

A key issue for all types of telecenters is funding. It is essential that telecentres have an economically sustainable strategy right from the beginning. In this context it is important to distinguish between demand-driven telecenters and telecenters established as part of special programs.

The latter model has been the dominant one in most industrialized countries. Typically, the establishment of telecenters in rural areas has been initiated through some sort of public funding — sometimes supplemented with grants from the national telecom operator. Public funding is usually given for a limited period of time. Either because they are given as part of a program with a limited lifecycle or because the intention is only to finance the start-up costs, keeping in mind that the centers ought to be economically viable in the long run. One exception is Australia where funds also are provided for operational expenses. Many countries have developed national programs supporting telecenters and at the international level the EU has initiated a number of supporting programs for both the EU and Eastern European countries. Telecenter programs have also been implemented in developing countries, usually with (time-limited) financial support from government and, in some cases, from international development agencies.

It is however difficult to make a sharp distinction between a demand-driven telecenter with a commercial orientation and a telecenter set up as part of a special program. Most centers have a commercial orientation and have been created as the result of a local initiative, but they have also received some type of financial support. In addition many commercially oriented telecenters generate a substantial part of their income through provision of services to the public or participation in public-funded projects.

The possibilities for funding are related to the services provided and the objectives to be met. Many centers have focused on creation of local coherence by creating a meeting place for economic or cultural activities, and many services have been provided free of charge. This type of center has been found to be the most eligible for public funding, but they have on the other hand found it very difficult to redirect their activities in a way that can generate a sufficient income to survive after funding has ended.

The general experience from Scandinavia and other countries with a long history of telecenters is that telecenters, which are independent and not integrated in a larger organization, have found it very difficult to survive on public grants alone. Sooner or later they have to generate their own funds in some way or another. One important exception is Western Australia, where telecenters can receive some public support on an ongoing basis.

Some of the most common business models applied are:
1) Integration with local institutions (training centers, schools, libraries, etc.).
2) Service provision for local authorities, e.g., training courses, cultural information centers, etc.
3) Service provision for (local) businesses, e.g., accounting, Web design, etc.
4) Provision of public access to IT and telecom facilities (Internet cafés, telecenters, etc.).
5) Telework facilities for a number of companies.

Centers operating as part of a larger organization do not have to depend on external funding if their operations contribute to the overall objective of their parent organization. They can, e.g., provide training courses for unemployed as part of a labor policy scheme if integrated in a training center; or they can provide electronic access to library databases if they are located in a library, etc. The drawback is that the activities may be limited by the objectives of the parent organization. Centers which are part of a public institution may, for example, be restricted in their supply of services to private businesses.

Centers can also operate as independent entities, but provide public-funded services. This model gives more flexibility as the center more easily can develop and supply services for other customers as well. Different models exist. Some centers can produce IT services or similar services, which also could be obtained from any consulting or accounting firm. Such centers are hard to distinguish from any small, local, private company. Others may provide training courses, public information and cultural services for which they receive some payment from public authorities. A mixture of the two can also be found. The drawback will often be lack of basic funding. A center will generate its income from production of services sold to public authorities, but the centers themselves are responsible for generation of a sufficient income to survive.

A third source of income is provision of business services to local businesses. This could be IT-consultancy, accounting services or services with some ingredient of IT. These services will often, after an initial phase, be carried out on a purely commercial basis. Again it might be difficult to distinguish such a center from any other local company — in particular if the customer base is broadened gradually to include companies from other regions or from abroad. One difference may be the structure of ownership, which may be a cooperative or a partnership instead of a personally owned company.

A fourth source of income is provision of public access to IT and telecom facilities. This is the most important activity of many of the telecenters located in many developing countries. They provide mostly basic services such as telephone and fax. However, also countries with almost a universal service coverage have centers providing IT or telecom facilities. An example of this is the many Internet cafés popping up in many countries, although these cafés usually are excluded from the definition of telecenters.

Another example is related to the fifth source of income mentioned above, namely provision of working places for teleworkers. This model is quite widespread in the UK, and such centers have also been established in the US. Small companies can rent office facilities or individuals can come occasionally to do their work there.

As computers in private homes have become more common, such centers can also act as providers of an IT hotline for people working from home and supply them with more advanced facilities such as high-quality color printers, etc.

Demand-driven telecenters in the industrialized countries are often difficult to identify and most often they are not termed as telecenters, but rather as Internet cafés, call centers, information centers, small consultancies etc. These types of companies may offer services, which are hardly distinguishable from those of the telecenters, the major difference being their commercial orientation.

In developing countries the demand-driven centers mainly focus on provision of basic communication services such as telephone, fax and sometimes also e-mail and Internet. Demand-driven centers can be initiated on a franchise basis as is done by the public telecom operators in Senegal and in Punjabe in India. A similar model is applied by WorldTel (Ravi, 1998). WorldTel is a private limited company established on the initiative of ITU. WorldTel is more ambitious with regard to the range of services they want to provide and has developed a commercially viable model for rapid penetration of the Internet in emerging markets through establishment of community Internet centers. WorldTel plans to establish Community Internet Centers in Indian towns and in other regions as well.

Another type of demand-led centers are those established on the initiative of local private entrepreneurs without being part of any type of special program, as we have seen in Ghana.

Although the primary services of the demand-led centers are related to basic telecommunication the centers may offer other business services like photocopying as well. So far these types of centers are mainly located in urban neighborhoods, close to a large population of customers without residential access to basic telecom services.

The telecenters in Ghana are commercially oriented and established with the primary objective of generating income for their owners. However, they provide an important contribution to the local community by providing telecom access in a country with a very low penetration of phone lines. The number of phone lines is rapidly increasing and so is the number of phone booths. It is therefore possible that telecenters will become less important in the future. At present the telecenters generate most of their income on phone services and it is a question whether the urban-based telecenters will remain viable. The survey (discussed earlier) indicates that the density of telecenters is smaller in the most advanced districts.

The long run viability of the telecenters will depend on their ability to upgrade their services into new areas, such as e-mail and other computer services. In this relation lack of human resources may be a major barrier. Most of the employees in the telecenters do not have the expertise to offer more advanced business and community services of the type offered, e.g., in Hungary. They also lack a telecenter association, which can support them in development of their business. It has therefore been proposed to establish a Ghana National Service Center by private investors (Community Communication Centers for Knowledge Exchange, 1999). Such a center should support telecenters in development of skills in telecommunications technology and management, and possibly other skills as well.

In the rural areas affordability by the local people can be a barrier against further development. Although there is an obvious need for access to telecommunication facilities in rural areas, it may not be possible to provide the service at prices affordable by the local community.

This question cannot been answered in general, as rural income levels can be very different in different countries and regions. Experiences from remote areas in high-income areas such as Australia, Alaska and Northern Canada indicate that demand in rural areas generally exceeds forecasts based on population density or per capita income (Hudson, 1998). Preliminary studies of the financial viability of telecenters in rural areas in low-income countries indicate that they could be attractive business cases, even as a "stand-alone" business (Uganda), or in small groups (of 12 telecenters in the Indian case), at least for local entrepreneurs, but this remains to be proven (Ernberg, 1998). The experiences from Suriname indicate that low income can be a barrier towards establishment of an economically viable telecentre (Goussal, 1998). In Ghana the rural based telecenters had on average a smaller turnover than in the capital, but the costs of operation were also smaller.

Telecom access and services are a major expense for all telecenters. It is therefore important for their viability that a fair arrangement is made with the telecom operator. For example, the Ghanaian district town Sogakope is not serviced by Ghana Telecom, but by Capital Telecom. Capital Telecom offers wireless local access. Although this may be a cost-effective solution, their charges are several times higher both with respect to monthly line rental and call tariffs. A part of the explanation may be the high interconnection charges paid to Ghana Telecom, but the higher charge rate significantly affects the viability of telecenters in the region.

In some countries (e.g., Hungary) preferential rates for telecenters are negotiated. Another more ambitious solution is considered by WorldTel, namely construction of a common backbone network for their telecenters. This will make telecenters operated on a franchise basis by WorldTel less dependent of the tariffs offered by the telecom operators.

The Community Impact
of Telebased Information Centers

Although the point of departure varies from region to region, and although the services provided by telecenters also vary, the overall objectives are basically the same. Namely upgrade of the local access to information and communication in order to generate economic and social development in a local area.

This overall objective can be addressed in different ways. At least five related and partly overlapping issues are addressed by establishment of a telecenter in a local community:

- To create regional cohesion (cultural or economic)
- Infrastructure (1): To provide access to IT and telecom facilities
- To promote diffusion of usage and knowledge of IT
- Training: To train local people — particular in IT-related qualifications
- Infrastructure (2): To provide access to IT-related business services
- To create local employment

The priority between these issues and the way they are addressed may vary, but most telecentres will address a number of them.

Regional Cohesion

The concept of telebased community centers was first developed in Scandinavia in the mid-1980s. Rural communities suffered from lack of employment opportunities and lack of an infrastructure of local services. Employment in agriculture had in many countries decreased to one-third of the level a few decades ago, local shops were closing down and rural communities had developed into satellite communities with decreasing populations and no internal infrastructure.

The objective of the first telecenters in Scandinavia was to use modern information technology to strengthen the cohesion of local communities in rural areas.

Also in Hungary and Western Australia, regional cohesion is addressed very explicitly through provision of community services such as issuing of a local newspaper.

Creation of regional cohesion is a very broad objective and all the subsequent objectives contribute more or less to this overall objective. It reflects the dualism of the objectives of many telecenters, which both have soft-value objectives, such as enlightenment and cultural development, on the one hand, and more hard economic objectives related to generation of income and employment on the other. Although many centres address both types of objectives, the priority between the two has been very different.

Provision of Access to IT and Telecom Facilities

This objective is the most basic objective, as it in some sense is a precondition for usage of IT and telecommunication services to create local development. The first telecentres established in Scandinavia did not focus on access to telecom facilities, as universal access to basic telecommunication facilities was already in place. On the other hand, provision of local access to computers was one of the primary objectives. It was often the ambition to network these computers, and also to provide other more advanced telecom facilities, but in the mid-1980s these technologies were not matured and only a few telecenters had the capability to overcome the technical problems related to provision of such services.

In Ghana telecenters compensate for the low penetration of phone lines through public access to telecommunication facilities. This is often the most important service, both with respect to the number of people served and to the income generated. The telecom service is used both for private and business communication, and is of major benefit for many local businesses.

As yet e-mail plays a very limited role in this respect, but the service has a great potential. E-mail addresses are offered free of charge and the use of e-mail is growing rapidly. However, Internet connection from rural areas is still very expensive and the growth is centered around the capital.

Promote Diffusion of Usage and Knowledge of IT and Training

Usage and knowledge of IT is promoted both through the mere access to facilities and by training. Training is an important ingredient in the activities of many telecenters. In particular in their initial phase, many telecenter programs have focused much of their attention to training activities.

Training can, if successful, be a major source of income for a telecenter. Training activities also contribute to establish the center as a central meeting place in the community, and strengthen other informational activities.

Informational activities are usually supported by public funds, and it is unlikely that commercially based telecenters (like those in Ghana) will engage in this without any support from donors or the district council. Very few of the centers have the skills to provide training or other informational activities.

Provision of Access to IT-Related Business Services

Provision of access to IT-related business services is important for promotion of IT as well as creation of local employment opportunities. Employment can be generated both directly at the community center and indirectly when access to business services for local companies is improved. An example of this double impact is a South African entrepreneur, who, after taking a two-day bookkeeping course, was able to generate his own income by marketing of business services from a community center to local shops. Such a service would be an obvious extension of the services provided by telecenters already having a computer and offering ordinary typing services.

Business services are provided within many different types of institutional setups and telebased community centers are only one out of many. In Poland more than 200 non-profit business service organizations are set up as public information centers serving local business companies. These centers may not be termed as telecenters, but the services they provide to local small and medium-sized businesses are basically the same. Similar centers are present in a smaller scale both in Russia and in the Baltic countries.

Creation of Local Employment

Local employment can be created both by improvement of the local business environment as described above, and through different types of telework or distance work. An example of the latter is the French Telergos, which provides administrative services mainly to insurance companies. Such activities do not directly contribute to local coherence through improvement of the local business environment, but they contribute to local development through creation of local jobs and income. It can be discussed whether this type of activity should be included in the concept of community centers. However, it may be difficult to draw a borderline between centers providing tele-services locally and centers with a more national or international orientation. Many of the Swedish telecenters act indeed as local companies, but have expanded their customer base beyond their own region, e.g., by networking with other centers.

Telework in its traditional sense will hardly be practiced in a major scale in developing countries. However, telecenters can enable local businesses to expand their market as they can receive orders and make offers by phone or fax. The direct generation of employment at Ghanaian telecentres is limited, as most centers only employ one or two persons. However, the revenue generated may constitute an important extra cash income in small, rural-based communities.

Conclusion

The development of telecenters has been different in various regions of Europe, and the number of telecenters differs widely from country to country. In Scandinavia where the first telecenters were established in the 1980s, the number of centers has declined in the 1990s. These first centers played an important role in the dissemination of information technology to a wider audience by provision of basic IT facilities and courses.

Today the Scandinavian countries all have a very high penetration of PCs in private homes and PCs are also available at public libraries. In Denmark IT courses are provided by the labor market authorities. For these reasons the needs for the services first provided by the telecenters have declined. In Denmark and Norway the result of this has been that very few centers exist today. Sweden and Finland however still host a fairly large number of telecenters. These centers are part of a very active movement for development of rural areas. The primary functions of these centers are not strictly related to provision of and training in use of IT. IT is rather a tool used to improve local infrastructure and to generate local employment. In both countries the centers operate on a commercial basis. Although they generate a part of their income by provision of services to public authorities, many of the centers can hardly be distinguished from other rural-based small businesses using IT in their operations. In Sweden this is reflected in the list of members of the organization of telecottages, which include many private companies and consultants using IT.

Networking has been a critical factor, particularly for the Swedish telecenters. By networking with other centers and businesses they have access to more customers and have the ability to share projects with other partners.

The experiences from Scandinavia as well as other parts of Western Europe indicate that telecenters must be able to generate their own income from commercially based activities if they are to survive for a longer period. Public funds can be used to set up new telecenters and train the participants, but it seems to be very difficult to survive on attraction of public funds in the long run.

Telecenters have therefore been forced to focus more on provision of services generating sufficient income to make the telecenter sustainable. Many centers can hardly be distinguished from other small firms offering consultancy or other business services. Others concentrate on creation of a telework environment and host a number of small business service firms. Although community services receive less attention in this way, the telecenters still contribute to generation of local employment and income.

Telecenters generating their own income can also contribute more directly to supplying noncommercial activities related to enhancement of local economic and cultural development in rural periphery regions, if local authorities out-source part of their obligations to the local telecenter. Telecenters can carry out services on behalf of local authorities and thereby generate income for themselves. The most successful model seems to be a center generating its own income, but with a relationship to local development movements—being private or public. Public support is reduced to initial funding and the buying of services from the local telecenter.

The Australian experience is somewhat different as public funding is provided on an ongoing basis. Although the grants given are limited, this has resulted in growing number of telecenters and a very high penetration compared to the number of inhabitants. This has enabled the centers in their strategy to maintain their focus on community impact rather than on generation of income.

The Scandinavian centers were established at a time when penetration of PCs was very low among private households and small and medium-sized enterprises. Therefore access to IT equipment and development of IT-based skills were given high priority. These services also play a role in Hungary today, but are less important in Scandinavia. In Ghana provision of access to telecommunication services is the most important. So far, none of the centers visited are able to offer computer training.

Both in Hungary and Western Australia, a very broad concept of services offered have been applied, the most important criteria being community impact rather than relation to information and communication technologies. This concept seems to have been very successful with regard to community impact, but it requires a strong ongoing commitment from local authorities or donor agencies to be economically viable.

The telecenters in Ghana have been set up without any type of financial support. This is reflected in the type of services provided, as they focus on the currently most profitable service, namely telephony services. Many of the centers complement the revenue from the telecenter by combining several types of businesses, such as su-su (a type of informal banking service), renting of video cassettes, gift shops, restaurants, etc.

The economic viability of the Ghanaian telecenters may be threatened by the growing competition from telephone booths and private phones. If they are unable to renew their range of services they will have to close down, just as many of the Scandinavian centers focussing on computer access did. Only few of the existing centers in Ghana will be able to manage such a reorientation of their activities without support for development of IT and other skills. If telecenters are upgraded to supply e-mail and other Internet services this will greatly enhance the rural communication infrastructure. Internet services can also be easily combined with informational services on market information, agricultural consultancy services, etc.

Informational services could partly rely on information provided by development assistance projects. In this way telecenters would increase the outreach of these projects. Telecenters make an important contribution to the information structure in developing countries. Limited support and national coordination can increase this contribution substantially and add to the economic viability of these centers.

References

Anyimadu, A. (1999). *Telecommunication Services in Ghana—A Sector Overview and Case Studies from Southern Volta Region.* ZEF/DETECON and German

Watch Workshop on ICT and Economic Development. Bonn, May 31-June 1, 1999.

Center for Tele-Information: *Telecottages in Estonia*. http://www.itu.int/ITU-D-UniversalAccess/casestudies/estonia.htm.

Center for Tele-Information: *Telecottages in Hungary*. http://www.itu.int/ITU-D-UniversalAccess/casestudies/hun_mct.htm.

Cogburn, D. (1998). *Knowledge in development: Multimedia Multi-purpose Community Information centres as catalysts for building innovative knowledge based societies*. World Bank Background Paper.

Community Communication Centers for Knowledge Exchange. (1999). Mission report of a collaborative mission of InfoChange Foundation and infoDev, United Communications Systems International and the World Bank.

Engvall, L. (1998). *CTSC International* (http://arla.rsn.hk-r.se/~engvall/CTSC/7.html).

Ernberg, J. (1998). *Universal Access for rural Development—From Action to Strategies*. ITU Seminar on Multipurpose Community Telecentres. Budapest, 7-9 December 1998. http://www.itu.int/ITU-D-UniversalAccess/johan/telecentremain.htm.

Gillespie, A., R. Richardson and J. Cornford (1995). *Review of Telework in Britain: Implications for Public Policy*. University of Newcastle Upon Tyne.

Goussal, D. *Rural Telecentres – Impact Driven Design and Bottom-up Feasibility Criterion*. ITU Seminar on Multipurpose Community Telecentres. http://www.itu.int/ITU-D-UniversalAccess/seminar/buda/papers/final/F_Goussal.pdf.

Hudson, H. (1998). *Rural Telecommunications: Myths and Realities*. National Telecommunications Co-op Association, Nov.-Dec., USA.

ITU (1998). *World Telecommunication Report 1998*. International Telecommunication Union, Geneva.

ITU/BDT (1999). *Report on the Regional Seminar for Central European Countries, Budapest, 7-9 December 1998 (organized by the ITU in partnership with UNESCO, CTSC International, the Telecottage Association of Hungary and the Communication Authority of Hungary)*. International Telecommunication Union, Telecommunication Development Bureau, Geneva.

Jæger, B., J. Manniche & O. Rieper (1990). *Computere, lokalsamfund og virksomheder (Computers, local communities and private companies)*. AKF Copenhagen.

Mansell, R. and U. Wehn (Eds.) (1997). *Building Innovative Knowledge Societies for Sustainable Development*. Working Group on Information Technology and Development, UN Commission on Science and Technology for Development.

Ravi, N. ITU Seminar on Multipurpose Community Telecentres. http://www.itu.int/ITU-D-UniversalAccess/seminar/buda/papers/final/f_ravi.pdf.

Sangonet Database. African Internet Connectivity. http://www3.sn.apc.org/africa/ghana.html

Simmins, I. (1999). *What is the difference between a "Telecottage" and a "Telecentre?"* http://eto.org.uk/faq/faqtcvtc.htm.

U.S. Aid (1998). *Our Telecottage – Rural Communities Towards Information Society*. Budapest.

Endnotes

1 Information provided by the French Distance Expert Nicole Turbe-Suetens.
2 I would like to thank Lilian Holloway, Telestugan i Ammarnäs (The Telecottage in Ammamäs) for information used in this section.
3 A telecenter in Nykvärn outside Stockholm was even established two years earlier with support from Telia. This center was, however, not a telecenter in the traditional sense, since its objective was to serve telecommunters working in Stockholm and not to generate local development.
4 I would like to thank Lars Tollet, partner in the telecottage of Taitoverkko who has kindly provided information for this section.
5 This section builds on information from Gail Short: "The Socioeconomic Impact of Telecentres in rural and remote Australia." The role of Community Telecentres in fostering Universal Access and Rural Development Regional ITU/UNESCO Seminar for Central European Countries Budapest, Hungary, 7-9 December, 1998. See ITU (1999).
6 Information from Mátyás Gáspár, Honorary President of the Hungarian Telecottage Association.
7 Elek Straub: "The Role of Multipurpose Community Telecentres in Fostering Universal Access and Rural Development." The Role of Community Telecentres in Fostering Universal Access and Rural Development Regional ITU/UNECO Seminar for Central European Countries Budapest, Hungary, 7 - 9 December, 1998, Reported in ITU (1999, p. 9).

CI Applications

Chapter XIV

Cafematics: The Cybercafe and the Community

James Stewart
University of Edinburgh, Scotland

Introduction

While mainstream industry and government focus on individual, home and business ownership and use of new ICTs, there is a quiet revolution going on as computers, and all their applications from games to the Internet, move into public spaces. There are commercial kiosk systems in the streets and malls, and many government projects to empower communities and stimulate the local economy, but perhaps the most important, overlooked and oft-derided development is the cybercafe.

The cybercafe is a cafe or shop open to the public, where a computer can be hired for periods of a half hour to access the Internet, write a CV or play a game. With the explosion in the use and profile of the Internet and personal use of new information and communications technology—'multimedia'—cybercafes have become part of contemporary culture, established among the public places of modern cities, towns and villages around the world. In December 1999 an on-line cybercafe guide listed 4,397 cafes around the world.[1]

There is very little research on what these cybercafes are used for, who uses them and why.[2] This study, conducted in 1998 (Stewart, 1998), addressed the use and users of three cybercafes in the same city, the reasons and manner they were set up and developed, and the role cybercafes play in the general development of use and knowledge about multimedia. What emerged was that cybercafes are not only sites for technical access and consumption and use of multimedia content and services, but also public, physical, community and cultural spaces. In this context I challenge the view that computers either undermine the community, or are only relevant to the formation and activities of 'virtual' communities.

The cybercafe is not a transitory phenomenon, but the evolution and extension of a very old and traditional institution, the cafe. Cybercafes may service and reflect the communication and information needs of people living in a global society, but they place this in a local context, providing a social space and a convenient and hospitable location for technology access: the 'human face' of the information society. If the city is our home, then the cybercafe is becoming an important part of our domestic life. Cybercafes bring IT into real communities, allowing people to use and learn about them in their own way. The managers and customers of the cafes are finding new ways to incorporate this global phenomena into the everyday life of the city.

The study looks at three different cybercafes in the same city: how and why they have developed and are being used, the business, the technology, the customers, and the staff. In this chapter I look at who uses the cafes, and why, highlighting the *triggers* that brought them in in the first place, and the reasons why they come back. The convenience, sociability, learning opportunities and games stand out as principal factors. I follow with a discussion of the importance of the cafe as a focus point and gateway for local, virtual and distant communities. Finally I argue that rather than becoming irrelevant, the cybercafe has a strong future, in ever more diverse forms as increasing numbers of people come to use and rely on the Internet, electronic entertainment, commerce, communication and information services as they go about their everyday life. People like to do things in public spaces and in social spaces, and if those things involve multimedia, then cybercafes are in a position to satisfy a growing market.

Cybercafes, Cafes and ICTs in City Life

Multimedia in the City

There is a growing body of work examining the way that the use of network technology affects or might affect contemporary city life. It includes ideas such as electronic commerce and government, exclusion and inclusion, virtual communities, and city life moving into virtual spaces, with the 'digital city' (Graham and Aurigi, 1997). Much of the literature focuses on on-line communities that are no longer geographically bound (Reingold 1994). It conceives of individual users locked away in their own rooms, offices or homes accessing these communities. However there is little interest in the points where IT use becomes public. This is reinforced by a dominant paradigm, supported by industry, of individual ownership, and individual use of multimedia and the Internet. However recent surveys show that many people access the Internet in public spaces (16% in the U.S., Spring 1998; 24% in a UK survey, The Guardian, Summer, 1998). The huge uptake of free Web based e-mail accounts also indicates that many people do not have their own Internet access, or frequently access their e-mail away from their own computer.

New media and communications technologies tend to go through a sequence of public and then personal ownership as they are simplified and become cheaper, and uses and knowledge develop among users. This has occurred for the television and telephone, photography, video games, and increasingly for the computer. It occurs in organisations as well, e.g., reprographics and computers. There are some technologies where the skills and costs make it very difficult for people to actually own and maintain the technology, but it is cheap enough to install in a very convenient local community location. In these cases we have seen the development of local provision of video hire, photocopying, DTP, pay-TV and film processing.[3]

Even with technologies that moved into the home or office, a commercial or public provision often develops locally to service the local market and provide a professional level of provision. In the case of the telephone and the television these technologies have become as common in public spaces as in private ones. One common thread though all of these developments is *convenience*—local, on demand, and pay-as-you-go services. Cybercafes fit into these trends of renting equipment, convenience, and need for expertise in maintenance and training.

Origins of Cybercafes

Cybercafes appear to have been developed in the USA in the early 1990s, often as an extension of existing attempts to democratise access to computers and to media in general. A trendy cafe with computers to surf the Net was a bizarre novelty, computers and the Internet being associated with a solitary occupation of 'anoraks' or with work. They did not seem to fit with the conviviality of a cafe atmosphere, where face-to-face contact, escape from work, etc., are central to the experience. However the relaxed, informal atmosphere of the cafe was precisely the aim: the cafes were promoted as a 'human' place to learn about computers and find information. However, as the Internet and computers suddenly became a widely diffused part of mainstream culture, why should cybercafes or public Internet points continue to exist? This paper shows that they have a very good reason to become an even more common and permanent part of our world.

Cybercafes are not the only public access points in the city. There are range of 'cyber' centres. There are schools with IT centres for local business (Micro-borough, FT 25/3/98), Libraries with Internet and CD-ROM facilities, business centres offering Internet access and computer facilities, 'telecottages' providing technology for business and cultural projects, video stores with computer terminals, and computer training centres. Even banks offer a chance to surf the Net. There are also initiatives to open government one-stop shops to provide on-line access to national and local government services through local telecentres.

Cafes

Computers and the Internet in cafes are a natural extension of existing facilities and uses of cafes and other public or semipublic venues. People have always met to eat, drink, talk and play games in places such as inns and taverns. The first 18th century 'cafes' were centres of community for informal discussion of politics, local affairs, and culture, frequented by particular social groups. Perhaps more than traditional drinking establishments, information was central to early coffee shops (Sennett, 1977, 81), some even publishing their own newspapers, others becoming financial institutions (e.g., Lloyds of London). Cafes are sites for learning, socialising, and playing. They are a place for travellers to find some home comforts, to write letters and find out about the area or meet others. They are places to do business or have a celebration. All these activities are characteristic of cybercafes. Many of the activities that people come to the cafe for, they could do at home, but we prefer to do them outside: cafes are a home away from home. Home is not always convenient, or even pleasant, and we like to be in the company of others. It is not only a home, cafes are also a public venue that is not a formal work or office space.

While most cafes have to operate in the market, they are also social centres: they often offer a focus for a particular social group or geographical community. The cafe is more than the physical space or the products it serves, it is the people who use it and work in it: a cafe is successful when it attracts customers back. Not that the physical aspects are not important—the decor, the drinks, the games, etc.—but they are there to mediate, facilitate and lubricate the experience and activities of people.

Social games appear to have a special place in cafes. Pubs, cafes, clubs, and of course amusement arcades and casinos all feature games as part of the activities. In many countries games are the central activity in cafe life. Electronic games made their way into cafes and pubs as soon as they were invented. Although some games attract a limited clientele, others such as the 'pub quiz' have a broad appeal.

The 1990s have seen a change in pubs and cafes in the UK. Many new cafes have been opened, attracting a different clientele who are looking for more modern design and ambience than can be found in the traditional pub or bar, including an interest in the 'continental' style cafe that is not primarily a drinking space, but also for socialising, relaxing, working, shopping, etc. This often includes linking the cafe with another specific function—a bookshop for example, or making it part of a Gallery or Museum.[4] The design of the new cafes is also a moving away from more traditional styles. This global style can be traced to ground-breaking cafes such as Philip Starck's 1984 design in Paris (Boyer, 1994). It is a modern, young and futuristic design, a model many of the cybercafes follow.[5]

Survey of Cybercafes

Method

Research was done by questionnaire to the cybercafe customers, observation and interviews with managers during repeated visits to each cafe over four months. Some information was gained from newspaper articles, but little previous academic work was found on cybercafes, except for some articles on telecottages and social experiments with computer and media equipment access centres in Denmark (Cronberg, Duelund et al., 1991). However, since then a number of people have started to do work on cybercafes.[6]

Case Studies: Three Cybercafes

The study looked at three very different cybercafes.[7] Cafe X is an upmarket city centre cybercafe, on a franchise from a London firm. It attracts tourists and people travelling on work or working in the area. The main uses are e-mail and the World Wide Web, and the cafe facilities. The cafe has a high profile in the city, and the users are 50/50 male and female. Cafe Y is a small, private Internet and computer access centre in a middle class residential/high street area. (It is not actually a cafe, although it offers free coffee.) Users are all people living locally, especially teenage boys, and short- and long-term immigrants. Main uses are networked games, office facilities and e-mail. Cafe Z is a local government-funded cafe and Internet access centre in an area of 'social deprivation' on the outskirts of the city. It is in a modern building and operates as a 'healthy eating' cafe. It has the broadest range of users of the facilities, from under age 10 and upward. These customers also use a broad range of services: WWW, chat, e-mail, Web page design, Web-camera and word processing. A key feature is the free access introduced during the study period which encouraged many more people to come in, especially children.

Similarities

The cybercafes have much in common, apart from the provision of computer access. They all have a regular customer base, with over 50% of customers coming in at least once a month and many more regularly. Users are very mixed, male and female, young and old, although there is a marked bias toward younger people using the cafes. The cafes, even Cafe Y, are social meeting points, and many of the customers considered the atmosphere and the chance to be with friends an important reason for coming in. The managers of the cafe are not 'technology' people, even though there is an important role for a technical manager. Two cafes (X and Y) are run on a day-to-day basis by young women with a background in hospitality and catering who do not have a technical focus or approach. The manager of Cafe Z is not from a computer background, sees himself as a 'people' person, and stresses the importance of the personal relationships in making his business successful. The managers and staff advise and train customers, but also learn from them. They have developed the cafe facilities over its lifetime, responding to demands of users and technical change from outside, acting as *intermediaries*, facilitating new uses and shaping the services offered. All of the cafes, but especially those outside the city centre, struggled to bring in customers, but the managers feel optimistic of a growing trade. Technical expertise is essential to expanding the cafe activities, which in itself has obvious limits to growth. They were all developing business outside the cafe, based on the expertise and public profile built through the cafe. These activities include developing Web sites, setting up computers and networks, and teaching and servicing the business and home computers of customers, and generally expanding into the local public and business community.

Differences

There are also key differences between the cafes, shaped by their location, the aims of the managers and owners, and the type of clientele. Cafe X, based in the city centre, has many visitors, who come in predominantly to use e-mail, but also the Web. The customers use these mainstream applications because their interests are information and communication with family, friends and colleagues. The staff help the customers, and they run some training courses, but there is little effort to develop new uses, because the customers generally know what they want, or are satisfied with what they are shown. There is not a great deal of interaction between customers either, except for groups who come in to drink coffee. Cafe Y, on the other hand, was almost completely dependent on locals who come at the recommendation of friends and family, and live within 10 minutes walk. The Internet connection in the cafe is not very fast, and e-mail and the Web are not the dominant uses. Games are very important, and encouraged, the staff being experts on them all. The games are the main interest of a key user group, teenage boys, who come from the local school. This is in contrast to Cafe X where they have few games, partly because of restriction on children by the licensing laws, partly because they rather contradict the 'trendy' image of the cafe.

Cafe Z is also largely supported by local residents, and many children use it, especially since access was made free (after the survey). The use of services is broad, as the managers have to work hard to find relevant uses for local people, many of whom have no wide social network which they would need e-mail for, and no general interest in 'information.' Games are discouraged, at least the wilder killing games, however a Scottish league football game is popular. Chat is encouraged, especially on a monitored site, and is very popular. There is much more interaction between users of Cafes Y and Z who often go in groups, or meet friends there, than in Cafe X. Cafe Z stood out for developing Web pages related to the local community and the users, and for encouraging customers to represent their interests on the Web site.

Tables in the appendix summarise the answers to some of the questions posed in the questionnaire.

Findings: The Human Face of Multimedia

The development and popularisation of personal multimedia services, together with the research on particular cybercafes suggest a number of key theses (Stewart, 1998):

1) *Cybercafes are convenient, on-demand multimedia access centres.* Cybercafes exist because they provide technical services to customers which are not readily available to them elsewhere, or when they are away from their own facilities. However it is generally not basic access to computers that is central to cybercafes—it is facilities such as networked games, the Internet, printers and scanners and the latest software and hardware. They are not necessarily for novices; even competent and regular customers do not have the resources, or do not see the need to own expensive new technology, but still want to use it occasionally.

2) *Cybercafes are traditional cafes.* Cybercafes are more than technology access centres, they are also cafes in the traditional sense—they are public spaces where many age-old activities can be conducted, but mediated by modern technology. These can be social and personal uses, but they are not new activities. In contrast to the dominant trend, computers do not have to be kept in private or in formal premises; they can be public and informal.

3) *Informal learning and appropriation space.* Cybercafes are also points of individual learning, the informal atmosphere makes it easier to learn and to experiment. They are the 'human face' of computers and the Internet, technologies and services that are frightening to many people. Local, cheap, community access, seeing other people using multimedia, and nonexpert help can act as a trigger to bring someone across the 'use threshold.' The technology itself, fast changing and new, needs this informal space in order to be appropriated and domesticated into city and community life.

4) *Diverse styles reflect different environments and goals.* Cybercafes, like normal cafes or other public spaces, come in different flavours and appeal to

different tastes. They are part of a developing cultural and media landscape that we exist in and use. The computers and the computer-mediated services are inserted into physical environments that reflect existing patterns of leisure, home and work activity, and existing aesthetic and social tastes. Just as public leisure spaces have diversified to reflect particular tastes, cultural or social milieu, geographical and market position or use, cybercafes will as well. However at the moment cybercafes appear to bridge many of the groups and interests with a common theme of multimedia facilities.

5) *Cafe managers are 'reflexive intermediaries.'*[8] The cafes, the technology and the use are mediated, but not dictated by the owner and managers. These people *facilitate* access to technology and content, but also shape use through *configuring* the computers, Internet connections, relations with customers, choice of the space, opening hours, food, prices, music, user clubs and marketing. Whatever their aims, commercial or otherwise, the managers are able to have very direct experience of what people are using and how. They also have time to look for new technology, talk to suppliers directly, find new content such as Web sites and games, and help create that content themselves.

6) *The cybercafe takes computers and Internet outside the mainstream paradigm of individual use and ownership.* The dominant industry paradigm is for individual ownership and/or use or computers, communicating in a 'virtual space' or community over a network. Use of the cybercafe undermines this, as it is based on people buying time to use the computer, not owning technology, and sharing them in a public space, not in a private space. They also favour the 'network-centric' model of information technology development: a virtual presence on the network, though a Web page or e-mail account, is more important than a physical presence, e.g., owning a computer.

7) *Cybercafes could be 21st century public spaces.* More speculatively, cybercafes are more than an extension of existing cafes, they are community centres that bring together a wider range of people doing a wider range of activities than most cafes or other public venues. Local communities are fragmented and divided; cybercafes, by appealing to a broad range of users and being more than casual centres of consumption, such as shops, create contact between customers that can form social, learning and sharing relationships around use of the technology and consumption of different media. They also allow communities to project themselves in 'cyberspace' and allow people to interact on global networks as part of a local community, rather than as isolated individuals. Cybercafes are community centres for the 21st century.

These theses will now be illustrated within a discussion of more detailed aspects of the empirical material.

Multimedia, Customers and the Cybercafe in the 1990s

A cybercafe, or any other local ICT access centre, privately or publicly run must attract customers. What services can a cybercafe offer that will entice people to come in and spend time and/or money using the services? There are three main groups of potential users—those people who may not otherwise be interested in computers and the Internet, or who perceive the services and information available through them as being irrelevant to them; those who are interested but do not have technical access; and people who already have access to computers and networks elsewhere. What sort of environment, services and promotion will attract these different groups? In the study the users of the cafes came from all three groups, and the research attempted to find why they used the cafes. The research gives some fairly clear answers from the statistics, but these can be enriched by the verbal responses. Table 1 gives the aggregate figures for the main reasons given by customers at all the cafes for using the cafe. Other answers such as for training, or the cost, or just 'good' were also given.

The main reason for going to the cafe was predictably the computers or Internet services, whatever the application used. However many people wrote that the main motive was for the atmosphere, staff, or to see friends, and the technical service was a secondary factor. Table 2 shows the diversity of use and motivations.

Triggers to Use

As intermediaries, the managers and cybercafes act as agents in the diffusion process, facilitating voluntary uptake of their services by customers. People encounter and engage in new technologies in many different ways subject to personal and local contingencies. In trying to make sense of this, I introduce the concept of *triggers to use* to examine what makes people start to use and continue to use the facilities of the cafe. Four main categories are suggested: 1) life events; 2) social push; 3) multimedia pull or instrumental need; 4) curiosity and interest in technology or content. They are all reversible.

These triggers are useful categories to understand not only why an individual might starting using a service such as a cybercafe, but also to understand the market that the intermediaries are addressing, and the tools they deploy to try to engage new customers.

Life events: These are changes in occupation or circumstance of an individual, or in that of friends and family and other personal social network members, that create 'need' or force a need on someone. The technology may be a solution to a new problem

Table 1: Main Reasons for Using Cybercafe

Provides a technical service	68.75%
Atmosphere and helpful or friendly staff	37.5%
Convenience of location	17.5%
Place to be with friends	17.5%
Total number of respondents = 80	

Table 2: Detailed Reasons for Using Cybercafes

How and why do people use cybercafes?

The reasons for using cybercafes are very varied:
A friend had gone away—need to contact them
Have friends who live abroad
Travelling the world, want to keep in touch
Living abroad—read home newspaper and keep in touch with developments in
    home country
Need to write a CV or letter
Need to send documents by e-mail
Want to download an application form
Want to find out about a particular subject for leisure interest
Want to do work-related research
Find out about possibilities of using the Internet
Want to 'improve' oneself
Want to have fun with friends
Own computer does not have facility, e.g. modem
Need two computers to play a game
Not allowed to use Internet at work/college
Cannot afford a computer and/or Internet
Do not want to buy a computer or Internet—rational calculation re. use/cost
Something to do to fill time
No longer a student with access at the university
Equipment broken down at work/home
Don't have a printer/scanner
Don't own a particular game
Enjoy using chat systems

---

or barrier, or a new obligation. Among life events one of the principal changes are those related to travel, which reflects the increasing global mobility of many people. More people travel today, and longer distances, and technology now allows us to keep in touch with personal and professional social networks across the globe. There is undoubtedly a large number of people whose participation in the 'global society' and 'global culture' is not just as consumers of goods and media from around the world, but as active members of global social network. Several of the customers who previously did not use e-mail, or even computers, and have no other use for them, found the cybercafe the only way to keep in touch. Many tourists and visiting business people came to Cafe X; all the cafes had long-term immigrants visiting to mail home or read home newspapers. Several people started to use e-mail and computers for the first time because they had made friends abroad, or a girlfriend or boyfriend had moved away. Other life events include changing jobs or leaving college and being deprived of an Internet connection or computer access.

Social push: The adoption and use of multimedia is a result of being introduced
    to the technology through other members of a social network, and using it with

or because of them. Many people come to cybercafes on recommendation from friends, or with friends to pass the time. Cafe Y had most of its customers through recommendation, and many of them came to play games with friends. Parents came in because of the their children, and grandmothers had heard about their grandchildren using the Internet. The owner of Cafe Y knows that almost all of his clients came in because of personal recommendation. In Cafe Z many of the customers only come with friends, to spend the afternoon on-line.

Multimedia pull/instrumental need: In these cases technology that a customer owns or has access to is no longer sufficient for the purposes they want it for, or they are restricted from using certain functions due to economic or external social constraints (e.g., restrictions at work). The trigger is primarily instrumental need. Many people using cybercafes have their own computers, but do not have Internet connections: either they can't afford them, or do not wish to spend money on something they use only occasionally. The gamers especially need more power than their home computer can offer, and get the network connection to play the games against friends. Some of Cafe Z customer's had computers in nearby offices, but did not have Internet connections, so came next door to search for information.[9] The cybercafe can provide an opportunity for people to explore uses of the Internet and computers that they might be restricted from using at school, college or work.

Curiosity and engagement: Some people develop an intrinsic interest in new technology or multimedia, a curiosity which they wish to satisfy, with no particular goal apart from developing knowledge about the system or particular content or application. Even without using it, multimedia has become part of most people's life today: the cybercafe offers a convenient way to convert that background presence into a more practical and informed experience.

Other Key Aspects of Cybercafes

Learning and teaching

Multimedia is still evolving as a phenomenon; it has not yet been packed into a 'black box'; it has bits hanging out all over the place. Computers and the Internet are complicated tools that we need training and learning to use. Many people need personal training and help to get to grips with and use computers and the Internet at home or in an office. They also need help in learning how to use multimedia interfaces and content, how to search the Web and play the latest computer games. For many people help is not needed once, but over and over again, as new functions are needed, upgrades installed and software and hardware faults emerge. Sometimes people will know in advance how to use the facilities, but almost by definition, many will have to learn. The cybercafe is one place where this help can be and is given— it offers a local, high profile, informal space for learning and receiving advice. All the cybercafes have a learning/teaching element, both formal and informal. These relate to formal business activities of the cafe business, and to informal social activities between customers and customers, and customers and staff. Some cafes turn teaching into a formal activity, everywhere the staff offer informal advice, and sometimes customers become informal members of staff (Cafe Z) or even get jobs there (Cafe Z).

The demand for convenience

Multimedia is fast changing, there is new technology available that many individuals and businesses cannot keep up with even if they own computers, but may have occasional need for: currently Internet, colour printing, scanning, etc. The cybercafe can provide these. Many people have very little need for the facilities of the computer or Internet, so are not interested in spending a great deal of money on the technology. However there are an increasing number of instances when one of the services could be useful at a particular moment, for example word processing, downloading a job application, or sending an e-mail. Some other applications may not be desired out of necessity, but are for entertainment or casual use. Cybercafes are useful and attractive places to do these things.

Communications media and technologies proliferate today: public phones, mobile phones, answering machines and faxes all provide the solutions to communications in increasingly mobile and unpredictable lifestyles. Even with facilities at home or at work, there are times that we need access to electronic communications and information when away from home and the office. Just as cafes provided phones in the early days of telephony, for people who did not own them, they now provide phones as convenience. Cybercafes provide access to communications in the same way, both for those who do not have the technology, and those for whom it is a convenience.

The sociability of games

Just as games are important in many cafes and pubs, they are so in the cybercafes, only now using new technology. Games can be played alone, or in a group: taking turns, watching and interacting around the game playing. Network games extend this. In the cybercafes that had games, group activity appeared to be very important. Groups of children and students played games together, or watched each other. The games become centres of social activity, just as other nonelectronic games always have, with people inviting strangers to join in and more experienced players offering advice to newcomers.

The Internet Chat is another social 'game.' In Cafe Z the members knew each other, and friends would be in the same room talking to each other on the chat system and to their other 'cyberfriends.' The chat is a favourite of the teenage girls and women in their 20s: social and escapist like the boys' video games; calling on particular skills; providing excitement and a continuing week-in, week-out experience of taking part in another world. Like the video games, the participants play in silence, then call out to each other, laugh, and make suggestions.

Lessons for Cafe Managers

The strength of the cybercafe is its visibility on the local street, and its flexibility in catering for a diverse range of customers. However cybercafes do not always succeed, many have closed or struggled to exist, especially in the early days.

The cafes in this study survived because they concentrated on the customers and satisfying them, rather than implementing the latest technology. This meant experimentation with access conditions and costs, the cafe facilities and ambience, changing services, and expanding outside the cafe business. It is not enough to assume that the mainstream technologies and content will appeal to all potential customers, or that the people who walk past are interested in communicating across the globe or looking up difficult-to-find information. Most people are just not interested. These sensible uses are not enough—cafes survive when they let people socialise, play, experiment and learn. Although customers can be encouraged to use things that they 'should' use , the cafe encourages people to follow their interests, whether it be chatting or looking up the pages of their favourite pop star.

Managers must provide a service that responds to the needs and interests of potential customers as they are triggered. They must also reach out to bring customers in through engagement with outside groups, and to continue to provide services to individuals, firms or community groups even if they adopt technology themselves. Like any business, a network of satisfied customers is key to continued success. This includes providing technical support, training, network services, Web page design, etc. Managers have to exploit social networks, and personal and professional recommendations to bring people in. They have a radical new product, and therefore must make sure the message gets out that there is something relevant in the cafe for different customers, be it a good atmosphere, helpful staff or a fast Internet connection. One of the benefits of the cybercafe is that there are no set uses, and nothing to sign up to. This is also a drawback, and some sort of structured or semi-structured help, such as introductory classes or niche-user clubs, should be provided that quickly gets them working by themselves. The experience from the cafes in the study shows that this does not take very long.

The main group of users of the cafes are those who already have technology knowledge and access somewhere. These people will form an important core group of users who can bring in new ideas and new customers. However, unless they satisfy the manager's goal (e.g., commercial success, serving the community), their uses and community should not be allowed to dominate the cafe.

Community and the Cybercafe: the Local and the Global

The cafe is a meeting point for community and neighbourhood, or members of subcultures. It enables members to meet and share a space on a 'real' scale, but also to extend links outside. An individual 'meets' their own social network in the cybercafe through e-mail or chat. Chat lines also open out to the community by bringing outsiders into the local space. The cybercafe is a social portal, whether it be face-to face or in cyberspace. This link can be on different levels, and each level serves a different community. The cafe reaches outside its limits in different ways, according to the configuration of the technology, the customers, both internal and external, the community and geographical location, and the manager's efforts in

Table 3: Service Domains of the Cybercafe: Individual to the World

Cafe Service User	Use
Individual	Individual customers coming into the cafe
Cafe groups	Subgroups of customers, e.g., game players
Cafe community	The general cafe clientele
Local community	Providing a service at community level rather than individual level; part of domesticating multimedia into the community
The City	Increasing awareness of multimedia, part of the city's multimedia facilities for locals and visitors; domestication of multimedia into the City
The Region	Serving the hinterland of the city, opening up a cyberspace gateway as well as physical use by individuals or groups
The World	Bringing in visitors and customers primarily in the 'virtual' from around the world, linking them to the individuals, the cafe, groups, the community and the city; links the global to the local

configuring the customers and the technology. These levels go from the individual to the World, as illustrated in Table 3 above.

The dominant paradigm of multimedia is individual use, at home or at work, which is closely linked to individual ownership: technology suppliers want everyone to have their own terminal (at least). However, while this may be very convenient for some individual users, it restricts possible uses that the cybercafe, being a social space, is able to exploit. These case studies illustrate some collective and social ways in which computers and the Internet can be used. To understand the different types of IT use, Table 4 divides them into uses of ICTs that are essentially individual, and those that are collective. It also distinguishes between those that involve access to global communications networks (Internet), and those that are purely within a local space.

As the case studies show, not all cybercafes exploit all these possibilities. Cafe X is solidly in the Individual use column, although they do occasionally try to include group use, such as in training, the kids club or in corporate events. Cafe Y is much more characterised by group use, but confined within the cafe walls. Cafe Z includes all four types. The local context , the needs and interests of the customers and the motivation of the managers, rather than the technology, is what makes the difference. These group uses are an example of appropriation and reinvention of the technology, outside the dominant industry paradigm, and also illustrate the development of a new use through a process of social learning between intermediaries—the managers and their customers.

Table 4: Dimensions of Cybercafe Use

	INDIVIDUAL	**COLLECTIVE**
CAFE BOUND USE	**A** An individual using multimedia within the cafe (or other location), e.g., using CD-ROM, office software.	**B** Group using multimedia within the cafe, e.g., network games, local chat.
'CYBERSPACE' USE	**C** Individual using multimedia to reach beyond cafe, e.g., sending e-mail to broad social network, surfing the WWW, publishing own WWW pages. Virtual groups. Can link the individual to the 'global.'	**D** Groups using multimedia to reach beyond cafe. Establishing group identity on line. Using chat or games on line as a group or a distinct community in the cafe. Producing group pages. Links the local to the global.

A. is the conventional use of computers, alone, without a network.

B. Requires (networked) machines in one place: cybercafe, office, terminal room; face-to-face group interactions run in parallel to computer-based interactions.

C. is a conventional individual use of networks or Internet. Can lead to the formation of virtual communities and groups.

D. Groups interacting on line. Need the conditions of B, but an application and incentive to project that group out of the cafe, and into a more global space.

The Future of Cybercafes?

A purely instrumental look at cybercafes, informed by a vision of eventual technological saturation, would see them as a temporary phenomena. Cybercafes provide access to computers and associated services such as the Internet that people can't afford at home. Eventually there will be no need for these as everyone will have personal access. They provide informal teaching and learning centres for a population getting to grips with this technology for the first time. They provide services to people away from the office and home until such time as mobile or personal terminals become commonplace.

These reasons for their existence now are certainly true, but that cybercafes are transitory is not at all certain. The previously stated findings would suggest there is a strong future for cybercafes. They also suggest there are more complex reasons why cybercafes exist today, and the part they are playing and important part in the process of the development and diffusion of multimedia. As use of multimedia becomes more commonplace, the number of applications relevant to our everyday life increases the need for an increased variety of means of access will increase too.

The pleasant and convenient cybercafe will continue to be popular. For many people it may continue to be the only place they can access these services. The cybercafes can continue to exploit and serve our increasingly mobile lifestyles and global relationships and cultural interests, and find ways to integrate these with our everyday local, community activities.

Conclusions

There are many dimensions to Community Informatics, but there is a tendency to concentrate on the serious aspects of community services—education, health, democracy, etc., rather than on the everyday activities of individual citizens, families and friendship groups, let alone entertainment and leisure. The cybercafe falls at the softer end of CI, and also stands apart as many cybercafes are commercial and cater to those with existing expertise, looking for convenience and conviviality rather than primarily for training and information. Many customers did not regard the technical services as the main or single reason why they visited the cafes. The atmosphere, the friendly staff, the chance to meet and spend time with friends, the music and decor. These are social or aesthetic factors that would influence the choice of spending time in any venue, be it public or private. The locality of the cafes were also important. The cybercafes are convenient local centres, either for residents or for travellers: most customers did not travel more than one mile to visit them, so they really serve a local community. Like other media services they are 'on demand,' just turn up and plug in.

The cybercafes today are slowly becoming specialised, catering for different groups and interests (e.g., surf and smoke dope in the Amsterdam cyber 'coffee shop,' work in a business cafe, compete in video game leagues in gaming cafes, check out wave conditions in a surfer's café, plan hiking trips in a outdoor-pursuits cafe[10]), but most of them cater to a wide range of intersecting members of the community. Like some other semipublic spaces (shops, leisure facilities), cybercafes bring together people from different backgrounds, and of different ages, engaged in different leisure, work and learning activities. A good cybercafe creates an ambience where they can all feel comfortable, and some can stimulate interchange between customers.

Cybercafes are also about learning, opportunity and access. They are important training and advice centres, for new and experienced users alike. In these cases, the majority of users have access to computers outside, and the cafe made up for the lack of access to the latest technology. The informal nature of cybercafes is very important as a way of providing a gentle introduction to the world of new ICTs for people who may not like a classroom atmosphere, who just want to find out a little, and are not initially interested in using the technology for any particular reason. For community projects that want to give opportunities to people previously marginalised from computer use, the cybercafe is probably the best way to bring technology and expertise into a neighbourhood and lower barriers to learning and experience.

There are important differences between cafes, even within the same, relatively affluent city: the city centre cafe is mainly for people with global social networks and interest in global information and cultural resources. The other cafes are much more based around local uses, and collective and social uses of the cafe and the technology and content. All the cafes are expanding their clientele and developing new areas of business, particularly those that are not resting on established communication and information uses, but are searching to expand the uses and relevance of new technology to new groups.

The role of cybercafes must not be overestimated. Many people learn about IT, have access to services, and encounter multimedia in private spaces and though private networks. However, as the manager of Cafe Y pointed out, only about 20% of the population are on line, despite a massive increase in computer ownership and free or cheap Internet access even since this research was conducted. Even though the Internet and computer services appear to be becoming an integral part of our cultural, economic and public life, many people have not had direct experience of the Internet. The cafes have played an important role in raising the profile of the Internet in the city through direct contacts or through the media. If the city is our home, then the cybercafe is becoming an important part of our 'domestic' life, and the managers and customers of the cafes are residents finding original and appropriate ways to incorporate this global phenomenon into the everyday life of their communities.

In a final message to policy makers, I would encourage local and national governments to include cybercafes in their ICT policy, as our local authority finally did after initial rejection. I hope this paper shows the diversity of the concept, and

Table 5: Cafe X (all values in percentage); Total cases = 35

Age	12-18	18-24	25-30	31-35	36-44	45-50	Over 50
	8.6	31.4	25.7	20.0	2.9	5.7	5.7
Gender	Male	Female					
	48.6	51.4					
Distance from home (miles)	<1	1-5	5-10	>10			
	37.1	48.6	2.9	11.4			
Education	Secondary	Vocational	Further	Higher			
	11.4	2.9	20.0	65.7			
Occupation	School	Student	Unem-ployed	Trade/skilled	Unskilled	Profes-sional	Self-employed
	2.9	22.9	0	34.5	5.7	25.7	5.7
Income (£UK'000)	<5	5-10	11-15	16-20	21-24	25-30	>30
	25.7	20.0	28.6	2.9	5.7	2.9	8.6
Number of services used	0	1	2	3	4		
	3	54.3	22.9	17.1	3		
Cafe visits	Daily	<3x/wk	<3x/mth	Few times/year	First time		
	14.3	25.7	22.9	2.9	34.3		
Reasons for using cafe	Services	Conven-ience	Friends	Atmos-phere/staff			
	71	17.1	8.6	40.0			

Table 6: Cafe Y; Cases=27

Age	12-18	18-24	25-30	31-35	36-44	45-50	Over 50
	3.7	51.9	14.8	14.8	7.4	3.7	3.7
Gender	Male	Female					
	81.5	18.5					
Distance from home (miles)	<1	1-5	5-10	>10			
	63.0	22.2	3.7	11.1			
Education (achieved /current)	Secondary	Voca-tional	Further	Higher			
	55.6	0	3.7	40.7			
Occupation	School	Student	Unem-ployed	Trade/ skilled	Unskilled	Profes-sional	Self-employed
	44.4	22.2	7.4	14.8	7.4	3.7	3.7
Income (£UK'000)	<5	5-10	11-15	16-20	21-24	25-30	>30
	59.3	7.4	7.4	11.1	3.7	3.7	3.7
Number of services used	0	1	2	3	4		
	59.3	25.9	7.4	3.7	3.7		
Cafe visits	Daily	<3x/wk	<3x/mth	Few times/year	First time		
	18.5	48.1	22.2	3.7	7.4		
Reasons for using cafe	Services	Conven-ience	Friends	Atmos-phere/staff			
	63.0	22.2	29.6	37.0			

Table 7: Cafe Z, Cases=18

Age	12-18	18-24	25-30	31-35	36-44	45-50	Over 50
	11.1	38.9	16.7	5.6	16.7	5.6	5.6
Gender %	Male	Female					
	66.7	33.3					
Distance from home (miles)	<1	1-5	5-10	>10			
	72.2	11.1	5.6	11.1			
Education (achieved /current)	Secondary	Vocational	Further	Higher			
	50.0	0	11.1	22.2	16.7		
Occupation	School	Student	Unem-ployed	Trade/ skilled	Unskilled	Profes-sional	Self-employed
	44.4	5.6	22.2	16.7	5.6	5.6	5.6
Income (£UK'000)	<5	5-10	11-15	16-20	21-24	25-30	>30
	61.1	11.1	16.7	5.6	5.6	0	0
Number of services used	0	1	2	3	4		
	0	27.8	27.8	27.8	11.1	5.6	
Cafe visits	Daily	<3x/wk	<3x/mth	Few times/year	First time		
	27.8	44.4	16.7	5.6	5.6		
Reasons for using cafe	Services	Conven-ience	Friends	Atmos-phere/staff			
	72.2	11	16.7	33.0			

why commercial and community entrepreneurs should be encouraged to develop this type of project as a resource for local citizens and as an essential service to visitors.

References

Boyer, M.F. (1994). *The French Cafe*. London: Thames and Hudson.

Callon, M. (1987). Society in the making: The study of technology as a tool for sociological analysis. *The Social Construction of Technological Systems*. Cambridge, MA: MIT Press, 83-106.

Cronberg, T., P. Duelund, et al. (Eds.) (1991). *Danish Experiments. Social Constructions of Technology*. Copenhagen: New Social Science Monographs.

Graham, S. and A. Aurigi (1997). "Urbanising cyberspace?" *City* (7), 19-39.

Latour, B. (1986). *Science in Action*. Milton Keynes, Open University Press.

Mitchell, W.J. (1995). *City of Bits: Space, Place and the Infobahn*. Cambridge, MA: MIT Press.

Molina, A. (1993). Sociotechnical constituencies as a process of alignment: the rise of a large-scale European information technology initiative. Edinburgh, PICT: 51.

O'Connor, J. and D. Wynne. (1997). From the Margins to the Centre: Post-Industrial City Cultures. *Constructing the New Consumer Society*. P. Sulkunen, J. Holmwood, H. Radner and G. Schulze. Basingstoke, Macmillian.

Østby, P. (1993). *Escape from Detroit—The Norwegian Conquest of an Alien Artefact*. The Car and its environments. The past, present and future of the motorcar, Trondheim, European Commission COST A4.

Reingold, H. (1994). *The Virtual Community*. London: Minerva.

Sennett, R. (1977). *The Fall of Public Man*. Cambridge [Eng.]: Cambridge University Press.

Silverstone, R., E. Hirsch, et al. (1992). Information and Communications technologies and the moral economy of the household. *Consuming Technologies: Media and Information in Domestic Spaces*. London: Routledge.

Sørensen, K. (1994). Adieu Adorno: The Moral Emancipation of Consumers. *Domestic Technology and Everyday Life—Mutual Shaping Processes*. A. Jorrun-Berg. Brussels, EC.

Sørensen, K. (1996). Learning Technology, Constructing Culture: Socio-technical change as social learning. *SLIM Working Paper*. Trondheim, Centre for Technology and Society, Norwegian University of Science and Technology.

Stewart, J. (1998). Computers in the Community: Domesticating multimedia into the city. Edinburgh: Research Centre for Social Sciences, University of Edinburgh.

Stewart, J. and R. Williams (1998). "The Coevolution of society and multimedia technology." 16(3).

Stewart, J., S. McBride, et al. (2000). "Cybercafes and cyberpubs, adding coffee to computers." *New Media and Society*, (Special Issue). Forthcoming

Williams, R. and D. Edge (1996). "The Social Shaping of Technology." *Research Policy*, 25, 856-899.

Endnotes

1 http://cybercaptive.com/.
2 One report is that by Frederico Casalegno, "Les Cybercafés," a study of cybercafes in Paris in 1995 where he develops the idea of the cybercafe as an 'outside living room,' where the 'real' groups can link to 'virtual' groups.
3 Even as we get 2Mb links into the home, there will always be a point when we want to use a 100Mb connection, e.g., to download a film. Even when we have advanced mobile systems, batteries run out, and memory and bandwidth will always be lower than a land line.
4 The Victoria and Albert Museum in London caused controversy in the early 1990s when they opened a new cafe and advertised with the slogan along the lines of, 'Nice cafe with museum attached.'
5 American readers may note the similarity of the development of modern cafes and cybercafes to the 'Googie' style cafes in the 1950s, where cafes were designed using modern technology (plastics, aluminium), and were based around the automobile. They also incorporated other contemporary icons and images such as space ships. As the computer replaces the car as the symbol of process, youth and freedom, and the cyberspace, the deep space frontier, it is hardly surprising that cafes, which are designed to reflect and stimulate new tastes, should now adopt the key themes late 1990s.
6 e.g., Wakeford, N. (1999). Gender and the landscapes of computing in an Internet cafe. *Virtual geographies: bodies, space and relations*. M. Crang, P. Crang and L. May. London: Routledge., research at the University of Sussex and University of Teeside in 1999.
7 A full account of the cases, including Irish cyberpubs, can be found in Stewart, McBride et al., 2000 (forthcoming).
8 The role of the managers as intermediaries, and the process of the development and management of the cafes is covered elsewhere (Stewart, 1998).
9 Until their offices were directly wired to the cafe server.
10 The closest cybercafe to my own home is a restaurant offering in-house and on-line help in planning outdoor adventures!

Chapter XV

Facilitating Community Processes Through Culturally Appropriate Informatics: An Australian Indigenous Community Information System Case Study

Andrew Turk and Kathryn Trees
Murdoch University, Western Australia

Introduction

This chapter discusses how community processes may be facilitated through the use of information systems (IS), developed via a highly participative methodology. It examines the utility of several approaches to modeling community information requirements. By way of illustration, it describes progress on the participative development of the Ieramugadu Cultural Information System (ICIS). This project is designed to develop and evaluate innovative procedures for elicitation, analysis, storage and communication of indigenous cultural heritage information. It is investigating culturally appropriate IS design techniques, multimedia approaches, and ways to ensure protection of secret/sacred information. Development of ICIS is being carried out in close cooperation with an indigenous community in Western Australia.

Developing Culturally Appropriate Community Information Systems

Community Information Systems (CIS) are ways of utilising information technology to address some of the needs of communities. They build upon, and coexist with, complex informal and formal (non-computerised) preexisting systems of communication and information storage. Such communication is often oral and much information is stored only within the memories of community members.

Computer-based CIS are becoming much more common because of the increased affordability of powerful hardware and software and the greater availability of relevant data sets, especially as government agencies convert their records to digital form. There has also been a massive increase in the interconnection of computer systems via client-server architecture, local and wide area networks, and especially the development of the World Wide Web. Usability of computer systems has been greatly enhanced making them accessible to a much broader range of users. As well as these 'technology push' factors, there has been 'demand pull' through the desire of community members to have access to technology enabling them to play a more effective role in decision-making processes. This has, in part, been fueled by demands for greater equity of access to information and by developments in participatory democracy.

The Information Systems (IS) discipline studies the way individuals, groups and organisations use information. This is generally in the context of computer-based IS, which can be considered to consist of the following five aspects: hardware, software, data, people and procedures. IS analysts seek to assist organisations to fulfill their objectives through IS interventions (usually termed 'projects') which seek to identify information processing requirements and to design, implement and maintain a suitable IS. Such projects are usually executed in accordance with a particular set of procedures, techniques and tools, collectively referred to as a 'methodology.'

Jayaratna (1994, p. 35) defines a methodology as:

"... an explicit way of structuring one's thinking and actions. Methodologies contain model(s) and reflect particular perspectives of 'reality' based on a set of philosophical paradigms. A methodology should tell you 'what' steps to take and 'how' to perform those steps but most importantly the reasons 'why' those steps should be taken, in that particular order."

Although IS development methodologies have traditionally focused primarily on the design of hardware, software and data aspects, newer (so-called 'soft') approaches involve more consideration of human factors issues (Avison et al, 1993; Checkland, 1981; Crowe, et al., 1996; Finkelstein et al., 1990; Flynn, 1992; Mumford, 1983). Socio-technical methodologies combine hard and soft approaches (Eason, 1988; Travis et al., 1996). They incorporate a higher level of participation by system users and focus on identification of culturally determined user needs and constraints.

Socio-technical methodologies should be employed in the design of CIS. Whitley (1998) cautions that IS practitioners should not rely blindly on any one methodology and emphasises the need to develop a thorough understanding of the information situation and the people involved. Bell (1998) reviews the range of available IS development methodologies and discusses two frameworks which can help practitioners select the appropriate methodology (or combination) for a particular project. Hidding (1998) states that cultural factors may play a critical role in whether or not a system development methodology is successful. Human-computer interaction issues and a user-centred design paradigm (Dix et al., 1998; Hix and Hartson, 1993) are of particular importance for CIS, especially where they are used for cross-cultural communication.

Even socio-technical methodologies are open to threats to their validity when they are performed by a system's analyst who has minimal contact with the stakeholders. In general, the more participation in (and understanding of) the system development processes by the stakeholders, the better will be the requirements determination and hence the system design. Potential users will thus feel more 'ownership' of the system and will be more supportive of its implementation. Where the number of stakeholders is small and/or they share similar needs, or where the system requirements are simple and/or easily represented in formalisms, the challenges in delivering equitable and effective participation may be minor. Where these conditions do not apply, the process will be more difficult and the analyst is likely to deal more adequately with a subset of the stakeholders who are tractable to the methodology the analyst normally uses. Hence, the views of other stakeholders may be undervalued and underutilised because of the limitations of the analyst's modeling tools and techniques. These issues are of special significance for CIS because of the broad range of stakeholders and the open-ended nature of system content and potential uses.

Culture, values and attitudes towards technology of the system users (and designers) are important factors which can be addressed through the use of a highly participative system development methodology. This should give the system owners more practical power over development processes (Dahlbom and Mathiassen, 1996) and provide effective mechanisms to manage the changes in the information situation (Brugha, 1998). Morley (1993) recommends that different modes of participation be adopted for different types of stakeholders, taking into account the social context and the project's goals. Guimaraes and McKeen (1993) emphasise the need to ensure that participating users understand the development processes and system design options. Effective ways of incorporating the needs of all stakeholders must be developed. One way of doing this is through examination of the use of potentially richer modeling approaches for systems development in complex cases.

Most importantly, the development of any IS should be considered a social, as well as a technical process (Kling, 1996; Ledington and Heales, 1993). McMaster et al. (1998) discuss a process model of IS development based on 'networks of association' which describe the social aspects of information use. They observe, "In IS development, the tradition has been to separate social and technical issues and to apply different treatments. This means that the due process is not enacted and the likelihood of the network becoming aligned is reduced. Stabilized networks of association can and do emerge in the absence of due process, but a visible due process would allow organizations to talk about and act on a projectory for an IS development, rather than being ineffectual observers of its trajectory" (p. 354). This applies most strongly in the case of CIS, especially IS for indigenous communities. A CIS must be culturally appropriate not only in terms of the design of the user interface (Preece et al., 1994) but also in terms of its deep structure (ontology and epistemology) and the procedures for development and use of the system.

Description of the ICIS Project

The Community

Ireamugadu (Roebourne) is a small town, near the coast, in the Pilbara region of Western Australia. It has a population of approximately 1,500 people, 95% of whom are indigenous. It is the place where many Ngaluma, Injibandi and Banjima people live, although the traditional 'country' of some of the people is up to 200 km inland. These peoples have been progressively displaced from their 'country' over the last 150 years as part of the colonial process. They have been obliged to make way for, initially, pastoral activities, and in more recent times, the mining industry (Rijavec and Harrison, 1992).

Health and living standards have been, and are still, poor. Most of the older indigenous people in Roebourne have little or no formal (nontraditional) schooling. However, children today attend primary and secondary schools and a small but increasing number go on to undertake trades courses or university degrees. In the 1990s strong indigenous leadership and a relatively united community has resulted in significant changes to the life of the community and a desire to investigate innovative ways of storing and using cultural heritage information. The information requirements of native title land claims have also generated interest in the development of CIS.

Basis of the ICIS Project

Since 1996 the authors have been working with the indigenous community at Roebourne to develop a cultural heritage information system utilising multimedia, geographic information system (GIS) and database technology. This project provides a very demanding application domain for the investigation of highly participative approaches to the development of CIS. Ethical issues are also foregrounded because of the culturally sensitive nature of the information.

The information system is for the storage of Ngaluma, Injibandi and Banjima peoples' heritage information and is called the Ieramugadu Cultural Information System (ICIS). It is aimed at addressing the fragility of cultural information by developing ways of preserving its richness. ICIS is being developed with, rather than for, its users. Its development requires a significant conceptual advance in the way that complex cultural information may be identified, linked and represented.

The aim is to provide a flexible information bank capable of producing convincing products in a variety of circumstances (e.g., education or negotiation) in line with community needs and not infringing on cultural constraints. Thus the project seeks to aid in empowerment of the indigenous community through highly participative, culturally appropriate CIS design and implementation. This will result in a system where data can be input, manipulated and output by members of the indigenous community themselves, hence appropriate training is also an important issue.

Indigenous cultural heritage information must express the integrated relationships between:

places—   not just an arbitrary configuration of physical locations but an assemblage of places connected by meanings associated with traditional belief systems;

people—   the specific group/s of people who possess the meaningful relationship with (and are responsible for) those particular places;

procedures— the laws and customs which link the people to the places and sustain their unique relationship to the land and each other;

presentations—the practices and physical manifestations by which the laws and customs and meaning relations between the people and places are expressed (and hence maintained), such as ceremonies and paintings.

ICIS is being developed on a large desktop computer fitted with an image scanner and high quality colour printer. The system involves an innovative integration of interactive multimedia, GIS and database software and will be accessed through an extremely user-friendly interface being designed for this purpose. Data sets include information about locations (overlaid on digital topographic map data); details of individuals, families, sub-tribal and tribal groups; and information relating to traditions, laws, ceremonies and cultural representations. The linking of these data sets in an effective way is a key facet of the development.

Research Progress and Plans

Work on the development of ICIS and the methodology is progressing very well (Trees and Turk, 1998; Turk and Trees, 1998, a, b). Two versions of an initial (requirements animation) prototype have been evaluated by the indigenous participants. Their responses were extremely encouraging and there is expanding community interest in the project. More cultural information (e.g., family trees) is being collected and converted into digital form. The first prototype of a purpose-built database program for family information was tested in 1998.

The researchers assisted in the establishment of Ngurra Wangkamagayi (the cultural training group), the development of cultural awareness courses and training of the indigenous presenters. Half-day and two-day cultural awareness courses are being run for staff from mining companies, teachers, school groups and others. This may be followed by development of cultural tourism activities. The developing ICIS is already playing a key role in the preparation of promotional and teaching materials for these ventures. While embracing these opportunities for employment and greater interaction with the non-indigenous community, it is important that there is indigenous control over what cultural information is provided and the form of its presentation. This is a key objective of ICIS and this research project.

In 1999 the authors are continuing with the development of ICIS in close collaboration with the indigenous community at Roebourne, through Minurmarghali Mia (Roebourne Education Centre) and Ngurra Wangkamagayi. Research aspects include investigation of:

- multimedia scenarios which reflect traditional narrative structures;
- the utility of different types of interface metaphors;
- effective utilisation of video and audio materials;
- ways to deal with restrictions on the use of names and images of people who have died;
- collection and storage of meta-data, such as who a particular story belongs to;
- ethical issues in the context of the highly participatory system development methodology.

The research project at Roebourne is being extended to incorporate on-line computer-mediated communication (CMC). Trials of some approaches are already under way (including a trial Web site for Ngurra Wangkamagayi), however, the particular forms of WWW-based communication will be determined in consultation with the community to ensure that it reflects their needs and does not infringe cultural constraints. Possible scenarios include:

- a rich Web site describing the activities of Ngurra Wangkamagayi;
- negotiations with organisations wishing to arrange cultural awareness courses;
- interactions with potential cultural tourists.

Use of Cultural Heritage Information in Education

Education is a critical area where more effective use of cultural heritage information is needed. This will assist in raising the well-being and self esteem of indigenous communities, as indicated in the recommendations of the Royal Commission into Aboriginal Deaths in Custody. It is important that cultural information previously collected by governments and non-indigenous researchers is made available. Empowerment of indigenous communities through culturally appropriate use of multimedia technologies will also assist in the engagement of young people in the learning process.

Some of the more significant ways that this information could be used in educational settings within indigenous communities include:

- development of culturally appropriate teaching materials/practices in schools;
- targeted cultural heritage study programs for young people;
- encouragement of inter-generational dialogue;
- maintenance and transmission of cultural heritage among older people.

Indigenous cultural heritage information can also play an important role in education of non-indigenous people. This can assist in the development of cultural sensitivity, reconciliation and meaningful negotiation. Relevant educational scenarios include:

- cultural studies programs for school children;
- efforts to raise cultural awareness among members of the general public;
- cultural awareness training for specific interest groups or professions (e.g. mining company employees, teachers, police, medical workers and magistrates).

The ICIS project addresses these issues through linkages with relevant existing educational programs in the Roebourne area and specific cultural awareness projects. The research program is being carried out in cooperation with Minurmarghali Mia, the Aboriginal Education Centre at Roebourne and the Ngurra Wangkamagayi culture group.

Use of Cultural Heritage and Other Information in Native Title Claims

The Australian High Court 'Mabo' case in 1992 found that although Australia is a settled colony it was not *terra nullius* and hence the common law doctrine of native title was applicable when British sovereignty was extended over Australia. The Crown could extinguish native title to land providing the appropriate legislation or executive act embodied a clear and plain intention to do so. The central question is whether the regime of land control established by the act in question, or the disposition/alienation of the particular land, was inconsistent with the continued enjoyment of native title. Hence a major aspect of the determination of a native title claim revolves around whether extinguishment has occurred with respect to the particular area of land under consideration. Information concerning the nature of the occupancy claimed by a group of indigenous people will also be necessary to prove that the claimant group is the one with native title rights.

A large variety of information types must be analysed and the collated results presented during the native title claims process. Much of this information will refer to locations in space and hence the efficient and equitable consideration of native title claims requires the use of spatial information in the most effective way possible (Turk and Mackaness, 1995). Integration of the information from various stakeholders is a potential aspect of claim mediation by the National Native Title Tribunal (NNTT) (or arbitration by Federal/State courts or tribunals). The procedures must ensure a proper separation between shared databases and information which must be kept confidential for cultural reasons as well as for the purpose of negotiation strategy. The required information systems must also be capable of including a wide variety of information formats and be accessible by various types of users from different cultural backgrounds.

If indigenous communities are to interact effectively with native title procedures they need to be able to present a coherent set of relevant information (Turk, 1996). This must be organised in a manner which gives them control over what information is used in any particular circumstance and the form of its presentation. ICIS is designed to investigate these objectives, although it has been inappropriate for it to be used directly in the legal processes. It is envisaged that information gathered for the native title court case in the Roebourne area will be incorporated in ICIS after the legal proceedings are concluded.

Multimedia as Heritage Narrative

Culturally appropriate technology developments must complement existing oral traditions and honour cultural constraints. The ICIS project is important for the way that it engages with specific cultural practices such as naming taboo—the prohibition on using a person's name after death. With the use of photography, film and multimedia in indigenous communities, the naming taboo has been redefined to take into account the use of images (Michaels, 1990). The use by the indigenous community of valuable films such as "Exile and the Kingdom" (a documentary video about Roebourne) (Rijavec and Harrison, 1992) is severely restricted because people in the film have passed away and showing their images would transgress cultural restrictions. In cooperation with the film makers, the researchers will digitise this film, other related footage, 1960's anthropological films and old photographs, then write the material to CD-ROM. Video segments will be classified to enable their use in interactive multimedia programs. A relational database will allow specific material (e.g., about a person who has passed away) to be isolated and not used in a culturally inappropriate way. Through this process films such as "Exile and the Kingdom" can be made even more useful to the indigenous community.

Such procedures raise not only technical issues but also involve the indigenous community in thinking about ways in which new technologies influence their cultures. They also raise ethical issues for producers of film and multimedia, extending the discussion within disciplines such as anthropology and ethnography, and building on the work of authors such as Michaels (1990).

Ethical Overview of the ICIS Project

This project addresses key ethical aspects in the context of postcolonial practice, critical ethnography, visual anthropology and GIS (Turk and Trees, 1999). It is impossible here to address in detail the ethical issues raised by this project. In general the ICIS project is being carried out in a manner which the authors believe reduces the opportunity for unintended outcomes; however, they do not claim that there is no room for improvement. The ethical principles being applied to the project are based on a general interpretation of postmodernism and postcolonialism and consideration of the particular indigenous GIS issues raised by Rundstrom (1995). The following list summarises the key points:

- The greatest possible project control by the indigenous community—they initiated the project (formal as well as informal consent was obtained) and can terminate it at any time.
- A highly participative, user-centred development methodology operationalises the ethical principles.
- Continual interaction with the community for data collection, system design and critical review of project processes and products.
- Use of multimedia and innovative design approaches to permit epistemological pluralism and to minimise degradation of information concepts.
- A highly reflective mode of operation—at the individual and project group level.
- Conscious effort to involve personnel from cultural studies as well as technical backgrounds.
- Training of indigenous participants.
- An absolute bar on project information being used outside the community (except for Web site, cultural awareness and negotiation materials, as authorised by the indigenous participants).
- An intricate system of security measures for different types of information is being developed.

Modeling Community Processes

The Need for Modeling

The development of an information system for any organisation needs to be a social as well as a technical process. This is especially the case where the client is a community, with diverse interests and goals which are neither well formulated nor stable. Where the community is primarily made up of indigenous people living in poor economic conditions within a dominant non-indigenous culture, the adoption of culturally appropriate practices and socially effective procedures is critical.

For an information system to be able to truly address a problem (or opportunity), it needs to reflect the real needs of the stakeholders involved. Hence the requirements determination phase of systems development needs to involve procedures which elicit and capture the different viewpoints about the problem/opportunity and how an IS might address them. One problem is how to obtain (and retain) the richness of the explanations from each stakeholder. Ideally this should be done in a way which permits the differing viewpoints to be integrated in an effective, efficient and equitable manner. An initial step, which is likely to retain as much richness as possible is the use of some form of narrative. This has the advantage that the stakeholders themselves are likely to be able to understand the narrative and can critique it. The next problem is to find an appropriate way to move from each narrative to some integrated model of system requirements, e.g., via representational techniques such as rich pictures, means-ends tables, and responsibility modeling (Cross and Turk, 1997).

Potential Use of Organisational Modeling Techniques

Michael Gurstein suggests in the Introduction to this book that "CI accounts for the design of the social system in which the technology is embedded. ... Thus technology is an extension from 'organisation' to 'communities' of the socio-technical approach to system design." In the ICIS project we see the need to move from 'community' to 'organisation,' i.e., facilitating the creation of organisational structures (and IS) to address specific opportunities and problems within the community. This should help to ensure that the community is "enhanced and enabled through the use of ICTs."

Information systems should be designed to suit the organisation in which they are situated. However, the information and communication technology which is available may well be a significant factor in determining the most effective structure for the organisation, especially since the advent of distributed systems and collaborative technologies (Karsten, 1998). Iivari (1992) discusses this requirement for 'fit' between organisation and IS and notes that the design issues are becoming more complex and 'reciprocal.' "This means that unidirectional causalities expressing the fact that either the organizational context (environment, organizational technology, structure, etc.) determines the characteristics of an information system, or that, vice versa, information technology and information systems determine organizational technology and structure, may be too simplistic" (p. 5). Hence there is an enhanced requirement to model organisational processes (technical and social, formal and informal) and to evaluate a range of socio-technical IS options, within a highly participative systems development methodology.

Espejo (1996) argues that effective participation is produced by the interplay between the ideas and activities of individuals and the organisational structures, which regulate their interactions. However, these structures are modified by the individual actors (in a formal reorgnisation or informally) to suit changing demands to produce what he terms a 'recursive organisation.' "Effective participation requires that all of us are involved in the invention and formation of self-constructed action spaces ... Whether or not we are aware, our actions are producing their embedding structures which are producing the spaces for these actions in a never-ending regression ... an organization is not something given to us but something produced by us in our moment-to-moment interactions. Once these interactions become stable and develop closure, regardless of any formal declaration, an organization is constituted with identity" (pp. 414-416). A community can thus be considered as a nested set of organisations constituted in this way and a CIS can be designed to service the information needs through modeling the interactions between the participants. Espejo applies Beer's Viable System Model (Flood and Jackson, 1991) to aid in understanding the communication requirements, to facilitate the "self-constructed action spaces" in a manner which fits the organisation's objectives and the demands of team work. This process involves the notion of 'organisational citizenship,' leading to cohesion and effective performance. "... As a participant I am both constituting the organization by my moment-to-moment interactions with other participants and observing the organization as I reflect upon it. ... Languaging these reflections is a crucial form of action" (p. 514/416). Organisational modeling can provide 'boundary objects' (graphical and/or linguistic representations) which facilitate the envisaging of new structures and discussion of their suitability and utility.

Hart (1997) emphasises the social and political aspects of IS development and use. He recommends modeling of stakeholder interests via the concept of 'information wards.' Such an approach can aid in identifying and minimising potential conflicts which may be caused by changes in power relationships precipitated by proposed new information access and custodianship arrangements. This can reduce political machinations between stakeholders with different interests and hence reduce the likelihood of failure of the IS development project. However, if a participatory methodology is applied it is important that the conceptual modeling of the system is represented in formats which can be easily understood by all the stakeholders (Patel et al., 1998).

ORDIT (HUSAT, 1993) is an example of the type of methodology (tool) which can be used to provide a more effective way to integrate organisational design with information system design. Conventional systems analysis focuses on defining information processing requirements, rather than IS in the wider context. Hence, systems do not satisfy the needs of their human operators, even when technically sound. In the past, designers have generally ignored organisational requirements, which come out of a system being placed in a social context. Sources of such requirements include: power structures; obligations and responsibilities; control and autonomy; and culture, values and ethics.

The objectives of ORDIT are to:

- identify organisational requirements;
- provide a set of techniques (and how to use them) which allow organisations to explore the implications of different potential system options; and
- summarise the requirements in a form usable by IS designers and problem/ opportunity owners.

ORDIT attempts to bridge the gap between technical developments and organisational systems. It is not a substitute for current methodologies—these are needed in parallel. The complete approach needs to emphasise the exploratory nature of systems development. It adopts a socio-technical systems approach to design of a system which serves organisational goals, acknowledging that the technical system needs to be well-integrated with the organisational structures and processes. ORDIT adopts a user-centred perspective (Karat and Bennett, 1991). A system needs to be designed as a tool to serve the needs of users, not the reverse.

A common problem of information technology is that it can unnecessarily impede work performance by placing arbitrary restrictions on the tasks users can perform or constrain users in the ways they perform tasks. ORDIT is about providing choices and creating new opportunities. System design should not confine itself to creating solutions to problems by merely computerising existing practices. Opportunity scanning is needed to identify a number of socio-technical options, not just to generate the solution, but to expand the problem space and explore wider system implications.

The analysts need to maximise consultation with the users, to ensure relevant user knowledge of the organisational structure and processes is incorporated in the new system. This will facilitate the building of consensus as users come to understand the proposed system and become committed to its implementation. ORDIT therefore seeks to identify these key users, their concerns and requirements. It attaches importance to facilitation of communication between problem owners and system designers.

ORDIT utilises the notion that an organisation is a network of responsibilities. Such an idea helps identify organisational requirements and represents them so that both users and analysts can understand and evaluate them. This is especially important since users may not have a clear understanding of their needs at the commencement of the project.

The key to ORDIT analysis is the idea of roles, agents and responsibilities. Within ORDIT, the focus is on understanding and describing an organisation as a set of related work roles, because consideration of human organisation is central to understanding the requirements that organisations will have of the technology. The core concept in understanding roles is the agent entity. These agents are the primary manipulators of information. An agent represents an office in the sense of a role holder and may be an individual or a group.

The key to modeling structural relationships is the realisation that they are basically describing responsibilities that agents have to each other. ORDIT analysis and modeling is based on the fundamental premise that the function of an organisation is manifest in responsibilities held by individual role holders. The structure of the organisation is manifest in the relationships between them. Therefore identification of different role holders and their responsibilities leads to identification of both functional and organisational requirements. In modeling an organisation we also have the elements of actions and resources, where an action entity is an operation that changes the state of the system and a resource entity is what enables the agent to do the action (see Figure 1).

The framework in Figure 1 can be used at three different levels of abstraction: responsibility, obligation and activity. The concepts at each level of abstraction include the three basic entities—agents, actions and resources—and the relationships between them.

The responsibility level is the level of most apparent importance. Structure and organisational relationships are implicit in the responsibilities of agents. When an agent performs a work duty this is a manifestation of the responsibilities that they have to other parties. Requirements infer rights, e.g., to access information required to fulfil responsibilities (see Figure 2). However, a person in an organisation may not be explicitly aware of their responsibilities. Typically they are manifest in their knowledge of what they have to do. The time that responsibilities are made explicit is often when something goes wrong and blame is apportioned.

ORDIT demonstrates how responsibility modeling can be used as a means of identifying and specifying requirements of an IT system in a way which is meaningful to both users and system designers. Organisational structure can be interpreted in terms of responsibility relationships. Thus we can model responsibilities, their associated obligations and subsequent activities. Obligations define what a responsibility holder must do and subsequently these can be divided into those actions undertaken by humans and those transferable to an IT system—thus identifying the functional requirements of that system. By listing what a responsibility holder needs

Figure 1: Primary Model of Agents, Information and Activities (After HUSAT, 1993)

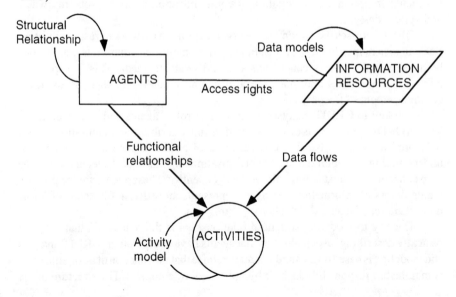

Figure 2: Modeling of Responsibilities and Rights (After HUSAT, 1993)

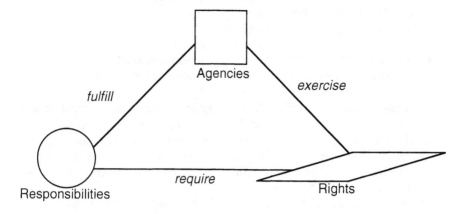

to know and needs to record, we can create lists of information requirements.

An example of such modeling for the ICIS system is the listing of responsibilities (and consequential information access rights) for an 'agent' responsible for the design of a particular cultural awareness course, e.g., as follows:

Responsibilities	*Information Access Rights*
• Understand the culture	- Discuss with elders
	- Read background materials
• Select cultural information	- Review cultural information repository
• Establish course requirements	- Access details re: participants
• Plan course delivery	- Review courses timetable / staff availability
• Prepare course outline	- Use document preparation functionality

The approach discussed here is based on socio-technical systems theory, which requires that any system is made up of technical and social (human) resources. These are interrelated so that any attempt to optimise only one of these resources may adversely affect the other set, with the result that utilisation is suboptimal. The objective is to capture and understand those requirements. However, there is a need to acknowledge the difference between technical systems description and the user's language, appropriate to the organisational context. What we need is some set of 'boundary objects' where these two worlds can meet. ORDIT's premise is that the concept of responsibility is one such 'boundary object'—that responsibilities may be regarded as the key to understanding requirements in implementable terms. This is because responsibility has attributes that can be appreciated in both worlds, though the language and implications differ.

The concept of responsibility as a 'boundary object' between users and designers should lead to a better understanding by designers of what the technical system should achieve (rather than how), and its context of use within the socio-technical system. It will also help support its acceptance within the organisation because the users can speak the language of implementation, at the boundary of responsibility.

In the planning stage of Ngurra Wangkamagayi, there were sessions about how to set up the organisation so it could:

- be an integral part of the community;
- involve elders as both teachers and advisers;
- involve members of other community organisations as advisers;
- facilitate members to be advisers, core group or casual workers.

As this structure was unfamiliar to many of the people an explanation was needed. The graphic in Figure 3 was produced in response to the culture group's need for a model (boundary object) to describe their proposed organisational structure. The authors were discussing this with the group members over morning tea and used plates of different shaped biscuits to physicalise the concepts. The graphic was then prepared for use in a community meeting later in the day. It operates as a 'boundary object' to facilitate participative CIS design, in the same way as do Soft System Methodology's 'rich pictures' (Checkland, 1981), or ORDIT responsibility modeling.

This 'boundary object' or 'rich picture' is not isolated from the group's experience, rather it is 'situated.' It

Figure 3: Ngurra Wangkamagayi organisational arrangements

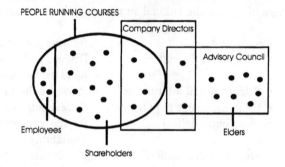

has the very tangible link of being derived from the morning tea biscuits. More importantly, as it was being devised the group was imputing names of possible advisory members, employees, etc. The 'boundary object' was not merely a theoretical model, it was part of a process and one which integrated the group and the wider community. In this way it fulfils the need to be part of a longitudinal rather than atomatistic approach. The 'boundary object' then serves not only as an organisational model but also as a demonstration of the already existing CIS which the computer-mediated system seeks to facilitate and operate from.

Social Modeling Approaches

In developing a CIS it is important that community social processes and requirements are clearly understood and that the IS fits in with and serves those needs. A socio-cognitive perspective on organisations is recommended by Hemingway (1998) as a way to facilitate "... the alignment of the semantic organization of information systems with the characteristics of effective action based on experiential knowledge" (p. 285). Hemingway stresses the need to understand the philosophical basis upon which CISs are implemented and the necessity to learn from experience.

He works from the premise that "stimuli are not passively recorded and passed to working memory, but are considered to be organised relative to an existing structure of knowledge, which reflects past experiences and any physiological aspects of structure" (p. 277).

Kaplan (1998) proposes a 'social interactionist framework' to guide socio-technical IS development so as to "... capture the complexity of interactions, interrelationships, and intereffects that occur during these processes" (p. 335). This framework can assist in modeling communication processes, tasks/services and control processes within a specific context. The relationship between these factors and information system design is crucial—"responses to the new technology are influenced by values, by communication patterns, and by individuals' places in a social network. All these influences are interrelated, they interact with each other, and are, therefore, part of an interactionist perspective" (p. 332).

The cultural fitness of IS is examined from an 'activity theory' perspective by Gobbin (1998). Because computers act as tools, mediating both individual and social processes, systems design needs to take account of organisational culture and patterns of social communication—"... the introduction of new computer systems can generate cultural rejection because whole areas of cultural, social and cognitive issues are omitted from current system analysis and design methodologies" (p. 109).

Gasson (1998) suggests that there is a tendency in discussions of IS design to "take a fragmented perspective of the nature of IS design by examining methodological, social and political issues in isolation from the design context" (p. 308). She concludes that this leads to a limited understanding of IS design and advocates adoption of a 'situated' design paradigm utilising theories and techniques from recent work on 'situated action' and 'social cognition.' This approach explicates the contingent nature of human activities arising from a given cultural and social situation. In a study of these phenomena, she found that a "'common vision' of process objectives, resolving the question of what constitutes cultural knowledge and how such knowledge is communicated and learned, is more important to successful design than a common vision of design goals" (p. 323). This is a similar argument to that made by Friere (1972). Friere, a South American sociologist concerned with the interrelationship of pedagogic strategies and oppression of marginalised people, suggests that even well-intentioned community project managers can contribute to the continued oppression of marginalised people. According to Friere this occurs because people, projects and community process are atomised rather than being treated as an integrated part of a larger process. The message is clear: developers of CIS have a responsibility to work with all relevant stakeholders in a way which ensures that such a holistic shared understanding emerges and that it truly controls the CIS design and implementation processes.

Need for Modeling Appropriate to the Particular Community

In his introduction to this book Michael Gurstein states that ICT is concerned with "providing resources and tools that communities and their members can use for local economic, cultural and civic development ... ." He goes on to suggest that CIS development needs to pay attention to "physical communities and the design and implementation of technologies and applications, which enhance and promote their objectives."

This is certainly the approach that we have taken when working with Ngurra Wangkamagayi on the development of ICIS. The computer-based CIS that we are concerned with providing is a manifestation of the cultural IS that already exists within the community. The Ngaluma, Injibandi and Banjima peoples have a complex sub-tribal ('skin') system. For instance, for the Ngaluma people the four 'skins' are called bananga, burungu, balyirri and garimarra. This system contains the rules which govern personal intra-group relationships and broader inter-group relations, including who may marry whom and the degree of respect that people must afford to each other. It also determines who is to care for 'country' and hand down cultural information. The purpose of the 'skin' system is explained in the video "Exile and the Kingdom" (Rijavec and Harrison, 1992) as being:

- to classify strangers;
- for forging alliances between groups, permitting them to survive in the long run;
- to regulate marriage;
- to organise law meetings;
- to organise talu (sacred) site operation and natural resource use;
- to ascribe character to individuals.

"Within the system all humanity is divided into four classes, including Europeans. All of the known universe and its phenomena are divided and allotted to these classes; plants, animals, thunder, rain, wind, ... even post-contact phenomena can be placed within the scheme. The totemic ideas bring man and nature into one moral and psychological system. It obliterates the division between the human and the natural. The section systems are unique to Australian Aboriginal Society, and in complexity are unparalleled in any other culture. Their 'rationale' is based on sympathies and antipathies. They also bestow power on initiates; power derived from the magical potency that flows from knowledge of vitally important secrets." (Rijavec and Harrison, 1992)

Each person in the community has a 'skin' resulting from those of their parents. However, there are complications because of the post-colonial displacement of tribal groups and their shift into town communities. A person may have one 'skin' in Roebourne and another when he or she is back in their own 'country.' The authors are working with a prototype database design to explore these complexities and to find effective ways of representing and explaining the 'skin' system. The ICIS information structure will utilise this approach, within the general structure reflecting the relationship between places, people, procedures and presentations, discussed earlier.

The use of the 'skin' system to help structure ICIS is an example of how CIS development should utilise preexisting community structures and principles of social organisation. It will take considerably more work to adequately understand the complex web of interpersonal relationships and to develop a CIS which faithfully represents them. This is only possible if a special database is developed for the purpose and if the development process involves a very high level of participation by the community.

Conclusions

A successful CIS should be 'situated,' i.e., based on a participative study of its organisational and community context. This chapter has used a discussion of the ICIS project to help explain the role of modeling in facilitating the development of a CIS that is well-integrated with preexisting community processes, structures and culture. This requires that the conceptual models be represented (as 'boundary objects') in a manner which is easily understandable by both analysts and potential system users. Such an approach facilitates effective and equitable participation in the CIS development project by all stakeholders and aids in ensuring that the resulting system is culturally appropriate. It operationalises the imperative that CIS development be postmodern and postcolonial.

ICIS was born out of a desire by members of the community to have cultural information stored in a computer-mediated form which would allow, for instance, children to access this material within the formal school system. To date it has been most useful as a means of producing cultural awareness course manuals for Ngurra Wangkamagayi. In the future it may be used as a repository for the cultural information being collected for the native title land claim. This information will then be available for a variety of purposes including computer-mediated interactive learning and sources of information for tourists, prospective clients, teachers, police, other indigenous groups. Importantly, its uses will evolve over time as community participation increases.

The authors intend to extend their research by further development of ICIS in close cooperation with the indigenous participants. The utility of CIS development techniques and tools will be evaluated in this extremely challenging application domain.

References

Avison, D., Kendall, J. E., and DeGross, J. I. (Eds.) (1993). *Human, Organizational, and Social Dimensions of Information Systems Development.* Elsevier Science.

Bell, F. (1998). Two Frameworks for Understanding and Evaluating IS Methodologies. Avison, D. and Edgar-Nevill, D. (Eds.) *Matching Technology With Organisational Needs—Proceedings of the 3rd UKAIS Conference.* Lincoln University Campus, University of Lincolnshire & Humberside, UK, 15-17 April 1998. pp. 321-330.

Brugha, C. M. (1998). Information Systems Development: Who Owns the Decision? In: *Proceedings of PAIS II, the Second Symposium and Workshop on Philosophical Aspects of Information Systems: Methodology, Theory, Practice and Critique.* University of the West of England, Bristol, UK, July 27-29, 1998.

Checkland, P. (1981). *Systems Thinking, Systems Practice.* Wiley.

Cross, R. and Turk, A. G. (1997). Quality specification of information systems: An integration of three socio-technical approaches. Ahluwalia, J.S. (Ed.) *Total Quality Management, the Proceedings of WCTQ 97, 7th World Congress on Total Quality.* February 1997, New Delhi, India: Tata McGraw-Hill; pp. 619-632.

Crowe, M., Beeby, R. and Gammack, J. (1996). *Constructing Systems and Information: A Process View*. McGraw Hill.

Dahlbom, B. and Mathiassen, L. (1996). Power in Systems Design. In: Kling, R. (Ed.) *Computerisation and controversy: Value conflicts and social choices (Second Edition)*. Academic Press. pp. 903-906.

Dix, A. J., Finlay, J. E., Abowd, G. D. and Beale, R. (1998). *Human-Computer Interaction (2nd Edition)*. Prentice Hall.

Eason, K. (1988). *Information Technology and Organisational Change*. Chapter 4, pp. 44-59.

Espejo. (1996). Requirements for Effective Participation in Self-Constructing Organisations. *European Management Journal*, 14(4), 414-422.

Finkelstein, A., Tauber, M. and Traunmuller, R. (Eds.) (1990). *Human Factors in Analysis and Design of Information Systems*. IFIP.

Flood, R.L. and Jackson, M.C. (1991). *Creative Problem Solving: Total Systems Intervention*. John Wiley and Sons.

Flynn, D. J. (1992). *Information Systems Requirements: Determination and Analysis*. McGraw-Hill.

Friere, P. (1972). *Pedagogy of the Oppressed*. Penguin Books.

Gasson, S. (1998). A Social Action Model of Situated Information Systems Desin. In: Larsen, T. J., Levine, L. and DeGoss, J. I. (Eds) *Proceedings of the IFIP WG 8.2 and 8.6 Joint Working Conference on Information Systems: Current Issues and Future Changes*. Helsinki, Finland, 10-13 December, 1998, pp. 307-326.

Gobbin, R. (1998). Adoption or Rejection: Information Systems and their Cultural Fitness. In: Hasan, H., Gould, E. and Hyland, P. *Information Systems and Activity Theory: Tools in Context*. University of Wollongong Press. pp. 109-124.

Guimaraes and McKeen. (1993). User Participation in Information System Development: Moderation in All Things. In: Avison, D., Kendall, J. E. and DeGoss, J. I. (Eds) *Human, Organizational and Social Dimensions of Information Systems Development*. Elsevier Science, pp. 171-192.

Hart, D. N. (1997). Modeling the political aspects of information systems projects using "information wards." *Failure and Lessons Learned in Information Technology Management*. 1, 49-56.

Hemingway, C. J. (1998). Towards a Socio-Cognitive Theory of Information Systems: An Analysis of Key Philosophical and Conceptual Issues. In: Larsen, T. J., Levine, L. and DeGross, J. I. (Eds.), *Proceedings of the IFIP WG 8.2 and 8.6 Joint Working Conference on Information Systems: Current Issues and Future Changes*. Helsinki, Finland, 10-13 December, 1998, pp. 275-286.

Hidding (1998). Adoption of IS Development Methods Across Cultural Boundaries. In: Hirschheim, R., Newman, M. and DeGoss, J. I. *Proceedings of the Nineteenth International Conference on Information Systems*. Helsinki, Finland, December 13-16, 1998, pp. 308-312.

Hix, D. and Hartson, H. R. (1993). *Developing User Interfaces: Ensuring Usability Through Product and Process*. Wiley.

HUSAT. (1993). *ORDIT Training Manual*. HUSAT Research Institute, Loughborough University of Technology, UK.

Iivari, J. (1992). The organizational fit of information systems. *Journal of Information Systems*, 2, 3-29.

Jayaratna, N. (1994). *Understanding and Evaluation Methodologies—NIMSAD: A Systemic Framework*. McGraw-Hill.

Karat, J. and Bennett, J. L. (1991). Working within the Design Process: Supporting Effective and Efficient Design. Carroll, J. M. (Ed.) *Psychology at the Human Computer Interface*. Cambridge University Press, pp 269-285.

Karsten, H. (1998). Collaboration and collaborative information technology: What is the nature of their relationship. In: Larsen, T. J., Levine, L. and DeGross, J. I. (Eds.) *Proceedings of the IFIP WG 8.2 and 8.6 Joint Working Conference on Information Systems: Current Issues and Future Changes* Helsinki, Finland, 10-13 December, 1998, pp. 231-254.

Kaplan, B. (1998). Social Interactionist Framework for Information System Studies: The 4Cs. Larsen, T. J., Levine, L. and DeGross, J. I. (Eds.) *Proceedings of the IFIP WG 8.2 and 8.6 Joint Working Conference on Information Systems: Current Issues and Future Changes*, Helsinki, Finland, 10-13 December, 1998, pp. 327-339.

Kling, B. (Ed.) (1996). *Computerization and Controversy: Value Conflicts and Social Choices*. Academic Press.

Ledington, P. and Heales, J. (1993). The social context of information systems development: An appreciative field perspective. In: Avison, D., Kendall, J. E., and DeGross, J. I. (Eds.) (1993). *Human, Organizational, and Social Dimensions of Information Systems Development*. Elsevier Science. pp. 455-473.

McMaster, T., Vidgen, R. T. and Wastell, D. G. (1998). Networks of Association and Due Process in IS Development. Larsen, T. J., Levine, L. and DeGross, J. I. (Eds.) *Proceedings of the IFIP WG 8.2 and 8.6 Joint Working Conference on Information Systems: Current Issues and Future Changes*. Helsinki, Finland, 10-13 December, 1998, pp. 341-357.

Michaels, E. (1990). *Bad Aboriginal Art: Tradition, Media, and Technological Horizons*. Allen & Unwin.

Morley, C. (1993). Information Systems Development Methods and User Participation: A Contingency Approach. Avison, D., Kendall, J. E. and DeGoss, J. I. (Eds.) *Human, Organizational and Social Dimensions of Information Systems Development*. Elsevier Science. pp. 127-142.

Mumford, E. (1983). *Designing Participatively*. Manchester Business School.

Patel, A., Sim, M. and Weber, R. (1998). Stakeholder experiences with conceptual modeling: An empirical Investigation. In: Hirschheim, R., Newman, M. and DeGoss, J. (1998). *Proceedings of the Nineteenth International Conference on Information Systems*. Helsinki, Finland. pp. 370-375.

Preece, J., Rogers, Y., Sharp, H., Benyon, D., Holland, S. and Carey, T. (1994). *Human-Computer Interaction*. Addison-Wesley.

Rijavec, F. and Harrison, N. (1992). *Exile and the Kingdom*. Video, Film Australia.

Rundstrom, R. A. (1995). GIS, Indigenous Peoples, and Epistemological Diversity. *Cartography and Geographic Information Systems*, 22(1), 45-57.

Travis, J., Boalch, G. and Venable, J. (1996). Approaches for the learning organisation: A comparison of hard and soft systems thinking. *Proceedings of EDPAC'96*, pp. 487-500.

Trees, K. A. and Turk, A. G. (1998). Culture, Collaboration and Communication: Participative Development of the Ieramugadu Cultural Heritage Information

System (ICIS). *Critical Arts*, 12(1-2), 78-91.

Turk, A. G. (1996). Presenting Aboriginal knowledge: Using technology to progress native title claims. *Alternative Law Journal*, 21(1), 6-9.

Turk, A.G. and Mackaness, W.A. (1995). Design considerations for spatial information systems and maps to support native title negotiation and arbitration. *Cartography*. 24(2), 17-28.

Turk, A.G. and Trees, K.A. (1998-a). Ethical Issues Concerning the Development of an Indigenous Cultural Heritage Information System. *Systemist Volume 20, Special Issue*, pp. 229-242. Reprint of Turk, A.G. and Trees, K. A. (1998). Ethical Issues Concerning the Development of an Indigenous Cultural Heritage Information System. *Proceedings: Second Symposium and Workshop on Philosophical Aspects of Information Systems: Methodology, Theory, Practice and Critique—PAIS II*. University of the West of England, Bristol, UK. 11 pages.

Turk, A.G. and Trees, K. A. (1998-b). Culture and Participation in Development of CMC: Indigenous Cultural Information System Case Study. Ess, C. and F. Sudweeks (Eds.), *Proceedings International Cultural Attitudes Towards Technology and Communication - CATaC'98*, Science Museum, London, UK (Published by University of Sydney, Australia), pp. 219-223.

Turk, A.G. and Trees, K.A. (1999). Ethical Issues Concerning the Use of Geographic Information Systems Technology With Indigenous Communities. *Proceedings of the Australian Institute of Computer Ethics Conference, AICEC99*. Melbourne, Australia, pp. 385-398.

Whitley, E. A. (1998). Method-ism in Practice: Investigating the Relationship Between Method and Understanding in Web Page Design. In: Hirschheim, R., Newman, M. and DeGoss, J. I. *Proceedings of the Nineteenth International Conference on Information Systems*. Helsinki, Finland, December 13-16, 1998. pp. 68-75.

Chapter XVI

On-Line Discussion Forums in a Swedish Local Government Context

Agneta Ranerup
Göteborg University, Sweden

This text describes experiences of four on-line discussion forums that are used in a Swedish local government context. The main issue is how aspects such as the implementation of the forums, functional features of the forums, and activities to increase access to Internet affect the on-line debate. Furthermore, the debate in the on-line forums is evaluated against the ideal of deliberative democracy. Lastly, three strategies for how the amount of debate in the on-line forums might be increased are outlined. One strategy would be to provide citizen groups with access to technology in order to involve them in the forum and the local government network as such. Another strategy would be to welcome a completely open debate in the forums, in the hope of getting a discussion that also includes issues that are of relevance to local government. Yet another strategy would be to more seriously involve local politicians in the discussion. This group seems to be of strategic importance when aiming at democratic effects of on-line forums.

Introduction

The main aim of this text is to present experiences of on-line discussion forums in a Swedish local government context. In particular, we will discuss various aspects that affect debate in on-line discussion forums, but also strategies that might be used to increase the number of contributions to the debate. There are several reasons why these experiences are of value to the field of Community Informatics. Using information technology to improve democracy, equal access to technology, and active citizenship can be defined as ideals that are of importance to this field. There are many connections between these ideals and the experiences of on-line forums that will be presented. According to writers like Buchstein (1997), it is more likely that the Internet will have democratic effects if it is used in connection with existing political institutions: "Here computer democracy would be based on an already existing community and used to distribute and collect information and to foster deliberation" (Buchstein, 1997, p. 260) he argues. This is the case in the experiences that will be presented in this chapter. Also, the on-line forums that will be discussed were introduced in local government projects aiming at improving democracy and increasing access to technology. This makes them particularly interesting to Community Informatics as a test of whether the J.S. J.S. J.S. Internet has had any democratic effects. Finally, in the small but growing literature on community networks and civic networking, to which this piece of research belongs, a number of issues and aspects are treated (Tsagarousianou et al., 1998). A common issue is for example how the networks as such are initiated ('from above or from below'), or more particularly to what extent the citizen groups have been active in the process of implementation. However, a more thorough investigation of the discussion in the on-line discussion forums is more seldomly seen, though there are a few exceptions (Docter and Dutton, 1998; Tambini, 1998). The following text will contribute with such experiences. This is another reason why the following experiences are of particular interest to Community Informatics.

There are several ideals in a discussion about how IT in general, and the Internet in particular, can improve democracy. One ideal is called quick democracy or plebiscitary democracy. According to this ideal citizens should make an on-line vote on all political affairs. Otherwise, opinion polls should constantly be arranged to keep the politicians informed about the views of their voters. An alternative ideal is deliberative democracy. Here, representative democracy is taken as a starting point when asking how it could be strengthened and made more participatory (Friedland, 1996; White, 1997; Åström, 1998). Consequently, a genuine discussion between citizens and politicians, or a deliberative process, is defined as being of utmost importance. For this reason, citizens must be allowed to form opinions with the help of IT, rather than to just express them. It is also important that there is somebody who can be held responsible for the political decisions (Street, 1997). As a consequence, a politician within, e.g., a local government council is highly relevant as a contributor to a deliberative process.

The choice between these alternative democratic ideals when studying on-line forums is almost an obvious one. First, on-line forums are designed to support a debate that goes on for an extended period of time, and not to give a quick hint of popular opinion. Also, due to the fact that new contributions and issues can be introduced over time, citizens cannot be expected to react as quickly and in such a structured way as would be necessary in an opinion poll. In conclusion, the focus of the following text will be on experiences of using on-line forums in a local government context with the aim of satisfying the deliberative ideal, rather than other democratic ideals.

The next section will treat some issues of interest that will be in focus in the following against a background of previous research. More precisely, we will find a description of aspects that, according to literature, are likely to influence the debate in on-line forums and its democratic effects.

Which Aspects Might Affect the Debate?

The Initiation and Implementation of On-Line Forums
In research about community networks and civic networks there is a common theme to focus on the initiators of the projects. There are several alternatives that might apply. Sometimes they are initiated by different voluntary organizations as, e.g., in Amsterdam (Francissen & Brants, 1998), Santa Monica (Docter & Dutton, ibid.) and Seattle (Schuler, 1996); by local government as in Bologna (Tambini, ibid.), or by researchers as in Athens (Tsagarousianou, 1998). The first alternative is sometimes described as initiatives 'from below,' and the others as initiatives 'from above.' If different kinds of organizations are involved in the project, they can be seen as stakeholders, and the process where they are introduced can be characterized as a negotiation of their diverse and often competing interests (Howley, 1998). However, the extent to which citizens are allowed to participate in the implementation process varies considerably. In connection with community networks and the like, all citizens can be seen as potential participators in the implementation process (Braa, 1996). Aspects such as these are according to researchers very likely to affect how various groups such as citizens and politicians are involved in the network as a whole, something which might influence, e.g., the amount of discussions in the on-line forums. As a consequence, *the impact of the initiation and the implementation process on the on-line debate* is the first aspect that is in focus in the following discussion.

Increasing Access to the Technology
Another common theme in research about civic networks and community networks is concern about how access to technology affects their democratic effects. High access is by many researchers (Tsagarousianou et al., ibid.), as well as by practitioners (Ranerup, 1998; Ranerup, 1999), seen as a prerequisite for getting genuinely democratic effects.

There is a risk that when information is increasingly provided by means of information technology this might also affect the access to information as well as to participation in the democratic processes *per se* (Tsagarousianou et al., ibid.):

> "[There is also] the fear that electronic democracy projects might be oblivious of the social and economic inequalities among the citizenry and, therefore, the differential distribution of the hardware and skills necessary to participate in them, has led many participants in the debate to argue that only public provision in information infrastructure and public subsidy for information (and more generally, for electronic democracy) services can ensure that the benefits of access to information will be distributed equitably and democratically" (Tsagarousianou et al., 1998, pp. 170).

As a consequence, access to technology should be increased by various measures. One example is that citizens could be provided with public computers (Tsagarousianou, ibid.); alternatively the municipal authorities could offer home access to the Internet at a reduced rate (Tambini, ibid.). In the following we will focus *on whether various measures have been taken to increase access to technology* to prevent negative

362 Ranerup

effects from limited access. According to previous research this might be considered as a factor that influences access to on-line information in general, as well as to participation in on-line discussion forums. Consequently, it will be in focus in the following as a factor that is *likely to influence the debate* in the forums.

The Functional Structure of On-Line Forums

In research there is also a discussion about the functional aspects of the on-line forums as such, and how they affect the discussion. First, there is the question of whether the debate is censured or not. In the U.S. there seems to be a strong opinion against any kind of censorship, both within the legal framework and in the public opinion (Docter & Dutton, ibid.). A milder form of regulation is when a moderator reads the contributions before they are published (Docter & Dutton, ibid.). He or she might be allowed to remove sexist or otherwise abusive contributions, etc. These are examples of how laws and other restrictions might affect the debate in on-line forums.

However, the functional or technical structure in itself might also affect the debate. This being so simply because technological artifacts such as on-line forums create a space of possible actions (Stolterman, 1998). This way of reasoning is of course inspired by Actor-Network Theory (Akrich, 1992). For example, one possibility might be to avoid censorship and moderators, and provide an open, unstructured on-line forum where the contributions are published in a long list. Another possibility might be to provide some kind of tree-structure where different issues in the debate can be separated from each other (Croon & Ågren, 1998; Benson, 1996). However, both alternatives enable and restrict the behavior of users. Another functional aspect is that the contributions sometimes are removed from the on-line forums after a certain amount of time. Alternatively, they are placed in some kind of archive for a longer period of time, and an interested citizen might easily get an overview of previous discussions (Croon & Ågren , ibid.). A last *aspect that might have influenced the on-line debate is therefore the functional structure of the forums*, as described above. This aspect will also be in focus in the following.

Issues of Interest

As has been described in the previous sections, the main research question in the following is how aspects such as the initiation and implementation process, measures taken to increase access to technology, as well as the functional structure of the on-line forums, have influenced the on-line debate. More particularly, it will be discussed whether the amount of debate in general has been influenced by these aspects. We will also discuss whether these aspects have influenced the amount of contributions from politicians and citizens respectively, as well as the discussion about political versus other issues. But there is also another research question with a special focus on the on-line debate itself. It concerns *whether the debate in the forums is sufficient as to satisfy the ideal of deliberative democracy*, as described in the introduction. This ideal emphasizes the value of a genuine discussion between citizens and politicians about political issues. Lastly, with previous research as a source of inspiration, there will be a more open discussion about *how the amount of debate in on-line discussion forums in a local government context might be increased.*

In the following text we will find experiences of on-line discussion forums in three districts in the city of Göteborg, and one forum in the city of Sölvesborg in Sweden. The author has conducted 23 longer interviews with civil servants and politicians in these two cities. Furthermore, shorter interviews have been held with local politicians (six) and members of the use groups (nine) in one of the cities. Also, the debate in the on-line forums has been investigated regarding its size, contributors, and issues in the debate. The study was conducted between January 1997 and December 1998 in Göteborg, and between January 1998 and December 1998 in Sölvesborg.

Experiences of On-Line Discussion Forums

On-Line Forums in Three Districts of Göteborg

The initiation and implementation of the forums

One important feature in the background of the on-line forums is the reorganization of the Local Authority of Göteborg in 1990 into 21 districts. These districts have their own councils, with the authority to decide how to spend their budgets within a framework of centrally set economic, political and legal limits. Decentralization has meant that responsibility for a number of elements of government—including schools, child care, libraries and social welfare—has been devolved to a unit with a comparatively small population and geographical spread. The district reform was also intended to increase the democratization of local government, as well as to result in increased participation of citizens in government (Ranerup, 1999). Despite these intentions the democratic goals have been attained to a much lower degree than the efficiency goals, according to recent evaluations of actual results (Johansson et al., 1998).

The on-line discussion forums in Göteborg were implemented as a part of a bigger project, the DALI-project, aiming at using IT to improve local government services and democracy in local government. The acronym DALI stands for 'Delivery and Access to Local Information and services.' The project was partly financed by the European Commission, partly by the city of Göteborg. At this point in time (i.e., in the first half of 1996), Göteborg had already started to renew its technical infrastructure by implementing Lotus Notes as well as Internet technology. A prerequisite for the districts in order to take part in the DALI project was that they had embarked on this project of technical renewal. This meant that three districts out of 21 were qualified to participate in the DALI project, and agreed to do so. The names of those districts are Askim with 21,000 inhabitants, Kärra-Rödbo with 9,000 inhabitants, and Härlanda with 19,000 inhabitants. Göteborg as a whole has 460,000 inhabitants.

The project group in Göteborg consisted of a technical consultant, a few civil servants from central and district levels, as well as a few others who worked with IT-issues in the local government administration. According to this group, the project should focus on implementing a homepage containing information and an on-line discussion forum in each one of the three districts. The group made a suggestion about how this homepage should be designed. Since December 1996, citizens of Askim, Kärra-Rödbo and Härlanda have been able to access the three homepages that are owned by the districts ('the DALI-system').

The European Commission supported the DALI project economically on condition that groups of potential users were to be engaged in the systems development process. As a consequence, various user groups have been involved in the development process from the autumn of 1996 to the spring of 1997. In Härlanda and Kärra-Rödbo the participants were recruited from the local political parties that are represented in the district council, whereas in Askim the participants were recruited from other local organizations such as the boards of private schools, child care institutions, and organizations of pensioners. There were several meetings between the systems developers and the groups of potential users, where the former delivered short presentations of the Internet and the DALI-system. Moreover, after a few months, the groups of users were able to express their reactions to the appearance of the system (late 1996).

As a part of the DALI project three public computers were placed in each of the three districts (see next section). This meant that the user groups also discussed what the instructions to the public computers should contain, as well as where they should be placed. Views of a more critical character were also put forward, e.g., on how to make the DALI project known to the citizens in the districts, which was considered important. However, when subsequently asked if their own organization used the Internet and the DALI, only two out of nine user representatives knew this to be the case. According to these experiences, at the time of the interviews (late 1997) very few of them used the Internet in local government politics apart from the politicians that were involved in the DALI project itself.

Increasing access to the technology

As mentioned above, the district homepage and the on-line forums of the three districts could be accessed from computers in private homes or from nine public computers. Furthermore, as a part of the DALI project, computers were distributed to four of the leading local politicians in each one of the three districts that took part in the project. There was also an intention of spreading Internet access to schools in the districts, which are very similar to schools in other cities in Sweden. This process progressed gradually in the three districts.

Parallel to this there was an objective in the city of Göteborg to spread the access to the Internet by other measures. Very late in the period that was investigated, in December 1998, an agreement was made with an Internet supplier about providing citizens with Internet access at no cost except for the communication as such. In February 1999 this offer had been accepted by 30,000 inhabitants according to official informants (Jurnell, 1999).

The functional structure of the on-line forums

As was indicated above, the DALI-system was implemented as a separate homepage owned by the districts of Askim, Kärra-Rödbo, and Härlanda in Göteborg, with a connection to the homepage of Göteborg. The functional structure of the homepages and the on-line discussion forums can be described as follows: there is an administrative information section that contains information about the opening hours, addresses, and activities of various municipal services, such as child care, schools, social services, libraries, and sports facilities, etc. There is also some information about how to apply for services. However, there are not facilities to make application for service via the Web. The section for current issues contains the political proposals of the district council, and revised protocols of its decisions. There is also local news of a general character, such as the menus of schools and the restaurants for elderly, information about cultural events, as well as special events in schools, etc.

The on-line forums themselves enable moderated, publicly accessible interactive discussions regarding current political issues. They are accessed by a button on the entrance page of the districts. In these discussions, citizens and local politicians can participate. The issues in the debate, as well as the headings under which the contributions are published, can be chosen according to the preferences of those who participate. If the contributions become too numerous, the moderators remove them from the forum after a couple of weeks, but they can easily be fetched from the archive. A citizen who wants to make a contribution must fill in his/her name, but can remain anonymous when their contribution is published. In each one of the three districts there is a civil servant that checks the contributions to the debate Monday-Friday before they are published. However, according to city informants, not many contributions have actually been censored. Unfortunately, two on-line forums, the ones of Härlanda and Askim, have been closed during summer vacations, when the moderator was not on duty. Lastly, the local government homepages contain lists of some, albeit a minority, of the local politicians and their electronic mail addresses. This way, direct contacts between citizens and politicians are made easier.

The debate in the on-line forums

Table 1 shows the contributions to the debate in the three districts of Göteborg, as well as those who contributed. In one district (Askim) a significant amount of contributions were made, whereas in the two other districts the debate was much smaller. Furthermore, the debate in the districts of Kärra-Rödbo and Härlanda almost came to a complete standstill in 1998. Citizens, as compared to politicians and civil servants, made the vast majority of the contributions. However, Askim had two politicians who made rather numerous contributions to the debate.

In Table 2 we can see the different issues in the debate. In Askim, most contributions treated traffic issues (the regulation of specific roads, etc.). Another important issue was whether or

Table 1: Contributions to the Debate in Göteborg

	Askim	Kärra-Rödbo	Härlanda
1997			
citizens	107	37	28
politicians	31	7	2
civil servants	2	4	2
1998			
citizens	91	13	8
politicians	20	2	3
civil servants	4	1	0

not the district of Askim should be a part of Göteborg in the future. This issue got its own heading in the autumn of 1997, but before this point of time, contributions could be found under various headings. Furthermore, in 1998 this debate transformed into a discussion about the rules according to which the districts get economic support from the central authority of Göteborg. One could also find other smaller debates on various issues, something that applied to the districts of Kärra-Rödbo and Härlanda too. Lastly, in Härlanda in 1997 there was a debate on primary school issues with

Table 2: Issues in the Debate in Göteborg

	Askim	Kärra-Rödbo	Härlanda
1997			
Environmental issues	23	3	3
Traffic/communication	56	14	3
Miscellaneous	25	5	9
The district council	0	7	0
Housing/physical planning	3	0	2
Choose a heading	8	0	0
Education	2	2	-
Primary school	6	5	12
Sports/culture	4	8	1
Child care	5	0	2
Local organizations	1	3	0
The independence of Askim	7	-	-
The disabled	0	1	0
1998			
Environmental issues	10	2	2
Traffic	21	0	4
Miscellaneous	6	1	0
The district council	5	2	0
Housing/physical planning	1	0	0
Choose a heading	4	0	0
Education	0	1	-
Primary school	6	2	4
Sports/culture	0	6	0
Child care	7	0	0
Local organizations	0	2	0
The independence of Askim	34	-	-
The disabled	0	0	1
Östra Trollåsen	6	-	-
The distribution of money to districts	15	-	-

several contributions. However, the politicians chose to answer through other means than the on-line discussion forum.

Experiences of an On-Line Forum in Sölvesborg

The initiation and implementation of the forum

Sölvesborg is a rather small town in the south of Sweden, with 16,500 inhabitants. Two features are especially important in the initiation process of the on-line discussion forum in Sölvesborg. The first is the fact that the municipality of Sölvesborg during the 1990s has been providing their citizens with two information offices where a broad spectrum of information about local government services and other kinds of services can be found. These offices have been using IT during their whole existence, but as new kinds of technology developed in the middle of the 1990s, they were replaced by modern systems.

Also, in Sölvesborg there have been several attempts to renew the technical infrastructure of the local government administration at large with the help of various sources of funding. As an example, Sölvesborg received financial support from the European Commission for a pilot study about technical renewal in the middle of the 1990s, but a few years later an application for a further phase of their project was turned down. However, Sölvesborg received economic support from Swedish sources (NUTEK and KK-stiftelsen) for a smaller project aiming at providing citizens with services via the Internet. As a consequence, by the end of 1997, the local government homepage of Sölvesborg contained an interactive service where parents could apply for child care through the Web, as well as an interactive on-line discussion forum. At that point in time, the municipality also got economic support for providing schools with Internet access.

The forum was designed by a small group of civil servants that had previously worked with the information offices, and by a technical consultant. The leading politician of Sölvesborg also supported the introduction of the on-line forum on the local government homepage. However, no other kinds of potential users were asked to participate in the process of design.

Increasing access to the technology

The citizens of Sölvesborg have access to public computers and the Internet at the information offices, as well as at the libraries. No public computers have been introduced in connection with the on-line forum. Furthermore, no politicians have been given access to computers and the Internet as a part of the process in which the forum was introduced, but four politicians already had access to the Internet at their offices. Consequently, the citizens have a means of communication with four of the politicians through Internet, and the local government homepage presents their e-mail addresses. Also, as was said above, there is a continuous process of implementing the Internet at the schools of Sölvesborg.

Furthermore, Sölvesborg wanted to continue with their project of technical renewal, but an application for a further phase was turned down by the European Commission. As a consequence, one idea that had to be abandoned was to provide all citizens with Internet access at a reduced cost.

The functional structure of the on-line forum

The homepage of Sölvesborg contains information about local government services, links to services that other public agencies provide, as well as information aimed at tourists. The on-line forum is accessed through a button on the entrance page, in a similar way as the forums in Göteborg. By contrast with the forum in Göteborg, the one in Sölvesborg has between December 1997 and December 1998 been open for discussion about only one issue at a time. In 1998 this issue was chosen by the project group that consisted of civil servants, with the assistance of the leading local politician. There has been one suggested issue for discussion during the spring of 1998, and another in October-November. In December citizens were allowed to discuss issues according to their own choice, as well as a suggested issue.

After the debate on a certain issue has come to a complete standstill, the contributions are removed from the homepage. They remain in the system, but can not easily be accessed by citizens. The issues for discussion are shortly introduced with the help of a text, as well as relevant maps, etc. Citizens wanting to contribute must provide their name, but it must not necessarily be presented on the screen. The different contributions to the debate are shown on a long list, with the latest on top. The date, the signature/name of the contributor, as well as 20 words from the contribution are also shown on the screen. In Sölvesborg there is no moderator that checks the contributions before they are published, except for an automatic control of whether they contain 'forbidden' words. Unfortunately, the on-line forum of Sölvesborg has been closed between July 1998 and October 1998, among other things because of the summer vacations and the lack of suggested issues for discussion. During this period, the public was asked to submit issues for discussion.

The debate in the on-line forum

The first suggested issue that citizens were invited to discuss was whether a big natural museum aimed at tourists should be built or not. In a period of six months approximately 50 contributions were made on this issue. Seven were made by politicians, and the rest by citizens. A third of the contributions dealt with pros and cons of building the museum, a third with how it should be financed, and a third treated the debate in itself. For example, the citizens wanted the politicians to be more active in the debate, and they wanted new issues to be brought into the debate. The forum was closed from July to October 1998, but after that a second issue for discussion was introduced. Now the forum welcomed citizens to discuss if some of the streets in Sölvesborg should be for pedestrians only. The result was a discussion with 19 contributions, most of them in favor of that suggestion. However, five of the contributions asked for new issues to discuss. This resulted in a completely open discussion, as well as a debate on where a new big road in the area should be built. In the open debate the citizens discussed environmental issues, as well as issues such as whether a special ceremonial arrangement should be introduced when immigrants received their Swedish citizenship. All in all, citizens made 81 contributions, politicians made 15 contributions, and a civil servant made one contribution to the discussion in the on-line forum.

Which Aspects have Affected the Debate?

The Initiation and Implementation of the On-Line Forums

In both cities the introduction of the on-line forums took place within a larger project with the aim of improving the technological infrastructure of local government. In fact, both projects aimed at improving information and service, as well as achieving democratic effects. These kinds of goals are by no means contradictory, because they could result in a technological infrastructure that is used by many citizens. Moreover, the provision of services and information through the Internet and a local government homepage might very well attract attention to the on-line forums as well, and as a consequence, result in a lively debate. However, during the period of investigation, this effect has most surely been rather modest, but could become a factor of importance in the future.

Furthermore, there have been very few activities in the two cities that have allowed various groups of, e.g., citizens and politicians to participate in the implementation process, and in that way become more deeply involved in the future of the homepage and the on-line forums. There was one exception to this. In the implementation of the forums in Göteborg, potential user groups were allowed to participate in the implementation process during a limited period of time. However, these groups were dissolved a few months after the homepages and on-line forums had been introduced to the public. Consequently, the project management did not take full advantage of this potential for creating a stable interest in the on-line forums. In other words, there have been no strategies for how an active interest in the on-line forum could be created with the exception of some very rudimental activities. This could be compared to previous research that states continuous user involvement to be a necessity in order to create a lively interest in a community network (de Cindio 1999; Howley, ibid.; Schuler, ibid.). In sum, it is here argued that the limited participation by citizens as well as politicians in the implementation process as a whole has influenced the amount of debate as such negatively for both groups. However, in the next sections we will find some more comments regarding the politicians.

Activities to Increase Access to the Technology

In both cities there has been an intention to provide various groups, as e.g., citizens, children at school, and politicians, with access to the Internet. The most ambitious activities to increase access have taken place in Göteborg, but Sölvesborg has not been far behind. All in all, both cities seemed to agree about the fact that increased access to technology is important when the Internet is used to improve democracy in local government. This, in turn, is in accordance with previous research (Tsagarousianou et al., ibid.) as well as practitioners (Ranerup, ibid.). However, what is most obvious is the comparatively small interest of the politicians in increasing their own access to technology. It is very likely that the limited access to technology in this group has led to fewer contributions from their part. In each district council or local government council only a few of the politicians had the most fundamental facilities for being able to participate in the discussions in the on-line forums (a computer, access to the Internet, etc.). Furthermore, if all politicians got access to the Internet they could publish their e-mail addresses on the local government homepage, and as a consequence, exchange messages with citizens more in private. In this way they will also be more experienced in using the technology in their daily work as politicians. As it is today, only between 10-20% of the politicians have published their e-mail addresses.

Also, the assumption that they are actually willing to discuss with citizens is very reasonable when considering their role as politicians (Johansson et al., 1998). As a matter of fact, this is a fundamental assumption in a report in which politicians themselves discussed how to improve democracy in local government (Johansson et al., ibid.). For this reason, their lack of the most basic infrastructure for participating in on-line conversation is of course very harmful to potential democratic effects. However, in the section on deliberative democracy further down we will find some more comments on this point.

Lastly, according to recent statistics, in May, 1999, 50.3% of the Swedish population between 12-78 years of age had access to the Internet, and the access rate is steadily rising (http://www.sifointeractive.com/index2.html). Thus, the problem of limited access in the population at large in Sweden, as well as in the other Nordic countries, Canada and the USA might become less important.

The Functional Structure of the On-Line Forums

A comparatively lively debate took place in the district of Askim where the choice of issues in the debate was left to the citizens. Moreover, late in 1998 the on-line forum of Sölvesborg also welcomed a debate about issues according to the choice of citizens. Therefore it seems likely that an open debate is a necessity, albeit not a guarantee, for a debate that attracts citizens. The need for openness might not be without exceptions though. Previous research describes the not very surprising idea that censorship and moderators limit the freedom of expression in a debate. Yet, all kinds of moderation does not have to be negative if one aims at creating a lively debate in an on-line forum. For example, in the forums in Göteborg as well as in Sölvesborg there was some kind of moderation in the form of a moderator and a list of 'forbidden' words. In spite of these arrangements citizens initiated a discussion of issues that local politicians most probably would not have chosen. The most obvious example is the discussion in Sölvesborg about ceremonial arrangements when immigrants receive their Swedish citizenship. Another example is the discussion about whether the district of Askim should be a part of Göteborg in the future or not. Moreover, a completely open choice of issues can be negative if it results in contributions that are considered to be offensive by some citizens, and they for this reason choose to withdraw from the discussion (Docter & Dutton, ibid.).

A more strictly functional aspect of a forum is if the contributions are published in a long non-structured list, or in some kind of tree structure. The former was the case in Sölvesborg, and the latter in the forums in Göteborg. Our experiences suggest that citizens should be allowed to make a rather open choice of issues in the debate. However, the open choice must be combined with some kind of structure that gives citizens an overview of the debate. A list that contains 50 contributions or more as in Sölvesborg might be of limited value here. The functional structure of the forums in Göteborg is more likely to meet this demand.

Lastly, in Göteborg as well as in Sölvesborg, the on-line forums were closed during summer vacations. This is negative as it interrupts the act of checking up on the debate that might have become habitual to some citizens. There are always limitations as to the forms citizens are allowed to participate in a democratic process. But limitations such as these are most likely to be harmful when one wants to create a lively debate in on-line forums. In addition, it ought to be fairly simple to solve this problem.

The Debate and the Ideal of Deliberative Democracy

In light of the experiences that have been described, can the on-line debate be characterized as sufficient to fulfill the ideal of deliberative democracy? In the districts of Härlanda and Kärra-Rödbo the answer must be 'most obviously not.' The debates came to what can be described as an almost complete standstill in 1998, which is reason enough for this judgement. But regarding the debate in Askim, the situation was somewhat different. Here we could find a discussion about various issues with contributions from citizens as well as from several politicians. Also, as was mentioned in the previous section, issues that the politicians most surely would not have introduced were introduced by citizens. This is also positive. Furthermore, the similar can be said about the on-line discussion in Sölvesborg. Here there was a debate of some size on various issues.

Another feature that is positive against the ideal of deliberative democracy is that the debate in all on-line forums contained discussion about local political issues as opposed to other more general issues. In Table 2, as well as in the description of the debate in Sölvesborg, many such issues were described (e.g., traffic issues, environmental issues, school issues). As a consequence, in the forums a debate on political issues has taken place rather than entertainment (nonpolitical) issues such as hobbies, 'chat' and other similar issues, which contrasts other experiences (Tambini, ibid.). This is also in accordance with the ideal of deliberative democracy.

But what about the reason for the difference between Askim on the one hand, and Härlanda and Kärra-Rödbo on the other? The result in the previous sections is not very clear on this point, except that Askim had at least two politicians that were active in the on-line discussion, which is more than the others. Also, the user representatives were recruited somewhat differently in the three districts: Askim chose the members of local organizations such as the boards of private schools or child care institutions etc., whereas the two other districts chose the members of the local political parties. However, no differences could be found between the districts regarding the limited use of the Internet in all these organizations in a post-implementation check-up. Consequently, the three aspects that have been discussed in this study cannot fully explain the differences between the size of the on-line discussion in the districts. In other words, there must be other aspects that are of importance here.

In a previous investigation (Ranerup, 1998), politicians and civil servants in the three districts in Göteborg were asked about what aspects they thought influenced the size of the discussion in on-line forums. Apart from aspects such as access to the Internet and the like, the local political climate was mentioned as a vital factor, i.e. if there were any 'hot topics' to discuss or not. According to the subjects' own experiences, this fluctuates significantly between various periods of time, as well as between districts. It seems likely that this is an aspect that to a certain degree can explain the differences between the districts. For example, Askim has had a more turbulent situation regarding its relationship with Göteborg than the other districts. Also, the district is more dynamic in the sense that more roads, etc. are being built, which in turn seems to have resulted in a greater interest in issues such as where, when and how the new roads should be built in the district. This difference is interesting as it emphasizes a dimension that is somewhat unfamiliar to systems developers (Ranerup, ibid.). However, similar aspects have been noticed in previous research, as e.g., when Schmidtke (1998) discussed the political climate in Germany by focusing on the attitudes toward grassroot activities among citizens. This is in turn said to affect the willingness to introduce on-line discussion forums.

However, there is another aspect that might be of value to explain the differences between the districts. In Askim several politicians contributed to the debate quite willingly as was mentioned previously. As a contrast, in the two other districts the situation was less fortunate on this point. For example, a discussion about school issues was started by some citizens in Härlanda in which the politicians chose to contribute by other means than the on-line forum. This indicates that the willingness of politicians to discuss with citizens by means of an on-line forum *as such* might be limited, which is an issue for further investigation. As a conclusion, as was said above, their role as politicians indicates their willingness to discuss with citizens (Johansson et al., ibid.), albeit not necessarily in an on-line forum. In other words, there is more to this problem than giving politicians access to technology to create a lively on-line debate. In the next section there will be some comments on how to increase the willingness of the politicians to contribute to the debate.

All in all, in Askim and in Sölvesborg there was a debate of some size, albeit not sufficient as to fulfill the ideal of deliberative democracy. To satisfy this ideal, a debate with more contributions from citizens as well as politicians is required. The debate as it actually was can be characterized as a reasonable beginning. In addition, experiences have been had that is of value to the process in the future.

Strategies for Implementing Successful On-Line Forums

All in all, some citizens seem to be interested in using the Internet to participate in a political debate, but the on-line forums have not yet been used to capacity. What strategies could make the on-line forums more successful?

A first strategy would be to introduce a certain quality of the U.S. community networks into this genuinely Swedish context, i.e., that they are implemented with the help of various citizen groups and local organizations (Schuler, ibid.). Moreover, in many community networks there is continuous user involvement in the implementation process (Ranerup, 1996), as well as afterwards. Therefore, with the community networks as a source of inspiration, local citizen groups, e.g., the parents of children at school or in child care, as well as the relatives of elderly that receive service from local government, might get access to the Internet at a reduced rate. As a consequence, they could build their own virtual community or network (Stegberg and Svensson, 1997). As a first step, the citizen groups might find it useful to share information among themselves, as well as discuss various issues online. However, they could also be offered an opportunity to participate in the implementation process in association with the local government homepage and its on-line forum. In this way, the common group interest of parents or relatives to participate in a virtual community that exists (IT Commission, 1997) will be used as a ground for activity. This is seen as opposed to a situation where a common place of residence is taken as a sufficient guarantee that a virtual community will be an active one (Croon and Ågren, ibid.).

A strategy like this might be criticized for being a way of colonizing citizen groups, so that they become a part of the political structure of local government (Castells, 1997; Montin, 1998). However, many citizen groups and non-governmental organizations are themselves eager to take advantage of modern technology (Hallam and Murray, 1998), which makes them very suitable as testbeds for how the Internet could be used in practice to improve democracy (Olsson, 1999). Also, this might be a way of including citizen groups that are working with school issues, child care issues, etc. in a democratic structure that counteracts the risk of fragmentation in situations where an organization focuses on only one kind of issue at a time (Lievrouw, 1998; Montin, ibid.). The latter is a risk that is noticed by current research in political science (Montin, ibid.).

A second strategy would be to make use of the interest in participating in a debate on other issues than local government politics that many people have, according to other experiences (Tambini, ibid.). Or put simply: we could try to sell the discussion forum as a local chat area where the choice of issues is left to those who participate. This is a way of using the pleasure of chatting without any limitations regarding issues that can be discussed, etc. (Jones, 1997). The many discussion groups that exist on the Internet as a whole might serve as an argument that these kinds of discussions are fruitful. However, the ultimate goal of on-line discussions is not necessarily that one arrives at a common point of view. Instead, as is argued in Robins (1995), the on-line forum might serve as an arena for a discussion between participators with different standpoints:

> "[...] we must recognize that difference, asymmetry and conflict are constitutive features of that world. Not community. [...] the ideal of common substantive interests, of consensus and unanimity, is an illusion. We must recognize the constitutive role of antagonism in social life [...]"
> (Robins, 1995, p. 152).

In other words, the on-line discussion might be of value in itself even if it does not result in the kind of common values and community spirit that are associated with, for example, community networks (Rheingold, 1994; Robins, ibid.; Schuler, ibid.).

Lastly, a third strategy might be to more seriously involve the politicians in the on-line forum as such. In an ongoing survey of all on-line forums that have been implemented in Swedish local government context (Ranerup, work in progress) the vast majority of on-line forums have been implemented without local politicians having been involved. In one-third of the approximately 40 on-line forums where closer links between the on-line forums and the politicians could be found, there seems to be more discussion about local political issues, rather than other issues, as well as more contributions from politicians, civil servants and the like. Also, the politicians have been involved in the implementation process to at least some degree, as a complement to local government IT-strategists, Webmasters, infomasters, etc. Furthermore, the importance of this kind of strategy was emphasized in the discussion in the on-line forums where many of the contributors wanted the politicians to become more active in the debate. This strategy also includes that all politicians should be given access to the Internet, as well as an e-mail address that is published on the Net and otherwise whenever there is a chance of doing so. Another part of it would be to emphasize how important it is, for the sake of democracy, that the politicians actually contribute to the debate in the on-line forums. In fact, the very participation by the politicians in itself makes it even more likely that the debate will actually affect the political process (Olsson, ibid.). Politicians could also use the Internet in general, and the forums in particular, as a tool for collecting the opinions on various 'hot' issues according to their own choice. Hopefully, several politicians will take part in such a process where ideas about issues are generated. In Sölvesborg, only one politician took an active part in the discussion about important issues that could be introduced in the on-line forum. Furthermore, the discussions about one single issue at the time cannot be character- ized as a success. Consequently, a conclusion that could be drawn from our experiences is that politicians from several political parties should contribute to such a process to create a discussion about the broadest possible range of issues.

In other words, one strategy might be to make the Internet at large into a political tool that is used extensively in local politics by the politicians and hopefully also by the citizens. As thus, this strategy would be a way of more actively using the Internet in an already established political context. As mentioned in the introduction, this has been defined as a critical success factor for the Internet to be able to improve democracy (Buchstein, ibid.). However, as we have shown above, in spite of the seemingly close connection between the forums and local government, this has not been done to a sufficient degree. But, is this not a defensive strategy, in which we only use a fraction of the immense potential of the technology? Mark Poster, a professor of History, answers that:

> "In response I can assert only that the 'postmodern' position need not be taken as a metaphysical assertion of a new age; that theorists are trapped within existing frameworks as much as they may be critical of them and wish not to be; that, in the absence of a coherent alternative programme, the best one can do is to examine phenomena such as the Internet in relation to new forms of the old democracy, while holding open the possibility that what might emerge might be something other than democracy in any shape that we may conceive it given our embeddedness in the present. Democracy, the rule by all, is surely preferable to its historical alternatives. And the term may yet contain critical potentials since existing forms of democracy surely do not fulfill the promise of freedom and equality" (Poster, 1997, pp. 214-215).

And the author is bound to agree.

Future Trends

The author is currently making a comparative case study of all on-line discussion forums that can be found on Swedish local government homepages. According to recent statistics, about 98% of the 289 local authorities in Sweden have their own homepage, all of which can be accessed through http://www.svekom.se/ adrindex.htm. As a contrast, only about 40 of them have what can be defined as a kind of on-line discussion forum (on the 30th of June 1999) (Ranerup, work in progress). In such a broad investigation as this, the exact number of on-line forums that can be found depends on how the concept is defined. At this stage of the investigation, only two criteria have been used: 1) An on-line forum must be accessible to all citizens through the local government homepage, and access must not be limited to those employed by local government. 2) Everyone who wishes to see contributions made by others must be able to do so. This is a way of excluding arrangements in which citizens are invited to send in views and suggestions by e-mail to politicians or to local government authorities in general. In those cases we cannot find an open discussion which is so important to the ideal of deliberative democracy. In other words, we cannot consider such arrangements to be a kind of on-line discussion forum. The main intention with this comparative case study is to increase knowledge about strategic success factors for creating a lively debate in on-line discussion forums. A second stage of the study will go closer into a few of the on-line forums and study more in detail the contents and the effects of the discussions as such.

The experiences that have been described above, as well as the suggested strategies, provide insights that hopefully are of value to Community Informatics in general, and to further research on community networks and virtual communities. Our discussion above has made clear that several aspects are of a strategic importance. For example, we should not only focus on aspects such as: who initiated and implemented the forums, whether sufficient measures have been taken to increase access to technology, or the effect of functional aspects of the on-line forums. Rather, we must more closely focus on the very groups of people (e.g., citizens and politicians) that have various kinds of relationships with local government as such, as well as various attitudes towards the technology. As a consequence, all strategies that are applied to increase on-line discussions must take these groups more seriously into account. For example, the first strategy that was suggested in the previous section includes involving citizen groups in the process where the local government homepage is implemented and managed. The most obvious question here concerns what groups of citizens should be included in the process, and perhaps as a second step get local government support to build their own network. And how is such a selective offer to participate in a democratic process related to the ideal of representative democracy? In other words, is it really fair to give some groups of citizens a more privileged position in the local democratic structure?

Furthermore, regarding the second strategy that was suggested in the previous section: the whole idea of on-line discussion forums is closely related to the ideal of representative democracy in that it supports deliberation, i.e., a genuine discussion, before a political decision is made (Buchstein, ibid.). Here we might ask ourselves if it is good for democracy to discuss absolutely *everything* that is in accordance with the freedom of speech. Maybe a completely open agenda for discussion can be negative when weak groups in society are taken as an issue for discussion, but do not themselves participate in the deliberative process as such. As an example, this might be the case when the policy toward immigrants was discussed, as in our experiences from Sölvesborg. In this, and similar discussions, the issues for discussion itself, and not the language that is used, can be characterized as being of an unclear democratic value.

Lastly, regarding the third suggested strategy in the previous section: is it possible to give the local politicians a more significant role in the on-line debate, and at the same time avoid that they dominate the agenda for discussion to such a degree that the citizens find it uninteresting? Or more precisely, how do we combine the freedom of citizens to discuss issues according to their own choice, with a situation where the politicians are allowed to introduce their issues of interest?

In sum, on-line forums in a local government context do in a very direct way affect how the local political institutions are structured and how our political rights are defined. Put simply, almost everything regarding on-line discussion forums touches upon our definition of democracy, freedom of speech, and the roles of citizens and politicians. This should not come as a surprise, but the author feels that this insight has not spread to a sufficient degree in a Swedish local government context, and similar contexts. Thus, to what extent such insights have been incorporated in projects where on-line forums are implemented is an important issue in the future.

Conclusion

Using on-line forums in local government politics is not an easy task that immediately gets a successful result. This is made clear in the experiences from Göteborg and Sölvesborg, where two of the on-line forums can be characterized as failures, and two as at least a reasonable beginning. According to our experiences, all forums were developed in larger projects aiming at improving the technological structure of local government, as well as attaining democratic effects. However, there where very few activities in the two cities that allowed various groups of citizens and politicians to participate in the process of implementation, and thus become engaged in the future of the on-line forums. Also, measures were taken to spread the technology to citizens, although not so much to politicians. Their lack of the most basic infrastructure for participating in the debate is most surely harmful to potential democratic effects. As to the functional aspects of the forums, there seemed to be almost no negative effects on the agenda of discussion when there is a certain amount of moderation. However, the arrangements must be made in a way that does not interrupt the debate as such. Furthermore, an open choice of issues for discussion seems to be advisable in order to increase interest in the debate among citizens. Lastly, the political climate, that is, if there are any 'hot topics' to discuss, was defined as an aspect that influences the amount of debate in some of the forums. Also, we should consider the attitude of politicians' discussion in on-line forums as an issue that needs further attention, and not only their access to technology, even though these aspects are interrelated.

As a conclusion, we might agree with Castells (1996, 1997) that the network society must include networking between big cities, regions and companies, as well as between people in their struggle to create reasonable living conditions in the information age. Research on how citizens and citizens' groups use IT for democratic purposes could be a way of safeguarding the interest of otherwise powerless groups in society. Thus, it might have an emanicipatory purpose (Habermas, 1968). Or at least it will have the effect that IT of a better quality is designed to support other sectors of society than private companies and public organizations. The experiences of on-line discussion forums in Göteborg and Sölvesborg might exemplify this kind of research.

References

Akrich, M. (1992). The description of technical objects. W.E. Bijker & L. Law (Eds.). *Shaping Technology/Building Society*. Cambridge Massachusetts: MIT Press.

Åström, J. (1998). Local digital democracy (Lokal digital demokrati, In Swedish). *IT and Local Government. An Overview*. Stockholm: Kommentus.

Benson, T. W. (1996). Rhetoric, civility, and community. Political debate on computer bulletin boards. *Communication Quarterly*, 44(3), 359-378.

Braa, J. (1996). Community-based participatory design in the Third World. J. Blomberg, F. Kensing, & E. Dykstra-Erickson (Eds.). *PDC '96. Proceedings of the Participatory Design Conference*. Cambridge, Massachusetts, USA: IFIP/CPSR/ACM.

Buchstein, H. (1997). Bytes that bite: The Internet and deliberative democracy. *Constellations*, 4(2), 248-263.

Castells, M. (1996). *The Information Age, Volume I. The Rise of the Network Society*. Oxford: Blackwell Publishers.

Castells, M. (1997). *The Information Age, Volume II. The Power of Identity*. Oxford: Blackwell Publishers.

Croon, A. & Ågren, P.O. (1998). Four forms of virtual communities, (Fyra former av virtuella gemenskaper, In Swedish). *Human IT*, 2, 9-20.

De Cindio, F. (1999) Community networks for reinventing citizenship and democracy. In *Papers for the EACN Workshop on Community Networking and the Information Society: Key Issues for the New Millennium*, European Association for Community Networks, Paris, 21-22 May, 1999. http://www.canet.upc.es/fiorella99.html [19990516].

Docter, S. & Dutton, W. H. (1998). The first amendment online: Santa Monica's public electronic network. R. Tsagarousianou, D. Tambini, & C. Bryan (Eds.). *Cyberdemocracy. Technology, Cities and Civic Networks*. London and New York: Routledge.

Francissen, L. & Brants, K. (1998). Virtually going places: Square-hopping in Amsterdam's digital city. R. Tsagarousianou, D. Tambini, & C. Bryan (Eds.). *Cyberdemocracy. Technology, Cities and Civic Networks*. London and New

York: Routledge.

Friedland, L.A. (1996). Electronic democracy and the new citizenship. *Media, Culture & Society*, 18(2), 185-212.

Habermas, J. (1968). *Knowledge and human interests*. Boston: Beacon Press.

Hallam, E. & Murray, I.R. (1998). World Wide Web community networks and the voluntary sector. *The Electronic Library*, 16(3), 183-190.

http://www.sifointeractive.com/index2.html [19990505].

http://www.svekom.se/adr/adrindex.htm [19990505].

Howley, K. (1998). Equity, access, and participation in community networks. *Social Science Computer Review*, 16(4), 402-410.

IT Commission (1997). Digital democracy. A seminar on technology, democracy, and participation. Report number 2/97 (Digital demokrati. Ett seminarium om teknik, demokrati och delaktighet. IT-kommissionens rapport 2/97, In Swedish). Sweden.

Johansson, L., et al. (1998). *The district reform in change*. (Stadsdelsreform i utveckling, in Swedish). Report to the City Council of Göteborg, 20th of August 1998. City of Göteborg.

Jones, S. (1997). The Internet and its social landscape. In S. Jones (Ed.). *Virtual culture. Identity and communication in cybersociety*. London: SAGE publications.

Jurnell, S. (1999). Director of the Internet supplier Utfors, private communication, 16th February 1999.

Lievrouw, L.A. (1998). Our own devices: Heterotopic communication, discourse, and culture in the information society. *The Information Society*, 14, 83-96.

Montin, S. (1998). *New forms of local democracy* (Lokala demokratiexperiment-exempel och analyser, In Swedish). Demokratiutredningens skrift nr. 9, SOU 1998: 155. Stockholm.

Olsson, A.R. (1999). *Electronic democracy* (Elektronisk demokrati, In Swedish). SOU 1999:12. Stockholm.

Poster, M. (1997). Cyberdemocracy: The Internet and the public sphere. In D. Holmes (Ed.). *Virtual politics. Identity and Community in Cyberspace*. London: SAGE Publications.

Ranerup, A. (1996). Participatory design through representatives (Användarmedverkan med representanter, In Swedish, with Summary in English). PhD Dissertation, Report 9. Department of Informatics: University of Göteborg.

Ranerup, A. (1998). Can Internet improve democracy in local government? R. Henderson-Chatfield, S. Kuhn, & M. Muller (Eds.). *PDC '98. Proceedings of the Participatory Design Conference*. Seattle, Washington, USA: CPSR/ ACM.

Ranerup, A. (1999). Contradictions when Internet is used in local government. R. Heeks (Ed.). *Reinventing Government in the Information Age*. London and New York: Routledge.

Ranerup, A. (2000). A comparative study of on-line forums in local government in Sweden. Presented at *Community Informatics. Community Development through the Use of Information and Communication Technologies*. University of Teesside, UK, April 26-28.

Rheingold, H. (1994). *The virtual community. Homesteading on the electronic frontier*. Reading, MA: Harper Perennial.

Robins, K. (1995). Cyberspace and the world we live in. *Body & Society*, 12(3-4), 135-155.

Schmidtke, O. (1998). Berlin in the Net: Prospects for cyberdemocracy from above and below. In R. Tsagarousianou, D. Tambini, & C. Bryan (Eds.). *Cyberdemocracy. Technology, Cities and Civic Networks*. London and New York: Routledge.

Schuler, D. (1996). *New Community Networks: Wired for Change*. New York: Addison-Wesley.

Stegberg, T. & Svensson, L. (1997). How to use Internet, when you don't have to. K. Braa and E. Monteiro (Eds.). *Proceedings of the IRIS 20*. Department of Informatics: University of Oslo.

Stolterman, E. (1998). Technology matters in virtual communities. Positioning paper for the workshop Designing Across Boarders: The Community Design of Community Networks. PDC '98/ CSCW '98, Seattle, USA, November 1998. [http://www.scn.org/tech/thenetwork/Proj/ws98/index.html].

Street, J. (1997). Remote control? Politics, technology and electronic democracy. *European Journal of Communication*, 12(1), 27-42.

Tambini, D. (1998). Civic networking and universal rights to connectivity: Bologna. R. Tsagarousianou, D. Tambini, & C. Bryan (Eds.). *Cyberdemocracy. Technology, Cities and Civic Networks*. London and New York: Routledge.

Tsagarousianou, R. (1998). Back to the future of democracy? New technologies, civic networks and direct democracy in Greece. R. Tsagarousianou, D. Tambini, & C. Bryan (Eds.). *Cyberdemocracy. Technology, Cities and Civic Networks*. London and New York: Routledge.

Tsagarousianou, R., Tambini, D. and Bryan, C. (Eds.). (1998). *Cyberdemocracy. Technology, Cities and Civic Networks*. London and New York: Routledge.

White, C.S. (1997). Citizen participation and the Internet: Prospects for civic deliberation in the information age. *The Social Studies*, January/February, 23-28.

Chapter XVII

Reinforcing and Opening Communities through Innovative Technologies

Alessandra Agostini and Valeria Giannella
University of Milano Bicocca, Italy

Antonietta Grasso and Dave Snowdon
Xerox Research Centre Europe, Grenoble Laboratory, France

Michael Koch
Technische Universität München, Germany

The aim of the Campiello[1] research project (Esprit Long Term Research #25572) is to promote and sustain the meeting of inhabitants and tourists in historical cities of art and culture. This overall objective is undertaken in two main steps: reinforcing the community bounds via collective participation in both creating community knowledge and optimizing access to it. Once the local community's sense of belonging has been reinforced, sharing its knowledge with outside people will become more natural.

In this paper first we present the various technological aspects, as well as where and how innovative technology can help local communities. Then we present the context of experimentation, future plans and current achievements in one of the two project settings: Venice.

Introduction

Local communities are declining under the pressure of globalization (mass media, delocalization of production, etc.). World-famous art cities are in a particularly difficult situation in this respect since on the one hand tourist flows take over the inhabitants' territory, and on the other hand the industrialization of tourism tends to transform art cities into a new sort of cultural Disneyland. In an art city like Venice, local communities suffer from a progressive and seemingly irreversible diminishing of identity. Can Information and Communication Technology (ICT) help local communities survive and revitalize their social lives? Can ICT help transform the relationship between tourists and local communities of cities of art and culture?

The Campiello project (Campiello, 1997-2000) — a European Community funded project — aims to answer these questions (Agostini, Grasso, Giannella, & Tinini, 1998; Grasso, Koch, & Snowdon, 1998). In fact, it experiments with the use of innovative information technologies as well as interaction paradigms in supporting the dynamic exchange of information and experiences between the community of people living in historical art cities, their local cultural resources, and foreign visitors.

The overall objectives are the following:

- *Supporting the dynamic exchange* of information and experiences between the communities of people who live in historical art cities and external (often foreign) visitors, using the local cultural resources and events organized in the local context as triggers.
- *Enabling people to interact and cooperate* in building a new and richer sense of community based on an exchange of knowledge about the cultural resources.
- *Interrupting the relegation of the local communities* to separate areas within historical art cities, allowing them to reconnect with their territory.

Two local communities were chosen as contexts for the project: Venice in Italy and Chania in Crete (Greece). Both cities nurture a culture heritage linked to their territory, while sharing it with people of different cultures from all over the world. In this paper we focus on Venice.

Due to its status, the city of Venice is the historical city where the critical condition of friction between local community and visitors is highest. The enormous number of visitors puts a heavy load on the city's weak physical structure and on its social structures too. Local people must share their cultural resources and public utilities, as well as the places where they meet and socialize, with visitors from other countries. There is a tendency to create separate meeting places for the local community, isolated from tourist areas. Moreover, new events organized within these cities are directed to tourists and not to the people living there; in fact, local people have no priority in participating in these events and rarely contribute to their organization. On the other hand, visitors — accessing exclusively the city's most visible and renowned resources — gain only a superficial understanding of the local culture. In order to enhance the Venetians' feeling of citizenship and visitors' opportunities to discover the deep and complex beauty of Venice, Campiello realizes a system developing new tools, technologies and solutions. Its main goals consist in making the local inhabitants active participants in the creation of the cultural information, and in supporting new and improved connections not only between the local community and tourists but also between the local inhabitants and the "official" managers of the town's cultural resources. This is achieved by supporting the bidirectional exchange of information about the town, its places and events. So the aim is to build not only a network for empowering existing communities but one that tries to bring different communities together or at least let them profit from each other.

One peculiarity of the adopted method is that we chose to involve in the experiment subjects of the community which already take an active part in the local culture and its resources (at present, a school and a magazine). Our idea consists in making such subjects the pivot for involving the whole community in local cultural life and for spreading knowledge about cultural heritage and resources to both tourists and the whole local community.

In terms of tools and technology, the project aims to develop and adapt a knowledge agent's architecture to support personalized information services; it also aims to design a novel set of interfaces to support ubiquitous computing (Weiser, 1993) and innovative interactions with the system. In fact, the Campiello system realizes a new concept of interactive participatory medium; it builds a dynamic shared knowledge diffused in the territory to facilitate and support the exchange of information between different communities in a local, physical environment. In particular, when shifting from office setting to community social life interaction modes are of primary importance and in Campiello have been designed starting from their invention and not from system functionality (Agostini, De Michelis, & Susani, 1998; Agostini, De Michelis, & Susani, 2000). Therefore, the user can interact with the system through a variety of interfaces that are, as much as possible, accessible throughout the physical places used by the communities to socialize. Both innovative (large screens and paper-based interfaces, see the following sections) and conventional ones (PC, Web access) have been developed. A strong emphasis has been put on selecting innovative interfaces which provide a high degree of accessibility of the system by the whole community (including, for example, elderly people). For instance, the use of paper requires no technical skill at all.

In focussing on the possible interactions with knowledge we envision three main user stereotypes. The first one, denominated *cultural managers*, is composed of the members of active groups in the community who create knowledge while performing their activities: in our case the students and teachers of the school "P.F. Calvi" in Venice and the editorial staff of a local magazine; in broader terms, all the people who—either out of personal interest or for work purposes—usually produce knowledge related to the community. Other examples include cultural associations, city administrations, etc. The second group, the *local community*, consists of generic members of the community who, in comparison to the cultural managers, produce new contents more occasionally since this is not part of their everyday activities. Finally, the *visitors* will mainly comment on and evaluate the knowledge. Therefore they mostly contribute to its enhancement, only rarely providing fresh information relevant to the community. While for cultural managers the PC interface seems the most suitable (due to its flexibility and efficiency), for the second and third groups of users the interfaces based on paper and large screens allow higher diffusion and accessibility in the physical territory as well as greater usability by non-technical persons.

The chapter contains two main parts: in the first we describe the system's key features; in the second, we present the context and the actors involved. We conclude the chapter with the first real experiment conducted and the lessons learned up till now.

Overview of Campiello Functionality

On the basic level the Campiello system acts as a repository of information related to places and events in the city. This information consists mainly in descriptions and comments. While every user of the system can contribute to the information space, there are some main actors (the cultural managers, see above) who provide an initial set of descriptions and keep on feeding the system with descriptive information. In structuring the information space and designing the access to it, we were faced with the well-known tradeoff between structured information system and unstructured 'communication area.' If the system is designed in a structured way it is easy for users to find information. On the other hand if users are given a lot of freedom in creating new information they are more likely to contribute, but this usually leads to a loss in structure and so can frustrate people when trying to find something in the network. This is why we decided to model the information services primarily around the recommendation function, while keeping search and browsing capabilities available for those who choose to use them.

The information space consists in a set of items. The major types of items are physical places (e.g., monuments, houses, restaurants, museums), events (e.g. traditional feasts, concerts, festivals) plus more abstract topics of interest like 'associations' and 'cooking.' In regard to search and recommendation issues, the items can be related and grouped into one or more categories. Additionally, the descriptions or comments of the items can contain links to other items. The system collects explicit and implicit user feedback and combines this with a given profile of interests to select information items that might be interesting for the user. This works in a proactive way (i.e., without action by the user) and in a reactive way (i.e., the user tells the system to give him items that match some attributes). More information about collaborative filtering can be found in Shardanand & Maes (1995); meanwhile Knowledge Pump, an existing collaborative filtering system from Xerox used as recommender engine in Campiello, is described in Glance, Arregui, & Dardenne (1998, 1999).

Beyond the information services devoted to improving the flow of information within and among the different communities, Campiello offers a layer of functionality to support better communication and facilitate contact. This layer includes services for finding people with similar attributes (e.g., interests) available for contact in preparing a visit to the town, for instance. The goal of Campiello is to provide more than just an asynchronous messages service (although we may well integrate synchronous communications support). We aim in fact to provide a dynamic community memory and actively push information toward people who might be interested in it. To this end we have been developing two novel interfaces; we have also been working on supporting services for managing the information contained in the community memory. Rather than relying mainly on users performing searches of the knowledge base or browsing it looking for items of interest, Campiello also actively tries to push information toward users.

In the next section we present the system architecture and explain how this "information push" is achieved. In subsequent sections we describe the user interfaces which make these services available.

The Architecture of Campiello

The Campiello system contains a number of components:

- *Storage agents* — These provide access to and specialized functions on the information stored in the Campiello knowledge base. Internally the information is stored in an SQL database but these modules provide a more convenient API hiding this fact.
- *Service providing agents* — These agents use the knowledge base to provide services to the user interfaces. Currently the most important of these modules is the recommender agent.
- *User interfaces agents* — Campiello provides a number of user interfaces which use the storage and service providing modules.

These modules are implemented as Java-based agents and can either run on the same machine or several different machines. This allows for some flexibility of deployment, as well as the ability to provide multiple instances of each user interface geographically dispersed throughout a city.

The Knowledge Base Objects

There are four important sources of information stored by Campiello. They are:

1. *Contexts* – A category which can be used to group related items. Since contexts can contain other (sub)contexts, it allows the creation of a hierarchical classification. Each context has a title in one or more languages and one or more descriptions in one or more languages. Contexts can also have context-specific information associated with any items related to them, allowing the knowledge base to be extended.

Figure 1: The Architecture of Campiello

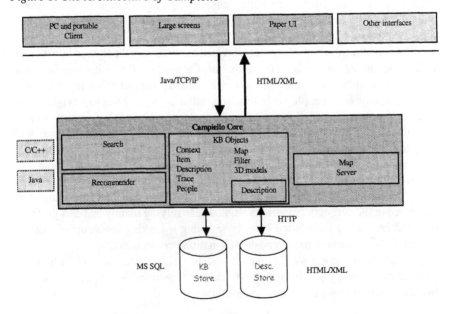

2. *Items* – An item represents a concrete or abstract entity about which information can be collected. Items can be monuments, streets, museum exhibits, cultural events, associations or any "thing" about which people might wish to collect information. Items have a title in one or more languages and one or more descriptions in one or more languages. Items are classified and linked to any number of contexts—for example, the item describing the "P.F. Calvi" school might be classified under:

Figure 2: A fragment of the context hierarchy[2] related to the Castello district of Venice

"Conosci **Castello?**"

eventi
e feste di quartiere

le case a Castello
abitare ieri e oggi

le strade, calli e campielli
ieri e oggi

i monumenti del quartiere

gli artigiani e i mestieri

fare la spesa e cucinare

le associazioni
ieri e oggi

- Education – it is a school!
- Architecture/Monasteries – the building hosting the school is a XVI century monastery (see Figure 3).

3. *People* – The users of the system. Campiello stores some basic demographic information about people, plus a user profile which is composed of 1) a static part indicating a user's preferences, such as which contexts s/he is interested in; and 2) a dynamic part composed of the *traces* (see below) the user has generated in the course of their interaction with Campiello.

4. *Traces* – As people interact with the system, they leave traces of these interactions, which are then used to find correlations between users and as implicit indications of a user's preferences and interests. Traces can be both

Figure 3: Pictures[3] of the courtyard of the "P.F. Calvi" school

Figure 4: An Example of a Hand-Written Comment; it has been Inserted by a User during the S. Pietro Feast

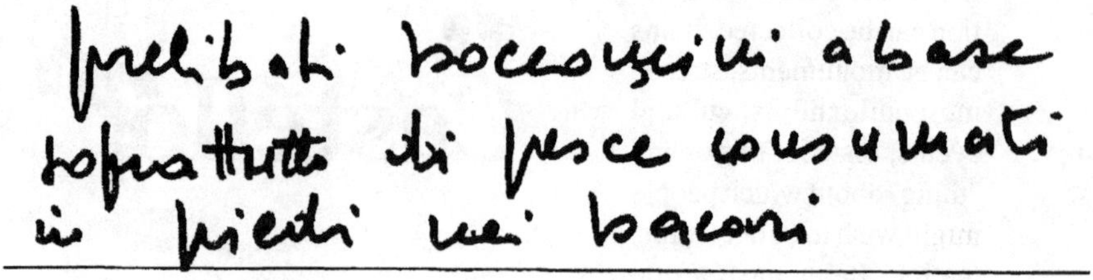

implicit and explicit. Implicit traces are generated by actions such as requesting information on an item, sending a message to another user, or writing a record of a visit to a particular place. Explicit traces are generated when a person explicitly wants to communicate information to the system and other people. Examples of explicit traces include the rating of an item (a score awarded on a linear numeric scale indicating like-dislike) and comments (a text or handwritten comment — see Figure 4 — indicating the user's opinion and intended to be communicated to other users). In order to maintain privacy, users have the option of marking individual actions as private. Traces marked as private will be shown to that user in their travel diary (a log of their visit) but to no one else.

With these four types of information, Campiello can provide a flexible knowledge base and allow users to share their opinions on items with others.

The Recommender

The recommender service provides two main functions:
1. Finding users with similar profiles (matchmaking).
2. Recommending items to users which they should be interested in, according to their profiles.

These two functions are interrelated. The recommender adopts a technique called *community-centred collaborative filtering* (Resnick et al., 1994; Glance et al., 1998) in which correlations are found between people in each context; then these correlations are used to predict a given person's interest in a given item based on the ratings of the users with which a person is correlated. These correlations are done on a context-by-context basis, since while two people might agree in one subject area (e.g., architecture), they might violently disagree in another (e.g., music). By restricting correlations to individual contexts we can provide better correlations of interest than by attempting to find overall similarities among users.

For example, suppose a number of users have visited various monuments in Venice. These users then rated the monuments and left comments that can be seen by other users. The recommender compares the ratings of the users for each monument and derives a value for the correlation between users using the Pearson algorithm (Resnick et al., 1994). Given the correlations between users the recommender can now predict user interest in monuments not yet seen. If two users tend to agree on their ratings of monuments and one has rated an historical building that the other has not, then the recommender can use the rating and the correlation factor to predict the other user's interest in this building. If the result of this computation is a score above a given threshold the system might recommend the monument to the user when s/he interacts next with the system or it might send her/him information directly, perhaps by fax or e-mail.

Complementing the PC with Paper and Large Screen Interfaces

As stated earlier, one of the goals of community networks is to offer widespread access in order to encourage utmost participation and to ensure that sections of the community are not excluded due to lack of access. For this reason Campiello has chosen to focus on novel user interfaces, in addition to using a PC-based user interface. The sections below describe these new user interfaces: paper and large screen displays.

Using Paper to Interact with the Communities

In Campiello we make use of some Xerox technology (Johnson et al., 1993; Xerox, 1998) for processing paper-based forms. We chose paper as one of our user interfaces because it has a number of advantages that make it an ideal medium for supporting dissemination and collection (see Agostini et al., 1998a) of information and feedback:

- It is portable (sheets of paper can be easily folded and carried in pockets).
- Paper artifacts are easily shareable.
- Paper can be easily annotated.
- Paper is already involved in many information gathering and annotation processes.
- It is cheap.

Using paper user interfaces could establish the integration of community support systems into the physical environments (Grasso, Koch, & Rancati, 1999). This is essential especially for local communities which have a close relationship with some kind of physical space. The technical background for such a solution involves storing encoded digital information on paper artifacts. Using Xerox Dataglyph™ we can do this in a very unobtrusive way (Johnson et al., 1993). This means that a printed information sheet can provide elements allowing users to use it to request more information, express ratings, add comments, etc. Such a sheet can be scanned and this action can activate complex functionality provided by the system. (See Grasso et al., 1998, 1999; Koch, Rancati, Grasso, & Snowdon, 1999 for more information in terms of technology.)

One outcome of our work is what we call an Active NewsCard. Active NewsCards are small (DIN A5 or DIN A6) postcards that show information on one or more items from the information space, enhanced with active and visible checkboxes and blank areas. Ticking any of the checkboxes distributed in the content allows the user to:

- express interest in more detailed or related pieces of information or
- give feedback (rating) for an item.

Checkboxes can have different shapes or can be attributed with attached icons to express the different functionality. Blank areas are used to give the possibility of adding free comments, which are scanned and saved. We are also considering using text highlighting on the NewsCards for input purposes. NewsCards are edited explicitly by "cultural managers." Therefore, at present Campiello provides an HTML-like language in which the content and various active elements can be expressed; a graphic editor is under development. Let us specify that, mainly for practical purposes (ease of printing, collection and scanning), at present the various prototypes of Active NewsCards have been designed in A4 format (see Figure 5).

We distinguish two basic types of Active NewsCards:
- *Static NewsCards*: These are Active NewsCards edited explicitly by cultural managers and distributed in large numbers around town. They resemble the information cards already found in several towns. According to the main usage of the feedback elements we distinguish two main subclasses:
  - *Information NewsCards*: The main goal of the NewsCard is to give information on one of the several items. Therefore, the NewsCard contains a lot of text and images. Feedback elements are mainly used for expressing interest in more information or for giving feedback (ratings and free text).
  - *Comment NewsCards*: The main goal of this kind of NewsCard is to collect comments and ratings on a specific event/place. It is assumed that the user already knows about the item, so there is little extra information, but large areas for writing comments or for giving ratings. Examples of such NewsCards can be feedback NewsCards at public events (e.g. concerts, feasts).
- *Dynamically Created NewsCards*: While static NewsCards are created in

Figure 5: Examples of Two Active NewsCards with Various Layout and Interaction Elements.

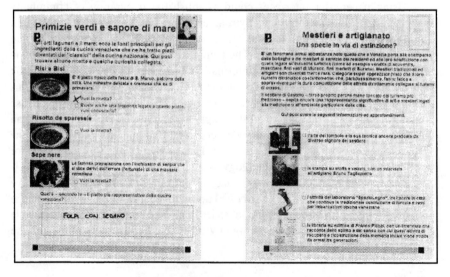

large numbers for generic users, there is also the chance to get a personalized printout of information (e.g., in response to submitting a filled-out static NewsCard). These NewsCards usually contain more information and are larger in size. That is why we also call them Active Newspaper.

Two possibilities for processing paper input have been implemented in Campiello: synchronous processing and asynchronous (batch) processing.

In regard to synchronous processing, scan/print stations are available at semi-public places where people can process their NewsCards and immediately receive the results (which can be a personalized NewsCard with requested information). Examples of such places are libraries, schools, tourist offices and hotels. Additionally, synchronous access is available through a fax server.

Alternatively, the users can drop the NewsCards in collection boxes set up in several places (e.g., newspaper kiosks, restaurants, hotels, and at public events) or send them by mail.

To give the users the possibility of identifying themselves each Active NewsCard can be associated with another paper "tool" called PID Stickers. Such a sticker (see Figure 6) contains a machine-readable ID of the user. Attaching a PID sticker to an Active NewsCard makes the system associate content and actions with the user. This creates conditions for producing useful information relative to recommending and collaborative systems, both essential in terms of giving valuable services back to the user.

Finally, it is possible to publish new information by either gluing it to a specialized comment NewsCard (where the context(s) to which the information belongs is (are) already specified)—or by adding a sticker to an existing piece of paper. With stickers the selection of contexts is done by choosing from different context specific stickers or by using stickers that include checkboxes. To further motivate users to publish new information and comments, publishing new information is modeled after the personal diary metaphor; in fact, this information is also inserted in an electronic personal diary that can be used like a paper one. In regard both to comments on existing items and to new items, we make use of the fact that the information we collect is mainly to be presented to users again. Therefore, it is not of primary importance to fully transfer the inserted comment/information into digital information (e.g. do recognition of handwriting, etc.). Rather the image is mainly stored and related to some basic contexts and/or users.

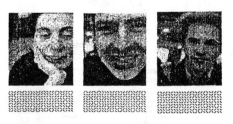

Figure 6: Examples of Personal Identifier (PID) Stickers

Making the Interface Available Where the Practice Takes Place

Up to now embedding recommender interfaces into practices has been difficult since the interface to the recommender has been available in a computer based way only, and the practice has taken place where no computer support has been available. Paper-based interfaces can close this gap and be available where and when the practice takes place.

People already collect a lot of information on paper and use paper quite naturally for marking feedback on items either for themselves or for friends. If we make this information gathered in paper form and written on paper artifacts available to electronic information systems, the users no longer have to make an extra effort and much more material and feedback will be available for the recommender system.

The CommunityWall

As described above, the paper user interface allows us to provide information on specific topics and lets users request more information and provide feedback. What is lacking in this interface is a general overview of the contents of the community memory, plus information about topics of current interest to community members. Eventually this information could be distributed via the paper user interface by providing forms which users can scan to receive updates or by automatically mailing or faxing information to registered users. In fact, this is something that we may try at a later date. However, our current approach is to use large screen displays to display broad coverage of topical information in a way complementary to the focused presentation of the paper user interface.

The purpose of our large-screen display, the CommunityWall (see Figure 7 for a draft design of the prototype for the local community in Venice), is to create an environment that fosters social encounters (conversation) using topical information and/or news as a trigger. It provides a focus for social activity in a way similar to existing notice-boards which display notices (ranging from formal printed notices to handwritten scraps of paper) concerning current community activities. Using the

Figure 7: The Draft Design of the CommunityWall Prototype for the Local Community in Venice (by Domus Academy)

CommunityWall, we aim to provide information on what are the interesting activities or topics of conversation, who is actively interested and what they are saying. If a topic displayed on the CommunityWall attracts someone's attention s/he can then request that more information be displayed on that item by touching the screen (or using some other means if a touch-sensitive display is not available) or that a NewsCard be printed on that topic on a printer/scanner situated near the display. Once the NewsCard is printed it can be used to comment on the topic. In this way the CommunityWall supports information discovery and asynchronous communication among members of the community. Furthermore, we hope that if a number of people use the display simultaneously this will help trigger conversations and allow people to meet others with similar interests (since it will be obvious which topics people are looking at).

Given the contexts, items and traces present in the knowledge base, the task of the CommunityWall is to select the topics most representative of the community at present and to display information about these topics in such a way that onlookers can see which items are of current interest to the community. Topics are selected for one of two reasons. First, privileged users, referred to as cultural managers (editors), can mark some items as being of particular interest. Second, the CommunityWall monitors the database in order to see which contexts and items are receiving most comments and ratings — these are assumed to be the items which are of current interest and which are therefore generating lots of "traffic." A set of customisable rules is invoked to prioritise items and decide which are the most interesting at present. We provide a collection of rules that can be combined to form more complex composite rules. For example, a rule exists that generates a priority according to the average rating of an item, another selects the highest priority returned by its sub-rules, another changes the priority of its sub-rules according to whether a cultural manager has commented on the item and finally we have a rule that decreases the priority according to the length of time that an item has been displayed in order to prevent a small set of items from monopolising the display. Items can be viewed as competing for screen real-estate and those that win are the ones that are generating the most interest, belong to contexts of other high interest items, or have been marked by a cultural manager as being of particular interest. We are still in the process of developing an appropriate rule set, but by way of example, we currently have composite rules that perform the following:

- If an item represents an event, then its priority gradually increases as the event draws near and drops to zero after the event has taken place.
- For items which are not events, the priority is determined by a combination of the average rating and number of comments that an item has received relative to other items. In addition a degree of randomness is added to ensure that every now and again items which might otherwise not have a high enough priority are given a chance to be displayed. Finally, the overall priority is modulated according to whether a cultural manager (privileged user/editor) has authored, commented on or rated the item.

Once selected, items displayed are grouped by contexts so that the display has some sense of consistency. Items are represented by a title, a brief description, *plus* user comments on the item and pictures of the users who have commented on that item. Onlookers can therefore see what is interesting, who is interested in it and what these people are saying about the item. We are exploring several different screen layouts to see which is most effective. In one version, in order to maximise screen usage while preserving context, older comments (and the associated images of the commenters) shrink gradually, and thereby progress from being full-size, to being small but legible, to being illegible but still visible, to finally vanishing altogether (see Figure 9). The motivation for this is to provide some indication of the volume of comments even if people cannot actually read them all. In another version, comments are displayed at a fixed size, and if an item has received multiple comments, the display cycles through them in series (see Figure 7). Currently, we favour the version in which comments are presented serial at the same size in order that they remain legible and we add some additional icons that indicate roughly how many comments an item has received. Optionally, ratings can be displayed alongside particular items to give another form of feedback.

We deliberately chose to display images of people making comments which were captured as part of the registration process. Although this might seem to raise privacy concerns we believe that there are good reasons for identifying people in this way. First, people are aware that the comments they are entering are intended to be seen by others. Secondly, we believe that some form of identification may help encourage responsible comments and avoid obscenity. Thirdly, we hope that people standing in front of the CommunityWall will be able to recognise people who have commented on a topic that is of interest to them and perhaps identify these people if they are also nearby. In this way the CommunityWall can help facilitate contact among members of a community.

We are currently experimenting with a number of different display and layout styles for the CommunityWall and also working to provide an interface that can be adapted for a number of different settings. Figure 8 shows an example CommunityWall display designed for an office setting and Figure 9 shows another prototype of the CommunityWall used for a demonstration at a conference of European projects (IST'98).

The Context: Venice

A Rough Picture of the Current Situation (Recent History)

Venice is an art city known worldwide for being built on water, having no streets and cars but canals and boats. It attracts some 15 million visitors per year, most of whom are excursionists staying in town for a few hours. Nowadays tourism is the major industry in the urban economy, as well as one of its main problems in terms of its impact on the town's physical, social and economic structure. This has changed dramatically in the last 50 years, so that it is difficult to understand what is actually going on without taking a look at the main transformations which have occurred. At the beginning of the 20th century (1919) the settlement of the

Figure 8: The CommunityWall for a Workgroup in an Office Setting

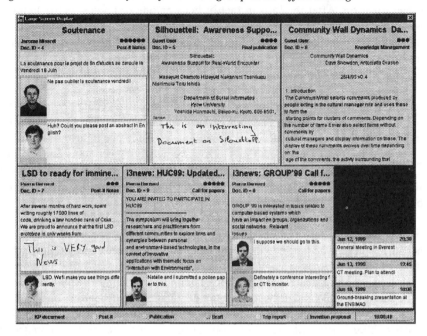

Figure 9: The CommunityWall Prototype Used for a Demonstration at the IST'98 Exhibition

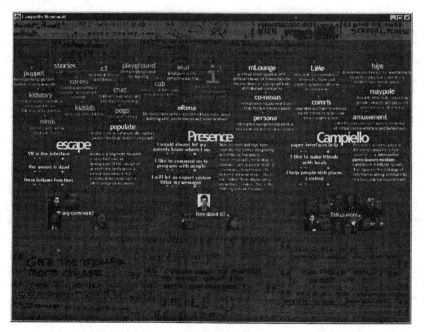

Marghera industrial district on the outskirts of Venice proved a major trigger for transformation. After the '50s in fact, it became one of biggest petrochemical areas in Italy, attracting a massive movement of workers with new employment opportunities and the prospect of a new and "modern" lifestyle. At that point insular Venice began suffering a constant, sometimes impressively fast drop in population, from 150,000 inhabitants in 1951 to the current 70,000. Speaking broadly about the last 50 years we can say that the economic heart of the city progressively moved inland. Meanwhile the historic center's economy specialised in tourism exclusively. Modern, industrialised economy entailed a major change in the city's relationship with its environment. While in pre-industrial times the lagoon was the resource to put to work to have a thriving economy and society, now the unusual environment became a penalty. To sketch it in a non-technical way, the local culture went from being a "water culture" to a "ground culture" but without the ground!

Throughout these processes the weakening of local communities has been impressive, changing lifestyles and culture. As often the case, the local culture adapted to stereotyped models in order to meet commonplace expectations. Nowadays culture in Venice is mainly a product to sell to tourists; authentic popular traditions like craftsmanship are lost. Local communities go through the storm losing cohesion, a feeling of integrity, the material foundation which it grows on and takes meaning from. In spite of everything, however, in this apparently irreversible process there are still signs of resistance against the city's mainstream commercialisation.

The Castello Area

"Sestiere di Castello" is one of the six Venice neighbourhoods. It is still very much a working-class and non-tourist area and certainly one of the few places in Venice where we can, while confusedly, feel some sense of reality. Not part of the sprawling Disneyland extending from Piazza S. Marco (Venice's main square) in almost every direction, it appears to be a quite lively and populated area. In fact Castello was somehow spared the blind development described above. Resident migration has been lower than elsewhere in Venice and a number of services devoted to residents have survived too. Tourist flows hardly touch this neighbourhood due to its marginal position in respect to major city attractions and its slightly longer distance from tourist terminals. These two elements have also determined a favorable condition for some small traditional handicraft activities re-proposing ancient Venice trades. Some Arts and Crafts Associations especially linked to the area still exist: a seasonal feast is held annually in a wonderful neighbourhood "campo,"[4] very well attended by local and town people; lastly, a few other community subjects (associations, committees, local schools) play a role in animating and reinterpreting the community's culture. However the district is not only a marginal, and working-class fringe with bits of the art and culture that in many respects make up the Venetian identity. To give just a couple of examples we can mention the Arsenale complex (the biggest naval factory of ancient times, well known worldwide) and the church of S. Pietro di Castello, built on one of the original settlements of what today we call Venice.

Thus we chose Castello as the context for our Campiello project experiments in Venice. Our hypothesis was that this is still a place where a local community is alive and can be usefully supported in its resistance against fragmentation and homogenisation to a stereotyped cultural model. To test this hypothesis, we selected one crucial subject, working with it to enter softly into community life and find occasions to observe its dynamics (as much as possible) from inside. This is obviously a major methodological point. An equally key point is the recognition that the existence of communities is not something natural and unquestioned; rather, it represents an effective choice, actually hindered by most contemporary social trends (Agostini et al., 1998a). Hence, if we agree that in our post-modern condition community life is one option among many others, then the direct observation of the way it is constructed through daily relationships becomes crucial.

The Local School

We selected a local school (pupils ages 11 to 13) as the privileged subject for our trip into the local community. In this school a group of enthusiastic and innovative teachers have carried out, for many years, a wide range of research studies strictly related to both past and present neighbourhood life. They think the school has to be an active subject in the neighbourhood, thus their educational goal is not limited to the children but extends to the surroundings as well. They are conscious and proud of their role and carefully consider every possible initiative to get involved in, in light of its prospective contribution in both school and neighbourhood contexts. In relating with the school, we have sought to be as unobtrusive as possible, trying not to modify their style of working (with the exception of consolidation of computer use). The implicit contract "signed" between the Campiello project and the school regarded on the one hand the possibility of entering the local community without bearing the label of "foreigners," thus gaining a closer and more accurate look at it; on the other, the opportunity for giving a wider diffusion and generally greater effectiveness to the work done in and for the community. Cooperating with the school has meant providing one more reason to explore local reality in its many past and present aspects; offering the chance to disseminate the work beyond the boundaries of school-related people (teachers, parents, plus a few others); and allowing for the collection of new contents by outside people belonging to the community or not.

The research studies carried out during the 1998-99 school year for the Campiello project included the design of various itineraries in the neighbourhood as occasions for exploring and giving accounts of physical, historical, cultural and social features of the area. Furthermore a video was produced, portraying episodes of students' daily life in the neighbourhood context. In the future the video will be linked to research materials, with the whole working as a hypertext document.

The Tourists Visiting Castello

We have already said that Castello is almost untouched by the year-round tourist flows to Venice. This is actually one of the constantly repeated complaints heard chatting with neighbourhood people; in a sense, both positively and negatively, it is part of the local identity. In fact there is a lot of grumbling about not participating in the rich banquet of tourism economy. On the other hand the same exclusion is portrayed as a positive difference, since that way the neighbourhood maintains its original character and community life. As evident from many empirical researches, being "different" can be seen as both a handicap and a resource for identity-building processes. All this means that in one way or another tourism is an important "topic" in the neighbourhood life: something desirable to be enhanced by all means, or rather, something quite dangerous to be handled with care due to the possibly uncontrollable consequences of its expansion. Given the crucial importance of this theme, we decided to spend some time observing tourist presence in the area from a standpoint we had missed: that of tourists. Since one of Campiello's main goals lies in the encounter and sharing of experiences between visitors and the local community, we tried to answer questions such as: Who are the tourists coming to Castello? Why and how do they come here? What are they looking for?

We thought the best way to get an answer to these questions was by posing them directly to tourists. Thus we decided to interview them during their visit. The vast majority of people we contacted were very keen on answering our questions. Only a very few refused to stop and relate with us. We interviewed people from a wide range of countries and most of them complained about the lack of accurate information or the difficulty of getting a hold of it. The set of questions we asked them concern four main points:

1. information obtained by tourists before arriving and once in Venice;
2. existing interactions between tourists and local people;
3. tourists' interest in meeting local residents and in learning about aspects of local life and culture while visiting Venice;
4. tourists' interest in a system allowing them to gather information about aspects of local life and culture.

We believe that this kind of information is essential for refining both the interaction mechanisms and the functionality provided by the Campiello system.

In the following we provide some hints on the results of the interviews. In regard to the first point, all tourists have problems getting information. Before arriving in Venice, tourists get information from guidebooks, friends, and tourist bureaus. They think that generally this information is incomplete and inaccurate; they sometimes even found contradictions between different guidebooks. Once there, a lot of tourists say they have difficulty getting more information about things to do, places to see, where to eat, cultural events such as exhibitions and concerts to attend. In particular, they complain that it is not possible to get all different kinds of information from one source (for example, places to visit and means of transport or cultural events and restaurants). It is difficult to get information about events that are not well known. Relative to the second and third points, tourists generally think that local people can provide useful information about visiting the city and are very kind and inclined to give information and chat with them. Moreover many tourists think that local people can offer them with accurate information about non-tourist areas, including strangest and remotest places. Various tourists say they would be interested in interacting with locals primarily for the following reasons: contact with locals may be very nice and moreover help in understanding the 'spirit' of the city better than a guidebook; locals can give accurate information about local life and culture not provided by tourist guides or bureaus; locals can give information about not well-known cultural events which otherwise may be difficult to notice.

One of our main purposes in interviewing the tourists was to check out their interest in a system like Campiello (fourth point). What we found from carrying on this task is that all types of tourists would be keen on having more information about the places they visit. Significant differences appear in regard to various types of tourists (see below). Thus it may be possible to specify different profiles of users connected to different options, services and so on. Tourists say they would be interested in such a system for several reasons. For instance, the system could allow them to overcome difficulties encountered in getting more information; it could provide information on any aspect of Venice, including those relative to local life and culture. It could be accessible on the Internet, there for the consulting from home, for instance; it could in advance provide up-to-date information about cultural events, so that people could plan better when to visit Venice. Moreover, it could include information about less well-known events which may otherwise be difficult to notice; it could collect commentaries about such events which may be provided by other people; it could facilitate interaction with other tourists and/or locals. Finally, it could support 'indirect' interactions by giving and receiving messages and commentaries, 'direct' interactions too by providing occasions for meeting people.

In elaborating the answers we got from the interviews, we figured out a typology of tourists visiting the area. This is based on their attitude exploring a place generally considered "of minor interest." The three profiles we outlined can be used as ideal types for putting the gathered information into meaningful pictures. Here are brief sketches of each type.

The *classic tourist* arrives in Castello quite casually after visiting one of the town's main attractions (S. Marco Square and some well-known monuments are not far away). S/he basically thinks there is not much (or even nothing) to see in Castello. The *curious tourist* often arrives in Castello by chance, yet with the idea of seeing more than the stereotyped image of the place s/he is visiting. This person is actually curious about local habits, although keeping very well in mind "s/he is here only for tourism." This means keeping a basic sense of detachment toward the "object" of the visit. The third figure, the *empathic visitor*, is split in two subclasses. There is the tourist who comes to Venice for just a few days but wants, nonetheless, to get in touch with as many aspects of local life as possible. This person looks for the genuine characteristics remaining in the town s/he visits in spite of mass tourism. S/he would also love to meet local people, especially in relation to specific interest s/he cultivates. Unfortunately the length of the visit often impedes this. Then there is the tourist who decides to live in Venice for a while. This individual is usually a foreigner who rents a house for some months with the intent to dedicate time to some personal interests and to a lengthy discovery of the city, its human culture, its rituals... It is interesting to note that Venetians normally love this kind of visitor. They usually have very good relations with her/him and are very open and proud to introduce the same to their own cultural heritage.

Summarising, we found that all types of tourists (for different reasons) would be keen on accessing a system providing information both before their arrival and during their stay in Venice. It is important to stress that in any case the kind of user of such a system would not be the typical customer of the mass tourist industry; rather, it would generally be someone with the chance to stay in town for at least a few days and to organise at least portions of the trip on his or her own.

The First Campiello Experiment

As evident in our approach, there is a particular vision of the community in current times. This needs to be both close enough to preserve cultural models and maintain local identities, and open enough to continuously reproduce culture and tradition without losing touch with reality. The Campiello project's main features allow for meeting both needs: community subjects are the direct producers of the system content, thus deciding what is relevant to themselves. At the same time the system knowledge base can be fed and enriched from users of all sorts (Venetians not part of the Castello community, tourists, scholars staying in town for short or longer periods, etc.). The demands for information the system can support are also manifold in that they involve various types of people: from those curious just to know more about real bits of local life, to those who want to actually get in touch with community subjects to gain a deeper feeling of the place through its inhabitants and hence trigger a sharing of cultures.

Since we were lucky enough to have at hand good content producers, Campiello had to find similarly good occasions for demonstrating the work done, collecting feedback and testing the general interest the project generates. The seasonal feast held every year (at the end of June) in one of the most enchanting and secluded islands of the neighbourhood was identified as the right moment to present the school work and the Campiello system through it. The idea proved right on mark: in fact, after two years of participating in this event, we can say the S. Pietro feast is very much a community celebration involving people of all ages in the effort to organise a 10-day event able to attract thousands of people. It mobilises the energy of school children, public officials, local entrepreneurs, young and elderly people, whole families who volunteer to cover the many different roles the organisation of such an event requires. The feast's character happens to be consistent with the approach described above: it is an event which the neighbourhood people both carry out and participate in. It certainly contributes to reinforcing the community identity, but is open to and draws meaning from participation by a much broader audience. Such consistency is confirmed by the prompt acceptance received by the Campiello project in the feast program. In the subtle play between closeness and openness, the presence of a tool speaking about the neighbourhood's many facets was deemed appropriate in that, *being at the same time a way for self and external representation.* Hence, amidst the slightly out-of-date style of the feast the Campiello kiosk's technologic presence was well and sometimes enthusiastically received in spite of the quite heavy contrast with the rest of the festivities. Or maybe precisely because of the same.

Here follows a list of subjects/contexts (also see Figure 2) covered by the school research studies which were available through paper or computer Web-based interface (see Figure 10) during the feast. All concern the Castello district: events and feasts; houses, present and past; streets, calli[5] and campielli; monuments, present and past (those no longer in existence are described too); artisans and craftsmen; shopping and cooking; associations. Even if within the first prototype of the Web-based interface used during the S. Pietro feast (see Figure 10) the innovative services (e.g., "pushing"-based access to knowledge, collaborative filtering of the relevant information, etc.) were still not available, some characteristics of the Campiello knowledge base are already visible. The list on the left shows the available contexts, as decided by the school; the second list from the left is an example of available items for a particular context (i.e., in Figure 10, events and feasts in the district); and one item description is shown (i.e., the feast of the Maries).

In addition, any comments inserted via the paper-based interface were also embedded in the Web page for an item (see Figure 11).

Applied Methodology and Lessons Learned

To conclude this chapter it might be useful to recap and make explicit some methodological aspects briefly reported above. Generally speaking, we would say that the Campiello project's major point of interest lies in the attempt to include the use of innovative technologies in a local community's everyday life and practices to support its potential. As already noted, Campiello moves from recognising the existence of a community to trying to be integrated into it. Community practices represent the basic point of reference in shaping the system and its performances; meanwhile the contents are almost entirely produced by community subjects. The system design and progressive definition grow through active participation in community activities and events; and its usefulness would eventually be measured on the appropriateness and relevance of the answers it will provide in meeting the demands of different kind of users (residents, community subjects, tourists and other type of visitors). Two questions are strictly linked to the methodology in use. We will highlight them here briefly, demanding further work and experimentation. We have made clear that the Campiello project came from the idea of involving the community subjects directly, using the contents they produce in their everyday activities to fill in the system. Two years down the line from this relevant choice we can draw some lessons from our experience so far and make adjustments for the future. First of all, it is important to recognise that working side by side with community subjects is a very good idea that is, nevertheless, quite difficult to put in practice. Everyone involved in multiple-partners projects are well aware of the effort needed to get different competencies in different organisations to communicate effectively. Adding to this complex interactive system the variable of real actors immersed in their "normal" everyday activity produces a great amplification of the complexity of the whole. This means that the time needed to go through the various steps of the project increases considerably. On the other hand, there is a consistent advantage in making the choice we did, particularly when working on the innovative use of information technology. One of the major problems we normally meet working in this field involves making nonexpert persons understand what the work is all about. We can

Figure 10: The First Prototype of the Web-Based Interface[6] Used During the S. Pietro Feast

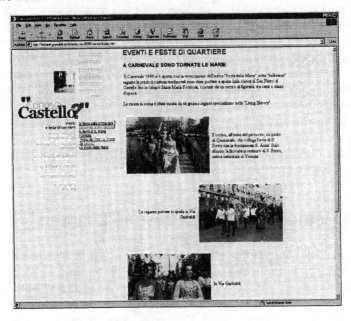

Figure 11: The First Prototype of the Web-Based Interface Showing the Comments Inserted via NewsCards

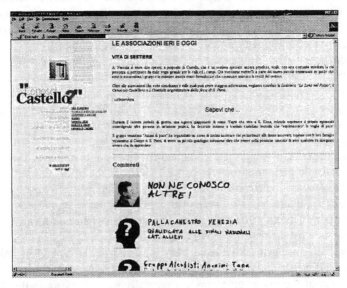

keep talking about the project, explaining its goals and making examples of its use, and still end up with very confused people who will raise the same questions upon the next occasion. In this respect the decision to have community subjects produce the system content makes quite a difference! The impact and understanding induced in nonexpert observers (of various ages and backgrounds) by real content—related to what they know and can connect directly with—is very impressive. A second question concerns how the work must be extended in the local context in order to go from the step of "demonstrating" the potential of the system to that of actually experimenting with its possible use. To reach this second step the number of community subjects involved should be much higher than what has been possible so far; moreover, the occasions for experimental use of the system must be broader. By means of this larger test-bed opportunity, the relevance of the system and its community supporting functions would be maximised. It is basically in this direction that we will focus our efforts next year.

In conclusion we can say that, in spite of all the difficulties encountered so far, we still deem the attempt of supporting existing communities in their double-face character of closeness/openness a fertile and innovative path worthy of exploration.

Concluding Remarks

The work presented here is the outcome of a project based on a three-tier methodology (see Agostini et al., 1998b, 2000) involving technology providers, interaction designers and end users. The result is a spiral refining both overall objectives and understanding of the needs of the local communities over the course of the project. The work presented here will continue until September 2000 and other experimentation will take place before the end of the project. We must stress that in regard to the local community support we chose neither an activist approach based on off-the-shelf technology, nor a purely visionary technical effort far from the real setting needs. At the end of the project we plan to evaluate not only the technology and its usage in real settings, but also the effectiveness of the proposed three-tier design methodology.

Acknowledgments

We would like to emphasize that the results of the Campiello project are due to the work and efforts of all partners involved, so our thanks go to them all. In particular, the Domus Academy Research Center (Milano, Italy) is in charge of designing the Campiello system interaction modes and user interfaces. The Laboratory of Distributed Multimedia Information Systems at the Technical University of Crete (Greece) developed the first prototype of the Campiello Knowledge Base. The authors would especially like to thank Giorgio De Michelis, who shared fully in the work experimentation of the system in Venice.

References

Agostini, A., Grasso, A., Giannella, V., & Tinini, R. (1998). Memories and local communities: An experience in Venice. In *Proceedings of the 7th Le Travail Humain Workshop "Designing Collective Memories,"* Paris, France, September, 14, 1998.

Agostini, A., De Michelis, G., & Susani, M. (1998, March). A methodology for the design of innovative user-oriented systems. In *i3 Magazine—The European Network for Intelligent Information Interfaces*, 2, 4-7.

Agostini, A., De Michelis, G., & Susani, M. (2000). From user participation to user seduction in the design of innovative user-centered systems. In *Proceedings of Fourth International Conference on the Design of Cooperative Systems*, COOP 2000. Sophia Antipolis, France, May 23-26, 2000 (to appear).

Campiello. (1997-2000). Esprit Long Term Research Project # 25572. Started in September 1997 and will last until August 2000 [Online]. Available: http://klee.cootech.disco.unimib.it/campiello.

Glance, N., Arregui, D., & Dardenne, M. (1998). Knowledge Pump: Supporting the flow and use of Knowledge in Networked Organizations. In U. Borghoff & R. Pareschi (Eds.). *Information Technology for Knowledge Management*, Berlin: Springer Verlag.

Glance, N., Arregui, D., & Dardenne, M. (1999). Making recommender systems work for organizations. In *Proceedings of PAAM '99*.

Grasso, A., Koch, M., & Snowdon, D. (1998). Campiello—New user interface approaches for community networks. In D. Schuler (Ed.), *Proceedings of the Workshop on "Designing Across Borders"* at CSCW '98. Seattle, WA, November 1998 (short version in *SIGGROUP Bulletin*, October 1999).

Grasso, A., Koch, M., & Rancati, A. (1999). Augmenting Recommender Systems by Embedding Interfaces into Practices. In S. C. Hayne (Ed.), *Proceedings of the International ACM SIGGROUP Conference on Supporting Group Work*. Phoenix, Arizona, November 14-17, 1999, pp. 267-275.

Johnson, W., Card, S. K., Jellinek, H., Klotz, L., & Rao R. (1993). Bridging the paper and electronic worlds: The paper user interface. In *Proceedings of INTERCHI '93*. ACM Press, April 1993.

Koch, M., Rancati, A., Grasso, A., & Snowdon, D. (1999). Paper user-interfaces for local community support. In *Proceedings of HCI International '99*. Munich, Germany, August 1999.

Resnick, P., Iacovou, N., Suchak, M., Bergstrom, P., & Riedl, J. (1994). GroupLens: An open architecture for collaborative filtering of Netnews. R. Furuta & C. Neuwirth (Eds.). *Proceedings of CSCW '94*. Chapel Hill, NC, October 22-26, pp. 175-186.

Shardanand, U., & Maes, P. (1995). Social information filtering: Algorithms for automating word of mouth. In *Proceedings of CHI '95*. Denver, CO, May 1995, ACM Press, pp. 210-217.

Weiser, M. (1993). Some computer science problems in ubiquitous computing. *Communications of the ACM*, July 1993 (reprinted as "Ubiquitous Computing,"

Nikkei Electronics, December 1993, pp. 137-143).

Xerox. (1998). Information on PaperWare and multifunctional devices [Online]. Available: http://www.xerox.com/paperware/

Endnotes

1 A small Venetian square is called *Campiello*; these squares are places where people meet and socialize.
2 The graphical representation of the Castello hierarchy has been designed by Domus Academy.
3 The pictures have been taken by students of the school during their research.
4 Campo and Campiello are the names used for denoting the squares in Venice (except Piazza San Marco).
5 Due to the peculiar morphology of Venice, there are dedicated names (both in Venetian vernacular and in Italian too) for the various typology of streets. These names in Italian are used only for Venice.
6 The homepage was designed by Domus Academy, while the various item descriptions have been collected and designed by the school.

Chapter XVIII

Academic-Community Partnerships for Advanced Information Processing in Low Technology-Support Settings

Rosann Webb Collins
University of South Florida, USA

What knowledge is required to enable individuals in community agencies to harness advanced information and communication technologies (ICTs) to promote their communities' development? Around the world organizations of all types use ICTs to significantly enhance and transform the quality and efficiency of work. Knowledge workers in community agencies, like knowledge workers in general, want to employ the full power of ICTs, both to provide services electronically and to improve their own access to information. There is an increased need to be able to "gather and interpret data efficiently and effectively into functional information for professional acting social work settings" (Grebel & Steyaert, 1995, p. 163).

There are a variety of roles in community agencies, including administrators, program developers, counselors, and teachers. All individuals filling these roles would be categorized as knowledge workers because their tasks require non-routine and complex work, they must apply their knowledge capital to these tasks, their work requires significant cognitive information processing, and their written and verbal outputs have information content (Davis et al., 1993). Their needs for more information include content information to help community agency employees provide service to clients more effectively, information on community and other resources that can provide additional help for clients, as well as information that helps employees assess their service programs and the management of their agencies. These needs stem from both the desire to provide better service to clients and increased pressure from funding sources, public and private, for more accountability.

What is unique about the community agency setting is that resources available to purchase and develop ICTs are more limited than in the typical for-profit organization. Not only are community agency administrators reluctant to divert funds from direct service activities to purchase ICT hardware, software, networks and expertise, but often there is limited knowledge about how to be intelligent consumers of the products and services offered in the marketplace. This chapter argues that academic-community partnerships can enable community agencies to take full advantage of the power of advanced ICTs. Community agencies are viewed as low technology-support settings that, given adequate assistance from a neutral agent, can successfully develop and use quite sophisticated information and communication systems to enhance their work.

The idea for these partnerships stems from the author's ongoing participation in two community agencies. This chapter will include short descriptions of the specific ICT problems faced by these agencies, the source of these problems, and how the academic-community partnership acted to solve the problems. One case study involves technology planning and the development of a database resource for enhanced program assessment at a public and privately funded Crisis Center. The second case study concerns the development of a program assessment system that includes a within-agency database and network combined with interagency data exchange; this system is for a charter school for children who live in a privately funded shelter for homeless families. Background information and a summary of specific activities with each agency are provided in Tables 1 and 2 at the end of this chapter. Based on specific examples from these case studies, the chapter will detail the challenges and problems in informatics inherent in such situations, as well as how an academic-community partnership offers a workable solution.

Background: What is Meant by *Low Technology-Support Settings*?

Non-for-profit organizations that provide community services increasingly use information technology to capture data about their services and to support service delivery. Typically these systems are based on microcomputers and local area networks, and the systems include both standard personal productivity software and specialized software products for social service providers. For example, Crisis Center staff use a suite of word processing, graphics, and spreadsheet programs, as well as a commercial product called IRis, from Benchmark Enterprises, for client tracking and referral information support. In the past volunteers from the MIS department of a local company developed a database for one of the neighborhood centers, but that system has been abandoned because it could not be maintained by the staff and the local company no longer works with the Crisis Center. In addition Crisis Center staff use their microcomputer workstations as terminals to larger state and federal systems in order to input mandated information about services they provide in certain areas, such as help for victims of domestic violence. Similarly, in the charter school the principal and other school staff use microcomputers for word processing, to access a county school system administrative system, and to some extent in the classroom.

Although ICTs are used in community agencies, not all agencies have even basic information technology support, or what support they have is not evenly distributed throughout the agency. For example, the Crisis Center has no full-time IS professionals and relies on what a few staff learned on their own and one outside consultant (who primarily installs hardware). In the case of the charter school, the administrative office of the parent agency has a small MIS group that provides systems for administrative and development purposes. However, little information technology is supplied to the service components, such as the health clinic, counseling service and school, and the MIS group provides no support to nonadministrative functional areas.

These two cases illustrate what is likely a common situation: a community service agency in which system development and support are done by one or a few "power users" within the organization, by volunteers, by a small IS staff, and/or by outside consultants. Such community service agencies are therefore considered *low technology-support settings*. Despite this low and uncertain level of support, users are able to accomplish a great deal when the technology solutions are well structured for their particular needs. However, users in these settings are typically unable, or less able, to create or maintain technological solutions for their problems using more advanced information technology resources.

The Challenge and Problems of Advanced Information Technology Use in Community Agencies

Users in community agencies realize a need for more powerful applications of information technology, and in addition they face increased pressure from government and other funding sources to provide and share better information about their operations. In particular there is a need to demonstrate program effectiveness and create information networks that can coordinate services across agencies. For example, many agencies, both public and private, provide services to the homeless. A homeless person may call the Crisis Center hotline to find out where to get help, and the Crisis Center may refer that individual to a shelter, a jobs program, and a clinic. The Crisis Center measures that interaction as a service event to the homeless. The agencies that provide the shelter, job training and health care will also record their services to that same individual. Since there is no way to combine the information across agency, the government and private funding groups have no way to identify gaps or redundancies in services, or even how many individuals are currently being assisted. In some locales, such as West Palm Beach County, Florida, IBM is partnering with local governments and agencies to build ICTs to provide such information.

However, there are several impediments to addressing such needs. *Firstly,* community agency users may not fully understand the power of the technology they already possess. For example, in the case of the Crisis Center, some program evaluation reports could have been generated by the existing IRis system, but the users thought they were limited to the reports designed by a consultant at initial implementation, and they do not have the training to design reports themselves. Complicating this particular point is that the users may also lack knowledge of exactly how to process the information as requested. For example, in both cases individuals expressed a need to provide longitudinal data analyses to their funding or government agencies. Neither agency had staff with the statistical expertise to adequately set up such analyses, even if they had access to the data and the data analytic tools.

Secondly, knowledge workers in community agencies are likely to have some, but not all the technology expertise they need to create solutions with advanced technologies such as data bases and inter-agency networks. For example, research by Grebel and Steyaert (1995) indicates that in almost all European Community countries, the curriculum for social workers includes at least basic computer literacy. However, many schools do not have adequate resources for building knowledge on how to integrate IT applications into social work, and in general there is a "gap between the education of new social workers and requirements of today's profession" (p. 162). Using information technology for assessment of services through analysis of data aggregations and data sharing across agencies, while needed, may seem impossible for community agency professionals who lack IT education. Even what an IS professional might consider basic procedures, such as regular data and system backup with off-site storage, may not be done in some community agencies because of lack of training. In fact one of the action items recommended to the Crisis Center in the technology assessment plan was to provide facilities for system backup (which were not available in all parts of the agency) and establish policies on data privacy (that did not exist at all).

Thirdly, the community agency users may feel at a disadvantage in negotiating for assistance in addressing information technology needs. Managers at both agencies studied reported bad experiences with consultants (expensive and not very responsive); the MIS staff of other, related organizations (unwilling to help create data exchanges); and volunteer help from local companies (not fully committed for long-term assistance). This is also a problem when, as discussed earlier, there are needs for help in deciding what data are to be analyzed and how they should be analyzed or processed, as well as in developing or acquiring information technology to support that analysis. Finding IS consultants, MIS staff from parent organizations, or volunteers who combine both sets of expertise (skills in statistical analysis and development of ICTs) is difficult, and if such help is not voluntary, it is expensive.

Fourthly, knowledge workers in community agencies find it difficult to get funding for information technology projects if they cannot articulate exactly what is needed and why it is necessary for the agency programs. In the Crisis Center agency case, the academic partnership began with a technology assessment of the agency's existing and needed technology infrastructure (hardware, software, data, networks, expertise) that was done by an advanced systems analysis and design class under the supervision of the author. This agency had already been turned down for one grant for additional information technology because it could not demonstrate what existing computer systems were available or what they needed. This agency had a particular problem in that it had grown over time as the county government merged several smaller agencies with the Crisis Center. No one had ever attempted to document the technology infrastructure of this new, larger agency, nor had anyone taken any action to standardize on common application suites or platforms. One part of the assessment document modeled common data, processes, software, hardware, and networks, and described a plan for moving all areas within the agency to standard applications, platforms and data. Once the technology assessment document was prepared it was used to support funding requests from both a government agency and a local company.

Outcomes for Both the Academic and Community Partners — "Win-Win"

The academic-community partnership provides benefits to both the community agencies as well as to the academic partners who provide the assistance, resulting in a real "win-win" situation. The academic-community partnership does not replace assistance from the IT marketplace, but rather helps knowledge workers be wiser users of existing technologies and consumers of new technologies. The argument for an academic-community partnership that assists knowledge workers in community agencies in developing advanced uses of information technology is that expert academic help can:

a) **assess** the current information technology capabilities (hardware, software, networks, data, skills) of the agencies, so that users understand the power of their existing tools as well as how to enhance them through training and acquisition. This was the focus of the technology assessment plan created for the Crisis Center. This was also needed for the charter school, which was the recipient of a variety of donated computer equipment. With some assistance some of this donated equipment was moved from the closet, in some cases upgraded, and installed in the classroom. This kind of assistance is particularly important when the best solution is not to simply buy more hardware, but rather use what an agency has more effectively. If an agency is using a consultant who benefits financially when new hardware or software is acquired or developed, then it is unlikely this consultant will focus on a lower-cost solution.

b) provide **training and support** where that is not available in the marketplace. Again, this training is likely to go beyond the use of an information technology, and into training on what data processing/analyses are needed for a task. Academics who combine expertise in both technology and common data analytic procedures are able to provide both kinds of training and support. Parts of the program assessment system developed for the charter school to meet needs expressed by the teachers, counselors and principals required knowledge of specialized educational measurement techniques such as readability analysis and evaluation of facial expressions, as well as the typical kinds of data collection and analysis.

c) serve as a **communication bridge** between the marketplace of information technology or the MIS staff of other, larger organizations and the community agency, so that agency managers do not feel at a disadvantage in negotiating for services and products or for common systems. For example, the charter school staff knew that valuable information about their students was collected in the county administrative system, but the school was provided with read-only access to that data. The school's request to the county MIS group to facilitate downloading selected fields was initially denied. At a second meeting that included the academic partner who could "speak the IS professional's language," the request for data interchange was accepted.

d) **create structured solutions**, in the form of prototypes that can be adapted in many agencies. Much of the base technology employed by community agencies is widely used: general-purpose products like Microsoft Office, as well as special-purpose products like IRis (which is used throughout the U.S. for crisis center-type agencies). The idea is to leverage this fact by creating procedures and designs that can be valuable to many communities. The prototypes include not just the technology solution for a particular agency, but also the processes and procedures necessary for creating the solution in most/ many low technology-support settings. The prototypes may be critical for agencies located in areas where there are few or no IT professionals in academia or business who are willing to provide direct assistance. In Australia Rochester and Willard (1998) found that all community groups realized a need for information to pursue their goals, but groups in rural areas sometimes felt disadvantaged in their access to information. The next phase of this research will examine how well knowledge workers in other community agencies can take advantage of one or more prototype solutions created for the two case study agencies.

In addition to these benefits, the partnership also helps knowledge workers who serve as academic partners. For example, a successful academic-community agency partnership developed a community information network for Decatur, Illinois. This project used students from the local university for tasks such as Web page design, preparation of documentation, and analysis of legal issues in information distribution. The project director notes the "win-win" nature of the partnership, since it provides a "benefit [to] the student's educational experience and [fills] the agency's need for volunteer help in the networking area" (Hale, 1996, p. 205). Two examples from the author's own experience reiterate this point. Recently Arthur Andersen supported competitions in which university student teams from MIS and computer science departments design Web pages for local community agencies who lack the expertise to create high quality Web pages for themselves or the funds to pay someone for this work. Even in cases where the student team designs just the opening Web pages, these serve as a template design that can be built upon more easily by the agency staff or consultants.

Similarly the advanced systems analysis and design students who completed the technology assessment project for the Crisis Center had to work in teams to interview users, examine computer equipment, create models, synthesize models across teams, and present results both orally and in written form. The students get valuable experience dealing with, as one student put it, "real users" and real problems. They get to practice fundamental systems development and assessment techniques, and at the end they have a nice work sample to show employers.

Table 1: The Crisis Center

Context

The Crisis Center of Hillsborough County, Florida, USA, is a multi-unit community agency that is funded by government, private charities, and individuals. Initially the Crisis Center provided 24-7 hotline assistance to individuals with a range of problems, including drug or alcohol abuse, suicide, and violence. This single-purpose agency was supported by a small group of administrators and developers (fundraisers), and most staff were telephone counselors. Over time the Crisis Center has expanded its services and been asked by the county to merge with other units, so that now the Center provides educational outreach services; several specialty hotlines (for teens, parents and seniors); ElderNet (daily phone contact with seniors who live alone); two neighborhood family support centers (a single source of social service information for low-income families); Travelers' Aid; a sexual abuse treatment service focusing on abuse, prevention, psychotherapy, and life education; a rape forensics service for the entire county; and an ambulance service for patients who need hospitalization for mental health or substance abuse treatment. There is a central administration and development group for the Crisis Center, but units continue to operate somewhat autonomously.

Each unit has ICTs, but there is no specialized information systems staff. Neither the county government (for whom the Center provides services on a contractual basis) nor the United Way (a primary funding agency) supply support, so all decisions about ICT and development are made by the staff. The only assistance prior to the academic partnership was primarily hardware installation by one local independent consultant and the development of a database for one of the neighborhood centers by volunteers from a local company. All ICTs have been purchased or provided by government agencies (such as the Federal Emergency Management Agency Database).

Activities of the Academic-Community Partnership

Technology Assessment Plan. No attempt had ever been made to understand the total ICT resource nor identify what data could and needed to be consistent and shared. In 1997 the Center's request for additional technology was turned down by a funding agency because they could not specify existing ICT resources. This need to be able to demonstrate what was needed and why converged with increased pressure from funding agencies, both public and private, to provide better and different service assessment data.

The technology plan consists of two main reports. First, the plan includes action recommendations for the Crisis Center and an overall assessment of the

Table 1: The Crisis Center (continued)

existing information technology capabilities for the entire organization. The purpose of this map of the technology infrastructure for the whole organization is to identify how existing capabilities and data can be shared; and to identify where new hardware, software, data, training, and other resources need to be added. Second, for each unit within the agency, there is an assessment of that unit's existing information technology capabilities and its needs for additional information technology. The assessment of current capabilities includes a detailed inventory of the hardware, software, data, procedures (e.g., backup and restoration procedures), user skills, and supporting materials (system documentation, manuals, self-help materials) currently available in each agency or functional area. The assessment of the unit's additional information technology needs is matched to the unit's goals or strategy, and prioritized. This part of the report includes a description of what hardware, software, data, training, and/or supporting materials will have to be purchased or developed in order to meet those needs.

Seven of the eight action recommendations in the plan have already been implemented, and the eighth is in development (a replacement client database for the neighborhood centers). One action item was the creation of a Technology Change Management Team comprised of both academic and Crisis Center staff; this team is a formal mechanism for ongoing partnership efforts. The team is currently involved in planning the network wiring requirements for a new building to house the Center. The technology assessment plan has been used to acquire additional funding from private sources and as base information for assessing Y2K vunerability.

Database for Program Assessment. A primary activity of the Crisis Center continues to be hotline services. Since 1996 data on hotline services has been collected by telephone counselors using IRis, a specialized database system built on FoxPro. Prior to 1996 some data was collected in an Alpha-4 database. The reports provided by the IRis system are valuable, but the director of these services would like to investigate trends, combine data and analyze it in more sophisticated ways. Students in the MS in IS program are currently working on reconciling and combining the data between the two systems and creating a data resource for the manager of hotline services. This project is critical to a goal for the Crisis Center: to have the Center's hotline become the central contact point for community services for the entire county. This is similar to the United Way's "Dial 211" program in Atlanta, Georgia, that is accessed by dialing 211 on any phone or via the Internet and that gives information on agencies and programs in the metropolitan Atlanta area. The Crisis Center would like to demonstrate its ability to collect and analyze meaningful data on these service calls as justification for becoming the main access point to find or give help in the community.

Planning and Testing for Year 2000. The Crisis Center board of directors mandated that the chief financial officer of the Center, who is a power user of technology and a member of the Technology Change Management Team, lead the Center's Y2K efforts. The technology plan's inventory was used as base information for deciding what had to be assessed. The board specifically requested that an outside consultant be used to certify that the approach being used was appropriate and to help with the testing. Both the author and a doctoral student in accounting information systems reviewed the plan and letters to vendors, and MIS majors participated in the testing of each workstation in Fall 1999.

Table 2: The Charter School

Context

The Charter School of Metropolitan Ministries in Tampa, Florida, opened in Fall 1998. This elementary school is provided for children whose families are temporarily living in the Metropolitan Ministries shelter for homeless families. The purpose of the separate school for these children is to address their unique needs while homeless. These needs include help in overcoming barriers to receiving educational services (no proof of residency, poor or missing education and medical records), reducing delays in receiving services because of family transience, and dealing with the social and psychological stresses of being homeless. A core activity of the school is the development of a long-term plan for the education of each child that is based on the charter school's testing program as well as records from previously attended schools. This long-term plan is transferred to the child's new school when the family moves to permanent housing. The only existing information technology support for this school is participation in the county's school administration program. Charter school staff enter attendance data and can access some student background data and prior attendance data.

Activity of the Academic-Community Partnership

The primary partnership activity with the charter school is the development of a program assessment system, although the academic partner also designed a local area network for the school's administrative office. There are also plans to initiate a job training program for parents at the school that is focused on development of ICT skills. The initial information requirements determination has been completed for the program assessment system and the cooperative arrangement with the county school administrative system has been established. The program assessment includes the following information on each child:

1. background, contact, and prior school attendance information on each child, downloaded when possible from the county administrative system;
2. individual, long-term plan for education;
3. scores on reading and phonetic awareness tests administered by charter school teachers and counselors;
4. score on a measure of problem-solving ability administered via computer;
5. assessment of student drawings for emotional content, administered via computer;
6. readability and content analysis of writing samples for students' attitudes toward themselves and learning; and
7. attitudinal data based on paper-based questionnaires for students and teachers.

The main database has been created in Microsoft Access, with supplementary data and content analysis done with specialized programs.

References

Davis, G. B., Collins, R. W., Eierman, M. A. and Nance, W. D. (1993). Productivity from information technology investment in knowledge work. In R. D. Banker, R. J. Kauffman, and M. A. Mahmood (Eds.), *Strategic Information Technology Management* (pp. 327-342). Harrisburg, PA: Idea Group Publishing.

Grebel, H., and Steyaert, J. (1995). Social informatics: Beyond technology: A research project in schools of social work in the European community. *International Social Work*, 38, 151-164.

Hale, C. (1996). Project MILLIKInet' becomes 'DecaturNet': A library-initiated community information network. *Illinois Libraries*, 78(4), 201-210.

IRis (Information and Referral Software), Benchmark Enterprises, 597 Caroline Ave., West Palm Beach, FL 33413. (407) 697-4230.

Rochester, M., and Willard, P. (1998). Community organizations and information: Results of a study. *The Australian Library Journal*, 47(3), 254-263.

CI and Development

Chapter XIX

Communication Shops and Telecenters in Developing Nations

Royal D. Colle
Cornell University, USA

Bai Yuxiong, a farmer from a poor area in northern Shaanxi Province [of China], traveled at least 500 kilometers to Yangling Agroscience town to learn about prices of Qinguan apples and the value of a small pumpkin variety he grew....At Yangling Information Center, for the first time he saw the computers and heard about the Internet. He saw the director of the information center typing some things on a computer. In a short time he got all he wanted from the computer.... He knew the price of the apple. And his small pumpkin was found to be of a very precious kind, an indispensable delicacy for Japanese state banquets....Computers opened his eyes and his new [export] venture. (Li, 1999)

Information and Development

This chapter deals with the importance of information for the development of rural areas, and especially with providing the institutional structures for helping people gain access to the communication technologies that are instrumental to information flow in the 21st century. We start with a reminder about the potential value of information, then move to the role of information and communication technologies (ICTs), and then concentrate on the emergence of community-based communication centers (telecenters).

A recent book by Fraser and Restrepo-Estrada (1998) provides excellent documentation using concrete case studies that show where a systematic approach to providing access to information has made a significant impact on the welfare of rural and urban people. Their examples range from family planning successes in Indonesia and immunization programs in Colombia, to agricultural development in the Philippines and Mexico. They also note the consequences of inadequate information. Referring to situations in the Philippines, they report that a mountain tribe is losing its land to "land-grabbers" because the tribe doesn't know about existing laws that could protect them; and a farmer traveled by sea and land for seven hours to buy seeds, not knowing that a fellow farmer three kilometers away had them for sale.

The 1998 World Telecommunications Development Report summarizes several situations related to the value of information to the population:

In rural southern Ghana, petrol stations are able to place orders with suppliers by telephone when previously they could only be made by travelling to Accra; in Zimbabwe, one company generated US$15 million of business by advertising on the Internet; in South Africa, lives have been saved since citizens have been able to call the police from strategically located community payphones; in the mountains of Laos and Burma, yak caravans employ mobilephones to call ahead and find the best route to take during the rainy season to bring their goods to market; and in China, a little girl's life was saved when her doctor posted her symptoms to an Internet discussion group and received an immediate answer. (International Telecommunications Union, 1998)

The central and vital role communication and information play in the lives of people was officially recognized by the UN General Assembly in December 1997 when it endorsed a statement on the Universal Access to Basic Communication and Information Services. The statement concluded that the "introduction and use of information and communication technology must become a priority effort of the United Nations in order to secure sustainable human development." The statement also embraced the objective of establishing "universal access to basic communication and information services for all" (ITU, 1998).

Traditional interpersonal means of providing rural people with information have long been acknowledged as being inadequate. Government extension systems, for example, have been plagued by the social, cultural and educational gap between the agricultural scientists and technicians and their clientele, the poor training of the technicians, and their lack of understanding of farmers' problems (Fraser and Restrepo-Estrada, 1998, pp.16-17). In recent years, governments typically have sought alternative solutions to labor-intensive and expensive face-to-face community contacts, including greater use of radio, television and other media.

Information Technology

The dramatic introduction of new communication-related technologies in the latter part of the 20th century significantly accelerated the exploration and trial of innovative information systems. These technologies included miniaturization, portability and consumer-friendly innovations in conventional media such as radio and television, as well as the linking of computers and telecommunications to mass and specialized information systems. Thus, at the turn of the century, we speak of a telephone in every village in India and China, and women operating telephone businesses in Bangladesh using cell phones and no copper wires. How these technologies are penetrating our lives is reflected in our language with words like tele-medicine, tele-education, tele-commuting, and "community informatics" becoming a common part of our vocabulary.

The World Bank's Willem Zijp (1994) has made a strong case for the role that information technology can play in agriculture and rural development. In regard to rural communities, Zijp says that information technology is "better" because:

- Farmers' organizations can use fax and/or e-mail to obtain better information for their members and speak more effectively on behalf of their members, strengthening their advocacy role.
- Rural groups can use video camcorders to present their needs and potential solutions more effectively to policy makers through the use of visual images.
- Farmers and other rural groups can easily use powerful multimedia training programs with touch screens, even if they are not literate.
- Vertical farmers' organizations can use telecommunications, radio, and/or packet radio to obtain and transmit price information in order to become integrated into production and marketing chains, providing them with greater strength in negotiating with buyers.

Zijp says communication technology is faster because:

- Farmers can get information about daily market prices on the radio or through e-mail notices posted at a local center before taking produce to market.
- Rural people get information via radio about impending weather threats.
- Rural midwives can get immediate information about particular health problems from a microcomputer at a local clinic.
- Small craftspeople in isolated villages can get information about transportation via radio.

And, Zijp says, information technology has cost advantages because:

- Rural advocacy groups can make video tapes cheaply with camcorders to describe their problems and suggested solutions through powerful visual and oral messages.
- Community groups can develop their own radio programs, addressing local needs.
- Rural people can get high-quality, consistent training at low cost via distance education or interactive training technologies.

Zijp makes comparable arguments for the use of communication and information technologies in regard to public and private intermediaries (Zijp, 1994, pp. 24-29).

The Effectiveness of ICT

Some observers are cautious and somewhat skeptical about what they see as the "euphoria" surrounding anticipated benefits of ICT for development (Gómez, Hunt and Lamoureaux, 1999; Heeks, 1999a). However, the critics inevitably address factors *associated with* ICTs rather than those inherent in the hardware itself. Some of the concerns about the effectiveness of ICT include:

1) *The diversity and incompatibility of equipment and program protocols.* This applies to conditions within countries and between the "least developed countries" (LDC) and more advanced nations (Mansell and Wehn, 1998, pp.104-5). Recently, for example, a correspondent in China informed me that he could not download a report I sent because, while traveling in the rural areas, he could not find a computer online larger than a 286 or 386 model. Similarly, I have frequently sent documents to colleagues via email using attachments in WordPerfect only to be requested that they be sent in Word because earlier generation word processing programs did not convert alien programs.

2) *The language and access system.* Much of the information available in storage systems on networks may be in a language different from the users'. Not only might there be a linguistic difference, but also a difference in conceptual content. Scientific language is not designed for lay people; and the language of America's Silicon Valley can be a mystery even to a college professor.

3) *Literacy levels.* Although ICTs may be multimedia, many information storage systems (such as electronic databases) depend on one's ability to read and to understand the way information is packaged.

4) *Relevance.* Much of the information available via electronic networks may not meet communities' needs for local information on agriculture and health and nearby markets. Any information system, including interpersonal communication, is headed for failure if it does not meet the interests or the perceived needs of the consumer.

5) *The value of information.* The "culture" of information, for example, understanding what knowledge about diseases, distant markets, and laws can contribute to their welfare may not yet be part of some people's "rhythms of life" (Mansell and Wehn, 1998, p. 108).

6) *Technophobia.* In a world where a majority of the people have yet to make their first telephone calls, keyboards, disk drives, Internet and multimedia may be large, bewildering, and forbidding steps into ICT. It is a phobia most evident in adults — those most likely to represent the cultural mainstream of a national community.

7) *Maintenance.* Technology, appropriate or not, often falls victim to parts that cannot be replaced, servicing that is no one's responsibility, and weather that is inhospitable.

8) *Resource deficiency.* ICT ineffectiveness may result from the lack of resources people need to have in order to do what is proposed in ICT messages. We often hear farmers, for example, complain that they learn about innovations in seed technology or animal husbandry, but are unable to obtain the seeds or the starter animals.

9) *Accessibility.* The 1998 World Telecommunications Development Report tells us that two-thirds of the world's 150 million households do not have telephones which, potentially, are a key to the flow of information to and from people in the least developed nations (LDC). For some, high tariffs stand in the

way. However, another 300 million could afford telephone service if there was a sufficient supply of telephone lines (ITU, 1998).

It should be noted that most of these obstacles to ICT effectiveness can be addressed by suitable planning, support and training. Many of the so-called failures of ICT in development can be attributed to inadequate management rather than to anything inherent in the technology or hardware system.

Universal Service and Accessibility to ICT

The issue of accessibility to ICT is the focus for much of the remainder of this chapter. The challenge that has been taken up by many international agencies is to bring the benefits of information and communication technologies to populations and nations that have not shared significantly in the communications revolution of the last five decades. While ICT may not be part of their vocabulary, many are like some peasants in China, where a survey of Hunan peasants in 1993 showed that 64% of the peasants expressed their need for production, technology and market-related information, and only 6% said that they did not need such information (Lee, 1997). Asked about their grievances, the peasants "ranked lack of information" fourth after "heavy taxation, chaotic markets, and crime" (p. 120).

Our initial focus is on the telephone, because in the present state of information technology, the telephone line provides the principal connection to computer-based information networks, and to two-way connection for demand-driven information services. However, in most of the world, the strategy of one telephone and one Internet connection for each household is not likely to happen in the next several generations (Mansell and Wehn, 1998, p.100), despite the rapid growth and diffusion of communication technologies. ICT such as cellular telephony has provided for significant leapfrogging of traditional telephone diffusion patterns, but cost is a major constraint for cellular to be a viable alternative for first-time telephone users in developing countries (ITU, 1998).

There is a great difference in ICT penetration between developed nations and the LDCs, and between urban areas and rural areas. For example, the Internet has spread widely in China. Commercial Internet access is available in more than 200 cities — with all provinces represented. Estimates put the number of e-mail accounts at 1.2 million (Goodman et al., 1998). Besides the Internet, both Wide Area Networks (WANs) and Local Area Networks (LANs) add to the computer-telecommunications picture in China (McIntyre 1997). However, access to Internet and other communication technologies is more prevalent in urban areas than in rural areas — to the point of suggesting that rural areas are significantly disadvantaged (Lee, 1997).

Telecommunications specialists use the term "universal service" to describe the one-telephone-to-one-household ownership pattern. The more viable strategy for developing countries is "universal *access*."

The concept is that a telephone should be within a reasonable distance for everyone. The distance depends on the coverage of the telephone network, the geography of the country, the density of the population, and the spread of habitations in the rural or urban environment. (ITU, 1998)

"Reasonable distance" in Brazil is to be within five kilometers of a telephone; in South Africa, the standard is a 30-minute traveling distance to the phone; and in China, it is "one family, one telephone in urban areas and telephone service to every administrative village in rural areas" (ITU, 1998). Canada's International Development Research Centre (IDRC), which is especially active in the field of information and communication technologies in Africa, uses "an hour's journey on foot" as the criterion for reasonable distance (Fleury, 1999).

Implicit in the concept of universal access is the idea of *sharing* a connection or facility as compared to each household having its own. The most obvious example is the public telephone booth or telephone calling center. Originally the concept of universal access was applied specifically to telephone use. However, the process of providing access and sharing facilities has been used in other areas of telecommunications. For example, community television viewing has been institutionalized in some countries such as India where the government provides a television set for groups of people (Singh, 1993). Some international and national agencies are applying the idea of access to other communication technologies, in particular, the Internet. One approach to providing universal access to information technologies has been the establishment of community-based communication centers, or telecenters, where the idea of sharing ICT facilities is a dominant feature.

The Telecenter Movement

In the last five years of the 20th century, momentum began building toward the development of community-based communication centers as a means of providing access to information technologies. The emphasis was particularly on reaching rural people in Africa and Asia. Some observers suggested that Africa had missed the Industrial Revolution and it should not miss the Information Revolution. For example, in an *International Herald Tribune* story headlined "Africa is Missing Out on a Revolution, the authors, a professor at the University of Minnesota and a U.S. Foreign Service officer, note that, while one in every four Americans is an Internet user, in Africa the ratio is one in 5,000 (Kapstein and Marten, 1998). And that example is an appropriate one, for it is the accessibility to the Internet and other information technologies rather than the beat of a drum that has become the focus of contemporary communication in Africa. So on the brink of the 21st century, communication centers of various kinds were springing up around the developing world, but particularly in Africa.

Several forces have been driving this movement. These include:

1) *Communication helps rural development.* The assumption that information and communication technologies contribute to development is widespread. Many influential people and organizations believe that both evidence and logic point to the important role that communication plays in rural development. We reported some of these earlier in this chapter. Linked to this

assumption is the *hypothesis* that communication centers contribute significantly to providing the information exchanges that contribute to a community's development. We use the word *hypothesis*, because, as we shall see below, some major international organizations such as the International Telecommunications Union (ITU) are approaching the establishment of these telecenters as a research-and-development activity.

2) *Being part of The Information Society is an important goal for nations.* Many leaders believe that access to ICTs is essential to becoming part of the Information Society, which some believe is a sign of a 21st century nation. For example, a recent Government of India Action Plan proclaimed:

The Government of India, recognising that the impressive growth the country has achieved since the mid-eighties in Information Technology is still a small proportion of the potential to achieve, has resolved to make India a Global IT Superpower and a front-runner in the age of the Information Revolution. The Government of India considers IT as an agent of transformation of every facet of human life which will bring about a knowledge-based society in the twenty-first century.

One of the ways to become part of the Information Society is to join computers, telecommunications, and the Internet—and telecenters represent one highly visible and public mechanism for doing this. As FAO's Van Crowder (1998) notes: "Telecentres are a venue in which new ICTs, such as the Internet, can interface with conventional ICTs (print, radio and video)."

3) *Information technology is a new opportunity for creating small telecommunications business enterprises.* Thousands of small entrepreneurs, some with hardly more than a telephone receiver and a small shelter, have set up businesses for selling telephone service. In Bangladesh, the Grameen Bank has funded a system that provides cellular phone technology and encourages low-income women to start their own telephone service enterprises. In some cases these small ICT enterprises are the forerunners of more comprehensive communication centers, and they introduce communities to the benefits of using information and communication technology. The IDRC's Gaston Zongo reports having counted 9,000 quite sparse telephone kiosks just in Senegal, where, he says, they have opened up job opportunities for "some 20,000 people" (Fleury, 1999). In some cases these small ICT enterprises are the forerunners of more comprehensive communication centers, and they introduce communities to the benefits of using information and communication technology. Much of this becomes feasible because of the dramatic technical advances in communication and computer hardware in the last decade, along with the interest of many governments in privatization and in fostering micro-enterprises (Ernberg, 1998a).

In an October 1999 meeting in Addis Ababa of a special interest group in the African Development Forum, concern was expressed over the larger "donor-funded telecenters" which are seen as too large and unsustainable. "Certainly," it was reported, "they are not a model that can be reproduced across Africa." The report went on:

There are many thousands of smaller phone shops...that are generally entrepreneurial and quite successful. One means to providing greater access to ICTs across Africa would be finding ways of supporting 'migration' of phoneshops to include fax, then computers, then email, then Internet, then other information services as required. (Discussion summarized by Peter Benjamin, email November 11, 1999)

Much of this becomes feasible because of the dramatic technical advances in communication and computer hardware in the last decade, along with the interest of many governments in privatization and micro-enterprises (Ernberg 1998a).

4) *The market potential.* The market for computers and other information technologies in Asia, Africa and Latin America is enormous. Computer and telecommunications companies are working with governments to promote the ICT invasion of the developing world. This builds, for example, on government policies such as those in China where the year 2000 was the target for having a telephone in each of its 730,000 villages (China Computer Newspaper, 1999).

5) *Universal access is a step toward universal service.* National governments and international agencies such as the ITU recognize the difficulty in providing each third-world home with a telephone or other telecommunications connection in the foreseeable future. However, the need for information services makes it vital for governments to seek the universal access alternative through such devices as telecenters. Even where the Internet is perceived as a potential political liability, governments such as China recognize that the Internet is too important to their economic modernization to cut it off (Crossette, 1999).

Van Crowder (1998) notes that services offered by telecenters are a part of the solution to problems that rural communities and partner organizations aim to solve. By providing access to accurate and timely information "telecenters offer communities opportunities to:

- reduce the isolation and marginalization of rural communities;
- facilitate dialogue between rural communities and those who influence [them], such as government...planners, development agencies, researchers, technical experts, educators, etc.;
- encourage participation of rural communities in decision making which impacts their lives;
- coordinate development efforts in local regions for increased efficiency and effectiveness;
- share experience, knowledge, and 'lessons learned' with other rural communities to address issues within local contexts;
- provide information, training resources and programs when needed in a responsive, flexible manner (including, for example, resources related to agriculture, health, nutrition, and small business entrepreneurship);
- facilitate ongoing development initiatives aimed at solving a variety of problems;

- improve communication among stakeholders, thus overcoming the physical and financial barriers that often prevent researchers, extension workers, farmers and others from sharing knowledge and competence" (Van Crowder, 1998, p.4).

In addition to the development-oriented roles suggested by Van Crowder, communication centers might provide contact with distant friends and relatives, and recreational opportunities through videotapes and other entertainment media. Also, a common function of telecenters has been to familiarize people with ICTs and train them in their use.

What is a Telecenter?

There is great diversity in what is called a telecenter. In fact, there is equal diversity in the labels given to entities that provide various kinds of ICT service (Gómez, Hunt, and Lamoureaux, 1999; Conradie, 1998). In addition, there is the ambiguity over what is included in ICT. One view is that ICT relates specifically to telecommunications and computers. Another view (ours) is that ICT includes electronic media such as audio and video recordings, films, and print duplicators, along with telecommunication links, computers, network services, and databases. In this latter definition, ICT might be considered "electronic multimedia."

We have encountered more than 30 different labels for information and communication technology centers. The diversity of names undoubtedly stems from how pioneers in this field have perceived key variables associated with these centers. For example, we can start with the following list of telecenter variables.

Narrow Focus ← Multipurpose
Some enterprises may be exclusively devoted to individual consumer's access to communication technology (such as a public call office) while others may provide access, offer ICT training, facilitate distance learning for individuals or groups, provide development information and serve as a medium for information sharing in the community.

Community-Based ← Establishment
By community-based we refer to a style of ownership and operation that is marked by participation of a broad constituency of residents and groups who represent the fabric of the community. Establishment examples would be "top-down" government or business organizations or NGOs.

Stand Alone ← Attached
Some telecenters are built as solely telecenters; others are created as part of other community institutions, such as schools, cooperatives or government units. The attachment may be simply sharing space or may also include integration with other programs (as where centers are used by school children by day and by the community at night).

Thematic ← Universal

Some telecenters concentrate on particular kinds information such as health, government operations and services, agriculture, etc. (thematic), while others may respond to the perceived needs of a broader community (universal).

Other variables that help describe telecenters include:

Independent ← Networked, Grouped

Public Sector ← Private Sector

Profit Oriented ← Service Oriented

Publicly Funded ← Privately Funded

Commercial (Fee-Based) ← Free

Urban ← Rural

A Typology of Telecenters

The kinds of variables listed in the preceding section can be seen more concretely in the typology of telecenters developed by a team at IDRC (Gómez, Hunt, and Lamoureaux, 1999). In the list below we have added descriptive material to that of the IDRC. It should be noted that existing telecenters may combine characteristics of these types.

1) *Basic telecenter*—Typically small, independently operated community enterprise with a small number of computers and dial-up connections to a wide area network and the Internet. Used by the general public for becoming computer-literate, sending e-mail, and searching Web sites. Minimal information marketing within the center, heavily consumer-driven. They are generally subsidized by government or non-government organizations and are intended to be a public service, with minimal or no fees.

2) *Telecenter franchise*—Similar to the basic telecenter, but with a group of them centrally coordinated and nurtured. Examples are *cabinas públicas* in Peru, the Universal Service Agency telecenters in South Africa (see below), and the National Urban League's Family Technology Centers in the United States. The National Urban League's Family Technology Centers are housed in churches, youth groups and other community organizations. The latter has established a network of 65 centers, with 50 more planned by 2006 (Mendels, 1999).

3) *Civic telecenter*—Housed in public organizations such as libraries, schools, and civic centers, whose main function is other than ICT. The computers, Internet access, and other related services are woven into or provide a supplement to the organization's own cultural, educational, recreational, or governance functions. Civic telecenters can be seen in community libraries in Singapore and in public schools around Beijing in China. In the latter case, the

basic concept is that schools would establish computer centers with Internet access for use by students and teachers in class. Then, after hours, the center would be open to parents and other members of the community, who may not have Internet access. The school thus becomes the Internet access point for the community (International Telecomputing Consortium, 1999). The Family Technology Centers mentioned in (3) above also are examples of civic telecenters.

4) *Cybercafe*—Similar to a basic telecenter, but with a strong commercial and profit orientation focusing particularly on selling e-mail and Internet services. For example, in a five-story department store in Khon Kaen, Thailand, each floor has an Internet facility and each is usually well-populated by teenagers exploring networks and sending e-mail. As the name implies, often food and beverage service is available. The IDRC team notes that cybercafes, which often appeal to tourists and business people, have become popular in countries where connectivity is very limited such as Vietnam, Haiti and Morocco, as well as in the capitals and other cities of Latin America (Gómez, Hunt, and Lamoureaux, 1999). We have seen cybercafes in such widespread areas as China, Scotland, and Rome.

5) *Multipurpose community telecenter (MPCT)*—Offers a broader array of ICT facilities, including, for example, telephone, Internet, and facsimile, but also provides public services such as tele-medicine and tele-education, and user-support and user training. The International Telecommunications Union (ITU) has promoted this model, with an emphasis on community development, and is testing it in several nations. (See Ernberg, 1998a, and the discussion of ITU below; also visit the Nagaseke, Uganda telecenter at www.nakaseke.or.ug.) Similar to the MPCT is South Africa's Multi-Purpose Community Center (MPCC), "a structure which enables communities to manage their own development by providing access to appropriate information, facilities, re-sources, training and services" (MPCC Research Report, 1998). Although many have telephones, telecommunications connectivity is not essential to establishing MPCCs. However they are encouraged to become telecenters by applying to the program of the government's Universal Service Agency. In South Africa, there may be as many as 9,245 MPCC-related organizations (MPCC Research Report, 1998).

6) *Phone shop*—Emphasizes connectivity and offers a quite narrow range of services, beginning with domestic telephone calls, but sometimes expanding into international calls and facsimile. Significantly influenced by privatization of public services in many countries, the government monopoly phone business in some countries has been transformed into small, private-sector entrepreneurships. One can see in many cities and towns in India public call offices which are frequently no more than small wooden single-room shelters. Its more automated relative is the telephone booth in the U.S., Canada, Europe and elsewhere. The significance of the personally run phone shops is their potential for becoming broader-service telecenters built on a business model (Ernberg, 1998). Vodacom, the first South African cellular operator, contracts with people to run a phone shop (as long as it results in employment for the

community), and envisions upgrading the service "into something like an IT center in the near future" (Ballardini, 1998).

Two additional models can be added to the IDRC list:

7) *Communication Technology Center*—Similar to the multipurpose community telecenter but without the community development emphasis. This center provides public access to a range of ICT facilities which go beyond connectivity and the Internet, with more emphasis on being a place where people can use technology for their work or studies. An early example was the "telecottage," which Holloway (1994, p.11) defines as "a manned local center that provides computer equipment and communications equipment for local residents in sparsely populated or rural areas or in poor suburbs....A telecottage helps private people gain access to telephones, tele-learning, data courses, and meeting facilities." Telecottages emerged to serve rural areas, largely with national or local government support, in northern Europe in the mid-1990s and spread to Asia and Latin America. Brazil is expected to have 3,000 of them by 2004 (Ernberg, 1998a). In some countries, communication technology centers have been developed to support "telecommuting" in traffic-heavy urban areas (City of Chula Vista, 1999).

8) *The Community Communication Shop*—Offers a wide-range of communication-related products and services, many of which are available for a profit-generating fee for the entrepreneur. Other products and services may be free or subsidized by government or non-government organizations. Consumers may obtain printed and tape- and disk-recorded educational, training and entertainment materials, as well as access to connectivity services such as phone, facsimile and Internet-type networks, and rental of audio-visual equipment such as cameras and videocassette (or disk) players. While similar to the MPTC and the MPCC, it is marked by its market-orientation, entrepreneurship, and attempt to build a partnership between private-sector enterprise (for sustainability) and community service (Colle, 1998). One approximation of this model are family planning shops in Shanghai, China, where income from conventional entertainment products help sustain the sales of family planning information, education and promotional materials. However these shops do not offer telecommunications connectivity (Zhong, personal communication, 1998).

Major Actors in the Telecenter Movement

Two major issues dominate the attention being given to information and communication technologies in Africa, Asia and Latin America. These are: the establishment of telecenters, and providing telecommunications (Internet) connectivity and access for underserved urban, rural and remote populations. While these issues are sometimes closely related they are not necessarily the same issue. As we concentrate on telecenters issues, connectivity will sometimes surface.

Among the major international actors in telecenter and ICT diffusion are the

International Telecommunications Union, Canada's International Development Research Centre, and the U.S. Agency for International Development. While they are the agents most prominent in the development of telecenters, they and others are instrumental also in building the telecommunications infrastructure that is essential to building telecenters. This infrastructure includes connectivity and infrastructure items and networking devices such as Internet Service Providers (ISPs) and telecommunication transmission channels.

1) International Telecommunications Union (ITU)

The International Telecommunication Union is a major leader in the telecenter movement with its own telecenter sponsorship. In addition, the ITU has helped put telecenters on the agenda of other international development agencies and national governments. It is setting up Multipurpose Community Telecenters in Africa, Asia and Latin America. As mentioned above, ITU's MPCT activities are based on the *hypothesis* that information and communication technologies can significantly contribute to rural development. Thus, it is taking leadership in the creation of Multipurpose Community Telecenters as *pilot projects*. Built into its approach is an emphasis on evaluation.

This ITU activity operates within the framework of the Valletta Action Plan, adopted at the World Telecommunications Development Conference, whose overall objective is "to develop best-practice, sustainable and replicable models of ways to

---

1. *Partnerships with international agencies.* MPCT are created in collaboration with other agencies. Principal ones include: the World Health Organization (WHO), the Food and Agriculture Organization (FAO), IDRC, Canadian International Development Agency (CIDA), UNESCO, United Nations Development Program (UNDP), the Swedish International Development Agency (SIDA), the International Red Cross/Crescent, and the Danish International Development Agency (DAVIDA).
2. *Partnerships with in-country agencies.* ITU involves national and local bodies — public sector and private sector — in MPCT planning, ownership, and start-up and operational phases. For example, in India the MPCT partners include the Department of Telecommunications and the Gujarat State Government National Dairy Development Board.
3. *Universal access.* ITU assumes that universal *service* to conventional telecommunications is not presently feasible so that universal access via MPCT is the better approach.
4. *Multi-purpose services.* MPCT *are telecommunications and computer oriented.* While they are designed to provide ICT consumer services (phone, fax, Internet), they are also expected to provide public services such as distance education, tele-medicine, tele-trading, etc. The mix of services responds to expressions of community needs.
5. *Funding.* There is a partnership emphasis in funding the pilot MPCT. Besides joint funding from international agencies, national and local partners provide 50% or more of the resources for implementation MPCT. Some funding comes from user fee revenue.
6. *Research.* The MPCT pilots are intended to test telecenter models and learn about the dynamics of ICT in rural development.

Sources: Ernberg (1998a & b) and http://www.itu.int/ITU-D-UniversalAccess

provide access to modern telecommunication facilities and information services, particularly to people in rural and remote areas" (Ernberg, 1998a).

In 1999, ITU pilot projects were being negotiated or built in Benin, Bhutan, Honduras, India, Mali, Mozambique, Suriname, Tanzania, Uganda, and Vietnam. The ITU is studying the feasibility of MPCT projects in Haiti, the Maldives, Romania and Senegal. However, telecenters are not a goal in themselves: they are part of an overall approach to providing access to modern telecommunication facilities and information services, and they are part of a cross-sectoral and multi-disciplinary effort for community development. Specific characteristics of the ITU approach to communication centers are shown in the box on the previous page.

2) The International Development Research Center

IDRC is providing leadership in two major projects related to communication centers. Acacia concentrates on Africa, and Pan Asia concentrates on Asia and Latin America. In addition IDRC has a connectivity project called Unganisha that will link IDRC program offices with local project partners through the Internet. We concentrate on Acacia because of its relevance to Africa. We borrow heavily from the IDRC Web page available at <http://www.idrc.ca/.

The purpose of Acacia — actually an international collaborative effort led by IDRC — is to empower sub-Saharan African communities with the ability to apply information and communication technologies to their own social and economic development. IDRC, along with various partners, expects to provide significant funding during Acacia's first five years (spanning the beginning of the 21st century) and help Acacia grow to financial maturity over the first quarter of the new century.

Acacia will test the proposition that ICTs can have significant transformational effects in the developing world. One aspect of this activity is support for telecenters in Senegal, Mali, Mozambique, Uganda and South Africa (Whyte, 1998). The IDRC believes that by using ICTs to their own ends, disadvantaged communities in Africa may be able to shift some decision making away from metropolitan centers and international development organizations towards communities themselves where challenges are most acute. IDRC was one of the pioneers in employing ICTs in Africa.

IDRC's experience and competencies reinforce its reputation as an organization that has responded to global development problems. Over its 25-year history, IDRC has made significant investments in research and capacity building in relation to information and communication throughout the developing world. In fact, IDRC was among the pioneers in the adaptation and use of ICTs in Africa, and Acacia is building on IDRC's existing and emerging networks, programs, and partnerships.

One key partnership is with the African Information Society Initiative (AISI), which unites African governments and donors in a framework to extend the use of information, communication, and related technologies for development. Led by the UN Economic Commission for Africa, AISI provides an African perspective on the opportunities and challenges of that continent in an emerging information age.

Acacia works mainly with rural and disadvantaged communities, and particularly their women and youth groups, because these communities often are isolated from the ICT network and resources enjoyed by others. Acacia aims to achieve three mutually reinforcing objectives that combine to promote equitable, sustainable, and self-directed development among disadvantaged and rural communities in sub-Saharan Africa:

1) to discover and demonstrate how disadvantaged sub-Saharan African communities, especially their women and youth, can use information and communication in solving local development problems;
2) to learn from Acacia's research and experience and to disseminate this knowledge widely;
3) to foster international interest and involvement in using ICTs to support rural and disadvantaged community development, thereby increasing community access to information and communication.

Acacia's activities fall into four categories:

1) policies to encourage access to ICTs by low-income urban and rural groups, particularly women and young people;
2) fostering innovative human and technological infrastructure;
3) tools and techniques that will make it easier for low-income groups to use ICTs; and
4) applications and services that meet community needs (Fleury, 1999).

Specific outputs from Acacia will include:

- pilot projects which test different approaches to providing community ICT access;
- models showing how ICTs can be used to extend the reach of community voices in local planning and in all levels of governance;
- on-the-ground applications at community sites to meet health, education, natural resources management, and other local development needs;
- technology (software, hardware and content) adapted for use in rural and disadvantaged communities;
- innovative infrastructure which extends networks at low cost;
- research into making ICT policy, regulation, and practice more friendly to those who are currently disenfranchised;
- new forms of partnerships in development; and
- more effective use of research results by communities.

The Evaluation and Learning System of Acacia (ELSA) will be a key element in implementation of IDRC's program (Whyte, 1998).

IDRC will invest up to 60 million Canadian dollars over five years (through 2004) and will seek to mobilize significant resources from other donors, technical agencies, and the private sector. But as the lead agency, IDRC is expected to play a key role long enough to ensure a firm foundation for the Acacia program. South Africa has already been through a telecommunications policy reform process, important elements of which were supported by IDRC. One outcome was the creation of the government's Universal Service Agency (USA) intended to define the meaning of universal access in South African terms and to initiate pilot community access projects. The Agency expects to create 150 telecenters a year for the next 10 years. For the country, universal access is a step toward a longer term goal of universal service. The Agency is supporting telecenters as pilot projects in providing universal access. The telecenters will be in disadvantaged areas, particularly rural areas. It works with schools, libraries, churches, and existing community centers as a means for sustaining the telecenters. The Agency's work is guided by explicit mandate in the nation's 1996 Telecommunications Act.

The Universal Service Agency is a key Acacia partner in South Africa. Acacia will invest in selected telecenters and ensure that the experience of different models is evaluated and fed into the Acacia learning process.

IDRC's other major ICT program is Pan Asia. PanAsia's Research and Development Grant Programme provides funds for original and innovative networking solutions to specific development problems in Asia. During its first two years, countries benefiting from support included Bangladesh, Bhutan, Cambodia, Laos, Maldives, Mongolia, Nepal, the Philippines, Sri Lanka and Vietnam. Where Internet access was not available (as in the cases of most of the above), the program assisted in setting up a national ISP. Support is directed also towards regional Internet information hosts, World Wide Web services and authoring, and networking applications. Pan Asia has also been active in supporting national groups, particularly non governmental organizations (NGOs), in Latin America in establishing telecenter projects.

3) United States Agency for International Development (USAID)

USAID has two major efforts that deal directly with the diffusion of telecenters. These are: The Global Information Infrastructure Gateway Project (The Leland Initiative) and the LearnLink project.

The Leland Initiative — Overview

The Leland Initiative is a five-year program designed to bring the benefits of the global information revolution to people in Africa. While the project includes "global information infrastructure technologies (GII), it especially focuses on extending Internet connectivity to 20 or more African nations. Approved by the U.S. Congress in 1995, it is aimed at:

- Improving Internet connectivity within Africa.
- Increasing access of Africans to people and information "for sustainable development."
- Enhancing African ability to find solutions to African problems.
- Making African-produced information available to the world.

The initiative has three strategic objectives (SO):

- SO-1: Create an enabling policy environment in project countries to facilitate electronic networking and access to information technologies.
- SO-2: Strengthen the local telecommunications infrastructure to facilitate Internet access and support a local ISP industry to ensure the local availability of reliable, accessible, and cost-effective Internet access.
- SO-3: Achieve broad-based use of information and global information technologies among USAID's partners to promote sustainable development.

The USAID program consists of substantial advocacy for the Internet, facilitating the creation of private sector enterprises related to the Internet, and training personnel who can further the interests of Internet diffusion and applications. The project works on the supply side (creating the Internet capacity) and the demand side (promoting awareness of uses for the Internet). USAID encourages the development of affordable, public Internet awareness and access centers to promote Internet use. In this approach, the telecenters could provide such services as: free public demonstrations; fee-based Internet accounts and Internet workstations for those without access to home or office computers; fee-based end-user training; fee-based Web page development and training; free institutional information and communication strategy consulting; fee-based publishing consulting services and training; and free proposal development consulting for Internet-related activities. It is significant that the Leland Initiative emphasizes a mix of free services and fee-based services, thus addressing social development and private sector business development goals. The Leland Initiative is also promoting associations of Internet Service Providers and Internet users — in some cases this may mean fostering national chapters of the Internet Society.

A substantial amount of the USAID effort involves identifying organizations in a country that have the potential to use the Internet system in their work. This includes examining their "readiness for effective use of the Internet" and assessing the barriers (policies, infrastructure, awareness) to sustainable and effective use. An initial step in the USAID work in a country involves support for a USAID-designed workshop called, "Internet for Development: Applications and Training." The workshop is designed both to brief USAID Mission personnel in a country and to help their development partners navigate through applications and resources related to the Internet and the Leland Initiative.

The Leland Initiative and other USAID country programs assist with infrastructure building and encourage and support the development of Internet pilot projects and other activities that help increase the awareness and use of the Internet.

The following nations have been identified as participants in the Leland Initiative:

Benin	Ghana	Malawi	South Africa
Botswana	Guinea-Bissau	Mozambique	Tanzania
Côte d'Ivoire	Guinea-Conakry	Namibia	Uganda
Eritrea	Kenya	Rwanda	Zambia
Ethiopia	Madagascar	Senegal	Zimbabwe

It might be useful to pause for a moment to emphasize that the efforts of ITU, IDRC and USAID are not focused exclusively on the construction of buildings and the installation of equipment. The establishment of telecenters requires a wide range of initiatives that relate to policy, promotion, training, and technology.

The LearnLink Project

USAID is also developing communication centers through its LearnLink program whose activities range from improving girls' education in Egypt to various applications of the World Wide Web for education. It is managed by the Academy for Educational Development, a not-for-profit NGO based in Washington, D.C.

[LearnLink] uses culturally appropriate communication and educational technologies to strengthen learning systems essential for sustainable development. This includes using technologies to link individuals, groups and organizations, and to build the capacity of people to access the resources they need to meet their learning needs, particularly those associated with basic education. (USAID, 1999)

Of particular significance to this chapter is LearnLink activity in the development of Community Learning Centers (CLCs) in Ghana and Paraguay, and Community Networking Services Centers (CNS) in Benin — all operated by NGOs.

In Ghana, the NGOs use the name Community Information Center (CIC). CICs provide public access to information on the Internet as well as training to use the computer and other ICTs. Each of Ghana's three pilot centers operate under the auspices of a community development and training NGO. One CIC is in Kumasi, operated by the Center for the Development of People. A second is in Cape Coast, operated by the Central Region Development Commission, and the third is handled by Partners in Internet Education, a group originally organized by teachers to promote the use of computers and the Internet for education. In its design, each CIC is equipped with a local area network and shared high-speed access to the Internet. Each CIC also contains a library of printed materials and a room for the public to improve computer literacy and learn about information resources available on the Internet. Staff is available to help visitors learn to use the Internet. NGOs are expected to operate CICs on a cost-recovery basis. The Centers organize training programs for the public and they organize public seminars on topics related to the Internet and ICTs. This consumer-oriented approach is important in developing a steady clientele whose use as customers will influence the sustainability of the Centers, which are expected to be self-supporting. By mid-1999, the principal sources of income were through e-mail accounts and access to the Internet (Fontaine and Foote, 1999).

In Benin, a CNS Center has been established in a facility for agricultural training. This center is sustained, in part, by the profits of a restaurant that is associated with the training center. Two additional sites will be established in Benin.

For a perspective outside of Africa, in Paraguay, the Municipality of Asunción is seeking to provide basic education, and communication and information services to less advantaged citizens through 12 CLCs housed in public buildings and municipal centers throughout the city. Built on the idea of simple business centers which offer a variety of electronic and communication services, the CLCs emphasize the educational and civic development benefits of computers and communication technology, and, more specifically, their ability to increase access to basic education resources, life-long learning opportunities, and information and services from municipal sources. All of the CLCs were expected to be open and fully operational in 1999.

Among the questions raised by project leaders in the LearnLink CLC/CNS activities are:

• How can the Centers' community outreach and training be made relevant and useful?

- How can monitoring and evaluation be most effectively carried out? What are the appropriate baselines and evaluation data? What impacts should be measured?
- What are the most feasible means for assuring economic sustainability of the Centers?
- What kind of training is appropriate for CLC/CNS staffs? (Steven Dorsey, personal communication).

Other Organizations in Telecenter Development

While ITU, IDRC, and USAID are major international leaders in the telecenter movement, a variety of other organizations operating regionally or nationally are also involved in the telecenter movement. We name and describe a sample of these, realizing that the list is incomplete.

4) *Universal Service Agency* — USA is an agency of the South African Government and is mandated to foster universal access to telecommunications throughout South Africa. For the country, universal access is a step toward a longer term goal of universal service. The Agency is supporting telecenters as pilot projects in providing universal access. The telecenters will be in disadvantaged areas, particularly rural areas. It works with schools, libraries, churches, and existing community centers as a means for sustaining the telecenters. The Agency's work is guided by explicit mandate in the nation's 1996 Telecommunications Act.

5) *Peruvian Scientific Network (RCP)* — This is an NGO in Peru fostering the development of a franchise network of telecenters called *cabinas públicas*. Approximately 190 of these are operating or are being developed.

6) *Association for Progressive Communications (APC)* — The APC is an association of 22 nonprofit computer networks around the world. It is working with IDRC to implement, operate and evaluate two community networking pilot projects in Latin America. Ecuanex in Ecuador and Colnodo in Colombia will work with IDRC telecenters' projects to establish network access points in community telecommunications centers. APC's intent is to encourage wider use of computer networking by individuals, and to develop the capacity of individuals to participate in community development through information exchange and collaboration with computer networking as a key tool. Research and evaluation are built into the projects.

APC also has a broader mission. It links thousands of NGOs worldwide to exchange development-related information. *SangoNet*, in South Africa, a typical example of an APC member, provides low-cost Internet access, training and other resources to hundreds of development organizations and NGOs in the Southern Africa region. In Asia and the Pacific, the PacTok network, closely associated with APC member *Pegasus*, provides low-cost basic e-mail service mostly to more remote and rural areas, including Papua New Guinea, Solomon Islands, Cambodia, Vanuatu, Fiji, Western Samoa, Indonesia, and Malaysia.

Major Supporters of Telecenter Development and ICTs

A wide range of organizations play supporting roles in the telecenter movement process. We identify some key ones in this section and indicate the role they play.

1) *Bellanet* is a consortium that arose from a conference at Bellagio (Italy) where participants struggled with the question of why development assistance and programs are not more effective. Bellanet activities are based on two basic assumptions: that more effective collaboration among development agencies will increase the impact of their programs, and that the use of information and communications technologies (ICTs) can create an enabling environment for such collaboration. (See http://bellanet.org.)

Bellanet includes IDRC (which houses the Bellanet Secretariat), UNDP, SIDA, the Canadian International Development Agency (CIDA), the Netherlands' development agency (DGIS), the Rockefeller Foundation, and the MacArthur Foundation.

Bellanet judges that to stimulate more effective use of ICTs within institutions, it is useful to support partnerships with a specific focus. These "Collaborative Initiatives" respond to a specific need for information sharing, or to an identified communication blockage. They are "best practice" strategies designed to serve as lessons learned for others to build on or emulate. Bellanet's goal is collaboration and learning. The consortium plays a role in the communication center movement by facilitating the sharing of information about telecenters and brokering partnerships. Organizing Web sites, databases, and e-mail discussion lists are means to that end. For example, Bellanet has been hosting a Web site for PICTA (see below) and for the World Bank's InfoDev on funded and proposed projects.

2) *PICTA (Partnership for ICTs in Africa)* is a forum for the collaboration of donor and executing agencies acting within the framework of Africa's Information Society Initiative (AISI). Prior to a donor meeting in Rabat in 1997, there were two principal international collaborating groups focusing on ICT development in Africa: the African Internet Forum (UNDP, USAID, Carnegie Corporation, the World Bank and others) and the African Networking Initiative (IDRC, ITU, the Economic Commission for Africa [ECA], UNESCO and others). PICTA was created through a merger of these initiatives. PICTA membership is open to all donors and executing agencies which have substantiative programs or projects in Africa and the neighboring islands.

3) *UNDP* has an Information Technology (IT) for Development Programme, and it has at least two ways of working with the communication center movement. The most direct way is through partnerships with other organizations through its Sustainable Development Networking Programme (SNDP). For example, UNDP/SNDP is an international partner in ITU's MPCT projects in Benin and Honduras. UNDP and partners in Egypt have joined to establish two "technology access community centers" which will provide the public in Sharkeya Governate with a direct gateway to the Internet (INFO21, 1998). UND/SNDP supports the creation of national nonprofit electronic networks for sustainable

development information in about 30 countries. Various countries in Asia (for example, China, India, Indonesia, Republic of Korea, Pakistan and the Philippines), as well as parts of the South Pacific, are already at different stages of network development. UNDP typically provides seed money for two years to build users nationally — expecting that they will become self-sustaining after that.

4) *UNESCO* is a partner in communication center projects led by IDRC and ITU. For example, it is a partner with ITU in Mozambique, Tanzania Uganda, Honduras, Suriname, Bhutan, India and Vietnam. Its contribution is in funding and/or in support for the programs of centers. UNESCO also sponsors workshops associated with operational aspects of communication centers such as telecenter business planning, product and service development, and evaluation. In 1999, UNESCO undertook a major project to develop materials for training personnel of telecenters.

5) *WHO (The World Health Organization)* is a collaborating partner with ITU in various pilot MPCT projects.

6) *FAO (The Food and Agriculture Organization)* contributes to funding for some ITU MPCT pilot projects, and collaborates in other ways in some of them. FAO has been the most active of the UN agencies in studying and publishing about Internet in relation to rural development. (See, for example, Richardson, 1998.)

7) *The World Bank* is involved with rural communications activities in more than 15 countries, focusing on policy, revenue and tariff arrangements and infrastructure development for rural telecommunications. Central to the Bank's policy work is access to communications by the poorest people, most of whom live in rural areas. The Bank has established the multi-donor InfoDev Fund. The latter has supported major distance education facilities in Ethiopia, Kenya, Uganda, Tanzania, Zimbabwe and Ghana. InfoDev is a program designed to provide developing nations' governments with policy advice and "best practices" information on the economic development potential of communications and information systems. In this context the World Bank operates as a knowledge broker for governments.

8) *CIDA (The Canadian International Development Agency)* supports Internet connectivity and content creation capacity building in five African countries, and supports IT industry development, for example, in South Africa.

Communication Centers
Around the Developing World

In this section we approach the issue of telecenters geographically. Telecenters have had a substantial start in North America and Europe, where they continue to emerge even though the national telecommunications infrastructures are highly developed, and universal service is a near reality. In 1999, the State of Texas and the Federal Government each committed USD10 million to support community-level ICTs, such as wiring community technology centers to the Internet (Cisneros, 1999).

Similarly, the National Association for the Advancement of Colored People, an NGO, and the American Telephone and Telegraph Company launched a new initiative to set up 20 community technology centers in cities across the United States. The U.S. Government's Department of Education, at the turn of the century, was actively involved in funding telecenters around the country. Canada, an early supporter of ICT and telecenter programs abroad, has aggressively supported similar programs on its own soil. These efforts, much like those in developing countries, are designed to bridge the "digital divide" that separates low-income people, minorities and isolated populations from the mainstream communication revolution. We have also seen how Western Europe, Japan, Australia and other industrial nations have witnessed the establishment of telecottages (Holloway, 1994). We concentrate now on a summary of the telecenter movement in developing nations.

1) Africa

In 1996, the World Bank said: "If African countries cannot take advantage of the information revolution and surf this great wave of technological change, they may be crushed by it. In that case they are likely to be even more marginalized and economically stagnant than they are today." Fewer than five years later, one of the problems that confronts Africa is the coordination of international agencies and donor governments that are sweeping through Africa with ideas, schemes and money for universal access, Internet service providers and community telecenters.

As we have documented, Africa has clearly become the arena for the most aggressive efforts to develop ICT and build communication centers. The importance of ICT in Africa is symbolized in the creation of PICTA and the 1999 BICA (Building the Information Community in Africa) conference in Pretoria. A rally in March 1999 further emphasized the determination of Africans to be part of the Information Age. Called The African Connection, the rally was a 16,000-km journey from Tunisia to South Africa designed to suggest to the world that Africa offers "new and dynamic frontiers for the development of telecommunications and information technology." During the trip a telecenter was to be established in each of the countries in Africa through which the three rally cars would pass: Tunisia, Egypt, Sudan, Eritrea, Ethiopia, Kenya, Tanzania, Malawi, Zambia, Zimbabwe and South Africa.

One of the important problems facing a communication center is attracting clients. For example, in Ghana, much of the population in the community around a telecenter is illiterate (Jonnie Akakpo, personal communication). However, potential partners such as health workers and agricultural extension people also have insufficient familiarity, skills and experience with ICTs to appreciate and use ICTs' potential in their work. A new training facility established in Kenya in 1999 addressed this situation. The Regional Information Technology Training Centre (RITTC) provides training for health professionals in basic information technology, including email, CD-ROM, Web, and Internet technologies. It began by providing a basic three-day course (offered three times in 1999 and four more times in early 2000), and a three-day Information Technology Trainers course. The Trainers course is designed to prepare those who complete the basic course to train co-workers at their home institutions. The programs are open to nationals of Eritrea, Ethiopia, Kenya, Tanzania and Uganda. Two organizations have established the center: they are SATELLIFE HEALTHNET KENYA and SATELLIFE (USA). Funding comes from the InfoDev program.

2) Latin American and Caribbean Region

Some countries in Latin America, such as Peru and Mexico, have made some progress in establishing telecenter service. Networks of telecenters have been created, or are underway, in Peru (190 telecenters), Mexico (23 telecenters created in 1995, but with only five operating in 1999) (Robertson, 1998), Paraguay (eight telecenters), and El Salvador (about 100 basic telecenters to be created in the early part of the new century).

Apart from these networks, other projects can also be observed in the region — such as the two telecentre pilot projects supported by ITU in Suriname and Honduras. Two other pilot projects supported by APC, IDRC and local organizations have also been implemented in Ecuador and Colombia. Further examples include telecenters found in the Dominican Republic and Guatemala. And other telecenter projects are being planned in Honduras, Costa Rica, Haiti, Guatemala and Trinidad.

Based on a survey conducted by IDRC in 1999, Gómez, Hunt and Lamoureaux (1999b) identify "certain tendencies" concerning telecenters in Latin America:

1. Latin American telecenters are primarily funded by the private sector, local communities and NGOs rather than by governments.
2. Students appear to be the main users of telecenters although there are efforts to target marginalized populations in remote and poor areas.
3. Telecenters depend on a variety of sources for funding, which, along with technical problems, is among their major obstacles.

3) Asia and the Pacific

The arrival of telecenters in Asia will be assisted by the relatively strong telecommunications network that is either in place, or soon to be in place in Asia. In Singapore, telecenters are being established principally in libraries and community centers. India, though heavily endowed with a Silicon Valley-style industry in the South and with a strong reputation as a pioneer in satellite television and rural radio, was only slowly moving beyond the policy statement quoted earlier in this chapter. China, though trailing other countries, made rapid advancements in infrastructure development in the 1990s and at the turn of the decade both Internet service and communication centers were beginning to emerge in towns well beyond the major cities (*China Computer Newspaper*, 1999; Hua, 1998). Elsewhere in Asia, a number of initiatives such as PanAsia are focused on networking NGOs and international organizations in order to increase communication of best practices and reduce redundancy (IDRC Singapore, 1999). With major funding from the International Fund for Agricultural Development (IFAD), Electronic Networking for Rural Asia/ Pacific (ENRAP) is working with eight countries (Bangladesh, China, India, Indonesia, Nepal, Pakistan, the Philippines and Sri Lanka) to bring the benefits of the Internet to rural development projects supported by IFAD. The three-year pilot project is funded through 2001 (ENRAP, 1999).

Major Issues Surrounding the Diffusion
of Communication Centers

Four Ps consistently emerge in discussions related to the development of communication centers: policy, partners, participation, and planning. These begin our list of issues that can serve as the basis for guidelines or lessons being learned. This discussion points directly to two major concepts in the diffusion model: adoption and discontinuance (Rogers, 1995). In development circles, the practical translation of these concepts is *sustainability*. The short history of telecenters (and particularly telecottages) makes us wary of discontinuance, and much of the international support for telecenters, as we have suggested in this chapter, is predicated on the need to discover models and best-practices that will increase the chances for telecenters' sustainability.

Policy and political leaders. It is useful to have a policy framework such as the one announced in India in 1998 which can serve as a support and reinforcement for senior political leaders at the national, state and local level. Their public endorsement may be critical to mobilizing government officials, NGOs, business groups, and important resources. Governments may need to create the regulatory environment for investment in ICT. An appropriate policy and a regulatory body that implements that policy are important preconditions for successful and sustainable telecenters.

People in rural areas are generally poor, and many of them cannot afford telecommunication services. This is a vicious circle — few potential customers means high prices, which further reduces the number of potential customers, and so on. New wireless technologies offer more cost-effective solutions. However, often, the incumbent telecom operator hardly has the capital required for investment in infrastructure to meet needs in more profitable urban areas and there are policy and regulatory barriers for new entrants who might be willing to invest in rural telecommunications. The support of the international community is useful in developing decision makers' awareness of the need to adopt policies which promote the building of an information society.

Many countries are now in the process of developing policies to improve telecom penetration in rural areas. Such polices include license obligations to serve rural communities; subsidies by means of rural telecom development funds; variations of build, operate and transfer arrangements; low-interest loans; etc. The World Bank is currently supporting regulatory arrangements which promote communications in rural areas, in particular, appropriate licensing, interconnection, revenue sharing, and tariff arrangements. It is trying to promote solutions that have the capability of attracting private investment and expertise into rural areas. USAID, ITU, IDRC and others have been organizing regional seminars and workshops to raise awareness among decision makers and policy makers about the potential of ICTs to promote economic and social development in rural and remote areas. In dealing with the connectivity issue, the Leland Initiative also emphasizes starting with the top policy makers. However, to be useful, a policy does not necessarily have to be explicitly on ICT; several years ago FAO began promoting national policies related to development communication (Fraser and Restrepo-Estrada, 1998). Such policies can provide a good framework for creating telecenters.

Partnerships. Partners come in various forms. There are the big international partners who can provide funding, connections, and related experience, and there are important large in-country partners such as national and state governments that can provide political and financial support. The major actors such as IDRC and ITU have found it productive to have big partners such as FAO and WHO — as well as local partners such as cooperatives, libraries, municipalities and universities. Another kind of partner is the "champion." Based on limited experience with ITU's MPCT projects especially, Ernberg says that "the need to find local champions, who are motivated and able to drive the project, cannot be overstated" (Ernberg, 1998a). Champions are the people who endorse the telecenters, help give them visibility, provide leadership for their establishment, and recruit other individual and organizational supporters in the community.

Community participation. While in some cases, the introduction of ICT services such as telephones in rural areas have been met with eager users (such as the

An example of regional statement of policy that helps focus governments on goals is the "Vision for the African Information Society Initiative." It includes the following points:

- Information and decision support systems are used to support decision making in all the major sectors of the economy in line with each country's national development priorities.

- Every man and woman, school-age child, village, government office, and business can access information and knowledge resources through computers and telecommunications.

- Access is available to international, regional, and national 'information highways' providing 'off-ramps' in the villages and in the information area catering specifically to grassroots society.

- A vibrant business sector exhibits strong leadership capable of forging the build up of the information society.

- African information resources are available which reflect the needs of government, business, culture, education, tourism, energy, health, transport, and natural resource management.

- Information and knowledge are disseminated and used by business, the public at large, and disenfranchised groups such as women and the poor, in particular, to make rational choices in the economy (free markets) and for all groups to exercise democratic and human rights (freedom of speech and freedom of cultural and religious expression).

Source: Mansell and Wehn, 1998, p. 113

Grameen Phone), it is generally acknowledged that would-be clientele's participation in communication centers is not always spontaneous. Frequently observers have suggested that "access is not enough." Efforts need to be made to make telecenters understood, valued and used by the community. Nor is the concept of participation in regard to telecenter development absolutely clear. It is important to go beyond "participation" as an rhetorical, fuzzy concept (Heeks, 1999b; Cohen and Uphoff, 1977). Telecenter personnel should strategically and explicitly address effective ways the community can participate, and they should build participation into the planning operation. Some participation should begin before "start-up" with members of *different constituencies* being involved in the planning, including, for example, anticipating gender, class, and locality obstacles to access. Other roles for the community to play need to be clearly identified.

Planning. Planning has multiple dimensions. Most effort goes into the technical and facilities planning. Likewise, much training of communication center staff members is hardware and Internet related. In most cases of establishing information and communication access centers, planning needs to include the following aspects:

a) *Marketing of telecenters*. While communities may need and benefit from the resources of a communication center, systematic efforts are required to market (more than publicity!) the telecenter to the community, including school officials, health people, agricultural agencies, small businesses, and young people, the most likely early adopters. This includes basic marketing research and needs analysis, identifying persons who can provide user leadership. This involves linking telecenter resources with the *communities' perceptions* of need, positioning ICT as a useful commodity, and providing training for a core of potential adopters. Ernberg (1998b) notes that most people are unable to imagine the potential of ICT until they see and actually try out the tools. Evidence suggests that some people are willing to pay for ICT services if they perceive the advantages of them and the pricing is market-oriented. International and indigenous partners need to include marketing and community relations in their strategic planning, and factor in timing and resources for this marketing activity.

b) *Business planning*. Most communication centers, whether they are private sector or public sector, are charging fees for some services. Some centers depend heavily on public subsidies. Whatever the arrangement, a business plan needs to go along with the marketing strategy. An important additional P that needs to be considered under business planning is the issue of privatization. There is strong sentiment among some telecenter experts that private sector entrepreneurship needs to play a more prominent role in financial planning for telecenters.

c) *Orientation and training*. Training is an important and ongoing requirement for keeping up the vitality, support, relevance, and sustainability of a telecenter. Training applies to various stakeholders in the telecenter environment. They include: members of the center's governing body, telecenter partners, telecenter professional and volunteer staff, users of the telecenter, and the community at large. For the staff, training goes beyond ICT technical competence: it includes, for example, developing skills in research and evaluation, community relations and community development, information management, information production, and entrepreneurship. Local partners and potential partners in the community such as educational organizations, health facilities, schools, and NGOs need to understand the benefits of such a center to their operations and how they can play collaborative roles; these are training considerations. Communities have to work hard to make information and communication technology work for them. As Fuchs states, "People need to be encouraged to become involved in 'information-seeking behavior.' Simply put, people need to come to learn that it is worth their while to take the time and trouble to find information to help solve their problems" (Fuchs, 1998). For the community, training may involve providing an understanding of the opportunities and benefits of being part of the information society, and overcoming technophobia.

A review of communication experience also suggests other aspects of the telecenter movement that are vital to sustainability.

Dynamic product mix. CCCs need to be flexible about the services they provide. Factors that will influence a service's inventory (and dictate changes) include, for example, changes in technological infrastructure such as the introduction of an Internet service provider and telephone services, an increase in *individual* access to some ICT hardware (such as computers), availability of new products (for example, Web site construction services or low-cost telephonic connections), increased sophistication of consumer demand (for example, for distance learning opportunities), and expanding opportunities (such as collaboration with a consumer cooperative or a traditional library).

Relevant and meaningful information materials. While the Internet provides an enormous quantity of information and diversion — some of which may be useful to communities — a center that is designed to support community development should be aggressive and creative in *localizing* its knowledge and information resources. This is where local partners such as universities, schools, NGOs, broadcast stations, farmers' cooperatives and others can help produce information materials such as Web pages, audiotapes, videotapes, and printed materials by local people on local issues in appropriate languages. Centers also need the skill to add value to database and network resources by repackaging them (translating, simplifying, editing, and interpreting) for the community.

On the establishment of centers:

Starting-up centers. Communication centers have different kinds of beginnings. In many cases they are new, independent institutions in a community. In other cases they emerge as part of an existing entity, or they are transformations of previous organizations. Local bodies that have hosted communication centers include schools, municipal government bodies, cooperatives, libraries and local businesses. Entities that have been converted into communication centers often include more narrowly focused ICT organizations such as post offices, telephone centers, and libraries. Experience is too brief to know if there are different sustainability successes for the different approaches. Some evidence suggests that "stand alone" communication centers take longer to institutionalize and are more vulnerable to early demise. While other factors influence these situations, some advantages to building a center onto or from an existing organization are: it has an already established core clientele, an established position in the community, a ready-made organizational structure, and often a prime facility.

Networking telecenters. Communication centers in a country and across borders often have common interests and can benefit from economies of scale related to both hardware and software if they join in some kind of association. Telecenters that can link together can also share resources such as training and innovative ideas. A variation of this model is the establishment of a collection of centers under one organization. Another variation is the franchise model where a regional or national entity fosters the development and continuing services for locally owned and operated community communication centers.

Financing telecenters. Evidence and comments from the field suggest that telecenters supported exclusively by government grants is not a good sustainability formula, financially or politically. Telecenters need to be demand driven, and demand should be reflected in the community's willingness to pay for some services. Further evidence suggests that people are willing to pay for information that is perceived as valuable. In addition to start-up grants, governments (and NGOs) can contribute regularly to the centers by contracting with them for services. The need of a telecenter to obtain support from the public provides an important incentive to respond to community needs. Nevertheless provision may need to be made for providing some subsidized or free services, such as those related to health and basic education.

Evaluation. Evaluation is linked to the objectives of a project. Where profit is the driving motive, evaluation is a relatively simple matter. However, most telecenter projects have a community development dimension to them. Profits and losses may significantly influence the sustainability of an establishment but more sophisticated means of measuring social outcomes are necessary to determine the viability and worth of a center. For example, social change in the community, including factors such as health, education, self-efficacy, and overall family and community welfare need to be considered if communication centers are initiated as part of community or rural development programs. ITU's "pilot project" (research) approach looks at evaluation in terms of a center's social impact as well as its financial sustainability (Ernberg, 1998a). The challenge is to place a value on nonmonetary outcomes.

A Concluding Observation

Those who are contributing to making telecenters a dynamic movement include a wide range of participants, from large multinational agencies to small communities. One of the striking aspects of the telecenter movement is its simultaneous development as a top-down phenomenon driven by some of the major actors we have discussed and the grassroots less-well-documented, small, private, micro enterprises that are likely to play a major role in the 21st century: for example, the husband and wife in Uganda, who, using their own resources, acquired some computers and a telephone line and now have their own business. It is a story repeated hundreds of times around the world that needs more analysis than we have been able to provide in this chapter. Lying therein are probably some of the answers about making telecenters sustainable and responsive to community needs.

While we are seeing more private sector corporate support for telecenter operations in the United States, in developing countries there is relatively little visibility of support by private sector telecommunications multinationals except for Ericcson (Sweden), which has offered competitive grants for ICT projects, and Nortel Networks which, in October 1999, gave IDRC and Acacia CDN$2 million to set up two "centres of excellence in telecommunications for Africa" (Fleury, 1999). Many corporations have entered the mobile and cellular phone market in developing countries, but their more active participation in the establishment of telecenters, and especially telecenter infrastructure services, would substantially accelerate the movement at the town and village levels. In addition, two prominent UN agencies with significant experience in communication and who play important community development roles are missing among the community telecenter supporters: these are UNICEF (the United Nations Children's Fund) and UNFPA (the United Nations Population Fund). Their considerable contribution to our progress in using communication for social development during the last two decades of the 20th century suggests a greater role for them in community informatics in the 21st century.

References

Ballardini, A. (1998). E-mail communication to telecentre-1@lyris.idrc.ca.

China Computer Newspaper, Beijing, China. [Online]. Available: http://www.ciw.com.

City of Chula Vista. (1998). *Telecenters: The Workplace of the Future, Local Government Guide for Telecenter Development.* Washington: Public Technology Inc. [Online]. Available: http://ci.chula-vista.ca.us/tc.

Cohen, J., and Uphoff, N. (1977). *Rural Development Participation; Concepts and Measures for Project Design, Implementation and Evaluation.* Rural Development Monograph No. 2, Center for International Studies, Cornell University, Ithaca, NY.

Colle, R. (1998). "The Communication Shop, A Model for Private and Public Sector Collaboration Sustainable Too." Paper prepared for the Don Snowden Program Conference: Partnerships and Participation in Telecommunications for Rural Development: Exploring What Works and Why, Guelph, October. [Online]. Available: http://www.telecommons.com/documents.cfm?documentid=43.

Conradie, D. 1998. "Using Information and Communication Technologies (ICTs) for Development at Centres in Rural Communities: Lessons Learned." *Communicare*, 17(1), 97-116.

Crossette, B. (1999). "The Internet Changes Dictatorship's Rules." *The New York Times Week in Review*, August 1.

Ernberg, J. 1998, a, "Empowering Communities in the Information Society: an International Perspective." Richardson, D., & Paisley, L. (Eds.). *The First Mile of Connectivity*. Rome: Food and Agriculture Organization of the United Nations, 191-211.

Ernberg, J. (1998b). "Universal Access for Rural Development: From Action to

Strategies." Paper prepared for the First International Conference on Rural Communications, Washington, November 30 - December 2. [Online]. Available: http://www.itu.int/ITU-D-UniversalAccess.

Electronic Networking for Rural Asia/Pacific. (1998). [Online]. Available: http://www.enrap.com.

Fraser, C., and Restrepo-Estrada, S. (1998). *Communicating for Development*, London: I. B. Tauris.

Fleury, J.M. (1999). "Internet for All, the Promise of Telecenters in Africa." IDRC Briefing #3. [Online]. Available: http://www.idrc.ca/media/TelecentresAfrica_html#Teleaddresses.

Fontaine, M., and Foote, D. (1999). "Ghana: How You Can Use a Computer Without Owning One." [Online]. Available: http://www.TechKnowLogia.org.

Fuchs, R. (1998). "Little Engines that Did." [Online]. Available: http://idrc.acacia.com.

Gómez, R. Hunt, P. and Lamoureaux, E. (1999). "Focus on Telecenters: How Can They Contribute to Social Development?" [Online]. Available: http://www.idrc.ca/pan/chasqui.html.

Goodman, S.E., Burkhart, W. A., Foster, Mittal. A., Press, L. I. Tan, Z. (1998). *The Global Diffusion of the Internet Project, Asian Giants On-Line*. Fairfax, VA: The Global Information Technology Assessment Group.

Holloway, L. (1994). *Telecottages, Teleworking and Telelearning*. Stockholm: TELDOK.

Heeks, R. 1998, a, "Information and Communication Technologies, Poverty and Development." Institute for Development Policy and Management, Manchester, UK. [Online]. Available: http://www.man.ac.uk/idpm.

Heeks, R. (1999b). "The Tyranny of Participation in Information Systems: Learning from Development Projects." Institute for Development Policy and Management, Manchester, UK. [Online]. Available: http://www.man.ac.uk/idpm.

Hua, X. (1998). "Farmers in Hebei Like Computers," *Travel China*, 10(4).

INFO21. (1998). Information and communication technologies for development. [Online]. Available: http://www.undp.org/undp/info21.

International Development Research Centre. (1999). Homepage. [Online]. Available: http://www.idrc.ca.

International Telecomputing Consortium. (1999). Homepage. [Online]. Available: http://www.itc.org.

IDRC Singapore. (1999). [Online]. Available: http://www.idrc.org.sg.

International Telecommunications Union. (1998). *World Telecommunications Development Report — Executive Summary*. [Online]. Available: http://www.itu.int/ITU-D-UniversalAccess.

Jensen, M. (1998). "African Telecenters as Models for Rural Telecoms Development." Paper presented at the ITU Regional Seminar on Multipurpose Community Telecenters, Budapest, December 7-9.

Lee, P.S.N. (1997). "Uneven Development of Telecommunications in China." Lee, P.S. N. (Ed.). *Telecommunications and Development in China*. Cresskill, NJ: Hampton Press.

Li, D. (1999). News report. *China Computers Newspaper*, May 13. [Online]. Available: http://202.99.18.34/ciw/822/D0101.htm. (Translation: F. Yang)

Kapstein, E.B., and T. Marten, T. (1998). "Africa is Missing Out on a Revolution," *The International Herald Tribune*, September 24.

Mansell, R., and Wehn, U. (1998). *Knowledge Societies; Information Technology for Sustainable Development*. Oxford: Oxford University Press.

McIntyre, B. T. (1997). "China's Use of the Internet: A Revolution on Hold," in Lee, P.S.N. (Ed). *Telecommunications and Development in China*. Cresskill, NJ: Hampton Press.

MPCC Research Report. (1998). [Online]. Available: http://www.sn.apc.org/nitf/mpcc/introduce.html.

Richardson, D. (1999). "The Internet and Rural Development." Richardson, D. and Paisley, L. (Eds.). *The First Mile of Connectivity*. Rome: The Food and Agriculture Organization.

Robertson, S. (1998) "Telecenters in Mexico: Learning the Hard Way," *Journal of Development Communication*, 9(2).

Rogers, E. (1995). *Diffusion of Innovations*. New York: The Free Press.

Singh, R. (1993). *Communication Technology for Rural Development*. New Delhi: B. R. Publishing.

United States Agency for International Development. (1999). [Online]. Available: http://www.info.usaid.gov/regions/afr/leland.

Van Crowder, L. (1998). "Knowledge and Information for Food Security; the Role of Telecenters." Paper prepared for the Seminar on Multipurpose Community Telecenters, Budapest, December 7-9.

Whyte, A. (1998). Telecentre Research Framework for Acacia. [Online]. Available: http://www.idrc.ca/acacia/04066/index.html.

The World Bank. (1999). About the African Virtual University. [Online]. Available: http://www.avu.org/avusite/about/index.htm.

Zijp, W. (1994). *Improving the Transfer and Use of Agricultural Information: A Guide to Information Technology*. Discussion Paper 247, The World Bank, Washington.

Endnote

1. For another typology of telecenters, see Conradie, 1998.

Chapter XX

Virtual Communities, Real Struggles: Seeking Alternatives for Democratic Networking

François Fortier[1]
Independent Researcher, The Netherlands

Introduction

There is a technology that was said to have the "power to disband armies, to cashier presidents, to create a whole new democratic world — democratic in ways never before imagined, even in America" (From Daniel Boorstin's *The Republic of Technology,* cited in Winner, 1996, p.20). This technology was none other than television, whose potential for low-density mental reformatting is today more widely recognised than its affinity with democracy — in America as elsewhere. In fact, "Dreams of instant liberation from centralised social control have accompanied virtually every important new technological system introduced during the past century and a half" (Winner, 1986, pp.95-96). Collective memory is short, and information and communication technologies (ICTs) are now on the leading float of the technophile carnival. For many, the new technological artefacts promise to end the alienation of labour and industrial apocalypse, to leapfrog the so-called Third World into post-industrial *informationalism*, and to cast the foundations of slave-less, gender-balanced Athenian democracy (see notably Cairncross, 1997; Burton, 1997; Negroponte, 1995; Bissio, 1996; Annis, 1991; Lipnack and Stamps, 1986).

Yet, beyond the hype of the so-called *Information Revolution*, ICTs are having other implications, more tuned to neo-liberal substance than classical utopia. Those implications call for a critical political economic analysis and precocious system planning and deployment. On the one hand, this chapter compares the overall political impact of the technology in relation to the immediate advantages it is said to confer. On the other hand, the analysis shows that the development and implementation of ICTs, far from serving democracy, does in fact consolidate social injustice through ideological homogenisation, restrictive controls, and an enhanced capacity for surveillance. In search of alternatives, the last section of the chapter focuses on the technological conditions and political strategies through which information systems could be more relevant to progressive social forces and grassroots emancipation.[2] A matrix of relevant political issues is proposed in an effort to construct strategies of progressive community networking.

New Technologies, Old Tricks

The Agenda Setting

Optimistic perspectives on technologies pay little attention to the processes that lead to the formation of characteristics inherent to technologies. As a result, those views are often incapable of understanding the interests that sustain these characteristics in the first place. Technologies never occur fortuitously, they are always created and implemented through funding, research and development (R&D), production, improvements, commercialisation, and support, with purposes that become integral part of what they are, and can or cannot do. Not surprisingly then, dominant social sectors, those who control both investment and most technological development, have the opportunity to select options that are best suited to their own interest—not to a broader, vaguely defined, social progress (Stewart 1978, pp. 110-111; Noble, 1984, p. 195).

It cannot therefore be taken for granted that ICTs are necessarily beneficial either to a society as a whole nor, in particular, to groups already exploited and oppressed within a society. Analyses ought to shed light on the hidden agenda of such an alleged technological determinism, unmasking how social structures and relations come into play to develop and maintain a technology, fostering specific interests in the process. In order to understand all social implications of technologies, and ICTs among them, it is therefore necessary to understand multidimensional processes, with ICTs being affected and shaping social relations throughout their development and deployment (Stewart, 1978; Noble, 1977, 1984; Sussman, 1983; Feenberg, 1991).

From their very inception, information and communication technologies have indeed benefited most those sectors that invest in their development. Computer networking was born of United States military funding to serve related academic research.[3] Not far behind were corporate investors that recognised, long before the advent of the public cyberspace, the paramount importance of data flows for their industrial, commercial, and financial activities. Data flows permit them to improve research and development, facilitate design and manufacturing, adjust to markets, reduce stocks, transfer funds, trade commodities and control the workforce. Networking also increases the global mobility of corporations, further consolidating their power in relation to states and civil societies over taxation (Cairncross, 1997, pp. 266-270), labour rights, environmental regulations, and other conditions of investment. Most crucially, financial operations have been greatly improved by information systems, "unifying capital, markets, and leading to 24-hour 'follow the sun' trading" (Cowie, 1989, p. 24; see also Hills, 1990, p. 76). In fact, financial institutions are responsible for about 80% of global data communication (Sussman, 1997, p. 38).

Of course, there are numerous examples of independent initiatives that foster grassroots communication, information exchange, and alternative news coverage. But as the Internet increasingly commercialises, the dices are henceforth heavily loaded in favour of well-endowed corporate runners. For example, Internet service providers "are moving toward an ability-to-pay pricing structure and shunning periodicals that do not bring in large revenues, especially politically progressive publications that are not in vogue or do not have sufficiently large subscription bases" (Sussman, 1997, p. 182). The result is that alternative information providers are today swimming in rough seas of hostile sharks, with CNN Interactive, *USA Today*, MSNBC, and *The Financial Times* filling the limelight. Most North American and European media organisations have Web pages, many with around-the-clock updates. Furthermore, the advent (forced-fed on Windows 98) of push technologies and so-called "active channels" compel a pace that only large media organisations can keep up with, let alone satisfy the criteria of gatekeepers. In turn, search robots, used on the Internet to locate information, are now renting keywords at high prices to insidiously route users towards their highest bidding client sites (Roszak 1996). Needless to say, the grassroots and anything on the fringe of the commercial mainstream are not going to shine, if appear at all, on the road maps of the information highway and the yellow pages of cyber-emporium (see also Lee, 1997, p. 171).

While the potential of ICTs for social change is great, current trends in the development of the technology are less than encouraging for progressive activists and their communities. Through directed technological development, cyberspace is unequivocally becoming the domain of prominent commercial interests and their liberal discourses. The potential of ICTs thus needs to be confronted to the actual insertion of such technologies in society if we are to understand how they affect the *overall balance of power* between progressive sectors and other social forces. In that balance, the benefits of ICTs for the military, bureaucratic and corporate apparatuses are overwhelming if compared to the gains offered to progressive sectors of civil society. Metaphorically, "using a personal computer makes one no more powerful vis-à-vis, say, the National Security Agency than flying a hang glider establishes a person as a match for the U.S. Air Force" (Winner, 1986, p. 112). Yet even more concerning is that, irrespective of how much ICTs actually benefit progressive social sectors, current trends in technological development also present a serious threat on free information flows and, ultimately, democratic rights.

New Threats

It has been argued that ICTs, and computer networking in particular, are rendering censorship and monitoring of information difficult, if not impossible. This has supposedly eroded the ability of states, corporations, and other ideological gatekeepers to repress dissent and control civil society activities (Rheingold, 1994, p. 14; Swett, 1995). As a result, computer networking is said to break workplace hierarchies, foster freedom of the press, and widen political participation and avenues of resistance (Negroponte, 1995, p. 158; Cairncross, 1997, pp. 25, 148). Indeed, many authorities have been taken by surprise through spectacular civil society use of computer networking, from Tien An Men and Moscow to the Chiapas.[4]

By now however, both liberal and authoritarian régimes are taking measures to prevent electronic networking from further threatening their prerogatives. A wide range of policies and technical measures is already being implemented to subjugate ICTs, translating into control of access, flows, and content. These include regulations on trade and use of networking equipment and software, on operation of Internet service providers, and the surveillance of private communications and censorship of broadcast information. So-called secure gateways and other *censorware*, which permit monitoring and restriction of what users access through computer networks, have been operated by ISPs in the United States and countries of the European Union for many years (Lewis, 1996; Beeson and Hansen, 1997; ACLU, 1997a; EPIC, 1997). The same countries are now adopting a number of regulations dealing notably with illegal content and cryptography[5] (Thoumyre, 1996, pp. 57-63; Sédallian, 1996; Akdeniz, 1996).

While many rejoice at the checking of criminal activities in cyberspace — particularly child pornography and hate literature — restrictions and surveillance in these areas also conveniently lay the ground for more politically motivated controls (Lee, 1997, p. 172; Madson, 1997). In practice, censorware renders invisible a large proportion of the material on the Internet, blocking whatever is summarily judged morally or politically questionable by authorities or services providers, while also masking untargeted material. In the United States, a search robot offering a 'family-friendly Internet' has been found to filter 90 to 99% of otherwise available material (EPIC, 1997). Even when intended for parental control, censorware remains "clumsy and blocks out a whole range of legitimate non-obscene speech" (ACLU, 1997a). Casualties include gay and lesbian sites; scientific information containing sexually or anatomically explicit language; sites advising on safer sex practices, contraception, abortion, or several forms of cancers; as well as resources for victims and prevention of rape or human rights abuses.

Beyond moral sanctioning, overtly political implications of the censorware technology is just a step away. Many countries forbid public Internet access altogether or impose very high tariffs to limit usage by civil society (Tech Channel, 1998; PoKempner, 1997). Since 1996, Singapore has enforced surveillance and severe penalties to compel private users to access the Internet through proxy servers, censoring culturally or politically controversial material (Arnold, 1996). China, having first imposed prohibitive tariffs and then police registration (PoKempner, 1997) has now resolved to segregated and censored networks (Deron, 1996; Usdin, 1997).[6] Similarly, and since the Fall of 1997, Vietnam imposes certification for all levels of Internet access, service, and information providers, as well as for users themselves. In addition, the government called for Internet access providers to operate secured gateways that block viewing of blacklisted Internet sites and prevent the use of several TCP/IP protocols, including remote SMTP (mailboxes), NNTP (newsgroups), and Internet telephony.

Besides permitting restriction of information flows, ICTs are also tools of control. Used in industrial automation, ICTs allow management to minutely supervise, fragmentize, and de-localise production (Clement, 1988, pp. 224-233; Spencer, 1996). These practices further de-skill manual labour but also threaten part of the information workforce, stripping yet more workers of their individual and collective power. The workplace is frequently subdued to surveillance and monitoring through remote cameras, tracking devices, activity and communication recording, and the restricting of workflow by hierarchically controlled *groupware* (Clement 1988, pp. 228-232; ACLU, 1997b; Cairncross 1997, pp. 194, 274). Exploitation of labour increases through new contractual arrangements, dislocation of unions, and individualisation of workers in isolated units or locally bound subcontracting companies subsumed to global competition (Robins and Webster, 1988, p. 56; Castells, 1989, pp. 29-32; Gurstein, 1998).

The commodification of private information also surreptitiously threatens citizens in relation to state and private bureaucracies (Sussman 1997, p. 283). ICTs enhance the domination of producers and merchants over consumers, allowing personalised marketing and tailored consumerism to manipulate demand and fit it to the needs of supply, way beyond the fiction of 'friction-free capitalism.' In the sphere of social reproduction, ICTs are powerful tools of control in activities of surveillance, commercial and political behavioural monitoring (data warehousing), and bureaucratic functions in social services, taxation, immigration, intelligence, etc. (Robins and Webster, 1988, p. 61).

Digital communication is in fact easier to monitor than analogue voice, printed, or facsimile transmissions. While transactions leave digital footprints and are easy to wiretap, content can be automatically recorded, scanned, logged, sorted, screened, and indexed by recipients and originators (Winner, 1986, p. 115; Sussman 1997, pp. 194, 212; Warren, 1997). In this respect, the United States and other OECD[7] countries have made persistent attempts to restrict the use of cryptographic software, responding to pressure from police authorities who wish to maintain their current and

Table 1: Summary of Critical Arguments on Computer Networking[8]

Dominant Discourse	Critical Analysis
Technologies, and ICTs among them, are neutral artifacts that benefit the development of societies as wholes, despite being sometimes used for anti-social goals.	Technologies and ICTs are social products, only reflecting the economic interests & political agendas of those groups that have invested in their development & deployment.
Computer networking contributes to the democratisation of information by broadening access and diversifying sources.	Computer networking is increasingly dominated by high value-added media oligopolies that reproduce control & streamlining mechanisms previously found in other media.
Computer networking prevents censorship and monitoring of information, eroding the ability of authoritarian states, monopolistic corporations, & other gatekeepers to control ideologies & repress dissent.	Both the technologies and informational content of computer networking are increasingly being censored, monitored, regulated, and controlled by dominant sectors to ensure their ideological and political superiority.
ICTs are presented as tools of productivity that encourage democratisation through the need that modern economies have for free information flows.	Irrespective of their impact on productivity, ICTs are first & foremost tools of control over productive & socio-reproductive processes. Digital integration of different ICTs is making surveillance ever more pervasive & efficient.

often not-so-judicial ability to wiretap a large proportion of electronic communications (Sédallian, 1997; Wright, 1998). Beyond the discursive monopoly being increasingly secured through information and communication technologies, akin to processes in other electronic media, ICTs therefore also raise the dreadful spectre of active panoptic surveillance. In fact, "It would be extraordinary if a technology as rich in opportunity for extending control as office automation were not seized upon to exploit this aspect of its potential, even at the cost of some loss of productivity" (Clement, 1988, p. 243).

As the dust settles over the hype of information and communication technologies, the analysis above indicates that they do not emancipate work or democratise politics as promised. On the contrary, ICTs have been created, and are evolving, largely under the impetus of financial, commercial, and political interests of dominant sectors—not only the capitalist classes of OECD countries, but also bureaucrats, technocrats, conservative and patriarchal ideologues, landlords, merchants, professionals, and labour aristocracies in all societies. Co-optation of processes, control of information flows, and the increasing risk of panoptic surveillance indicate that, far from revolutionising social relations, ICTs mostly reflect and reinforce existing power structures. Under both authoritarian and liberal régimes, power-holders not only dominate the ICT agenda, but also have the means to obstruct alternative options, making it increasingly difficult to use technologies for emancipation beyond a narrowly defined ideological and political legitimacy. Attempts to develop civil society networking in many countries have led to mixed results, serving mostly foreign agencies, development workers, media correspondents, and corporate bureaux (see also Koert, 1998). Alternative strategies must therefore focus not primarily on the technology *per se*, but on the political relations that shape its social insertion—that is the actual control of both information systems and content.

Reaching Beyond Access

Many analyses of ICTs focus on the gap between the 'haves' and 'haves-not' of information and knowledge. They raise concerns about the *exclusion* of certain groups, nations or social sectors, which are left behind the so-called Information Revolution. The problem is examined internationally, as a North-South disparity, or sociologically, as the result of economic marginalisation. In either case, it calls for the rich (North, power-holders) to include and *empower* the poor (South, excluded), who risk further impoverishment if standing by global changes.[9] Strategies then seek improved accessibility through national information technology programmes and donor-driven development projects of public awareness, businesses support activities, infra-structural works, etc.

An approach increasingly taken to secure grassroots access to ICTs is to provide public facilities, such as multipurpose community telecentres (MCTs) with special interest to the grassroots. Yet, the conceptualisation and deployment of MCTs often ignores the needs of the alleged beneficiaries, and addresses priorities that are not necessarily theirs. Mostly located in town centres, MCTs are not likely to reach the most disadvantaged, who are often illiterate, geographically isolated, stereotypically excluded from fashionable innovations, and too poor to pay for research or training fees. Worst, the current form of ICTs, their implementation through MCTs, and the informational content they offer may systematically bypass grassroots sectors. For example, an MCT project officer once commented that the landless do not have land to apply information on irrigation, that the poorest peasants do not have assets to implement extension advises, and that poor communities cannot take advantage of micro-generator technology for lack of capital. From the officer's angle, there is therefore little point in providing those sectors access to information systems in the first place.[10] In other words, the poor are said to have inadequate needs for ICTs, rather than seeing how existing information systems provide inadequate content, through an inadequate medium, to the deprived.

With this kind of approach, merely providing access to yet another tool of information, without questioning its forms and content, only consolidates social inequalities. It further reinforces the power and wealth of merchants, landlords, patriarchs, and civil servants, at the expense of landless and small producers, women, and exploited age, ethnic, or racial sectors (see also Schuler, 1996, p. 274). Despite the best of intentions of much developmental work, providing access to technologies and content geared towards the needs of other sectors may be of little help. Worst even, ICTs being tools of control, they may also contribute to intensify the exploitation, oppression, and the ill development of many social sectors. The argument made here does not call for more rejoining onto the informational bandwagon by merely seeking ever-greater access to and integration in the Information Society. Rather, the analysis calls for resisting the forms this process and its technological support currently take, and to promote alternatives based on a progressive re-definition of what communication and information technologies should do.

Alternative Networking

Building Alternative Strategies

Despite the reservations expressed above on the democratic implications of ICTs, domination and repression are never laid in a social vacuum. Attempts by power-holders to direct the development of ICTs and their use for control necessarily beget resistance and new forms of social conflicts. While progressive civil society networking organisations increasingly suffer from marginalisation, they remain pivotal if an alternative development and use of ICTs are to be possible at all. For being too faint in a sky of largely commercial ISPs, alternative providers can barely challenge the hegemony of mainstream media, but still circulate critical information among those willing to listen and act upon it.

There are in fact many Internet and information service providers working for progressive causes, such as member networks of the Association for Progressive Communication and countless initiatives for community co-ordination and progressive advocacy. However, beside OECD countries, much of this work is done at the level of non-governmental organisations (NGOs) and professional activists, and still often bypasses progressive grassroots sectors. Following the argument made above on the overall political implications of ICTs, computer networking will remain little more than a sophisticated tool of communication with a limited impact for as long as it only serves organised professional sectors. To have any positive significance on that balance, networking technologies need to both reach and organically serve progressive sectors of the grassroots. As indicated above, we understand the grassroots as those organisations that emerge from and work in the interest of those social sectors that are both economically subsumed and politically oppressed.

Most importantly then, and instead of using the technology in its current forms and hoping that this will suffice to democratise information, progressive networks must also emphasise the appropriation and redesigning of technologies. This has been the strategy of some progressive developers and ISPs for well over a decade. They have invested into the research and development of appropriate low-cost and cooperative tools of communication, such as the robust Fido and Unix-based mailing and electronic conferencing software (see notably Afonso 1996). Some organisations have been able to carve inroads into the dominant agenda of technological development, and serve civil society with cheap and open fora that better match its specific needs and political objectives. Without these efforts of technology appropriation and alternative analysis, left to dominant interests and so-called market forces, computer networking could have evolved into something very different, likely to be much less interactive and co-operative than it has managed to become. This shows that, despite the grim political economy described above, resistance and efforts of alternative development can partly reshape ICTs to better assist progressive movements.

But a democratic option for computer networking remains to be bitterly struggled for. The work already done must be pursued and reinforced as an enduring effort of appropriation that meets the specific and changing needs of progressive social movements, while taking advantage of their resources and strengths. This work requires the development of a coherent information and communication strategy, based on a critical analysis and a clear political agenda. As we have seen, both the forms given to ICTs, and control over them, mostly mirror social relations of a given historical moment. Strategies of technological appropriation cannot therefore be taken in abstraction, underlining that a progressive use and appropriation of ICTs should only be understood, planned, and emerge as one of the many organic instruments available to ongoing social struggles. While others (notably Schuler 1996, pp. 253-375) have looked at the organisational and technical aspects of grassroots or community networking, the sections below propose some of the political economic issues to consider for the consolidation or establishment of such networks.

Progressive Grassroots Ownership

To the extent it is logistically possible, alternative computer networking strategies should focus on appropriation by the progressive grassroots themselves, who are best placed to define the specific design, use, and informational content most appropriate to their own agenda of political and economic emancipation. This is particularly true for the most deprived sectors. Both urban and rural poor, for example, have elaborate survival strategies that already include non-electronic but community-based information networks. For activists, researchers, and development workers alike, implanting ICTs over existing information processes prescribes the adoption of participatory methodologies throughout the elaboration and implementation of information systems, including identification of current forms of oppression, technical and informational needs, constraints, resources, and organisational or institutional alternatives. It also implies grassroots ownership and self-management, and participatory monitoring and evaluation of those systems, without which adequacy to actual needs and the danger of co-optation remain problematic. This of course means the rejection of top-down approaches and externally induced brands of empowerment, often based on dubious technical assistance with more or less hidden industrial and commercial agendas.

Grassroots ownership also means that ICT systems must seek to reach the weakest sectors *within* non-governmental or community-based organisations. Those organisations are far from devoid of internal power struggles, mostly replicating larger social structures of gender, ethnic, age and class relations. Management and accounting staffs are often the first to obtain computers within NGOs, while ICT support units always retain much control over the use and development of the technology. Differences are also prone to appear between peer organisations, where the better endowed in capital and staff can dominate less fortunate ones (Fortier, 1996b, pp. 149-151). These risks, internal to the dynamics of social movements, need to be considered carefully in the implementation and use of ICTs to ensure that progressive social objectives are met not merely benevolently but also systematically.

Designing Technologies for Actual Needs
To be of relevance to progressive grassroots and activists, ICT-based information systems must clearly identify the informational needs of those sectors and address them with appropriate content and forms, while preventing the co-option, ostracism, and subjugation these technologies imply. In a way similar to what is done for conventional media, a computer networking strategy needs to design relevant information audits (a practice and needs assessment) and constant monitoring and adjustments of both technology and content. The audits need to take into account, however, that computer networking offers new tools and forms that will open avenues for new needs to be addressed, often of more interactive and grassroots-controlled nature.

Furthermore, a progressive networking strategy must raise awareness, improve access and training facilities, and provide digital documents and archives, alternative information fora, and autonomous utilities for searching and indexing these resources. Much of this is already being done by some progressive networks and should be further expanded. Examples of initiatives addressing at least part of these issues include communication bridges such as rebroadcasting through community radio, street theatres (Huyer, 1997), and public libraries (see UNESCO, 1996, ¶ 14-15; Kole 1998b, pp. 354-355). Support centres that provide informational, technical, and financial services are also being deployed (Agre, 1997). In addition, there are initiatives and proposals for progressive distance-learning education, virtual cross-boarder labour councils, and early-warning human rights networks (Lee, 1997, pp. 174-186).

One important step in developing and providing adequate forms of technologies and content, for example, is the ability of computer networks to carry messages and documents in languages other than English (Schuler, 1996, p. 289). This was not the case for at least the first decade of telematics, largely due to the lack of investment in the necessary appropriation of software. This posed serious difficulties for many progressive organisations. Activists of *Rede Mulher*, a feminist organisation based in São Paulo but active all over Brazil, emphasised back in 1992 their need for clear and aesthetic documents used for public newsletters. Not being able at the time to transmit Portuguese diacritics made the use of Alternex, the then Brazilian NGO-based ISP, unpractical for most of these communications.[11] In contrast, another progressive ISP, the Rome-based *Agorà Telematica*, had prioritised multilingual features in the development of its DOS-based interface, and provided services for users across Europe in seven languages, including Russian in the Cyrillic alphabet. The proper use of multiple languages has now been addressed by corporate investment into more flexible operating systems, such as the increasingly ubiquitous Windows.[12] Yet the issue illustrates how appropriation can increase the responsiveness of ICTs to the needs of progressive sectors, without lagging on the trail of corporate-driven initiatives for possible beneficial fallout.

Addressing Constraints Through Appropriation

Many obstacles in acquiring and using technologies the way they are, and transforming them to better suit progressive civil society organisations, remain to be addressed. Besides responding to needs, appropriate ICTs must therefore also consider the constraints of low-income, polyvalent, time-stretched grassroots users and progressive civil society activists. There are financial, technical, organisational, political, and legal issues to overcome, while bypassing current and increasing controls over equipment, software and information flows. To address these conditions, alternative ICTs require several characteristics shared by other appropriate technologies.

Financial Resources

While computer networking is a relatively inexpensive technology, both in terms of initial capital investment and in operating costs, it remains out of reach for quite a few small NGOs and most grassroots organisations of low-income countries. At least until very recently, the price of a microcomputer and peripherals adequate for networking could range from US$1,000-2,000. Moreover, markets, taxes and import regulations could significantly affect retail or black market prices, while many labour and peasant organisations, among others, often cannot afford a phone line, let alone computers for accounting or other tasks of office automation. The operating costs of phone lines, which usually cross-subsidise local by long-distance calls, also penalise rural users who need to call urban centres where most computer hosts are located. An additional and potentially important cost is of the Internet service provider. Except in affluent OECD countries, charges are normally calculated on the basis of logged time or quantity of transmitted data, which can add up very quickly beyond affordability (Fortier, 1996b, p. 95).

Taking all these charges into account, the opportunity-cost of the technology depends on the overall telecommunication needs of institutional users, both for e-mail and information gathering or publication (see notably O'Brien, 1989; Koert, 1998). Large and medium-size NGOs seem to have both sufficient needs and available capital to justify the use of computer networking. Organisations with little or no communication abroad, on the other hand, might find the cost unnecessary, even if already equipped with computers, since the sole value of the information on

Web sites, fora, and databases is often not obvious to potential users. The information available on these media thus has to be better known and address more closely the needs of those organisations. In a similar logic, this information has to be cheaper and more convenient than through other media, also taking into account eventual research and translation costs associated with on-line retrieval.

Financial strategies are obviously crucial to the viability of progressive ICT initiatives, but must address particular difficulties. Information systems need much capital to invest in research and development, infrastructures, equipment, training, management, etc. (funding of community networks is extensively discussed in Schuler 1996, pp. 366-374). Yet, unlike most dominant sectors and ideological gatekeepers, progressive and grassroots organisations do not usually have the luxury of large-scale funding or market demand to finance their activities. Squeezed between the limited purchasing power of their constituency and the need to remain economically viable, alternative networking initiatives must then control costs by making cheaper technological choices. This implies that systems should be flexible to various hardware and software platforms, do not necessarily require state-of-the-art equipment, use free or cheap software, adopt intermediate technologies with long life cycles, and piggyback on existing infrastructures, using the Internet but also direct phone, data lines, and airwaves whenever appropriate (see notably Lee, 1997, pp. 85-87). Corporate monopolies in telecommunication, software, and equipment must also be challenged through revised intellectual property rights (Verzola, 1998) and diversified research, development, and supply (as with the recent burgeoning of *Open Source Software* [OSS] like the freely available Linux operating system).

Most crucially, the dominant liberal discourse largely favours project sustainability based on user fees and market viability. Yet such basis can only lead, sooner or later, to the development of institutional interests—such as self-reproduction, expansion, and career development—that may come in contradiction with the initial political objective of progressive grassroots networking. This is true, for example, of the contradictions between user fees and universal access, between support services to commercial élites and political struggles from deprived sectors, or, often, between sponsorship by dominant community leaders and women's reproductive rights groups. Progressive networking thus needs to be financially supported by organised solidarity, based on the social and political value of their work rather than a necessarily balanced self-financing dictated by market profitability.

Information Infrastructures

The increasing sophistication of Internet infrastructures in the OECD often leaves users in low-income societies with difficult access to on-line information. The World Wide Web protocols, for example, require a live connection between users and an Internet host, and greatly benefit from a wide bandwidth. Much networking beyond the OECD is not, however, of that nature, still offering mostly store-and-forward e-mail services, or being limited by narrow data lines. By January 1997, there were an estimated 71 million Internet users worldwide, but only half of those with full access.[13] Since low-income countries often suffer from expensive, incomplete, narrower, or geographically circumscribed connectivity, their future access to the newest protocols of multimedia information exchange will also necessarily remain problematic.

In this particular case, an example of how technologies can more adequately address the needs of civil society is the deployment of automated Web-to-email servers, which permit store-and-forward systems to retrieve, albeit cumbersomely, information provided on World Wide Web sites and other Internet media such as newsgroups. By sending specific commands and site addresses through e-mail messages, the automated servers of systems like GetWeb of SatelLife retrieve the requested Web pages and return them as text messages through e-mail. This is necessarily time consuming, but still permits Web searches despite limited connectivity (Richtel, 1998). In this way and through other ICTs, SatelLife has been able to provide medical information to areas where infrastructures were deficient or even declining. Here again, however, it remains to be seen to what extent such systems address the needs of the most deprived sectors of these areas. Critics have underlined, on the one hand, that the medical information made available on such systems may be of little relevance to basic community health in hygiene, sanitation and drinking water, while diverting precious resources (Pruett and Deane, 1998). On the other hand, and as with the case of community telecentres cited above, the approach here also requires users to adapt to a top-down Web technology, rather than develop information systems that provide multidirectional flows irrespective of Internet constraints.

Beyond the nature of connectivity itself, other infrastructure considerations have to be taking into account. For example, most civil society users in low-income countries access computer networking through public or organisational rather than private facilities. This implies that the location and availability of these facilities is a determinant factor of overall accessibility and usefulness, and needs to be considered when designing services relevant to the intended beneficiaries. Staffs of progressive organisations are also often overloaded with multiple functions and insufficient resources for seeking support. In other instances, the time of non-professional activists must be shared between their political, income-generating, and household work. The design of alternative computer networks has to take into account, for example, limitations on time and mobility that parents, most often women, are facing when caring for children (Marcelle, 1998, ¶ 65). This may imply developing tools and practices that reduce the time needed for managing e-mail inboxes, searching and retrieving information, or maintaining programmes and hardware. It may also imply the provision of additional services such as transportation or kindergartens (APC, 1997).

Intermediate and Robust Technologies

Progressive civil society and grassroots organisations are, more often than not, working in less-than-optimal conditions, with power and communication infrastructures of variable quality. This implies that appropriate ICTs need to be simpler, more robust and flexible, rather than performing but sometimes fragile. Appropriate technologies should then function properly in hard-working conditions, under intensive demand by several users at once or sequentially, and endure unreliable and fluctuating power supply, noisy analogue or multiplexed digital phone lines, extreme conditions of dust, humidity, heat, etc. (see notably Marcelle, 1998, ¶ 65). For example, electronic equipment and data files can now be relatively protected from electrical surges and cut-offs, while communication software permit data transmission even on lines of very poor quality or with significant satellite or switching delays. Some of these technical improvements have been done by grassroots developers, such as the Fido store-and-forward e-mail and conferencing system that allows connectivity in many areas of poor infrastructures.

Yet, intermediate technologies may not necessarily be the most appropriate in all circumstances. Particularly in a field such as ICTs, the rapid pace of innovation implies that newer technologies may provide more robustness and flexibility, along with performance and cheapness, despite a lesser diffusion, than older technologies. A judicious choice thus needs to be made between proven, well-known and diffused technologies that can be manufactured and maintained easily, and newer, more elaborate, often more expensive options that may however, in some cases, be more robust and flexible. In all instances, the development and deployment of alternative ICTs should be directed by transparency rather than obscure technological sophistication. ICTs can then be distributed rather than centralised and permit autonomy rather than create technical dependency. While this may incur trade-offs in efficiency in the short term, the benefit over time will be a better socio-technical integration.

Skills and Training

Networking necessitates basic computer literacy and additional specialised training. Many organisations, despite owning computers and telephones, may have little knowledge of the potential of networking, or lack the time for their already overworked staff to acquire the necessary skills. Furthermore, and until recently, communication peripherals and software have generally been less than user-friendly, requiring much devotion and patience on the part of staff, both in the learning process and in facing the unavoidable teething problems of new systems. Too often, progressive alternative service providers have neglected user-friendliness and support, relying on the personal initiative and perseverance of users to overcome technical and managerial obstacles. This negligence might have been the single most important impediment to the spread of computer networks until at least the commercial expansion of the Internet, and needs to be given due consideration.

In specific contexts, computer networking also needs to adapt to the modality of communication used by given social sectors, rather than impose its linear format to them. Some individuals, or entire social groups, may be illiterate, experience difficulty with abstractions, interactivity, or multidirectional links. Others, including many activists, simply remain uncomfortable with written texts (Lasfargue, 1989, p. 10; Fortier, 1996b, p. 103). Yanomamo communities, for example, mostly utilise oral forms of communication and have made significant use of video recording and facsimile transmission technologies. In contrast, few of their organisations were among the early comers to computer networking.[14]

Work then needs to be done to improve accessibility and system relevance, for both collection and dissemination of information, for users with no or little literacy or inaptitude with linear text argumentation. This can be addressed through research, development, and deployment, for example, of more explicit Graphic User Interfaces (GUIs), robust voice recognition and electronic speech devices that can operate in several languages, or proper use of touch-sensitive screens (Marcelle 1998, ¶ 64; M-Powa, 1997). Specific and adapted training also needs to be devised, addressing the particular constraints and aptitudes of trainees, including attitudes towards technological processes that may differ between sexes (APC, 1997), age, class, ethnicity, or other sectors, depending on each society. Training should also account for constraints such as physical mobility or time limitations due to professional or family obligations (APC, 1997). Research and development needs to be conducted as a systematic effort by progressive social movements, since market-driven corporate investment will not likely pay attention to the needs of marginal sectors that can only express a weak commercial demand.

Information Management

Through computer networking, the mass of information, rather than its scarcity, often becomes problematic. The lack of organisational preparedness for absorbing, managing, and contributing to information flows can then become a serious obstacle to the adoption and usefulness of ICTs. This should be addressed with adequate training and set-ups for demand-driven information retrieval, doing away with older, supply-driven, practices. In addition, and beyond information overflow, there remain difficulties in gathering, analysing, and providing knowledge generated at the grassroots level and among activists. This is particularly true for organisations that may have only one Internet account for many employees, or only a few computers, most likely reserved for accounting and word processing. As a result, many workers (particularly support staff, that are women more often than not) may be marginalised and left without effective access (Kole, 1998b, p. 350; Fortier, 1996b, p. 105). E-mail and forum messages, as well as Web searches or postings, must therefore be properly circulated to and from individuals concerned (Frederick, 1993, p. 202).

The lack of such information management practices could entail significant wastes and perverse political implications to organisations with limited resources and progressive social objectives. Alternative system developers and service providers should emphasise the provision of training and guidance for information management whenever appropriate which include adequate bidirectional routing mechanisms. The latter are still insufficient but yet necessary to ensure that ICTs are geared towards multidirectional exchanges rather than top-down broadcasting.

Security

Both potential and actual users of computer networking have expressed concerns about the confidentiality of the information being circulated through the technology and the security of systems they come to rely upon. Although fears of eavesdropping by service providers themselves are often unjustified, we have seen that there are significant grounds for suspicion. In fact, progressive civil society organisations are likely targets of political surveillance occurring beyond service providers, or of repression and sabotage from a range of antagonists. Feminists and reproductive rights activists, for example, have long been victims of abusive e-mail messages (APC, 1997), while political organisations have suffered significant financial losses or system failures due to massive e-mail bombing (Mason, 1997; Fortier, 1996a, p. 240-241). In turn whatever the actual level of that surveillance, which remains difficult to evaluate due to its concealed nature, the concern it generates has had a significant impact on the development of civil society networking (see also Schuler, 1996, p. 265).

In this context, the use of cryptographic software should be facilitated among progressive organisations, ensuring confidentiality and facilitating exchange. By shifting the location of information resources under restriction, progressive service providers can also contribute to baffle monitoring and circumvent censorship, secure gateways, and the blacklisting, screening, tracking, identification, and repression of political dissent.

Taking Advantage of Resources and Strengths

Beyond paying attention to needs and constraints, alternative ICT developers and service providers can also take advantage of some specific resources and structural or organisational strengths characterising progressive social movements. The latter often enjoy broad-based constituencies, which may imply a large number of decentralised nodes and participants. This may provide an invaluable network for logistical support or information gathering and dissemination. This is particularly true of indigenous knowledge on agriculture, health, appropriate technologies, social relations, etc., that represent a still largely unused asset of alternative networking (Kole, 1998a). Furthermore, and due to the voluntary nature of much of the progressive and grassroots activities, human resources are likely to be diversified and dedicated, much to the contrast of the information systems of dominant sectors.

The very constraints that progressive networks often have to endure may also prove to be positive incentive for alternative uses (Pruett and Deane, 1998). For example, and despite capital, training, and operation costs, it is revealing that organisations with fewer economic resources are often the most enthusiastic about the appropriation of computer networking. Where information and telecommunication expenditure are not compressible, new ICTs offer quantitative savings and qualitative access improvements whose marginal implications are commensurately more important to low-income organisations than their richer counterparts (Kole, 1998b, p. 356). In fact, delays in appropriation have often been caused by OECD organisations that lag in their embracing of computer networking.[15]

Example of an Alternative Networking Strategic Matrix

Table 2 presents, in the form of a matrix, the hypothetical example of issues that may be raised while planning for an information network addressing adolescent reproductive health and rights. Deployment of such a system can address a series of political objectives and identify needs, constraints, and strengths, while proposing relevant strategies for adolescent emancipation through reproductive rights and health concerns.

Table 2. Alternative ICT Strategies for Adolescent Reproductive Health and Rights

Political Objectives	Needs	Constraints	Strengths	Strategy
Allow adolescents to gain power & autonomy in relation to adults, so that they can both access information & make choices on their sexuality.	Multidirectional flows of information relevant to adolescent sexual & reproductive health (SRH).	Resistance of conservative sectors, and risk of co-optation by sectors opposing reproductive rights (RR) to further consolidate their discourse and repressive practices.		Provide networked information through adolescent content building relevant to SRH and RR. Select adequate software for messaging, conferencing, chatting, and databases. Provide training in information management for adolescent organisations, ensuring bi-directional information flows within them. Provide technical support:on-line assistance to beneficiaries; search engine facilities; software packages on CD-ROM & dedicated sites, support centres or telecentres, etc.
	Relevant infrastructure set-ups for access points, reaching the youth in their own spaces & practices:			
	Accessible locations.	Little or constrained mobility of teenagers.	May enjoy significant freedom of movements within small geographical areas.	Possible location in each social context: youth centres, libraries, schools, public canteens, health centres & hospitals, markets, sport centres & playgrounds, re-education & transition centres, closed houses, sex shops, bars, night clubs, stores, places of cult, refugee camps, etc.
	Adequate hours of operation.	Schedules constrained by other (often institutionally imposed) activities.	May enjoy more leisure time.	Adequate hours of operation for youth in their cultural context. Providing acceptable waiting time.
	Adequate privacy.	SHR and RR are topics for which showing interest can be embarrassing.		Ensure private workstations and provide anonymous accounts. Ensure, if necessary, discrete access to facilities.
	Accessible prices.	Very limited financial resources of adolescents	Availability of voluntary labour and group solidarity.	Establish acceptable service pricing, subsidised if necessary and possible. Consider public sources and civil-society fundraising.

Table 2 continued

Political Objectives	Needs	Constraints	Strengths	Strategy
	Sufficient support.		Constituency is likely to be quick learning and self-teaching.	Sufficient support staff. Establish in conjunction with other services?
Allow female adolescents to gain power & autonomy in relation to male adolescents, so that they can make & enforce choices about the reproductive & health consequences of their sexuality.	Overcome imbalances that may prejudice against female use and benefit of computer networking.	Consider the relative weight of specific constraints on female adolescents: literacy, education, skills, languages, availability, peer pressure, financial resources, & mobility.	Computer network may be able to overlap with existing networks among female adolescents.	Establish female-friendly ICT strategies as needed, based on the elements listed above for adolescents of both sexes (some of these gender-specific elements are considered in Karelse, 1998, and Huyer, 1998).
Allow adolescents of marginalised social sectors (ethnic & religious minorities, lower casts, landless peasants, informal sector labour, etc.) to gain power in relation to politically and economically dominant sectors.	Overcome imbalances that may prejudice against the use & benefit of computer networking by marginal sectors.	Consider the relative weight of specific constraints on marginal groups: literacy, education, skills, languages, availability, stereotypes, physical proximity, age, etc.	Group identity & sectoral political consciousness may be useful to organic networking.	Sector-sensitive networking strategy based on constraints and strategies elaborated for adolescents in general.

Table 2 continued

Political Objectives	Needs	Constraints	Strengths	Strategy
Facilitate the advocacy work of progressive activists, NGOs, governmental services & international agencies in the programming of adolescent SRH and RR services.	Resource sharing & other managerial logistics of network operations.	Bureaucratic mechanisms		Integrating of programme activities with larger informational services to the youth and other sectors, notably within telecentre programmes.
	Research & programming.	Limited funding	Professional resources	Establish content building of databases and structures of fora and chat rooms, and secure ownership of content by reproductive rights activists. Selection of user and server software, CD-ROM packages, support centres, etc.
	Public information that valorises progressive SRH & RR discourses & criticise conservative ones.	Obstacles to relevant information flows.	Available sources	Extend information flows outward from network, through the media, re-broadcasting by community radio, street/school theatres, etc. Reinforce dialogue with opposition.
Monitoring & evaluation of SRH & RR programmes.	Participatory monitoring & evaluation through interactive mechanisms between network users & programme administrators.	Range & sampling limited to network	Ease of bi-directional access & open communication.	Indicators for monitoring & evaluation of Adolescent SRH projects could be significantly improved though ICTs: "There are new techniques available for such rapid assessment, including the collection, analysis & dissemination of anonymous questions & comments from adolescents. The participation of young people in planning & carrying out such research is crucial to its success" (UNFPA 20 April 1998, p. 23, ¶ 64).
System Reproduction	Need for system sustainability?	Costs are likely to be much higher than revenues	May have many avenues of reducing costs	Secure adequate mechanisms of management & funding.
	System monitoring & evaluation.		Participatory monitoring can be built in the system.	Secure proper technology & processes for participatory & continuing M&E.

Conclusions

Technologies are shaped by, and inherently serve, the social interests of those groups that have made such technologies possible. It cannot therefore be taken for granted that information and communication technologies will necessarily assist progressive activism and grassroots emancipation. In fact, the overall political implications of ICTs, in their current forms, clearly are detrimental to social justice. The technology is increasingly co-opted under dominant discourses, while alternative information can be restricted or completely censored, and electronic surveillance is ever more refined and ubiquitous. While corporations are fencing the common pastures of cyberspace, liberal democracies and authoritarian régimes alike are taking measures to prevent computer networking from breaching their ideological dominance.

These current trends call for much precaution in the deployment of ICTs, and the urgent promotion of alternative strategies. Information and communication technologies can be less oppressive, and have the potential of addressing the information and communication needs of the grassroots. Progressive organisations that establish computer networks have a crucial role in this play. They granted an edge, albeit temporarily, to on-line activism over the previous 15 years. By now, and despite the strength of mainstream media, they continue to provide alternative information, crucial to the logistics of progressive movements. Most important for the overall political balance, and through technological appropriation, different forms of ICTs and networks can more widely provide critical and alternative analyses, loosen information flows, and resist against pervasive surveillance.

But a networking strategy by progressive sectors must go beyond merely improving access for a wider public. To appropriate ICTs and computer networks to the needs and constraints of progressive social movements, many technical, organisational, and political obstacles must be addressed by users, developers, and service providers through adequate human and capital investments. This process has and must continue to transform the technology in close interaction with the social movements it seeks to strengthen, basing its practices on a demystifying political economy of technologies. If progressive information and service providers are able to develop new communication systems, responding to the needs of landless peasants, prostitutes, street children, secluded women, and other exploited and oppressed groups, then ICTs may start to benefit them more than perpetuate their alienation.

References

ACLU (American Civil Liberties Union). "Internet Free Expression Alliance Counters Censorware Summit". 5 December 1997a. http://www.aclu.org/.
ACLU. "Wayne State University Prohibits Net Use for Non-University Related Work". 5 December 1997b. http://www.aclu.org/.

Afonso, C.A. (1996). "The Internet and the Community in Brazil: Background, Issues, and Options," *IEEE Communications Magazine*, 34(7), 62. http://www.cg.org.br/artigos/artigos.html.

Agre, P. (1997). "The Next Internet Hero," *Technology Review*. November-December.

Akdeniz, Y. (1996). "Pornography on the Internet," *L'Internet Juridique*. http://www.argia.fr/lij/articleJuin11.html.

Annis, S. (Fall 1991). "Giving Voice to the Poor," *Foreign Policy*. (Washington). (84), 93-106.

APC (Association for Progressive Communications). (1997). "Global Networking for Change: Experiences from the APC Women's Programme." http://www.gn.apc.org/gn/women/temp/survey3.html.

Arnold, W. (1996). "Asia's Internet Censorship Will be Easy to Circumvent," *The Wall Street Journal Interactive Edition* (Asia). 11 September.

Beeson, A., and C. Hansen. (1997). *Fahrenheit 451.2: Is Cyberspace Burning? How Rating and Blocking Proposals May Torch Free Speech on the Internet*, Wye Mills, Maryland. ACLU. http://www.aclu.org/.

Bissio, R. (1994). "Cyberespace et démocratie," *Le Monde Diplomatique*, 41(484), 16-17, July.

Bonchek, M.S. (6 April 1995). "Grassroots in Cyberspace: Using Computer networks to Facilitate Political Participation." Political participation project, MIT Artificial Intelligence Laboratory, Working Paper 95-2.2: Presented at the 53rd Annual Meeting of the Midwest Political Science Association in Chicago, IL.

Burton, D.F. Jr. (Spring 1997). "The Brave New Wired World," *Foreign Policy* (Washington), 106, 23-37.

Cairncross, F. (1997). *The Death of Distance: How the Communications Revolution will Change our Lives*. London: Orion Business Books.

Castells, M. (1989). *The Informational City: Information Technology, Economic Restructuring, and the Urban-Regional Process*. Oxford: Basil Blackwell.

Castells, M. (1998). *End of Millennium*. Oxford: Blackwell.

Cleaver, H. (Summer 1994). "The Chiapas Uprising," *Studies in Political Economy*, 44, 141-157.

Clement, A. (1988). "Office Automation and the Technical Control of Information Workers." Vincent Mosco and Janet Wasko, *The Political Economy of Information*. Madison: University of Wisconsin Press.

Cowie, J.B. (December 1989). "Entering the Information Age: Implications for Developing Countries," *IEEE Technology and Society*, 8(4), 21-24.

Deron, F. (30 September 1996). "Pékin met la multimédiatisation sous haute surveillance," *Le Monde*, Supplément Multimédia.

D'Orville, H. (1996). *Technology Revolution Study: UNDP and the Communications Revolution, Communications and Knowledge-Based Technologies For Sustainable Human Development*, New York, UNDP. http://www.undp.org/comm/techn.htm.

EPIC (Electronic Privacy Information Center). (December 1997). *Faulty Filters: How Content Filters Block Access to Kid-Friendly Information on the Internet*, Washington, EPIC.

Frederick, H.H. (26 June 1991). "Breaking the Global Information Blockade Using The Technologies of Peace and War," Conference on Computers for Social Change: Tools for Progressive Action Hunter College, New York, NY.

Frederick, H.H. (1993), *Global Communication and International Relations*. Belmont, CA: Wadsworth Publishing Company.

Feenberg, A. (1991), *Critical Theory of Technology*. New York: Oxford University Press.

Fortier, F.. (1996a). "Living with Cyberspace: Vietnam's Latest Dilemma." Dan Duffy. (Ed.). *North Viet Nam Now: Fiction and Essays from Ha Noi* (Viet Nam Forum 15), Yale University Council on Southeast Asia Studies and Yale Center for International and Area Studies.

Fortier, F. (1996b). *Civil Society Computer Networks: the Perilous Road of Cyber-Politics*, Toronto, Ph.D. dissertation, York University.

Greenberg, L.T., and S.E. Goodman. (July 1996). "Is Big Brother Hanging by His Bootstraps?". *Communications of the ACM*, 39(7), 11-15.

Gurstein, M. (1998). "Information and Communications Technology and Local Economic Development: Towards a New Local Economy." Gertrude MacIntyre. (Ed.). *Community Economic Development*, University of Cape Breton Press.

Hills, J. (April 1990), "The Telecommunication Rich and Poor." *Third World Quarterly*. 12(2), 71-90.

Huyer, S. (18 February 1997). "Supporting Women's Use of Information Technologies for Sustainable Development,". Submitted to the Gender and Sustainable Development Unit, IDRC.

IDRC (International Development Research Centre, 1997a). "Stratégie Acacia au Sénégal," Ottawa, IDRC. http://www.idrc.ca/acacia/.

IDRC (International Development Research Centre, 1997b), "Communities and the Information Society in Africa: Overview," Ottawa, IDRC. http://www.idrc.ca/acacia/.

Kole, E.S. (1998a). "Whose Empowerment? NGOs Between Grassroots and Netizens." Draft available at http://dkglobal.org/crit-ict.

Kole, E.S. (1998b). "Myths and Realities in Internet Discourse: Using Computer Networks for Data Collection and the Beijing World Conference on Women." *Gazette: The International Journal for Communication Studies*, 60(4), 343-360.

Lamprière, L. (14 January 1998). "Microsoft s'attaque au téléspectateur américain," *Libération*.

Lasfargue, Y. (October 1989). "Technologies nouvelles, nouveaux exclus? Changements technologiques et évolution du travail." *Futuribles*, 3-13.

Lee, E. (1997). *The Labour Movement and the Internet: The New Internationalism*. London: Pluto Press.

Leiner, B., et.al. (1997). "A Brief History of the Internet." *Internet Society*. http://www.isoc.org/internet/history/.

Lemos, R. (30 June 1997). "Will China Squash Hong Kong's Net Freedoms?" *ZDNet*, http://www.zdnet.com/zdnn/content/.

Lewis, P.H. (6 September 1996). "On-line Service Cracks Down on Junk E-mail." *International Herald Tribune*, 14.

Lipnack, J., and J. Stamps. (1986). *The Network Book: People Connecting With*

People. New York: Routledge and Kegan Paul.

Madson, W. (1997). "Cryptography and Liberty: an International Survey of Encryption Policy." GLIC (Global Internet Liberty Campaign). http://www.gilc.org/crypto/crypto-survey.html.

Mason, M. (1997). "IGC Fights Digital Censorship: Basque Website Attacked by Internet Mailbombers." IGC Netnews.

M-Powa (1997). "Exploring a Basic Illiterate Web Access System." Midrand, South Africa: IDRC Study/Acacia Initiative. http://www.idrc.ca/acacia/.

Negroponte, N. (1995). *Being Digital*. London: Coronet.

Noble, D.F. (1977). *America by Design: Science, Technology, and the Rise of Corporate Capitalism*. New York.

Noble, D.F. (1984). *Forces of Production: A Social History of Industrial Automation*. New York: Alfred A. Knopf.

O'Brien, R. (1989). "Networking and Computer Networking."Major Research Paper, Faculty of Environmental Study, Toronto: York University.

PoKempner, D. (1 Aug. 1997). "Encryption in the Service of Human Rights," Briefing Paper, Human Rights Watch. http://www.aaas.org/SPP/DSPP/CSTC/briefings/crypto/dinah.htm.

Pruett, D. and J.Deane. (April 1998). "The Internet and Poverty: Real Help or Real Hype." London: Panos Institute. Briefing No. 28. http://www.oneworld.org/panos/.

Rheingold, H. (1994). *The Virtual Community: Finding Connection in a Computerized World*. London: Secker and Warburg.

Richtel, M. (27 September 1998). "Accessing the Web via E-Mail," *New York Times*.

Robins, K. and F. Webster. (1988). "Cybernetic Capitalism: Information, Technology, Everyday Life." V. Mosco and J.Wasko. (Eds.). *The Political Economy of Information*. Madison: University of Wisconsin Press.

Rohozinski, R. (1998). "Mapping Russian Cyberspace: perspective on Democracy and the Net." Paper presented at the UNRISD Conference on Information Technologies and Social Development, Geneva, June 22-24.

Schuler, D. (1996). *New Community Networks: Wired for Change*. Reading: Addison-Wesley.

SDNP (Sustainable Development Networking Programme). (1997). "Home Page News," New York, UNDP. http://www.sdnp.undp.org/.

Sédallian, V. (23 October 1996). "Controlling Illegal Content over the Internet: The French Situation." Presented at "Censoring the Internet: A Lawyer's Deceit," Media Law Committee, 26th International Bar Association Conference, Berlin. Also on: http://www.argia.fr/lij/control.html.

Sédallian, V. (1997). "Cryptographie: les enjeux et l'état de la législation française", *L'Internet Juridique*. http://www.argia.fr/lij/etatcrypto.html.

Spencer, G. (1996). "Microcybernetics as the Meta-Technology of Pure Control." Sardar, Ziauddin and Jerome R. Ravetz, *Cyberfutures: Culture and Politics on the Information Superhighway*. London: Pluto Press, 61-89.

Stewart, F. (1978). *Technology and Underdevelopment*. London: Macmillan Press.

Sussman, G. (1983). *The Political Economy of Telecommunication Transfer: Transnationalizing the New Philippines Information Order*, PhD Dissertation,

University of Hawaii.

Sussman, G. (1997). *Communication, Technology, and Politics in the Information Age*, Thousand Oaks, California: Sage Publications.

Swett, C. (17 July 1995). *Strategic Assessment: The Internet*. Washington: Office of the Assistant Secretary of Defense for Special Operations and Low-Intensity Conflict (Policy Planning), Pentagon. http://www.fas.org/.

Tech Channel. (15 February 1998). "High Rates Limit Net Access in Sudan."

Thoumyre, L. (1996). *Abuses in the Cyberspace: The Regulation of Illicit Messages Diffused on the Internet*, Master of Arts, ESST, Université Louis Pasteur, Strasbourg, Facultés Notre Dame de la Paix, CRID, Namur. Available at: http://www.argia.fr/lij/telechargement/file1.doc.

UNFPA. (1998). "Report of the Round Table on Adolescent Sexual and Reproductive Health and Rights: Key Future Actions." New York, United Nations Fund for Population Activities (17 April). http://www.unfpa.org/.

UNESCO (1996). *Information and Communication Technologies in Development: A UNESCO Perspective*. Paris, UNESCO Secretariat.

Usdin, S. (1997). "China Online: Behind The Great (Fire)Wall." http://www.zdnet.com/.

van Koert, R. (1998). "Bustling and Sprawling Cities: a Natural Environment for ICTs." Draft available at http://dkglobal.org/crit-ict.

Verzola, R. (1998). *Cyberlords: The Rentier Class of the Information Sector*. http://www.dkglobal.org/.

Warren, H.M. (23 October 1997). "Implementation of the Communications Assistance for Law Enforcement Act (CALEA)." Washington, Statement Before Subcommittee on Crime, Committee on the Judiciary, United States House of Representatives. http://www.fbi.gov/.

Winner, L. (1986). *The Whale and the Reactor: A Search for Limits in an Age of High Technology*. Chicago: University of Chicago Press.

World Bank. "InfoDev: Information for Development Program, Background and Introduction." Washington: World Bank. http://www.worldbank.org/infodev/. (For a broader public discussion, mostly with the same emphasis on information gaps, see Global Knowledge at: http://www.globalknowledge.org/english/index.html.)

Wright, S. (1998). "An Appraisal of Technologies of Political Control." European Parliament, Directorate General for Research.

Endnotes

1   Independent researcher on the political economy of information and communication technologies currently based in The Netherlands. François Fortier can be reached by e-mail at ttff@antenna.nl.

2   The grassroots are here understood as organisations that emerge from and work in the interest of those social sectors that are both economically subsumed (in relations of production) and politically oppressed (through the existing order of social reproduction). This inherently political definition draws attention to the very purpose of grassroots activities, i.e., political

emancipation.

3   For more historical details on the Internet, see Leiner et al., 1997.

4   On the use of computer networking surrounding events of the Tien An Men massacre, see Bonchek ,1995; Frederick, 1993, p. 236; Lee, 1997, p. 163-164. On the Moscow failed coup d'État, see Frederick, 1991; Rohozinski, 1998. On the Zapatista rebellion, see Cleaver, 1994.

5   Cryptographic tools permit encoding information into unintelligible formats, which can only be decoded by the intended recipient.

6   Recognising the need for specific information flows, the Chinese government is willing to let business networks develop, as long as they keep off politics. From the point of view of commercial ISPs, this is not problematic. To avoid political controversy with the government and secure a share of the market, corporations like Compuserve are willing to restrict content and limit civil society access by focusing on lucrative commercial clients and steering clear from serving the potentially problematic broader public (Lemos, 1997).

7   Organisation for Economic Co-operation and Development.

8   This table was inspired from Kole's work (August 1998, p. 358), which synthesised the "Myths and Realities in Internet Discourse," emphasising NGO and gender aspects.

9   See notably Annis, 1991; Greenberg and Goodman, 1996; World Bank Infodev, D'Orville, 1996; Cairncross, 1997, p. 252-253; IDRC, 1997a, 1997b; SDNP, 1997; the Web site of the International Institute for Communication and Development (http://www.iicd.org/index.ap), etc.

10   IDRC Panconsultation, message 13:17 by Miguel Saravia Lopez de Castilla, 25 November 1997.

11   Interview with Liana Karabossian and Moema Viesser, Rede Mulher, São Paulo, 29 October 1992.

12   Although it should not be forgotten that many grassroots organisations are still using slower, DOS-based computers with no multilingual ability.

13   Pruett and Deane, 1998, citing Matrix Information and Directory Services.

14   Interviews with François Beaudet, anthropologist, Rio de Janeiro, 27 September 1992, and Eduardo Leão, CIMI, Brasilia, 16 November 1992.

15   E-mail correspondence with Bruce Girard, AMARC, 24 February 1992.

Chapter XXI

Linking Communities to Global Policymaking: A New Electronic Window on the United Nations

John Lawrence
United Nations Development Program

Janice Brodman
IT Strategies Consultant

Introduction

The 1990s have been marked by extraordinary changes in many of the fundamental elements of human existence, among the most powerful, the introduction of a global networking system. Indeed, it is difficult to consider thoughtfully any major aspect of our socio-economic-political circumstances, current and future, that are not in some way profoundly affected by this revolution. For those of us with Internet service, even a few keystrokes on a laptop computer can now put us in touch with friends, family, colleagues, or strangers almost anywhere in the world, certainly on all seven continents including Antarctica. Business can be conducted, money transferred, medical records evaluated, books/papers jointly written and edited, inventions created, ideas shared. The unprecedented ease and speed of access to knowledge and experience, and increasingly commerce, is at the heart of the promise of the new technologies for cyberconnectivity. Communities in all parts of the world are finding ways to make the Internet serve them, and becoming energized, organized and activated as a result.

Two factors, however, contribute to a sobering backdrop that frames further exploration of these exciting new frontiers. First, access to the underlying technologies is severely constrained in developing countries, and in poorer communities of industrialized countries. Differential access to key resources, such as capital, electricity, telephone service, exacerbates gaps between the haves and the have-nots.

Furthermore, even for those who gain basic access, other constraints, such as predominance of "colonial" languages, limit their ability to take advantage of opportunities offered by the technology. Second, the glitter of cybertechnology tends to divert us from addressing broader problems of inequities in social and economic development, and their associated ecological consequences. These have been sharply documented in the UNDP Human Development Report Series. (The most recent of these 10 annual Human Development Reports, that of July 1999, can be found at: http://www.undp.org/hdro/99.htm.)

This chapter presents the results of an experiment to bring together these two contemporary forces — the Internet explosion, and a sense of growing inequality in economic and political power — to create a new channel into global decision making fora, particularly for communities that seem increasingly to be left behind. The context for this effort was the United Nations, and a series of global conferences that focused attention on the major social, environmental, and economic issues of our time. The objective was to explore ways to use new electronic networking to link communities around the world more directly to top level decision makers.

The global conferences have sought to raise international consciousness of the growing gap between rich and poor, and to set guidelines for future action. They have provided a forum in which Heads of State and national governments offer formal commitments through broad declarations and specific global action plans. They have also helped to engage private industry in new efforts to achieve more equitable, sustainable and benign development for all people. (See, for example, UN Wire 6, July 1999; also see Michael Hopkins, 1999.)

To be effective, however, action to redress problems of inequity must engage people at the level of the village, the community, the family. If new initiatives launched in global conferences are truly to permeate to local levels — to bridge the global-local chasms — and ultimately benefit individual livelihoods, then far greater citizen participation in the process is essential. Enabling that participation was the objective of the approaches discussed in this chapter.

The objective was rooted in the fundamental principles of the United Nations. For more than 50 years, the UN has provided a global forum for its member states. The UN Charter, in its Preamble, calls for the use of "international machinery for the economic and social advancement of all peoples." More recently, the UN system has reiterated the principle of working not only with governments, but also with broader dimensions of civil society. This resolve, while underlying much of the original intent of the UN charter, embodies an important new emphasis that reflects widespread disenchantment with relying solely on government to redress inequities, and, instead, turns to broader 'governance' as an instrument of progress.

The purpose of this chapter is to report the results of three projects conducted by the United Nations Development Program (UNDP) and the Education Development Center, Inc. (EDC) to explore use of the Internet as a means for creating a "virtual global meeting" that could have input into global conference preparations. The chapter describes ways in which electronic communications technologies contributed to the functioning of the UN system in the middle and latter part of the 1990s decade. It examines the ways in which electronic fora enabled individuals and communities who had little or no knowledge of, still less experience with, the UN, to engage in the UN deliberative process for the first time and in novel ways.

The forum was provided via a moderated Internet List discussion that offered an unprecedented new (electronic) window on the UN deliberative process during three of the major conferences of the decade: the World Summit for Social Development (Copenhagen, 1995), the Fourth World Conference on Women (Beijing, 1995), and the Global Knowledge Conference (Toronto, 1997). The first discussion List addressed issues facing the preparatory committee for the World Summit for Social Development (Social Summit). Subsequently, the List refocused to play a similar role in preparations for the Beijing Conference. The Beijing List drew greatly expanded membership, reflecting increased interest in this medium as a useful discussion and communications device for participation into a global forum.

The Global Knowledge for Development List was created for the Global Knowledge '97 Conference, in an effort to further refine the principle of virtual participation in a global conference. This List, still ongoing, formed the backbone of the "virtual conference" at Toronto in June 1997. Perhaps even more importantly, it continues to stand as a major and continuing source of breaking information on communications and information technology (CIT) and development worldwide.

This chapter includes four sections. The first provides a brief background on the conferences and rationale for the project effort. The second presents a brief examination of the experience gained and "lessons learned" from each List. The third section delineates some conclusions drawn from experience with these projects. The fourth offers some thoughts on future trends in this electronic medium for the policies of the UN and other international development agencies.

Background

The United Nations was established on 24 October, 1945 by 51 countries committed to preserving peace through international cooperation and collective security. Today, almost all nations (185 countries in 1999) have become members. Among the best known aspects of the UN are its peacekeeping and security functions, humanitarian assistance in moments of acute human crisis (whether natural disaster or human conflict), and the highly visible deliberative process of the General Assembly and Secretariat in New York. Far less well known are the longer term development efforts aimed at strengthening national capacities especially in developing countries, with the longer term goal of reducing — perhaps even eliminating — the need for either peacekeeping or humanitarian aid. More than 30 organizations and agencies together constitute what is termed the United Nations "system," which collectively is engaged in addressing major global issues in areas such as human rights, justice, international law and standards, and broader concerns such as poverty, health/nutrition levels, and environmental degradation.

The UN Charter outlines the mission statement for the system as a whole, but each agency has its own particular focus. In the forefront of efforts to bring about social and economic progress is the UN Development Programme (UNDP). The UN's largest multilateral provider of grants for sustainable human development, UNDP works in over 170 countries and territories to facilitate technical cooperation and eradicate poverty.

During the 1990s, a round of world conferences was held to address specific problems facing all countries, and to identify and seek consensus on practical solutions in a range of areas such as education (Jomtien, 1990), environment and development (Rio, 1992), human rights (Vienna, 1993), population and development (Cairo, 1994), social development (Copenhagen, 1995), the advancement of women (Beijing, 1995), human settlements (Istanbul, 1996), and food security (Rome, 1996). This panoply of global discussions could have been dismissed as an indulgent talkfest. However, they proved to serve an important function in identifying areas of wide consensus among nations on the key issues of our time, and providing a unifying focus for the UN system as it moves forward into the new millennium.

The 1990s high-profile global events have broken new ground in many areas, including the direct involvement and presence of Heads of State, and the initiation of special sessions of the General Assembly for monitoring various aspects of followup (see the UN Web site material on global conferences). In particular, however, these conferences have been notable for their inclusion of thousands of nongovernmental organizations (NGOs), citizens, academics, and business people, in both official and unofficial meetings, to create true "global forums." The UN has encouraged this trend, recognizing that the support of a wide spectrum of society is needed to implement the policies being discussed.

This principle of encouraging grassroots engagement was paramount in the effort to find a new, more participatory arena where the crucial subjects could be widely debated. The major means identified for accomplishing this goal was the Internet. Upon initial presentation of the approach, there were obstacles and substantial opposition, on grounds of protocol, precedence, etc. Of particular concern was the risk that the positive relationships between UN organizations and governments, which have been built carefully over the years on the basis of trust, might be undermined in the face of opening new and untested communication lines. Yet those who championed the projects felt that, given sufficient safeguards, the new window was too constructive and offered too much promise to be ignored. Those in support of the new approach argued strongly that:

- conference deliberations could be greatly enhanced by the broader involvement of organizations/individuals unable to be physically present, but able to engage "virtually;"
- wider knowledge of the events could be promoted through 'railhead' strategies that were tailored to each cultural or geographic situation, i.e., documents and other information could be disseminated to a terminus point electronically, then widely distributed via conventional means by individuals or organizations credible in the local areas;
- follow-up, monitoring, and continued engagement of civil society in implementing the agreements could also be strengthened by establishment of these kinds of electronic networks.

These arguments prevailed, and the United Nations Development Programme, in August 1994, initiated a pilot project which, with the help of the Education Development Center, designed and conducted the first in a series of Internet discussion Lists directed towards expanding participation in UN preparations for the global events.

Three Conference Internet Lists

The Social Summit

The first project took place during the preparatory phases, as well as the actual convening of the March 1995 World Summit for Social Development in Copenhagen. It continued after the Conference to examine the implications of the Summit outcomes and the requirements for implementation of Summit agreements. The Social Summit addressed the problem of global poverty from the perspectives of governance, employment and livelihoods, and social inclusion. Its major contribution was to commit signatory governments, for the first time, to a consensus that eradicating poverty is "an ethical, social, political and economic imperative."

The Internet had gained considerable popularity, particularly in North America and Europe. Several pioneering efforts suggested the value that electronic networking could offer to a global conference. One involved the early efforts of UNDP and others at the Rio (Earth) Summit in 1992 to share documents electronically with diplomatic missions as soon as they became available. Another was the notable contribution of the Earth Negotiations Bulletin in providing daily summaries of deliberative sessions so that delegates could have access to the previous day's progress (or lack thereof) before the next day's session started. These summaries were also available on the Internet. There were also efforts on the part of various organizations to provide access to UN conference documents electronically. These and other efforts broadened knowledge of conference committee deliberations, but still among a relatively exclusive few, with limited interactivity.

Goals

The UNDP/EDC initiative differed from previous efforts in three significant respects. First, it sought to establish an Internet discussion List that was designed to involve a broad spectrum of civil society, worldwide, in an active interchange of ideas and concrete policy suggestions around the global Conference topics. Second, the discussion List explicitly and vigorously sought Southern participation in an interactive global dialogue on conference issues, especially from organizations that might serve as catalytic to expression of community opinions on the issues being addressed in the global conference. Third, the project contained a specific analytical component that examined the experience of the Internet discussion group in order to advance our understanding of the role such a forum can play in these kinds of global events.

The project had three major goals:
- To widen the interactive dialog on Social Summit issues, and expand access to information on proceedings of the Summit, particularly among those who might otherwise not be involved.
- To expand access to information and discussion of Summit-related issues by less technologically advanced countries and constituencies.

- To examine systematically how access to electronic networks affects information sharing and exchange of ideas, as well as impacts, if any, on the deliberations of the Summit delegations.

Fundamentally, the project aimed to shed light on the process involved in conducting this type of global forum, and the impact such a forum could have on a more traditional global conference. A detailed account of the design, method, and impact of the Social Summit Internet discussion List can be found in two reports to UNDP (Education Development Center, 1995, 1996). These reports examine five questions that were central to the project:

- Can an Internet electronic forum expand useful participation in discussion of global conference issues, especially from Southern countries and other typically under-represented groups?
- What kind of discussion does an Internet List generate, and by whom?
- What are the unique contributions such a mechanism can provide?
- What is required to make such a List successful, and what constitutes success?
- What lessons can we draw from the experience for the future?

Unique Requirements

The approach was to design a process that created and promoted the Internet List, while clearly delimiting its goals and function as required by the specific constraints of the situation within which it operated. Three requirements were critical. The first responded to the very high visibility that the Social Summit gained. With 117 Heads of State and Government attending, it became the largest gathering of world leaders that had ever been assembled. In light of the complexity of Summit negotiations, and the diplomatic and political nuances surrounding the negotiation process, it was essential that Summit delegates and discussion List members understood the experimental nature of the initial UNDP-sponsored Internet List discussion. The UNDP and EDC List hosts had to convey in very clear terms the demarcation between what the List could and could not do.

The novel relationship between the List discussion and the formal negotiations was presented in the Invitation to the List discussion. The Invitation encouraged those interested in the issues to join the discussion, while explicitly explaining the limited nature of the relationship between the List discussion and the formal negotiations. Although UNDP could not use the List discussion to provide a direct channel for List participants into the delegates' deliberations, Summit delegates could participate in and obtain information from the List. Discussion points from the List were transmitted informally to Summit delegates. List members were also encouraged to contact their delegates through formal channels, and the List explained the process for doing so.

The second requirement was that the List discussion offer substantive value. All of those involved in the pilot had observed other Internet Lists that degenerated into meandering discussions completely without focus and/or marred by name-calling. It was determined at the outset that the List must be carefully and actively moderated in order to ensure that messages were relevant to the discussion and did not include personal attacks ("flaming"). A moderated List also was able to provide services to the participants, as well as to pay close attention to each candidate message, which was necessary for the analytical side of the project.

The third requirement related to the desire to include developing country participation to the extent possible. Project managers had lived in developing countries and had first-hand experience with the costs and other constraints affecting developing country use of the Internet. Special measures were taken to design the List to facilitate participation from developing regions. In addition, extensive efforts were made to publicize the List through UNDP field offices and other country level contacts in the South, as well as a range of NGO networks, including EDC's worldwide network of NGOs and universities, and others such as those of the Association for Progressive Communications (APC).

Before the List was launched, five functions were identified as being central to successful List operation/maintenance:

- Provide technical expertise and troubleshooting — It was anticipated that some technical troubleshooting would be necessary to help List participants resolve problems they encountered. In the early stages of the project, however, the technical demands were far more extensive than had been anticipated. Technical problems arose primarily with regard to maintaining contact with participants from developing countries. For example, many of the List messages to those countries bounced back to the List, and technical problems had to be resolved on an individual basis.

- Promote the List — One of the most time-absorbing functions was to reach out to those who might be interested in the List discussion, with particular emphasis on reaching those in developing countries. The outreach process involved identifying a wide range of channels to potential participants (organizations, conferences, other Lists) and communicating with them to disseminate information about the Social Summit List.

- Moderate the List — Project staff decided at the outset to make the List highly service-oriented. As a result, the moderators' jobs became far more complex than simply logging, screening, and posting messages. For example, if a message were deemed inappropriate (e.g., it was irrelevant to the List discussion or contained a personal attack on another participant), a moderator responded personally to the author and explained why his/her message was not being posted and suggested ways to make it "postable." Project staff felt that this level of personal service was important in order to convey a sense of service to the participants and to avoid any appearance (or instance) of unilateral censorship. Moderators also posted "facilitation messages," which helped to stimulate discussion and to keep the discussion on topic. Moderators also posted draft Conference Documents to the List, and suggested particular sections that might be especially relevant to List participants, given the interests and concerns they had expressed.

- Analyze the discussion — Data were calculated on numbers and geographic location of subscribers, numbers/themes of messages, and flow rates. Content analysis was conducted on more than 1,000 pages of messages.

- Maintain the List functions — Daily monitoring and activity record maintenance were conducted.

Results

In the course of the project, it became clear that the project budget could not cover all of the time that was required to carry out all of the "central" functions described above. In particular, the "service" element added far more time than could be compensated. EDC project staff met to discuss the situation and options for resolving the problem. It emerged that the staff were unwilling to sacrifice any of the "central" functions, or cut back on the service component. Instead, they decided to contribute their personal time in order to keep service levels high. As a result, EDC staff contributed extensive time (around 400 hours) to the project, for which the contract did not compensate EDC. Moreover, a large proportion of that time was personally donated by the EDC moderators, who were not compensated for that time at all. The result was one of the most unexpected outcomes of the effort: a high level of public/private cooperation in an activity that captured everyone's imagination.

The pilot project produced a number of other important results. For the seven-month period during which the List was active, about 600 subscribers were formally logged in as subscribers, representing 54 countries from all world regions. Although, today, these levels of participation are not noteworthy, at the time they were exceptional. Most importantly, half of the countries represented on the List were developing countries. Half of the participants were outside North America (18% from Europe, and the remainder from other regions and international networks) at a time when most Internet users were in North America. Nevertheless, during this first pilot, it was clear that the bulk of subscribers were from Northern (i.e., industrialized) countries. Yet with 21% of the members from Latin America, Asia, and Africa, the Social Summit List was extremely unusual for an Internet List. Furthermore, it was clear that many of the subscribers using North American servers were either students from Southern countries studying in Northern schools, or Southern subscribers dialing into Northern nodes. This level of participation from the South was especially impressive because neither UNDP nor EDC could offer any direct support for the (often substantial) costs for Internet service borne by Southern subscribers. Certainly this engagement extended the "inclusiveness" of the discussion of Summit issues beyond the normal bounds of the Preparatory Committee process and the Conference itself, and hypothetically beyond that practically available through any other means. A message from South Africa reflected the role that the List played in expanding participation in the Summit:

> 'We are not formally registered [in the Social Summit], but would like to get information on doing so; until recently we were not aware of the Social Summit.'

Content analysis was conducted to determine the themes covered in the messages. Each message was coded for each theme it discussed, and the results analyzed both qualitatively and quantitatively. A few themes emerged as central: information technology (IT) in developing countries, UN system reform, and the role of education/training in reducing poverty and increasing employment. The theme of increasing productive employment (one of the stated Conference themes) was the issue most frequently raised in messages.

The analysts noted that one of the most distinctive qualities of the great majority of messages was the high quality of the discussion. Participants wanted to get to the heart of the substantive issues. On the other hand, List subscribers largely appeared to be unfamiliar with the intricacies of Summit negotiations, and in general expressed little interest in the niceties of contested wording (and consequent wordsmithing) that was central to the work of the Summit Preparatory Committee. Nonetheless, in at least one documented case, the List had a significant, tangible impact on Summit documentation, as described below.

Impacts

The Social Summit List achieved two important impacts. The first was the acceptance of the List process among senior UN agency management. The List not only avoided being perceived as threatening or disruptive to the fragile Summit negotiation process. It actually achieved status as a productive, innovative forum through which to share ideas, engage in advocacy, and obtain a sense of the wide spectrum of public opinion on the issues central to the Summit. The new legitimacy gave indirect impulse to several initiatives which continue to this day to use the Internet to pursue Summit issues effectively, e.g., Social Watch, an international citizens' coalition monitoring implementation of the world governments' commitments to eradicate poverty and achieve gender equity. (See http://www.socwatch.org.uy/.)

Second, NGOs used the List to help them collaborate on a worldwide basis, and to raise awareness of the Summit issues. Through the List, NGOs disseminated proposed Summit documentation in draft form to more than 1,000 people, with List members distributing the information to others who did not have email. List members posted the formal positions of delegations. In several cases, members requested (and were given) names and addresses of their official delegates, so as to be able to contact them and suggest changes. A message from Cameroon is illustrative:

> I am from Cameroon. Do you happen to have any [information] on delegates from Cameroon? I will really appreciate any help you can give me....

NGOs used the List to develop strategies for collaborative efforts for pursuing their own agendas at the Summit:

> I think it is important for us [NGOs] to exchange our plan for lobbying before the Summit... it's much better if we can coordinate our strategy or at least exchange our ideas.... please tell me [Japanese subscriber] about what your organizations plan to do at the Summit.

List members also noted when representatives were not supporting document language they had previously promised to support. The dissemination of this type of information on the List led to activities among NGOs and others that affected document language, and thus the direction/strength of the commitments made at Copenhagen. Indeed, NGOs learned to use the List to make themselves more "visible" to Summit delegates.

Take, for example, one NGO's efforts during the Preparatory Committee (PrepCom) deliberations. An important part of the PrepCom process is to finalize language that has been contested in earlier meetings. Such "bracketed" language has important implications for the obligations to which governments commit during the conference. During the PrepCom meetings, one NGO posted a message to the Social Summit List registering deep concern about the position on certain bracketed language by one Northern country. In particular, the respondent was concerned about the fact that the country's delegates consistently refused to meet with the NGOs. The NGO urged all List members who shared their position on the bracketed language to contact the delegates in question. The *next* day, the NGO posted a message thanking "all List members who had contacted their representatives" and announced that the national delegation in question had responded to those messages by meeting in good faith with the NGOs to discuss the wording. The national delegates explicitly mentioned the messages from List members as a powerful impetus leading them to invite a meeting with the NGOs.

Key Factors

Several factors contributed to outcomes achieved. The number of subscribers was far higher than had been anticipated. Nonetheless, List moderators believed that the subscriber level could have been higher were it not for the List's late start. The List began just before the second PrepCom, several months before the Social Summit itself. However, given its novel objectives and role, it took time for the List to gain recognition, credibility and interest. The cost of Internet service also greatly influenced the distribution of participants. Extremely low direct costs associated with North American Internet access were an important factor contributing to the high level of participation in the List from that region. Messages from List members in the South made it clear that entry costs, both capital (cost of setup) and recurrent (access/management) were a substantial barrier. Indeed, the moderators endeavored to ensure that messages were relatively concise because they recognized that List members in the South often paid for every byte they downloaded. Yet Southern members of the List also suggested that cost was not necessarily the major constraint they faced. Indeed, many noted that the cost of exchanging information via electronic networks was actually lower than the alternatives, sending faxes, mail, or calling by telephones. A participant from Indonesia reflected the sentiments of many other Southern participants:

> I strongly disagree with people who believe that the Internet cannot be afforded or used by south[ern] countries. Although it is still in limited departments, in my country there are a lot of networks and they are most important to development.

Active participation from the South was generally limited. Language was no doubt an important delimiter. Although postings were encouraged (and were posted) in other UN languages, most messages were written in English. Resources did not permit simultaneous translation, which undoubtedly inhibited participation. Although many members requested translations of messages, limited resources made compliance with the request impossible.

The impact on the Summit deliberations was affected by a number of factors. First, given its novelty, the List began too close to the PrepCom and the Summit itself. Most Summit delegates were unfamiliar with an Internet discussion list, and had little time to become familiar with it. As a result, there was little participation in the List by Summit delegates, and the impact of the List on the Summit outcomes was limited. Yet, importantly, some key decision makers (delegates, agency staff) did participate, conferring legitimacy to the List and opening a channel into the preparatory process that hadn't existed earlier. The participation by these decision makers in the List discussion was important in conveying to the List participants the sense that someone "from the in-crowd" was listening, and generating enthusiasm and interest among participants. In turn, the active interaction among List members led many to communicate with delegates, which did have an impact on certain decisions that were made. Nevertheless, List participants strongly criticized the lack of direct articulation of the List into the formal process of UN business. Had the List been beyond the initial pilot stage, and thus more central to the deliberative process, it would likely have had greater impact.

The Beijing Fourth World Conference On Women

The Fourth World Conference on Women (Beijing, 1995) was convened to assess the advancement of women since 1985, particularly with regard to the objectives of the Nairobi Forward-Looking Strategies for the Advancement of Women to the Year 2000. The Conference also aimed to adopt a "Platform for Action" with steps to be taken in "critical areas of concern" that obstruct the advancement of women in the world, and to mobilize people to take action around the strategic objectives described in the Platform. Given the global nature of the Conference, the UN faced the same need to extend constituency engagement as had existed for the Social Summit. Once again there was an important opportunity for electronic networking to help expand participation in the discussion of conference issues. Moreover, it was evident that NGOs and the wider public would play a key role in this Conference.

In light of the effectiveness of the Social Summit List, and the pressing need for interactive channels linking NGOs and the public to the Beijing Conference, UNDP decided to reorient the Social Summit List to focus on preparation for the Beijing Conference, as well as the relationship between the Summit and the Beijing Conference. This decision by UNDP in itself was evidence of the success accorded the original pilot. EDC agreed to manage/moderate/facilitate the revised List, exemplifying the good public/private partnership, which had provided a practical, worldwide service.

An extensive outreach effort was conducted by EDC, which had a global network of NGOs and others involved in women's issues. The outreach contacted thousands of individuals, universities, women's groups, human rights groups, and other NGOs around the world. In addition, invitations describing the objectives of the List, with instructions for joining, were posted to regional Internet lists in industrialized and developing countries, and were sent to network administrators throughout the world, with the request that they post the invitation to their networks/ bulletin boards. The outreach, and the List discussion, benefited from other on-line efforts related to the Women's Conference, including UNDP's own Web site, other discussion lists such as the Association for Progressive Communications (APC) Beijing95-women List (a moderated, women-only discussion), the Women's Environment and Development Organization (WEDO) electronic conference on the 180 Days Campaign, International Women's Tribune Center Global (IWTC) Net-FaxNet activities, and the Earth Negotiations Bulletin.

Levels of Participation

The response to the outreach was huge. Membership in the List — renamed Beijing-conf — expanded immediately. By the end of the Conference, there were more than 1,600 subscribers, with an average of 1,000 participating from around the world at any particular point in time. Even more noteworthy, the geographic distribution of members became wider and more inclusive, with 65 countries represented, half of them developing countries. At least half of the subscribers were outside North America — an impressive achievement at a time when well over 90% of those on the Internet were in North America. An analysis of List members' e-mail addresses determined that more than 20% of the members were in Africa, Asia, South/Central America/Caribbean (SCAC), the Middle East, or Oceania.

Importantly, the representation from the South was undoubtedly greater than the numbers suggest. First, about a fifth of the members had global network addresses, making it difficult to identify their geographic location; however, many of those participants were certainly in the South. Second, many participants in developing countries were engaged in the discussion through Northern servers, since information 'railhead' strategies were abundant. Finally, a new phenomenon became apparent. List members served as hubs for local networks and bulletin boards, and provided information to, and transmitted messages from, a great many others who were not formally subscribed to the List. For example in the Philippines, a subscriber posted Beijing-conf List messages to a local bulletin board where others could access them locally at low cost. In South Africa, a similar function was performed by a local postmaster.

The implications of this experience are far-reaching. It is inconceivable that a physical gathering at one location, or that any other medium, could have brought together as many people, from as wide a geographical spectrum for an interactive discussion lasting several months, at such a small cost. Even more importantly, the membership was largely composed of individuals who could not possibly have participated in the Conference at all, were it not for the List. For many in developing countries, as well as in "bypassed areas" of industrialized countries, the List offered immediate access to information, direct links with others, and above all, a sense of active participation in a critical global event that illuminated and shaped women's issues worldwide for the rest of the century.

Contributions of the List

The Beijing-conf List served several functions: it provided a platform for discussion and debate; it offered a forum for information exchange; it also served as an alternative source of news.

Discussion and Debate. The List served as a forum for a vibrant, often passionate, discussion of issues, with virtually every position represented on each topic. Most of this debate and discussion revolved around controversial topics, such as the link between women's rights and human rights, abortion, roles and rights of sex workers, and the right of women to be free from intimidation, coercion, harassment, and physical threat. Members seemed to express their views on these issues frankly, without reserve, yet, for the most part, with respect for others who differed with them.

Information Exchange. The List provided a useful channel for NGOs, UNDP and other development agencies, and the wider public to exchange information about the Conference. For example, UNDP and other development agencies, such as the FAO, Australian AID, International Development Research Center (IDRC), and Canadian International Development Agency (CIDA), used the List to provide notifications of meetings in which NGOs and other members of the public could voice their views. The List moderators also obtained and disseminated other information that NGOs and individuals said they had difficulty obtaining, such as lists of NGOs accredited to attend the Conference.

The List also provided a means for NGOs to exchange information in three major areas: logistics, notification of activities that NGOs were undertaking, and key information affecting Conference outcomes, such as the positions of various governments. This information was often crucial to those attending the Conference, or wanting to affect the Conference outcomes. For example, there was considerable confusion before the Conference about registration and visas requirements for those planning to attend the Conference or the NGO Forum. For a time there was even uncertainty about whether the Conference would be held at all. Many hundreds, if not thousands, of people who expected to attend the Conference could not get information about their applications, and the procedures they had to follow. NGOs used the List to provide up-to-date notices of key logistics information needed by those seeking to attend the Conference or Forum.

NGOs also used the List to support collaboration and coordination of activities, posting descriptions of the kinds of activities they planned to conduct and the types of collaboration they sought. During the Conference, NGOs posted updated schedules of events that were held during the Conference, including sudden changes in plans and new offerings. Messages to the List also informed people of innovative activities. For example, an early alert was sounded by a German subscriber who described an embryonic new initiative that subsequently became a bellwether for development:

Inspired by the success of small loans to Bangladeshi women by Grameen Bank, NGOs at the Women's Conference have announced an International Micro-Credit Summit scheduled for November 20-22 in Washington, DC.

The wider public also generously shared information. Individuals who had been to Beijing or Huairou posted information about what to expect on arrival. They described the registration requirements, and offered advice about how to ease the process. They sent directions about accommodations and access for handicapped participants in the Conference. Those attending the Conference shared "eyewitness" accounts that brought the Conference "home" to those who were not able to attend.

Providing the News. The third major function of the List was to provide news items, editorials, and speeches, often from "alternative media," such as the *Earth Times*, but also from local media in China and other Southern countries, as well as the North. News items included speeches by delegates, news stories from within China, and positions by government figures at the Conference or at home. Most of this information was never carried by the mainstream news media in the North, even respected U.S. newspapers such as *The New York Times*, or public radio. Thus, List members received a tremendous amount of information and news they could not receive from any other media source.

Impact and Value

The value of the List for participants was reflected in the hundreds of messages that moderators received describing its importance to local subscribers. Take the following message from an Australian participant:

> Thank you for all the news and correspondence you posted over the past weeks. We operate a free public access INTERNET site for women... the Beijing-conf provided information for multicultural communities [in Australia] inspiring women here to make use of these tools for communication, exchange and mobilization.

The best evidence of the value of the List, however, was stumbled upon, counterintuitively, and quite by accident. Faced with budget constraints, UNDP announced to the List in December 1996 that the agency could no longer provide funding. The responses were unexpected in their intensity and scope. They shed much light on the way that members relied on the List for a variety of local community-oriented purposes.

The Zambia Association for Research and Development in Lusaka captured the feelings expressed by many, while providing further evidence of the railhead concept:

> We in ZARD are extremely sorry to hear that this wonderful, inspiring, invaluable service ends so soon. Women's groups in Zambia have benefited immensely from the information downloaded from the list which was repackaged and distributed to interested individuals and groups.... for us in Zambia the loss of service is a big blow....

Appreciative and supportive messages flowed in from around the world, from South Africa to Sweden, the U.S. Virgin Islands to Japan, as well as from NGOs in the U.S. and Canada. Several subscribers offered to pay a subscription fee and invited others to do the same if the List could be saved. It was clear from this outpouring of concern that the List had affected peoples' lives in important, useful and constructive ways. Eventually, an anonymous benefactor (a private firm) emerged from the List subscribers and offered to cover the funding of the List for a period. As a result, the List continued as a follow-up forum for issues raised in Beijing.

Global Knowledge for Development Conference

The Global Knowledge Conference (Toronto, 1997), hosted by the World Bank, the Government of Canada, UNDP and several other public and private organizations, focused on the challenges facing developing countries and the international community at the end of the 20th century and the dawn of the information age. The Conference explored the vital role that information and knowledge play in achieving sustainable development, the ways in which the information revolution affects the development process, and possible responses by the international development community to these new challenges. It examined the

484 Lawrence & Brodman

new opportunities that information and communications technologies offer for participation, partnership and dialogue; the equity and access challenges that new technologies pose; and the ways in which information and knowledge can serve as tools of economic and social empowerment.

Early in the planning process, during a meeting of potential sponsors, the question of participation arose. Who would be able to attend, how many, from what organizations, and what support would be forthcoming to help those eligible but less able to afford the cost of participation? At this juncture, UNDP suggested, on the basis of experience with the other two global events outlined above, that there should be some opportunity for virtual participation in preparation for the Conference, as well as in the Conference itself. Thus the 'Virtual GKD Conference' was born.

One of the central activities in the Virtual Conference was the GKD List, similar to the earlier Social Summit and Beijing Lists, supported by UNDP and moderated by EDC. GKD's primary aim was to gain deeper understanding of the answers to five major questions:

- How can we further encourage broad participation, especially from Southern countries and other under-represented groups?
- What kinds of discussions does a List encourage, and by whom?
- What unique contributions can such a forum provide?
- What is required to make such a forum successful?
- What lessons can we draw from the experience for the future?

Although some of these questions had been considered in light of the experience of the earlier Lists, it was clear that the entire landscape of information/communications technologies (ICTs) was rapidly changing, along with their implications for developing countries.

Outcomes

By the end of the Global Knowledge Conference, the GKD List had become a rare phenomenon: a true learning community committed to building better understanding of the key issues: how ICTs can and do affect development. GKD offered a new model for global fora. The value and power of the List discussion led to its continued support by UNDP, the World Bank, and UNESCO. As this chapter is written (December 1999), GKD continues to serve as one of the foremost voices on ICTs and development, providing a vibrant and rich source of cases, information and experience on those issues.

Over 2,000 people have joined the GKD List discussion, with membership continuing to grow, and an average of 1,600 people around the world participating at any single point in time. Even more noteworthy, an analysis of members' e-mail addresses indicates that the distribution of members has been far wider and more inclusive than had earlier been imagined possible. Members hail from 90 countries—51 of them developing countries—representing every region in the world, including Africa, Asia, Europe, the Middle East, North America, South/Central America and the Caribbean (SCAC), and Oceania. Even more impressively, well over half the subscribers reside outside North America. One-third to one-half are in Africa, Asia, South/Central America/Caribbean (SCAC), the Middle East, or Oceania.

This level of participation represents an extremely strong response from the South, both in comparison with other Internet discussion Lists and even compared with the number of NGOs from the South who were able to attend the face-to-face Global Knowledge Conference. Furthermore, as was the case with Beijing-conf, the data underestimate Southern membership, and for the same reasons. In addition, many List members living in Northern countries are Southern nationals working or studying in the North. Thus, they represent a Southern perspective on the List.

GKD List participants represent a wide range of organizations and positions. Many members are part of the general public, ordinary individuals who are concerned about the issues. Still others are members of NGOs, professors and students at a university, and professionals working in the field of development. There are also members from various development organizations, including UNDP, World Bank, USAID, ADB, FAO. Finally, many List members represent newspapers or other media.

GKD has become an important forum for a vibrant, substantive discussion of issues related to ICTs and development, with viewpoints representing virtually all points on the spectrum. For example, debate arose when some List members criticized the role the U.S. and other industrialized countries have played in using ICTs to help bring about "development" in the Southern countries. GKD members debate issues ranging from the opportunities ICTs offer indigenous communities, to the particular constraints women face in using ICTs to serve their own needs.

Members also use GKD to seek, and provide, technical information. The Cuban Center for Information on Africa and Asia offers another example. They explained the difficulty they faced in obtaining information, and noted that they could not reach the WWW. Shortly thereafter, they were receiving information from organizations throughout Asia and Africa. A small company in Kenya is also illustrative. They complained, "In developing countries E-Commerce and the Internet services remain a Chinese puzzle, even to the local Internet Solution Providers." Within days they had received information to help unravel the e-commerce problem they faced.

GKD also serves as a channel for news items, editorials, and speeches from media in Southern as well as Northern countries. Often, these articles present information about initiatives in the South that relate to the use of information/communications technologies. As a result, GKD members learn a great deal about specific conditions and initiatives in Southern countries, which are not discussed via other broadcast media.

Not only the level of participation, but also the density of GKD message traffic has been impressive. The number of messages fluctuates, rising after a facilitation message is posed by a GKD moderator. Just before the GKD Conference, message traffic reached an incredible, all-time peak of about 2,000 per week. In general, GKD continues to receive about 250 messages per week — an extremely high volume. These include substantive messages that are posted to the List, requests for help or information, and messages that are not posted but receive a response from the Moderators (the Moderators' role is described below).

What is important, of course, is not the quantity of messages, but the quality. This is where GKD is preeminent. Messages posted to GKD are highly substantive and provide concrete information on actions being taken to use ICTs to support sustainable development. It is eminently clear that people on the GKD List value the information they are receiving. Many have sent messages thanking

GKD moderators for the List. Perhaps most telling is the large membership that has been sustained for almost three years.

Impacts

The GKD List has made important, unique contributions to our understanding of the role of ICTs in development. It has also created a new way to build a global learning community. It has further demonstrated that an Internet List can serve as a substantive, highly inclusive worldwide forum, with substantial Southern participation, for focused, value-added discussion of key development issues. Seeking to emulate the success of GKD, many organizations have subsequently decided to use an Internet discussion List to support a physical conference, discuss development issues, and share information. At the launch of its Development Forum, for example, the World Bank acknowledged GKD as a model, inspiring their use of Internet discussion lists to gain substantive insights into development issues. GKD has made three key contributions that are particularly noteworthy:

- Affected Conference outcomes through improved participation
- Provides communication opportunities that are otherwise impossible
- Improves the quality and quantity of information disseminated

Affected Conference Outcomes. It is difficult to document the impact of the GKD List on the Global Knowledge Conference in any precise manner. A large number of recommendations and cases were presented during the GKD List discussion, providing an extremely rich resource of information. Nonetheless, presentation of these cases and recommendations at the physical Conference could only be done informally, when and if individual Conference facilitators decided to recognize GKD moderators during a session.

Yet GKD did have a direct, innovative impact on some elements of the Conference. For example, when GKD moderators posted the GK Conference agenda, GKD members recognized the dearth of women speakers. They responded by strongly encouraging Conference organizers to involve more women as speakers, and provided the names of a large number of women who are outstanding in the field. As a result, many of these women were invited to the Conference.

GKD participants expressed a number of ideas that seemed particularly germane and helpful to the Conference goals. The sponsoring agencies responded by offering support for several of these participants to attend the physical Conference in Toronto. Panels were organized around the issues presented by these GKD members, who represented a wide spectrum of constituencies, from Quechuan communities in Peru, to Sub-Saharan African municipalities and districts. GKD thus made a contribution to bringing about more balanced panels and stronger participation from women in the Conference.

In considering the value of the GKD List, it is illuminating to review the external evaluation report of the Global Knowledge Conference. Although the evaluation examined all of the Lists and other "virtual conference" activities, it singled out the GKD for special recognition. The evaluation team admitted that initially they were rather skeptical about the value of the GKD List. Their hypothesis was that GKD probably involved only a small number of atypical individuals, mainly from North America, and, as such, could be viewed as a tangential activity to the actual conference.

Once they began to monitor the GKD List discussions, however, they quickly realized that the reality was quite different. They reported:

> The [GKD] List represents a large number of people from around the world, from developed and developing countries, representing many diverse viewpoints. It was clearly addressing the major themes of the conference, with many thoughtful interchanges, the provision of many examples, and consideration of the practical realities involving ICT and development. As we indicate later, the List may represent the most significant aspect of the entire Conference...
>
> ...the List constitutes a rich database, which complements and expands upon the issues raised and discussed at the actual Conference. *(Evaluation report of the Global Knowledge 97 Conference, World Bank, EDI)*

The external evaluators also reported on interviews with GKD List participants who attended the physical Conference. These reports revealed unanimous agreement that the nature and quality of the GKD List discussions were quite different from, even superior to, that of the physical Conference:

> At the Conference, the assumption was that ICT expansion was a priori great, with little critical analysis of who benefits and who does not. On the [GKD] List...while many people indicated the potential of IT, they and others also expressed concern about how IT could retard development and increase the gap between the haves and the have nots.. The presentations at the Conference tended to deal in generalities...and rarely reflected a grassroots...perspective. In contrast, the List discussion reflected a development perspective and provided many concrete examples of various approaches which have been tried and their effects... The style of presentation was totally different. At the Conference, it was..."experts" talking, and little audience participation... The [GKD] List, in contrast, involved actual discussion of issues among the participants.

Martha Davies, a GKD List participant who spoke at the final plenary of the physical Conference, captured the feelings of other GKD members when she remarked that the GKD "virtual conference" was more productive than the physical conference. She noted that: "We [the GKD participants] ...preferred the virtual conference" because GKD provided an opportunity for discussion and for attempts to find solutions, which could not take place at the physical conference.

Provides New Communications Opportunities. The GKD List has provided a unique communications channel. As mentioned above, GKD differs from most Internet Lists, whose members are predominantly North American. In contrast, GKD hosts an active discussion linking over thousands of people from more than 90 countries, half of them developing countries. In doing so, it provides a vast forum in which people from different places and cultures can "meet" and confer on the issues related to ICTs and development, over the course of years. List members thus have an opportunity to discuss these issues with people representing vastly different experiences and cultures. Many GKD members have been involved in the GKD discussion since its inception almost three years ago. Together, they have created a phenomenon that is often discussed but rarely achieved: a true learning community that has produced a deep, extensive knowledge base on the complex issues of ICTs and development.

This communications channel has contributed to a wide range of policy and program decisions as well as development activities. Take a message sent from an advisor with the UNDP:

> Last week, the Prime Minister of Tuvalu came to Fiji to give the keynote speech for a meeting on Governance for Sustainable Human Development. He asked me for advice in setting up Tuvalu's first Internet Service Provider. He is thinking of doing it through Tuvalu's telcom corporation, since they have relevant expertise. But even telcom specialists need guidance on the various tradeoffs...Where can the Prime Minister turn to for independent, knowledgeable advice?

GKD members from NGOs, universities and private companies responded with information on free software, Web site resources, and even pro bono assistance. Members also described the experience of other nations. For example, one discussed the lessons learned in many countries, including Haiti and Peru, which emerge from strong conflicts between local telecommunications companies and people working on Internet service provision.

GKD has also helped NGOs in the South to build their base of activities, and generally assisted in extending networks among educational institutions, companies and NGOs. Take Quipunet, an NGO that serves rural communities in Peru. Quipunet's Director uses GKD to obtain technical information, learn about funding sources, build collaborative networks, and disseminate news about Quipunet's efforts.

Improves Information. Many GKD members use the List to obtain information on a wide range of subjects, including technical information about "appropriate" ICTs, project designs, and references to useful resources. For example, one GKD member, who works with a rural NGO in Ecuador, sought help in his efforts to provide computer access to villages. The NGO promotes environmentally sound technologies, administers microcredits, and sets up computer kiosks on market days in the villages they serve. He queried the group:

> [Does anyone] have experience of using a car battery to power a computer...it's the cheapest power source available for us to run our mobile computer kiosk...

GKD members offered practical advice, based on their own experience. Some contributed detailed instructions in alternative techniques. Still others suggested alternative approaches, including powering the computer by hand cranking, a technology that had been described earlier on the List.

The value of these types of contributions is likely to be even greater for those in developing countries, and in "bypassed" areas of industrialized ones, than for those in prosperous areas of the North, for the former have far fewer opportunities to participate in such global fora than do those who are in more advantaged situations. Thus, it can be argued that GKD offers benefits that may be more important to members in the South than those in the North.

Key Factors
GKD's accomplishments rest on three key factors:
- Outreach
- Responsiveness
- Careful moderation

Outreach. List moderators have devoted enormous time and effort to reaching out to NGOs and others in the South. The moderators have sent invitations electronically to thousands of NGOs, universities, and other organizations whose members might be interested in joining the GKD. This level of outreach has been essential to expanding the number of countries represented on the List, and is particularly important in drawing members in the South. The outreach effort is ongoing, and has persevered throughout the life of the List.

Responsiveness. Early in the life of GKD, the moderators decided to make the List unusually responsive and "service-oriented." Moderators receive numerous requests for information or assistance on a daily basis. All requests receive a response, and moderators have tracked down information, provided technical advice and assistance, and helped disseminate special information. For example, on request, Moderators have queried government officials regarding policy statements related to ICTs and development, have helped GKD members resolve trouble with their computers or communications lines, and have obtained information from Web sites for those in the South who do not have Web access. GKD moderators also help those in the South who have important and valuable information to offer, by assisting them in presenting the information so that it receives the recognition and attention it deserves.

Active and Careful Moderation. As was the case for the Social Summit List and the Beijing List, GKD Moderators carry out typical moderation roles but with added service. They review every List message before it is posted to ensure that messages are substantive and relevant. Moderators generally do not post about 5-10% of the messages sent for distribution to the List, because they are irrelevant, insubstantial, use profanity, or "flame" (personally insulted) other List members. The moderators return the message to the sender and suggest ways in which the message can be modified to make it "postable." The importance of this moderation task cannot be overestimated. Those observing other Internet Lists often note that they easily degenerate into irrelevance or name-calling. It is clear that the GKD List could have suffered the same fate were it not for careful moderation — a very tiny number of voices could have undermined the value of the List for the vast majority. Yet, by returning the message to the sender along with suggestions for changes or an explanation of why it wasn't posted, moderators also avoid alienating the senders.

Another important role of Moderators is to help keep discussions productive. They develop and post a weekly summary of the discussion, provide documentation and reference resources, and ensure that messages are not redundant. Given the large number of messages, this role requires ongoing tracking of different strands of the discussion and past arguments.

A third role undertaken by Moderators is to stimulate discussion. Moderators pose questions, raise issues, and post information that generates discussion. They have also arranged and held panel discussions, e.g., an expert panel on the Y2K bug. In doing so, the moderators help maintain strong interest in the List, retaining current members and continually drawing new ones. Much of this role relies on "behind the scenes" research to identify documents and facts that will stimulate discussion.

Conclusions

We have described an etiology of what began as a new window on the UN and global conferencing efforts. We have shown that this window permitted not only unprecedented participation in the global events, but also affected lives at grassroots levels. As is the case of any innovation, it was difficult and at times frustrating to detect and address the various bureaucratic inhibitions which surfaced during formal consideration of the early proposals. Many were aware that the Internet has many sides, some quite unattractive. Much of the resistance was justified, on the grounds of concern for these darker sides of human expression, and for the hazards of exposing the UN system to them in untested ways.

Today they are tested, and have proven useful and practical, as well as relatively low cost. Fears that a well-managed List might undermine negotiation processes were not founded. There is a robustness to the functioning of the UN system which we believe is easily able to bear the light of scrutiny which this new window sheds upon it.

Experience with these three discussion Lists illuminates the questions posed at their inception. One key concern involved broadening participation as widely as possible. The Lists' demographics reflect the fact that interest in Internet discussions is expanding rapidly in developing countries, even among NGOs working in the field. The number of List members, and of countries represented on the Lists, grew at a startling rate between 1994 and 1997. Although this upsurge in List membership reflects a number of factors, including access to the technology, level of interest in the topic and awareness of the upcoming event, a "critical success factor" was the significant commitment of human resources employed to conduct outreach. Each time a concerted outreach effort was launched, the List membership grew. Establishing and using networks with large Southern representation played a major role in extending the List membership in developing countries.

There are, however, still areas for strengthening. The spread of the Internet has been explosive since the initiation of this idea into UN system functioning in 1994. Yet, as the current GKD List continues to document, penetration into poorer communities is still slow. This enormous, and some fear, growing gap is one of the most profound challenges facing development, especially at community levels. National capitals and elite institutions may be among the first to be wired, merely transferring the gap issues down another level. In addition, the twin pincers of postal telegraphic monopolies and censorship limit access even when the technology may be available.

Once the List membership is well established, the question of content arises. The vast majority of those who venture into an Internet List discussion experience frustration — or worse: mailboxes full of messages that meander off in diverse, unrelated directions, personal opinion unanchored in any facts or experience or downright nastiness.

Perhaps the most important contribution of the three Lists has been to demonstrate that an Internet List discussion can be highly substantive, value-added, and civil. The key to the quality of discourse is active moderation. A central role of Moderators is to uphold certain standards: that messages are directly relevant to the goals of the List, provide substantive information, are rooted in research or experience. Yet screening messages is by no means the Moderators' only task. Perhaps even more important, and certainly more difficult, are Moderators' efforts to help members craft interesting, informative messages. The interchange conducted "offline" directly with individual members absorbs considerably more time than screening messages. Add to that task several others: soliciting useful information from members and others who seem to have valuable experience (e.g., based on the survey that members are asked to complete); obtaining and posting useful information from the Web and elsewhere; creating informed, provocative facilitation messages; editing materials submitted by members; and regularly summarizing the key points presented on the List. The Moderators are pivotal to creating a lively, content-rich, informative and enjoyable discussion.

The Lists' experience suggests that discussions that benefit from this type of moderation emphasize three types of interchange:

- Exchange of information, ranging from strategic "lessons learned" to technical instructions
- Debate of the issues, some of which can generate considerable passion
- Presentation of "news," often from media sources that are not readily available to most members of the List

These types of content are described in greater detail above. What becomes clear is that a well-moderated List can garner important information that is otherwise very difficult to obtain, e.g., descriptions of NGO projects in the field. List members, with support from Moderators, can create a unique and valuable knowledge resource. Feedback from List members, as well as their ongoing participation, attest to the important role this type of forum can play in the development field.

A discussion List can also help to "democratize" a global event and make it more inclusive. Many people who lack sufficient resources to travel to a global conference venue manage to get access to email and can participate virtually. It is also evident that List members present their views frankly and (with occasional nudging from the Moderators) evince respect of others' views. Unlike many face-to-face fora, the List discussion is comparatively egalitarian. Every List member has a chance to express his/her position on a topic.

Of course, stubborn barriers remain. The dominance of English, and other so-called "colonial" languages, continues to obstruct participation by those who have important contributions to make to, and benefits to gain from global discussions on development issues. Although messages were posted to the three Lists in any language that uses the Roman alphabet, most were not translated, and thus reached only those who could read the language of the message. As a practical matter, List members must be able to make their way through an English email in order to glean most of the value from the List.

Yet another threat to the potential impact of a List lies in the response of the host agencies. Although a List can provide value for its members, its impact on a conference — or on other decision-making processes — depends on the host organizations. Formal links between the List and the conference/process that it aims to influence are essential to ensuring that the List makes a contribution and fulfills its stated goals.

Governments and NGOs alike are grappling with these issues, but it is essential that practical solutions be sought by a wide coalition of interests, including the private sector and, of course, local people and communities. A broader concept of "governance," involving active coalescing of these constituencies, has been promoted for some time by UNDP's Human Development Reports series. One of the major tests of new governance approaches will be the extent to which they address gaps in access to technology, as well as utilize technology, through such vehicles as Lists, to expand meaningful participation in development decision-making processes.

Future Trends

Electronic discussion Lists are now a fixture in the UN system, as well as among other development agencies. Among the recent additions is an internal staff discussion List on UN system issues (http://www.onelist.com/subscribe.cgi/unsystem). It seems aeons ago when Internet Lists were not a part of daily desktop/laptop life. Yet it is important to learn from past experience with these types of fora in order to fulfill their promise for the future.

The experience with the Lists described above presents both promise and warning. Internet Lists can deliver tremendous value, to members, to host agencies, and to the wider development field. Technically, setting up an Internet List is trivially easy, further affording opportunities for valuable applications. Yet therein lies the threat to its potential value. Development organizations, once loath to sponsor an Internet discussion, now eagerly rush to join the "information revolution" by setting up a List. Yet they often ignore the central role of skillful moderation and the need for a clear channel into the decision-making process. Without talented and experienced moderators, extensive outreach efforts, and useful approaches to utilizing and archiving discussion inputs, Lists become *pro forma* exercises that mimic but do not embody genuine, valuable participation.

Yet, apart from the human, technical and cost considerations of running value-added Lists, despite the limitations and the threats, the overriding prognostication must be one of hope. Even in contexts where swift political and ideological cross-currents swirl around the discussion platforms, Internet discussion Lists — when skillfully managed — can provide invaluable input and information exchange. If more sustainable, human-centered development policies are to result from today's global political deliberations (such as future world summits), a well-managed electronic dimension must be paramount.

Communities must learn how to adapt these technologies most usefully to the goal of practical participation and the shaping of more livelihood-sustaining policies at global and national levels. Development organizations must be willing to invest the resources required to make Internet Lists a truly inclusive, substantive and value-added forum. They need to heed List input and incorporate it into their decision-making processes. The Social Summit and the Beijing Conference, along with other Internet-based fora, formally acknowledged the *unsustainability* of our current growth patterns. The GKD List warns of unsustainable trends while offering concrete alternatives. Major problems addressed by all of these summits — disempowerment and discrimination (sometimes of majorities by minorities), acute poverty, social exclusion, and steady, uninhibited environmental degradation — remain with us. But the experience of the three Lists described above provides evidence that the Internet can offer an avenue for more open, participatory involvement in the search for solutions to these issues. The UN system has once again demonstrated its capacity for tapping into the spirit and energies provided by thoughtful, cooperative engagement and action by ordinary people. Hopefully, this trend will continue, and the medium will flourish to the point of becoming central, rather than peripheral to national policy debates worldwide.

References

The Beijing List Message Archives. (1996). gopher://gopher.undp.org/1/unconfs/women/fwcw.

Beijing95-women List, Association for Progressive Communication (APC). http://www.igc.org/beijing/post/post.html.

Earth Negotiations Bulletin. http://www.undp.org/fwcw/dawngo.htm.

Education Development Center. (1995). *Exploring the Use of the Internet in Support of the World Summit for Social Development.* Interim Report to UNDP.

Education Development Center. (1996). *The New Global Forum: Expanding Participation in UN Conferences via the Internet.* Final Report to UNDP.

Fourth World Conference on Women Web site material. http://www.undp.org/fwcw/> and <http://www.undp.org/fwcw/dawngo.htm.

GKD List message archives. http://www.globalknowledge.org/discussion.html.

Hopkins, M. (1999). *The Planetary Bargain.* New York: Macmilan.

International Women's Tribune Center Global Net-FaxNet activities. http://www.undp.org/fwcw/dawngo.htm.

Social Summit List archives. gopher://gopher.undp.org/1/unconfs/wssd/wssd and ftp//:ftp.edc.org/pub/mailing-lists/beijing-conf-digest.

Social Watch. http://www.socwatch.org.uy/.

United Nations Human Development Reports Series. <http://www.undp.org/hdro/99.htm.

United Nations system issues. http://www.onelist.com/subscribe.cgi/unsystem.

United Nations Web site material on global conferences. http://www.un.org/News/facts/confercs.htm.

UN Wire. (6 July, 1999). ICC joins UN Global Compact.

Women's Environment and Development Organization electronic conference on the 180 Days Campaign. http://www.undp.org/fwcw/dawngo.htm.

World Bank/EDI. (1997). *Evaluation Report of the Global Knowledge '97 Conference.* Washington, DC.

Chapter XXII

Community and Technology: Social Learning in CCIS

Roger S. Slack
University of Edinburgh, Scotland

Introduction

This chapter outlines the development of a community information service in the Southeast of Scotland — the Craigmillar Community Information Service (CCIS). It develops a socio-technical analysis of the development of the service and draws out some potential lessons regarding the development of community identity through the use and application of ICTs. What we present here is not intended to be a model which others should follow — it is our belief that each implementation of ICTs in the community is unique, and that the same is true of the solutions. Rather we aim to highlight some problems and their local solutions in the hope that community information services can find something of value for their work.

The Social Shaping Thesis

It is important to examine the ways in which society and technology exist in a relationship of mutual shaping — no technology, however powerful, completely configures its users. Rather technology is a component of a larger picture that includes the efforts of promoters, users, regulators, legislators and others to both extend and constrict the uses and meanings accorded a technology within a given societal context. As Williams and Edge point out:

> Every stage in the generation and implementation of new technologies involves a set of choices between different technical options. Alongside narrowly 'technical' considerations, a range of 'social' factors affect which options are selected — thus influencing the content of technologies, and their social implications. (1996, p. 866)

Williams and Edge note that social shaping should not be taken as if it were an inversion of the technological determinism — social shaping attends to 'the complexity of socioeconomic processes involved in technological innovation' (idem). The trajectory that technologies take does not reflect a unitary discourse —

they are driven by a multiplicity of competing and often contradictory discourses and attendant choices. 'Significantly, these choices could have differing implications for society and for particular social groups. The character of technologies as well as their social implications are opened up for enquiry' (idem).

Using the social shaping thesis, we can see that technologies such as CCIS are *negotiable* — that is, what a technology comes to be is the result of a complex series of interactions originating in different groups (suppliers, users, managers, community groups, other projects). No one group has primacy, although some discourses might be more powerful than others (regulators, funders) and will shape the technology in certain ways. Williams and Edge point to the ways in which the choices made earlier in a technologies' history serve to constrict the potentialities that are available later. For example there are extant models of what a community ICT service can be (in terms of scope, technology, democratic intent and content — see, for example Schuler, 1996) ; there are regulations as to what is fundable by grant awarding bodies and also timetables in which remits have to be attained. As we shall see, each of these plays a part in constraining what CCIS *can* become. This does not mean that we should substitute a weaker form of determinism in terms of 'lock in,'[1] but that we have to appreciate the impacts that choices (which are themselves products of socio-technical ensembles[2] [Bijker, 1993]) have.

Social Learning

The term 'social learning' originates in the constructive technology assessment approach advocated by Rip et al. (1995). It stems from an attempt to link evolutionary accounts of technological development with those from a social shaping perspective. The term attends to reflexive relations between technologies and social actors and the manner in which they come to practical choices about the uses of technologies in particular contexts. We have used the term to focus on the negotiated character of technological development wherein there is participation from a number of sources, and thereby a learning process which encompasses a diverse range of players in any given socio-technical ensemble. Social learning is praxiological — that is, people learn through doing. Sorensen (1996) has drawn out a series of categories of social learning upon which this paper draws. He notes that social learning can take place in terms of efforts to regulate, experiment and so on. We can see how the players in any given setting learn from each other and how this has an impact on the technology and what it can/will become. The studies from which this case was taken were made across the EU, and focus on the various forms of regulation and development of technologies and actor constituencies together with the ways each impacts on the other.[3]

We look at the development of CCIS as an example of social learning in that the extant models of digital communities, the work of local community groups, users of technology and funders were brought into play in the development of the service. That is to say, the service did not simply develop for example as a result of the availability of the technology or the will of society, but was the result of a reflective dialogue with various actors and technologies as to what the service could be which changed over time. Social learning directs us to a diachronic and dialogic development of socio-technical ensembles — new questions, new problems, new possibilities and new solutions open up over time. New knowledge accrues as a result of these processes and shapes the services offered as well as the notion of the service per se. Projects such as CCIS bear the marks of learning — what they are and what they can be is a result of the extant knowledges in play.

In this way, social learning extends the social shaping thesis to include a temporal dimension that attends to the always-preliminary nature of socio-technical ensembles. In what follows we show how the service came to 'learn to be fundable' and to marshal some discourses over others in its development; we show how the service took on new user constituencies and how this was achieved practically and rhetorically; we show how the service developed notions of civil rights and how these were employed over time and by various groups. In short we supply a history of the ongoing struggles to develop a community information service.

What can those readers interested in developing their similar services learn from our history? As we have said, all cases are unique and it would be wrong to look to cases such as this for a blueprint on how to build an information service. It is our intention to shift away from the notion of an ideal-typical blueprint to a series of elements that could be articulated as part of a configuration elsewhere. From the perspective of social learning, we see how CCIS emerged from attempts to emulate earlier examples of community ICT projects. We can also say that this case could be one of the many elements in a constellation of factors impacting on the development of future projects. The experiences of those involved have something to say to both present and future projects — these materials are useful to think with. This applies to the people concerned in the project as well as those outside — as a part of our work we discussed the case and the evolving findings with those involved. This was not a revelatory exercise but enabled the participants to reflect on the history that they had produced and to consider (again) the ways in which the service was heading. The reflective practice (Schön, 1983) also allowed the researchers to expand their understanding of the situation and to correct misunderstandings.

We do not propose that our work is a model of reflective practice or that what we did had any great impact—it is after all only a part of an ongoing process. However, it is important for the conduct of social science research and especially for research in community organisations. The sociologist does not have the solutions, she is not normally in the business of producing revelatory materials, but the chance to reflect cooperatively is important. Although at perhaps the 'softest' edge of activism, we believe that reflection is important in that it allows all concerned to stand back and examine what they have made and what can be made of it in the future. We hope that in what follows there will be some cause for readers to reflect on their own practice.

Background: Craigmillar

Craigmillar is on the periphery of Edinburgh, both geographically and eco-nomically. The industries on which the area traditionally depended—brewing, brick making and mining—have either relocated or been closed completely, leaving the local population of more than 11,000 largely underemployed. With unemployment rates several times higher than for Edinburgh as a whole, over 80% of those living within the area receive some form of welfare benefit.[4] The area has acquired an unenviable reputation, being considered a locus of social problems (underemployment, declining population, crime, drug abuse and the like).

The development of Craigmillar has been described as follows:

In Edinburgh, as in most cities, the socially marginalised are also spatially marginalised. From the 1930s to the 1970s, a number of housing estates were built on the edges of the city and many of these are now officially classified as areas of multiple deprivation. (Rose 1996, p. 88)

Composed largely of local authority housing built after 1940, Craigmillar is roughly four kilometres east of the city center. The area that surrounds it is largely disused factories or 'waste' land. Some of the larger factories have now been renovated and are rented as units to small businesses.

Despite being perceived as a locus of social problems, there is a strong sense of community upon which many local groups build. There is, however, a marked difference between those groups that are a part of the community, often being organised from within, and those that follow the 'social work' model and are mainly located outside the community. The latter groups tend to be regarded with some suspicion by residents. Groups such as the Craigmillar Festival Society, on the other hand, tend to be well thought of and to have a large base of participants in their various initiatives (in this case community arts and employment initiatives). In our analysis of CCIS, it is important to appreciate that there is a tradition of community activism within the area, and a substantial number of community groups, including CCIS, which actively build upon this. Yet CCIS is not a part of other community groups, it is a hybrid that while encompassing enterprise promotion and social welfare, exists between the two domains (and as a connection between them in the sense of providing communications as well as ICT skills).

It is important to consider the synchronic and diachronic dimensions of projects that we discuss. It is also correct to suggest that the history of the area and the local culture (a term used here in preference to 'subculture') should be taken into account when explicating the uptake and development of projects. In what follows we aim to examine social and cultural factors as they impact on the development of CCIS. It should be noted that in doing this, it will be necessary to look to a wider context (i.e., outside Craigmillar) since this is in part the basis for the development of CCIS.[5]

The Prehistory of CCIS

Against this background, the project of connecting community groups based in Craigmillar to each other and to a wider network was initiated in 1994 by Community Enterprise Lothian (CEL), a nongovernmental body located in the voluntary sector, and funded by the Scottish Office to encourage 'enterprise' within communities regarded as deprived. There is also an element of serendipity[6] in that the Scottish Office was at the time looking for a location for such a project. The original application for Urban Aid funding states that:

The profile of the area...suggests there are widespread needs in terms of information, skills development and employment opportunities which

existing local organisations are attempting to meet. A project which develops a more coherent and strategic information package is therefore likely to compliment the work of these organisations and help them become more effective.
(Craigmillar Community Information Project, Urban Aid Submission. Section 3, § 4; 'The Need for the Project')

The plan submitted to the Scottish Office envisaged a community information service connecting voluntary and community groups within Craigmillar with the intention of providing information on funding available for welfare initiatives. Within the established infrastructure the majority of agencies already had some kind of computing capabilities. CCIS aimed to connect these groups through the provision of modems and 'bulletin board' software (BBS). The service was originally modelled on the Manchester Host electronic village halls. However, the cost of such a large project would have exceeded the funding available. In this new model the structure of CCIS was based around what we might think of as the 'welfare geography' that existed in Craigmillar. CCIS mirrored the community welfare structure of Craigmillar and traded in part upon the pool of information available from the welfare organisations. The aim of CCIS was, as has been noted, to link community groups together—not to provide citizens with a point of entry to the developing information infrastructure. In this sense, it would be correct to regard CCIS at this point in its development as an information service for community welfare agencies as opposed to a community information service *per se*.

The original remit of the project was directed at producing a community information service that was a part of a broader host system that enabled SMEs and other institutions to gain access to information networks in order to take part in the so-called 'information technology revolution.' As we shall see, this is substantially different from the final project. This is important here since it forms a central part of the translation terrain for CCIS. In short the original aim, as outlined above, was to have a community host project connecting Craigmillar's voluntary groups and local businesses. The aim was to enfranchise the local citizens through training in IT skills and provision of information.

CCIS has three employees and is located in the old high school in Craigmillar. The aim of CCIS was to provide a BBS connection for community groups within Craigmillar and a center for generation of an ICT skills base within the community (to promote employment opportunities). As we shall see, this remit has changed over the course of the project, and there is now an additional concern to enfranchise citizens through the provision of Web-connected terminals. Under the rubric of 'cyber-rights,' CCIS can be seen as both a technology project (in terms of connection provision) and a social welfare project (for the same reason). In discussions with the project manager, it has been stated that the technology is secondary to the provision of Web connections for the community (which the project manager regards as a 'major civil rights issue').

Discussing what they call 'frameworks of obduracy,' Law and Bijker note that:
Bijker explains action by relating it to the way in which actors are shaped by and implicated in a network of relations. There are commitments,

explicit or otherwise, to economic investments, normal practice, and skills. (1992, p. 303)

As we will show, this is certainly the case for CCIS—the service was originally 'shaped by and implicated in' a distinct set of relations: those of promoting local enterprise and the generation of IT skills in the community. It was only later that these networks became somewhat problematised by the lack of uptake of the service by community groups based in and serving Craigmillar[7] (including the withdrawal of the original project planners [see below]). Thus the original client base for what was to become CCIS was attenuated even before the service had the opportunity to obtain funding,[8] leading the management group to search for new constituencies (a *leitmotif* in the history of CCIS).

A History of CCIS

To summarise, the concept of the Edinburgh Host had been developed by Community Enterprise Lothian(CEL), based on the Manchester Host. City Council advice was sought in preparing a bid to the Scottish Office for funds. The aim at this point was to cover the Craigmillar area as opposed to the wider coverage envisaged previously for the Lothian Host. After the usual six month review process, the Scottish Office gave funds for the project. During this time CEL undertook a review of their own work and it was decided that the Craigmillar Host was not central to their work. At this point funding was awarded and the project was transferred to the Craigmillar Initiative which took on the then Edinburgh Host appointing a project manager.

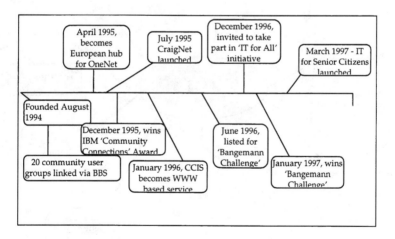

It is useful to think of CCIS as *a stable organisation employing changing technologies*. That is to say, the core group of personnel and the attendant manage-ment structure were stable — the project manager and the board have been in place for some time. There has been a migration of emphasis, so to speak, from the original BBS server-based concept to a Web-based service (although the BBS has been retained). Some users still employ the BBS, and others have moved to the W3 server. Within CCIS there has also been a number of developments of service, especially the establishment of the CraigNet (a community BBS and e-mail server intended to be used by members of the local community dealing with issues relating to Craigmillar as a community) and a growth of use by distinct community groups (such as seniors and school students). Recently CCIS became a limited company with charitable status, enabling the sale of Internet accounts from Scotland On-Line in addition to the existing services provided by CCIS.

A Possible Mapping of CCIS

A brief description of the diagram below is in order at this point. CCIS has been chosen as the focus since it has been the point of orientation for the research. Readers

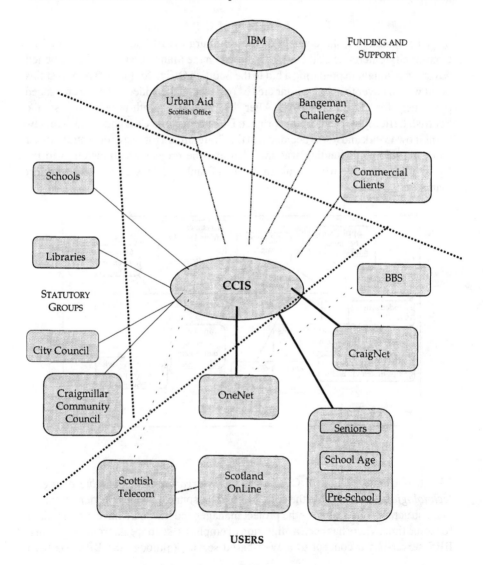

should note that CCIS does not have a large organisational structure, the manager is responsible for the day-to-day running of the project. However, the sum of the parts (shown here to the left of the diagram) constitutes CCIS as an operational entity. On the whole the use is grounded in the community, as shown by the heavier weight lines.

Since CCIS has become a limited company with charitable status, it has been able to take on commercial operations such as Scottish Telecom and Scotland On-Line (in terms of reselling leased lines and accounts for Web services). In addition, CCIS has an input to One Net—being the European hub of one of the largest free bulletin board services in the world.[9] These relations are shown in dotted lines to the lower part of the diagram.

To the left we have located a number of statutory groups, such as the libraries in Craigmillar, the City Council and Craigmillar Community Council. CCIS works with these groups as a part of its *raison d'être* — placing terminals in the library, obtaining funds from the city and community councils for further work, and so on. CCIS is not directly regulated by any of these groups — rather they constitute a realm of what we might call 'interest groups;' groups which it is politically expedient for CCIS to work with (both to promote the service and to involve new user constituencies).

Those groups located at the top of the diagram, linked to CCIS by chain dotted lines stand as funders or supporters of the project. In short, the project was founded with money from the Scottish Office and later won awards from the Bangemann challenge and from IBM.

It is important to note that the diagram does not cover all the dimensions of CCIS in that there are a number of personnel who can be found in various capacities in a number of organisations connected to CCIS. To represent the complex links within such a community would render the diagram unreadable.

Recurrent Themes and Tensions Within CCIS

Having given a detailed description of the history and development of the project, we would like to move on to examine some of the recurrent themes and tensions within the project. These can be seen to originate in the original conception of CCIS and the subsequent addition of new user constituencies. The discourses around these new constituencies turn on differing articulations of the original remit and the rhetorical use of the term 'community service.'

Development of CCIS

CCIS was originally conceptualised as being a part of the landscape of community groups—a local BBS service through which community groups could share information. The user constituency implied by this model did not develop in the manner expected. That is to say that while a number of groups were connected (indeed the targets for connection set out by the original plan were more than met), the service did not fulfil the evolving discourses of access extant around the project. At the same time, the growth of the Web meant that, to an extent, the BBS became unfashionable and hence less than attractive for new user constituencies. Thus, at least in part because of the simple availability of the Web (and its apparent popularity), a decision was made to base some of the operations of CCIS on the Web.

The original user constituency, still being catered for through the older technology, regarded this as a move towards a service that was in but not of the community—a service that was not grounded in the 'needs' of Craigmillar. One community group interviewed formulated this feeling as follows:

> Speaker One: "It would be difficult to say off the top of my head exactly what the original remit was. It was Community Enterprise Lothian (CEL—RS[10]) who came up with the idea and put the proposal forward. I imagine that it would have been more locally based than it has turned out to be. (. . .) I would be interested to hear what CEL's opinion is on the way the project has developed."
>
> Speaker Two: "Given the name of the project, it might be viewed as something of a misnomer. There is in my opinion a contrast between what the name suggests and what it delivers, at least thus far. I do wonder whether the 'grassroots' of the 'business' have been compromised by too great a focus on the bigger IT picture."
>
> (Interview 8/20/97; Amended 6/3/98)

There are three currents within the CCIS study; the first of these is ongoing and concerns the development of the remit; and the establishment of a user community of the project. Secondly, there is the development of that user community over time and the eventual reconceptualization of the user community in light of the uptake of the system. Finally there is the development of technology over the time of the project. The trajectory of the project is determined at least in part by the interaction of the three themes along with their independent/autonomous operation.

The Users and the Community

As noted above, the first part of CCIS's development was undertaken around community groups. As the manager of CCIS has pointed out (in informal discussions), the target set for a user constituency among the voluntary groups was met and exceeded in a relatively short time. While the target was relatively modest, we can see this fact as an indication of the success of the service in its own initial terms. There was a concern within the original project to get information from community groups about their remit, opening times, staff, location and the like—to create a common base of knowledge for those working within the community. The manager of CCIS, notes, "The paradox is that when we went to groups [in the community], the original idea was that the architects had a fetishism of databases and it was like 'get your information on' (. . .) but now I think it's about access, it's not necessarily who's got the information" (interview, CCIS Manager 7/8/97,extract from p. 3).

The use of CCIS by new users was achieved in part through a redefinition of the notion of the community to be served — accomplished in part by enhanced definitions of access and civil rights. The community was redefined to include groups outside Craigmillar (often community groups or voluntary organisations) enfranchised by the operation of the OneNet hub. CCIS *qua* hub for One Net was able to partially redefine its remit as a community service. The users of CCIS still include a number of the community groups originally targeted, but there is also a substantial user community drawn in by the OneNet service and the provision of Internet terminals in CCIS offices and the local library. Such a diversity of users creates a variety of tensions, some of which are outlined in the comments above.

Obviously, CCIS was intended to be a community service, serving the Craigmillar area. The original intention was to enfranchise local citizens, who were seen as deprived of opportunities to gain familiarity with Information and Communication Technologies (ICTs) and thus disenfranchised with regard to skills required by employers. Despite (or perhaps because of) recent additions to the services of CCIS, there is still a feeling that CCIS is a representative of Craigmillar and that it stands as evidence that the area can improve its image through the presence of IT-based projects in the community.

The definition of a user by the central actors is contingent on which users and which actors are being discussed. The question seems to turn on actors' uses of the term 'community' and the ways in which they define users in relation to that. Some users (e.g., those on OneNet) are regarded as being neither in, nor of the community, and thus a symptom of the ways in which CCIS has diverged from its original purpose. Others, such as local Web users, are regarded as being in, but not of the community—being too outward looking.

In short, the terms 'user' and 'community' are *rhetorical resources* employed by a variety of actors, variously to criticise the project from moving from its community base or to gain further funding for it through the expansion of its scope. The following sections describe the dynamics of technology and users in more detail.

Technology

Following the original establishment of the user constituency, users' needs have developed on an *ad hoc* basis. It is perhaps useful to think of the development of user needs as a *dialectic* of the available technology and the perceived user base. By this we mean to draw attention to the relations between technology and users as they are managed in situ by CCIS. CCIS has to manage the tension between the available technologies and the user groups. In practice this means that CCIS tends to use the Web and BBS servers. The service could be solely Web-based, however this would be problematic for some of the users of the BBS (whatever their means of accessing the OneNet, for example) who would then be disenfranchised by the adoption of alternate and incommensurable technologies. Disenfranchisement is, of course, an important issue for a community-based BBS provider: there is a need to be seen to provide certain services whatever their status vis-à-vis the current state of play in available technologies. In sum, a community service has to provide (to support in some cases) a range of modes of access that a more commercially oriented provider does not. They do this because of their status as community providers.[11]

The development of CCIS took place at a time of radical change in the field of ICTs. When CCIS was envisaged, the Web had not taken hold to anything like the degree it has today, and this in part accounts for the choice of a BBS as opposed to any other interface. It should be noted that CCIS was based in part on models of similar developments available at the time, and that it was only in what we have called the 'establishment' and 'redefinition' phases (see above) that other technologies and models of organisation took hold. Of course, these were in part shaped by the available technologies: discourses surrounding these technologies drawn from the media and more local sources of opinion entered into a dialectic with notions of what a community service should do and what its coverage should be. As we have seen above, the repertoire of technologies (the BBS, OneNet and the Web) have invoked a number of discourses from the management committee and from the wider community: some of these have been hostile and limiting (often suggesting that CCIS should not undertake new services because they 'remove' the organisation from the community base); others less critical (indeed the fact that the services are in existence at all within the operations of CCIS is testament to the agreement, albeit tacit, that the service should evolve).

Some of the difficulties that CCIS has encountered may turn on the fact that their service emerged at a time of paradigm shift and migration. In a sense, then, CCIS was a service whose provision was grounded in technology largely incommensurable with the Internet. This was to prove problematic in that the popularity of the Web grew exponentially (despite the potential of the first class-based service offered by OneNet). In order to enrol more users, it was seen as important to provide some Web service. Yet, as the following quote illustrates, the project manager regards the Internet as just part of the wider picture regarding enfranchisement through ICTs:

> (W)hat we do know for a fact, and this has been verified by what's interesting listening to other community experiences throughout the UK and other agencies is that email is king. It's the BBS, it's the email, it's the FTP stuff...It's not the multimedia, the kids come in on a Friday afternoon and look at the Spice Girls and the football team pages, and then they want to access their mail box. I take them to a Web café and they say 'can we no get on the CraigNet?' because they are getting email from South Africa (. . .) and Boston and that excites them. It's just bread and butter stuff and the Web is just king, it's like a social convenor, you can sit in front of it and amuse yourself for an hour.... (Project manager, August 7, 1997. Modified June 5, 1998)

There is, then, a realisation that both locally available (such as the CraigNet) and global services (the Internet) need to be available if the population of a deprived area is to be fully enfranchised through ICTs. There is a dialectic of the local and the global that turns on the discourses of enfranchisement and community service. However, within our study, some participants were critical of the perceived move away from the community through the adoption of a Web interface.

Community as a Rhetorical Resource Embodied in Interfaces

Among some of the people with whom we spoke during the study, there is an apparent tension between the BBS and the Web interface. We argue that notions of community can be seen as manifest in the interfaces used by the various services. That is, the BBS is regarded as being rooted in the community and offering a community service while the Web has been regarded as in but not of the community. The project manager appears aware of this point, and regards the critiques as expressions of parochialism:

> Our market is not just the local community (. . .) we're the biggest free
> BBS at the moment in the UK (. . .) and we don't realise the community

in terms of just being parochial—about, you know, down your alley in Craigmillar (. . .) like "what about Craigmillar?".(. . .) *(T)his is a world-wide phenomenon, and it's this thing about local depth and global breadth* (. . .) it's an international global phenomenon, so where does it start and where does it stop? (. . .) *(I)t's not just about Craigmillar, although Craigmillar's our primary remit.*
(Interview 7/8/97. Emphasis added.)

However, a significant number of critical comments that we have encountered in the research have been predicated on the premise that in moving away from the BBS, CCIS has turned away from the community and begun to look outward. The conservatism of those in community groups finds resonance in the formal conservatism of the BBS interface. The original remit involved *inter alia* the need to enhance the cohesion of the local community and to present what had been seen as a 'problem area' in a more positive light. In other words critics of the current trajectory of the service use the BBS as a rhetorical resource: there is a perception that the new service is not required since the BBS is adequate 'for Craigmillar.' The CraigNet BBS is taken as a benchmark in concentrating on the local dimensions of the service.

From the above, we can argue that technology is bound together with notions of community and with ideas of what CCIS should be doing. These elements are combined with the perceived growth of a diverse user base to make an assessment of how far CCIS is fulfilling its remit. Put simply, the prevailing view within the community groups appears to be that there is no need for anything other than simple technology as envisaged in the original remit. The argument runs that in using new technologies there is an implicit move away from CCIS's 'roots' in the community by moving away from the apparently easy-to-use community BBS.[12] We noted above that there was a conservatism with regard to the notion of 'community' espoused by those within community groups and that this is reflected in the discussions regarding the interface—those who regard CCIS as 'moving away' from the community couch their critiques in terms of the transition to the web (and the growth of other services based on the web interface).[13]

We might summarise here by saying that CCIS began using a closed standard technology—the BBS—and has now developed using a more open technology—the Web. In effect, CCIS has learned (and *had* to learn) how to employ emerging technologies so as to extend its remit to make the organisation viable. Questions of technology, community and viability, at the level of a user base, are all linked as components of the social learning situation.

There is in all the above a question regarding degrees of what one might term 'openness.' Again it is related to the technology. In a sense CCIS was originally a 'closed' system, akin to an intranet which could be accessed by those in community groups and in effect stood proxy for the interests of that community (at least as far as welfare and voluntary groups were concerned). The move to the OneNet technology was still rather constrained, but at a global level—in effect it was still a BBS, which by and large required one to be a member of the community to use it. The move to the Web led to a very different concept of what CCIS was about and its constituency; in effect CCIS was being opened to the world, with all the accompanying critiques.

Social Learning in the Case of CCIS

The main dimensions of social learning in CCIS have taken three forms:

- The establishment of the service within the local area from a model derived from other cases, namely the Manchester host. This led to a service in search of constituents, so to speak: while the groups to be connected within the plans drawn up by the originators of the service existed, there was a need to enrol users and connect them. One might say that there was a mutual shaping process in that the service was shaped by the practical exigencies of its implementation from the original plan, and that the putative users had to learn that there was a new way to communicate. This process might be thought of as a reflexive shaping of service providers and user constituencies. Over time there developed a number of user constituencies (those located in the local community, users of the OneNet BBS and those employing CCIS as a means of gaining access to the Web) during which this reflexive shaping process continued. Indeed one might point to the questions around CCIS's role (discussed above) as a part of this reflexive shaping process: CCIS attempted to ensure that it was usable by each group (and therefore fundable) while each group in some sense shaped CCIS itself (e.g., through the continued advocacy of BBSs as a means of 'communication in the community' where other clients used the Web).

- As CCIS developed over time, the technology available has also developed. There was a transition from the use of bulletin board services to web based provision. Within the original user constituency, this has meant that there have been competing discourses around the notions of community and service provision. It has been argued by a number of those with whom we have discussed the project that as a community service, the BBS is in and of itself enough—further, there is an undercurrent of feeling that suggests a move away from the BBS constitutes a 'move away' from the community basis of the service. The construction by community-based users of a discourse regarding the nature of an appropriate community communications system appears to be opposed to the use of Web interfaces as a means of communication. In this sense, there has been a failure within the reflexive shaping process discussed above. It is important to note that the discourses of community service and appropriate interfaces are intimately linked (moves to a Web-based interface being seen as moves away from the grounding of the service in the community and towards other user constituencies). In sum, CCIS has learned to employ new interfaces for communication, but the community user constituency has not learned that the use of these interfaces (often by new constituencies) are a part of the development of CCIS that maintain it as a viable community communications service.

- With regard to this last point, CCIS had to learn to be fundable through the provision of services to other user constituencies. One might contend that if CCIS had been restricted to the originally defined user constituency, it would have ceased to exist. Given the original remit, one can contend that the manner in which CCIS was expected to develop envisaged a wider use of the system *within* the community. As discourses surrounding the notion of access changed,

CCIS came to depend for its viability on other user constituencies; these were part of the social process of learning to be fundable through the enhancement of its user constituency. CCIS has successfully learned how to exist in a niche between its originally proposed remit and the space taken up by larger service providers such as AOL and CompuServe. To be sure, CCIS is still grounded in the community, but one can see that the community is a (substantial) part of the service not CCIS *in toto*.[14]

Having suggested an outline for the dimensions of social learning, we would like to fill in some of the details below.

CCIS has evolved from a community-based BBS through to a Web-based service running alongside a BBS server for both local and European groups. From our research, it is difficult to say how far the users have been involved in the design. Users were involved in the steering group of what was to become CCIS, and their vision for the system, as discussed above, was based on the Manchester Host. Users, insofar as they can be regarded as members of the community have been more involved in the management of the service, and that the service itself has been assembled from available technologies. That is, the origins of the service as a BBS reflected the technology available at that time—more accurately, it represented the technology available to a community group at that time (with regard to considerations of cost, usability and the like).

The main motivation for the service was that connection to the information society was an intrinsically good thing, and that the connection should be made; *however* it was made. Following from this, the community should be enrolled, notably through existing groups within the area and the provision of open terminals for community use. In short it might be said that the social-technical was the key configurational aspect of the project; users were enrolled as a result of the project's existence—as opposed to being the drivers. However, a caveat should be made to this statement in that the steering committee was made up of possible users, and they must have influenced the options (if only through the amount they were willing to pay).

It is fortunate for CCIS that the Web has evolved at the pace that it has, and that computer equipment capable of being connected to the Web has become significantly cheaper over the last three years. To be sure, CCIS has evolved some unique solutions to the environment in which it is located (the CraigNet for example). These have been facilitated by the availability of the technology at reasonable cost—there was no *a priori* need for CCIS to move from a BBS to Web, the availability of the technology might be seen as a key driver.

Interviews with management committee members indicate that there is a degree of difference as to the appropriate trajectory for CCIS. Some management committee members have said that they feel the system is moving away from its base in the community, becoming too outward looking and thereby eroding the remit of serving Craigmillar. Others have voiced similar concerns, but noted that technology has developed at such a pace that it was inevitable that the older community-based philosophy would be problematised; especially in the migration from BBS to Web technologies. The key tensions within CCIS appear to be grounded in the community-wide use dynamic. The notion of community is used as a rhetorical resource by those who critique CCIS's current trajectory (hence the charge of conservatism). The growth of the Web does not appear to have been taken as a possible reason for the dual focus of CCIS.[15]

Importantly, from discussions with members of the management committee, it would appear that these sentiments are kept out of the running of the project—the main purpose of the committee being to work through such conflicts while maintaining CCIS as a viable entity. Since the researcher has not had access to management committee meetings it is difficult to give any idea of the dynamics of management in CCIS.

The notion of user enrolment is key to an understanding of CCIS: it is a feature of CCIS that they seek to enrol distinct groups, such as the Senior Citizens. These people, together with the community groups, form the bulk of the user constituency. Some of the most interesting user constituencies are constructed firmly within the community, and what regulation there is takes the form of addressing presumed areas of interest for those groups. An example might be of use here: senior citizens have used the Web as a means of sharing community memory. However, their use is not limited to this niche — during observations at CCIS, the 'Cyber-Grannies' were shown how to use the Internet and were enfranchised as a user constituency within CCIS. Other groups such as parents and students of the local high school have also been successfully enfranchised within user constituencies. It is important to see this enrolment of new user groups *within* the community as a key component of CCIS's strategy as a community service. If there is any regulation within the user constituencies it is at the level of ensuring enfranchisement—in the majority of cases discussed with the project manager, users have to be brought to CCIS by the activities of 'entrepreneurs' either working in CCIS or in organisations closely associated to it in the community.[16]

We might question the relationship of CCIS to the traditional conception of a social experiment with technology in Scandinavia (for a discussion of the Scandinavian approach, see Cronberg et al., 1991). To be sure, CCIS is an experiment in that there is clearly a great amount of learning going on and that there is a service made available to an array of user constituencies where no service existed previously. In that CCIS has overtly social as well as technical aims, it might be regarded as a social experiment: the sheer fact of its aim to connect a depressed community to the information infrastructure must count for something here. One could conclude that CCIS is a social experiment within the fabric of existing community welfare provision—an embedded experiment with information and communication technologies—as opposed to a Scandinavian type of social experiment. Certainly, the nature of the debates and articulations of 'community' and 'community service' indicate that CCIS is by no means an uncontested terrain—as noted at the start of the paper, CCIS is only now undergoing a redefinition phase.[17]

Conclusion: Where is the Social Learning? What Can We Learn from CCIS?

If social learning is going on, it is to be found inter alia at the level of learning adapted after formal implementation (the promotion phase). We would contend that one might discern two distinct strands of social learning within CCIS's history: that taking place at the level of becoming an organisation within the voluntary and community group network in Craigmillar; and, more interestingly for our purposes, the stage *following* consolidation wherein CCIS redefines a part of its services. This redefinition can be seen to parallel the growth of interface technology (i.e., the move from BBS's to the Web) and the potential for funding inherent in the growth of ICT's in the public domain.[18] The reflexive link between the various user constituencies and the service is the prime site for this social learning (and for the evolving situation to be contested, as we have seen). Perhaps the main dimension of social learning has taken place in the development of CCIS as a number of fundable services: in the UK at least, services such as CCIS have to learn to be fundable since there is severe difficulty in obtaining funding for continuing projects.

Further, through this report it has been our contention that there is a mutual shaping process at work within the various discourses of community. As the term 'community' has been employed within and around CCIS and its operations we can say that there has been concerted activity to increase the community that CCIS serves. At the same time there has been a putative shaping of CCIS in and through the discourses *within* the community: some of these discourses have employed 'community' as a critical resource, suggesting that in taking on new constituencies CCIS is in some way moving 'away from the community.'[19] This shaping of CCIS as a community service is reflexive in that CCIS in some senses defines and is in turn defined by the community it serves.[20] In sum, as well as learning to be fundable, CCIS has to establish itself within the discourses of community both in Craigmillar and the other user constituencies that it serves; we would contend that this shaping process is at the heart of social learning for an enterprise such as CCIS.

What can we learn from CCIS? As noted at the start of the chapter, the first and most important lesson is to reflect on the development of a project and the marshalling of the socio-technical ensembles that constitute it. The case of CCIS shows us how a service can learn to articulate itself in novel ways and thereby develop as it takes on new user constituencies. However, we have shown that this is not an unproblematic exercise and that the community outside the project might react adversely to the redefinition, bringing into play alternative versions of community and community service.

Perhaps the most important lesson is the notion of being fundable. In the majority of government-funded projects we can conceptualise the life cycle of a project as akin to a package on a conveyor belt — when it reaches the end it falls off, irrespective of the value of the contents. When a funding cycle is completed, the project can fall off the end and close — the achievement of CCIS was to move beyond this cyclical development and to work with new user constituencies to further the work of the project. The development of new user constituencies means that one can expand the scope and remit of the project. How this development impacts on the original remit is, as we have seen, a contested point. Yet the fact remains that in an environment where a project can be ended just as it begins to develop due to the exigencies of funding, it makes sense to expand into new areas and user constituencies whatever the potential disputes.

Clearly, there is a message for funders here too. Projects have a life cycle that really only begins when the funding is largely used up (hence the need to learn to be fundable by developing new initiatives), and to cut off the supply of funds at this point is to condemn a large number of projects to failure. This also means that there is a sense of 'another first time' for some projects in that while funding may cease, the problems that it addresses are obdurate. All projects have their own trajectories and the traditional funding terms of calendar years should be more flexible. Funders have to establish mechanisms whereby demand from projects can change over time and in accord with points in project life cycles. This is not to say that there should be unrestricted 'demand valve' funding, but that a realistic funding program must critically address claims for funds outside the usual cycles that such organisations normally maintain.

Beyond this, we can point to the utility of the social learning and social shaping perspectives as they afford a means of reflection for those involved in the development of community information services. Emphasis on the negotiated character of technological developments and organisational forms enables analysis of important issues (such as accessibility of interfaces, location, charter, involvement in community activities and so on) that constitute social inclusion and exclusion. That a development is negotiated means that it is potentially open to renegotiation. Participants cannot take the status quo as natural and inevitable — this engenders a sense of openness and might serve to make changes thinkable over time. Organisations and technologies, like ideas, have histories, and the social shaping thesis enables these to be traced. For those interested in the discourses of civil society, a history such as this is a valuable resource. Critical assessments of a service should proceed with regard to the negotiated character of the enterprise, and not as if some ideal-typical formulation was within easy grasp. Critics are a part of the social learning milieu also. The central message is that all involved can, in concentrating on the negotiated character of technological development, bring about some changes — it is perhaps the duty of the analyst to enjoin reflection as a means to change.

Afterword

Since the writing of this paper, CCIS has undergone another transition — in an email dated December 14, 1999, the project manager told me that there had been an expansion, and that the envisaged 'Teleport' scheme had been realised. He noted that

Quite literally, you would not now recognise CCIS Teleport. We have opened up our old office as a (...) Learning Centre with eight multimedia PCs. We now have the conference theatre & the Mac training suite and business is literally booming.

I am happy to add this information as it shows that CCIS has achieved another marked success — it is an exemplar of the concept of learning to be fundable — with great benefit to the user constituencies (both new and existing) that are enfranchised thereby. This development and others like it emphasise the importance of social learning as both a practical and an analytic exercise.

Acknowledgment

This chapter is based on research funded by the European Union's Targeted Social and Economic Research programme grant no. 4141 PL 951003. The author wishes to record his thanks to the manager of CCIS and members of the community of Craigmillar for their assistance in the research. Thanks are also due to Dr. Robin Williams of the Research Centre for Social Sciences, University of Edinburgh, for his comments on various versions of the paper.

References

Bijker, W. E. (1993). "Do Not Despair: There is Life after Constructivism." *Science, Technology and Human Values.* 18(1), 113-138.

Brosveet, J. (1997). *Frihus 2000: Public Sector Case Study of a Norwegian IT Highway Project.* STS Working Paper, 1/98 (SLIM working paper No. 3): Center for Technology and Society, Norwegian University of Science and Technology. ISSN: 0802-3573-147.

Cronberg, T., Duelund, P., Michael, O. and Qvortrup, L. (Eds.). (1991). *Danish Experiments — Social Constructions of Technology.* Copenhagen: New Social Science Monographs.

Law, J. and Bijker, W. (1992). Postscript: Technology, Stability, and Social Theory. In Bijker, W., and Law, J. (Eds.). *Shaping Technology/Building Society: Studies in Sociotechnical Change.* Cambridge, Massachusetts: MIT Press.

Rip, A., Misa, T. J. and Schot, J. (Eds.). (1995). *Managing Technology in Society.* London: Pinter.

Rose, G. (1996). "Community Arts and the Remaking of Edinburgh's Geographies." *Scotlands,* 3(1), 88-99.

Schön, D. (1983). *The Reflective Practitioner: How Professionals Think in Action.* London: Temple Smith.

Schuler, D. (1996). *New Community Networks: Wired for Change.* Reading, Massachusetts: Addison Wesley.

Sorensen, K. H. (1996). *Learning Technology — Constructing Culture.* Available at http://www.ed.ac.uk/~rcss/SLIM/public/phase1/knut.html.

Van Bastelaer, B. and Lobet-Maris, C. (Eds.). (1999). *Social Learning Regarding Multimedia Developments at a Local Level: The Case of Digital Cities.* Namur: CITA, Facultes Universitaires Notre-Dame De La Paix.

Williams, R., and Edge, D. (1996). "The Social Shaping of Technology." *Research Policy,* 25, 865-899.

Endnotes

1 That is, the adoption of one solution above and beyond all others.
2 Sociotechnical ensembles draw our attention to the symmetrical reflexive shaping

Appendix: A Pre-History of CCIS

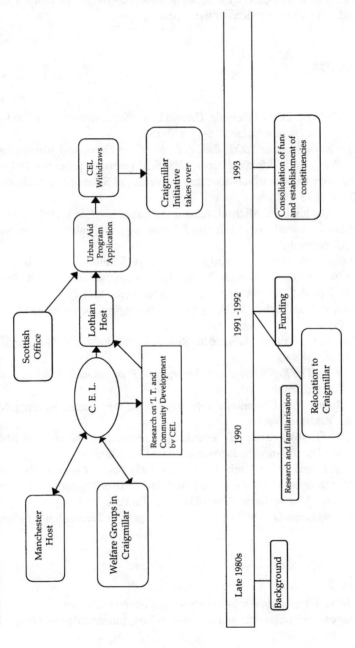

processes going on between the social and the technical. As Bijker points out, "the technical is socially constructed, and the social is technically constructed—all stable ensembles are bound together as much by the technical as by the social" (1993, p. 125).

3 See van Bastelaer and Lobet-Maris (Eds.) (1999).

4 Source: CCIS Urban Aid grant application, 1993.

5 Indeed it is might be thought of as the impetus for the development.

6 I.e., being in the right (fundable) place at the right time (to be funded).

7 Community groups here include those dealing with rights for unemployed, disabled, or otherwise disadvantaged citizens; those dealing with regeneration of the area; welfare rights groups; groups promoting arts, adult literacy, play schemes and other activities within Craigmillar; local community social work initiatives.

8 It should be noted that the original planners of the project withdrew in the time between the submission of the application and its funding.

9 CCIS receives all the traffic from Europe from the OneNet free BBS. Other hubs are located in the U.S. and Canada.

10 Community Enterprise Lothian exists to promote SMEs in the South East of Scotland, and was a part of the original development of what would become CCIS.

11 Arguably *qua* community-based providers they have to deal with parallel services that are incommensurable.

12 In interviews a number of respondents noted that while they had not used the BBS greatly, they regarded it as being rooted in the community.

13 In our transcripts and notes, no recognition of the changing dynamics of electronic communication is made by respondents when they discuss the idea of CCIS moving away from its community base.

14 The establishment of CCIS as a limited company with charitable status confirms the grounding in the community, but the place of CCIS as a re-seller of Internet accounts indicates that it has learned how to underpin its community roots through participation in commercial service provision.

15 It is useful to think of the situation thus: Web—outwith community and original remit—problematic; BBS—fulfils criteria of community service and original remit—worthy of support. Obviously this heuristic is over-simple, but it does enable those less familiar with the situation to understand basic dynamics of the argument.

16 This association is often at the level of members of the management committee—certainly this was the case for the senior citizens who were brought into CCIS by a local community education worker who was a member of the management committee at CCIS.

17 After having established itself as a fundable agency within the community (or at least the funders' perception of the community). Only now that the experimental-implementation phase is largely over can CCIS establish itself as a fundable body (especially with regard to providing Web connections for established service providers such as Scotland OnLine).

18 By this we mean that as the Internet comes to public attention, more people

become potential consumers—both at the level of using CCIS's existing services and those developing (for which CCIS acts as a mediator).

19 Formulations such as 'in but not of' the community have been used by interview respondents. While such formulations are significant, they are not widespread: readers should not regard CCIS as besieged by critics, yet there is a significant shaping process taking place within these discourses and others less critical vis-à-vis CCIS's role in and as a part of the community of Craigmillar. Further, in discussions the project manager rejected the notion of CCIS being 'in but not of' the community—although it is significant for the purposes of reflexive social shaping that he articulated the intention to further establish CCIS within the community.

20 Indeed the notion of 'service' is itself contentious and the object of ongoing negotiation, as we noted above.

Community Informatics
Case Studies

Chapter XXIII

Community Participation in the Design of the Seattle Public Schools' Budget Builder Web Site

Chris Halaska
Social Design, USA

This chapter provides a case study of the development of an Internet-based budgeting tool for the Seattle Public Schools, known as the Budget Builder. In particular, I describe the ways in which community participation affected the design and final outcome of the system. The Budget Builder project was unusual for a technical project because of its major focus on community participation. Although participation was stymied to some extent, the project can be seen as a success for community access. In the case study, I summarize the use of the Budget Builder over its first two years; describe the community participation and user input present in the design process; examine the social structure surrounding the Budget Builder, especially the division of power among the three main groups working on the project, and how those power relationships affected the final version of the project; and discuss some technical issues that appeared during the course of the project.

Project Overview

This chapter examines the development of the Seattle Public Schools' Budget Builder World Wide Web site. This site was designed as a tool to simplify and explain the process of budgeting at the local school level. A large team with members from the Graduate School of Public Affairs at the University of Washington, the Cross City Campaign for Urban School Reform, the Seattle Public Schools, and Seattle residents combined to produce the current system. As of this writing, the project is continuing; this chapter covers the first stages of the development of the Web site, from Autumn 1996 until the first official use of the Web site in April 1998.

The Players

The Seattle Public Schools (SPS) refers to the public school district in Seattle, Washington. SPS is a complicated enterprise consisting of several different groups, including the school administration, the principals, the school staff, the teachers, and the teachers union, the Seattle Education Association (SEA).

The Seattle Schools is one of the few school districts in the United States moving forward to decentralize their funding and to give schools control over much of their budgets. SPS is more progressive than most districts in making budgets public, particularly through the Budget Builder. After a presentation of the Budget Builder to the Teachers Union Reform Network (the progressive wing of the national teachers unions), several union members expressed amazement that the Seattle School district would freely provide so much information.

Within the school administration, the Budget Builder was considered primarily a principals' budgeting tool. For that reason, the district's budget office was the department most involved in developing the system, although people from human resources and information systems were key participants at various times.

The Cross City Campaign for Urban School Reform (CCC) is a non-profit organization based in Chicago that has school reform projects in several cities in North America. Cross City's interest in this project was furthering the movement towards school-based budgeting in Seattle. Cross City has supported school-based budgeting for many years, most actively in Chicago. From a community activist's point of view, school-based budgeting can provide an opportunity for communities to have a say in how their local schools are run, by moving financial decision-making authority to the schools. Communities can learn more about the details of school budgets and take part in linking budgets to goals for academic achievement.

The University of Washington Graduate School of Public Affairs (GSPA) is devoted to public policy training and practice. Its academic programs are geared towards Masters-level students heading into careers with public and nonprofit organizations. The faculty in the department are connected with organizations outside the university, and consequently, their research focus is often directed towards public institutions. While GSPA did not provide any official funding for the project, it sponsored the project indirectly by providing office space, equipment, personnel time, and official support from the university president.

The Scenario

The Seattle Public School District dramatically changed their school funding system in the fall of 1996. Before then, almost all of each school's resources were determined by the district's central administration. Using a method referred to as "staffing standards," each school's staffing level—both administrators and teachers—was determined centrally, based on the projected number of students. The district switched the resource allocation to a new plan, known as the Weighted Student Formula, in which resources were allocated directly to each school in the form of dollars. The dollars were driven by a formula that prescribed different weights for students based on educational characteristics such as grade level, special education level, or poverty level.

The district assumed that this move would create a dramatic change in the *process* of budgeting at the school level, as well as a dramatic change in the *mechanics* of creating budgets. Principals would be responsible for managing budget orders of magnitude larger than those they previously had to deal with. For example, under the previous system a high school principal might have had discretion over $50,000, while under the new system, she would control $3,000,000. Because of the magnitude of the changes, the Seattle Schools administration decided to provide a technical tool to support the principals' new budgeting task: the Budget Builder Web site. This decision was viewed by some as simply an attempt to placate the school principals, although the ramifications of the project were much more complex, as described below.

Joseph Olchefske, the school district's chief financial officer at the time, approached Anne Hallett, executive director of the Cross City Campaign, for assistance in creating the new tool. By chance, Anne met Andrew Gordon, a professor at the University of Washington, on an elevated train platform in Chicago. Andy and Anne have known each other for many years, from when they both worked on community development projects in Chicago. Anne described the Seattle Schools project to Andy and it intrigued him enough to draw him into the project. Andy and I were working on various projects together, so upon his return to Seattle, he drew me into the project as well.

Use of the Budget Builder

The first version of the Budget Builder was finished in the Spring of 1997. It presented users with a simplified view of building budgets that did not take into account many of the complexities of the Seattle Schools budgeting system. Because of that, principals were unable to use it for developing and submitting their budgets that year. Also, since it was the first year of the new Weighted Student Formula system, the district administration didn't want to push too many changes on the principals and decided to stick with a set of paper forms.

Throughout the following year, we demonstrated this simplified version of the Budget Builder around the country and made it available via the Internet. Many people saw the potential in the system and wished to have something like it for their own organizations. (People from large organizations such as Boeing and the United States Department of Defense seemed especially interested. I believe that they were responding to the open information and budget simplification represented by the Budget Builder qualities missing from their own organizations.)

For the Spring of 1998, a modified version of the Budget Builder was produced, which incorporated much of the complexity of the Seattle Schools budgeting system. It was presented to the principals as an option for developing and submitting budgets, although the paper forms were still available.

Although 46 of the district's 95 schools signed up to use the Budget Builder for their budget planning process, only five schools (all elementary schools, with smaller, simpler budgets than middle or high schools) went through the full process and submitted their budgets online. The big question, of course, is why didn't the others finish?

We tracked down answers by talking directly to people who used or failed to use the site, gathering feedback from the budget analysts who worked with principals to develop their budgets, and reading e-mail surveys of several school principals.

First, there was misinformation floating around which implied that schools would receive new computers or Internet access if they signed up to use the Budget Builder. Neither of those were true, and some people dropped out once they discovered that signing up didn't mean free technology. Next, there were many problems, including:

- Most schools (estimated at 80-90%) were not wired for Internet access, requiring that most principals who wanted to use the site had to do so from home, if they even had a computer of their own.
- Many of the schools did not have computers available to principals that were capable of connecting to the Internet or using the World Wide Web.
- Many of the principals were not computer literate and were intimidated by the site.
- The Budget Builder was a new system to learn, and in the last-minute crunch of building budgets, many people decided to stick with something they already knew—paper and pencil.
- The site had several problems: it was too slow, it was hard to get questions answered, it wasn't always clear where information was located, there was too much / too little information available, and there were confusing discrepancies between the electronic and printed materials.

On the other hand, we also heard many positive comments:

- All of the information necessary for completing a budget was in one place.
- The site made the budgeting process seem simple and straightforward.
- The computer dealt with the drudgery of calculations.
- Running totals of various budget balances were always available.
- The school-based budgeting explanations and examples were very helpful.

Within the school district, the focus was on principals using the tool to submit budgets acceptable to the budget office, so I was somewhat surprised and pleased to see and hear that people were using the Budget Builder for planning purposes, as those of us outside the district had hoped they would. Some principals said that even though they turned in their budgets on paper, they used the site to work on preliminary versions and as a way to look up information. While maintaining the site, I checked the server computer regularly and occasionally looked at the activity. I noticed one day that there were a several budgets with "scenario" in the title. On another day, I saw that someone had created a number of budgets named "Month 1 Budget," "Month 2 Budget," and so on. Those were obviously not going to be used for a final submitted budget, but were more likely planning documents for the schools.

Our only methods of gathering feedback from the community outside the school district were through the Discussion Forum on the site or via electronic mail. Unfortunately, there were only a few messages left on the forum. However, one of them praised the site and asked if we would develop a version for her city. The few e-mail messages we received were also positive.

Although use of the Budget Builder has been slow to take off, it will certainly increase in the future as more people become aware of it and have the tools to access it. It is also likely that the Seattle school district administration will require the use of the Budget Builder to submit budgets sometime in the next couple of years.

The rest of this chapter describes the development of the Budget Builder project by:

- detailing the community participation and user input present in the project,
- describing the power relationships I saw between the major players in the project and analyzing how those relationships affected participation,
- presenting technical issues we learned about the system design, and
- outlining plans for the future.

Community Participation

The potential for community participation in the design process was the main reason that Andrew Gordon, the Cross City people, and I joined the Budget Builder project. School-based budgeting can be empowering to communities if set up correctly, and all of us saw an opportunity to include community input in the technical development of the Budget Builder Web site.

The discussion of community participation is divided into several sections:

- a description of the different ways in which community was defined in this project,
- a discussion of each organization's (SPS / CCC / GSPA) effect on community participation and how they interacted, and
- a list of the participatory design and community-friendly steps we took in the design process.

Defining Community

One important factor in attempting to gather community input for this project was the difficulty in defining who was the "community." Each of the three main organizations had their own successively broader definitions of community. The school district administration viewed school principals and budget staff as their community. Cross City included those people, but also added parents, teachers, educational activists, and businesses. Andy and I had an even larger view, which included the previous people, plus staff, students, and anyone else interested in school performance. These different definitions caused conflicts about who to develop the tool for, as well as difficulties in how to think about including such a diverse number of people into the design process. Complicating matters even further is that the decision to make the Budget Builder available on the Internet broadened the definition of community to include people none of us imagined at the beginning, such as people from school districts in varied cities across North America, or people from large organizations such as the U.S. Department of Defense who found out about the site in one way or another.

An important question to ask is: Why do those of us working on the project have the authority to make decisions about community participation? A simple answer is that all of us on the design team are community members of our own cities, interested and affected by the public education system. Furthermore, Cross City staff, as educational activists, were deeply involved in grassroots organizing around school budget issues in Chicago and other cities. That gave them the standing to speak about the importance of including people from outside the district administration in budgeting decisions.

As a way to test that belief, after the Budget Builder had been officially introduced, we demonstrated the site and its features to many groups including the progressive wing of the American teachers unions, people working on school-based budgeting in many cities, and interested people in Seattle. Universally, the project was received with excitement, admiration, and sometimes envy. (People expressed great surprise that we had such easy and open access to the district's budget data.) Other people who had heard about the site visited it on the Internet and have asked that we develop a version for their school district. The positive response we have received from people who fall under a broad definition of community indicates that our community focus has been in line with the people we are trying to represent. The following sections discuss how each of the three major organizations involved in producing the Budget Builder handled community involvement in the design process.

Seattle Public Schools

One of the people with the greatest amount of influence over the project was Joseph Olchefske, the Seattle Schools chief financial officer. Joseph had originally conceived of the idea to create a simple, technical tool for school principals to build budgets. One of Joseph's strong points is that he always put his cards on the table, so it was clear from the beginning that his agenda was to have a tool built for the school principals, period. If other people were able to use it, that was fine, but he did not want to spend time or resources worrying about other audiences. With the introduction of the Weighted Student Formula, he was shifting a large amount of work from the central office to the principals, so he needed to offer them something in return to make their job easier.

Discouraging Community Participation

There were several times in the early development of the Budget Builder when Joseph discouraged any outside participation. The first came after Andy and I met with a group of school activist parents and later, with a group of principals who weren't always in agreement with the administration. We were trying to use a participatory model of developing the tool, so we started talking to potential users of the system about relevant issues.

When Joseph and Geri Lim, the manager of the SPS budget office, found out about the meetings, they got quite upset and reacted as if we had gone behind their backs to undermine them. At the next joint meeting with Joseph, Geri, Anne Hallett, the head of Cross City, Andy and I, Joseph demanded that we not meet with anyone without running it by him first. At the time, Andy, Anne and I couldn't believe he was asking for this but did not confront him.

The second time was in a later planning meeting at the SPS central office, with several SPS budget staff, Joseph, Geri, Diana Lauber of Cross City, Andy, and I. When we once again pushed the community input idea, Joseph once again reacted strongly, claiming that it was inappropriate. His rationale was that the Weighted Student Formula and all of its uncertainty for schools put principals in the awkward position of not fully understanding a new system and of having to explain it to questioning parents. By our talking to parents, we were letting the parents "get out in front of the principals," potentially embarrassing them. He said that he wasn't opposed to community input, but that it would be more appropriate once the controversy over the Weighted Student Formula had settled down and principals understood it better.

The last time he directly rebuffed our community attempts was after the Weighted Student Formula system had been modified by a committee of principals, teachers, and administrators, and approved by the Seattle School Board. During another planning meeting, we again brought up our desire to include community in the Budget Builder design. Once again, Joseph said that it wasn't the right time, that we needed to get the system working well first for the principals and then look to community concerns.

Soon afterwards, Anne, Diana, Chris Warden, also of Cross City, Andy and I met about this and agreed to focus first on the principals' version of the Budget Builder. One of the reasons we agreed is that Joseph and SPS were voluntarily offering to make budget data publicly available on the Internet. School-based budgeting can't proceed without data about teachers' salaries and other costs. Although the data is public information, in practice it can be difficult to obtain.

Because Joseph could have probably turned to other sources for the creation of the budget tool he needed, but we couldn't turn to other sources for the school district's data, he was in a better position than us to make demands.

Lack of Concern for Community Issues

Besides the specific incidents above, Joseph regularly demonstrated a lack of concern for community issues. One example of this was the way he categorized the various areas of the Web site. The Budget Builder is primarily divided into two planning sections and two budget-creation sections. After the four sections were fleshed out in early 1997, Joseph took to referring to the planning sections as the "magazine"—the sections with supplementary information you would go to if you had extra time on your hands. He declared that the important part of the Budget Builder was for the "real work" of creating budgets, since that's where principals would spend their time on the site.

Because people outside of the district administration would be those least familiar with budgeting, they would be the ones most in need of an explanation of school-based budgeting and suggestions for how to get started. Their "real work" would include using the planning sections as well as the budgeting section.

Furthermore, it became clear that the principals were generally not big fans of public input, although there were major exceptions to the rule. Many of the principals would have preferred to build their budgets in their office without bothersome distractions from outsiders. In early meetings with principals, we found that they were very uncomfortable with the idea of having their budget "drafts" available on the Internet. We ended up adding a password feature so that budgets could be private until the principals submitted them.

We also found the Seattle School District very conservative about when they felt that budgets switched from being internal planning documents to being public data. The official word from Geri Lim was that the numbers weren't public until the School Board had officially adopted the entire district's budget. Of course, this is long after it is possible for community members to have any meaningful input into a school's budget.

Proactive Decentralization

On a positive note, the transition the Seattle School District made to school-based budgeting, and their willingness to make budget data public on the Internet were significant steps. Only a handful of school districts around the country had attempted anything like that, and in many respects, SPS was ahead of most public school districts in its willingness to decentralize.

Cross City Campaign for Urban School Reform

The staff of the Cross City Campaign brought two major skills to the Budget Builder project: school-based budgeting expertise and fundraising connections. Anne Hallett, Diana Lauber, and Chris Warden had all been part of the movement to decentralize public school budgeting in Chicago. Anne had many local connections in Seattle from her earlier educational work there, as well as from her national work through Cross City.

A Unique Funding Relationship

Andy and I thought that Anne's work generating funding for the Budget Builder put us in a unique position. In Andy's experience with several technology projects involving both community groups and public institutions, it was unique for the community group to be bringing the funds to the project.

We believed that this funding relationship would allow those of us with a wider definition of community input to direct the design process more than if the funding had come through the school district. That belief was borne out to some degree, described below in the sections on participatory design meetings and community-friendly Budget Builder features that we included.

However, as detailed above, we agreed not to meet with community groups until the first version of the Budget Builder was finished. We followed this counter-community path because the district administration was providing public access to budget data that was crucial for effective school-based budgeting. Without that data, a community-focused budgeting tool wouldn't be of much use.

There was also a personal aspect to the decision to go along with Joseph Olchefske's no-meeting request. Anne, as director of Cross City and provider of funds, had the main authority to confront him and declare that meeting with community groups was essential to the project. However, she did not want to start a contentious debate, particularly in a community where she had built many long-term connections that she wished to maintain. This was not surprising given that various people described her strengths as a "diplomat" and "facilitator."

Part of Cross City's appearing to turn away from their community-focused roots may simply have been from being in a new situation. Diana Lauber explained that this was the first school-based budgeting project in which they were working cooperatively with a district administration.

Missed Opportunities

There were a few other ways in which Cross City could have pressed for community participation more actively but didn't: at several points, Cross City declined to put additional information about school-based budgeting on the Web site. Most of this information was in a printed form that they didn't want so publicly available. This seemed to be an issue of ownership of information upon which they relied for income for their organization.

In the Fall of 1997, I pushed to begin a series of community meetings to demonstrate, get input on, and train people to use the Budget Builder. I was repeatedly asked to wait to set up the meetings, first because Cross City was planning for their annual meeting in Seattle, and later because they were in the process of lining up a permanent staff person in Seattle. Andy and I went ahead and organized a (poorly attended) meeting anyway, but it lacked the "official" support of Cross City and so was not as useful as it might have been. It was not until the Spring of 1998 that we were able to set up the first Cross City-sponsored community meeting. By then, most of the features of the Budget Builder for that year were fixed.

Cross City staff were not aware of the extreme importance of getting community input as early as possible in the design process, since they were relying on Andy and I as technical experts, and we did not educate them appropriately on that point.

Graduate School of Public Affairs

Andrew Gordon and I had a large amount of influence over the development of the Budget Builder since we were in charge of the technical side of the project. We both recognized the importance of gathering community input early and often. For that reason, we set up meetings with interested parents and independent principals at the very early stages of the project. As discussed in the Seattle Schools community participation section, these meetings upset Joseph Olchefske, who feared they would undermine the groundwork he was laying for the Weighted Student Formula.

Difficulties in Gathering Community Input

In addition to that most obvious impediment to community participation, there were several other challenges and missteps in including community: despite our recognition of the importance of early input, Andy and I did not communicate this understanding very well to our Cross City partners. They looked to us for the technical details, without understanding the interaction of the technology with the social environment. We should have made the point very strongly that decisions made early on and features agreed upon in beginning stages of the project would be very hard to reverse later, if we found that they didn't support wider community goals. This might have caused Cross City to be more adamant about meeting with community members early, in opposition to Joseph's wishes.

One of the most important reasons for not including community is that my role switched from managing the overall content of the Web site to actively working on the budget-building code, when we abandoned a Java version of the budgeting section of the site for an HTML/database version. Since the budget-building section was acknowledged as most important once we accepted Seattle Schools' priorities, that's where most of the development time was spent. This caused my technical work load to increase, decreasing the amount of time available for community-related issues.

In addition to the practical reason for not focusing on community concerns, there were also personal reasons related to the change in technical focus. As I became more intimately involved with the workings of the budgeting section, I became more aware of all of the shortcomings and problems still to be fixed. I spent most of my time focused on fixing those problems instead of thinking about community input. Unfortunately, I was in a no-win situation (which I did not recognize at the time). Because the HTML/database version presented several inherent disadvantages, it would never work the way I wanted it to.

Andy offered at times to find someone else to work on sections of the code, but it was difficult to separate out independent pieces of work for others to do. Also, we had previously had difficulty finding competent technical help, and one person who was supposed to help with a section didn't come through. All of those caused me to feel uncomfortable with giving up control of the program I had already put so much work into.

Related to the previous point, the Budget Builder project turned out to be a far more enormous and involved project than we imagined at the beginning. In some ways, the amount of work overwhelmed our ability to do it and the resources we had available. There wasn't enough time to do everything we wanted to do. At the beginning of the project, when we were excited about the possibility of an Internet-based system, we were seduced by the technical talk of how wonderful Java would be as a programming language, and that it would allow one to easily accomplish complicated tasks on the Internet. Unfortunately, the hype was overblown. The rudimentary state of the Java language, compounded by having an inexperienced Java programmer, made the programming task nearly impossible. It would be much more feasible at the time of this writing. We spent quite a bit of our limited time following a path that was more challenging than it first appeared.

On a more social level, we found that trying to coordinate aspects of the project with a group in another state led to many difficult communication problems. In addition, the logistics of arranging meetings and worrying about travel costs were complicated by the distance, all of which made it difficult to set up community meetings.

Another practical social problem is that it was very difficult to find many interested people to give input. For our community meetings, we contacted all of the educational activists we could find—community education organizations such as Powerful Schools and the Alliance for Education, principals who were known to work with their local communities, and others. Our meetings were still sparsely attended. Since school-based budgeting was new to Seattle, and relatively uncommon throughout the country, it's not surprising that few people knew about the topic. Furthermore, budgeting—although extremely important—is not an exciting topic. The community organizing task is still in its early stages.

Community Surrogates

Because direct participation of community outside of the school system was minimal, Andy and I were effectively acting as community surrogates in our work on the Budget Builder. Our overall principle was to maximize access to information for as many people as possible.

Were we effective surrogates? At first look, it might seem that we weren't, since we acceded to Seattle Schools' requests to postpone community input until after completing a principals' version of the Budget Builder. However, it is likely that pushing for community participation at that point could have ended our involvement with the project.

It then must be asked if it was it better for us to be present, with whatever we could offer, than to leave and have a technical team without community concerns replace us. In other words, what would the system have looked like without us? It almost certainly would have had fewer community-friendly features and narrower participation in the design than it did have. It also likely would:
- have been more centralized, with either a spreadsheet-based system distributed only to principals or an Internet-based system that was accessible only to principals;
- not have had as much data publicly visible; and
- not have placed the budget-building tools in a context of site-based budgeting information, examples, and assistance.

Participatory Design

As a result of our participatory motivation, we included people from several communities in the design of the Budget Builder, which resulted in many community-friendly aspects of the design.

Participatory Design Meetings

Below is a list of design meetings with various participants in the process, in order of the number of meetings held with each group:
- SPS Budget Staff (20+)
- Cross City Campaign—School-based budgeting experts (10+)
- SPS Systems Staff (10+)
- Principals Steering Committee (6)

- Interested Parents (3)
- Activist Principals (1)
- Teachers Union (SEA) (1)

- In addition, there was a fair amount of unstructured user testing of the site by several individuals.

One can see that the design meetings were heavily weighted in favor of the Seattle Public Schools administration (36+) over those with groups who might have had different interests (15+). For that reason and others, much of the design, particularly in the budget building section, is heavily geared towards SPS administration interests.

Community-Friendly Budget Builder Features

However, many additional features were the result of requests from the non-administration design meetings, and of the site designers keeping a wide view of the potential user community:
- budget data publicly available on the Internet instead of only accessible via paper or Excel spreadsheets internal to the school district;
- budget forms and guidelines publicly available on the Internet instead of only available to school administrators;
- school-based budgeting expertise provided through an on-line handbook;
- school-based budgeting examples, ideas and case studies provided online;
- cross City Campaign commitment to a multi-year process of training the Seattle educational community about school-based budgeting;
- passwords or registration not required to use the site unless a user specifically wanted to keep a budget private; and
- site was designed for users with limited equipment such as small monitors, slow Internet connections, and older computers.

While limited in some significant ways, community participation and concerns had a direct effect on the design of the Budget Builder.

Power Relationships

To more fully understand the development of any technical system, it is always useful to examine the social structure surrounding the technology and the places where power is concentrated in that system. In the case of the Budget Builder, there were three significant types of power: the project funding, control of information, and technical expertise. Each is discussed below, followed by an analysis of their interaction.

Funding

Most of the funding for the initial stages of the project was raised by Anne Hallett of the Cross City Campaign. Because she and her organization brought significant financial resources to the table, their principles strongly guided the direction of the Budget Builder. The fact that community participation was raised as an issue at all, that people outside the district administration were considered important users of the system, and that Andrew Gordon and I were brought into the project all sprang from Cross City's focus on strengthening community control of public schools.

However, although it was unique to have the project funding controlled by an organization different from the one which wanted the technology, we found that it didn't provide enough leverage to regularly include users from outside the school system in the development process. Several times, our attempts to gather community feedback were resisted by Joseph Olchefske, the Seattle Public Schools chief financial officer. Anne, as power-holder because of her fundraising, acquiesced to Joseph's requests partly because of diplomatic concerns, partly because she did not realize the importance of early feedback from all users, and partly because her power was overridden by the power held by Joseph.

Information

Joseph's power arose because the Seattle School District offered two important features to Cross City, Andy and me: firstly, it provided a chance for Cross City to interact cooperatively with a major public school district while advancing its school-based budgeting goals. In other cities where Cross City works, the districts have not always been so cooperative, particularly about sharing data. Secondly, the district budget office was instructed to quickly provide all necessary data for the project. Without the data on staff costs and revenues available to schools, it would be nearly impossible to develop accurate budgets for schools. While it would have presented some difficulty for Joseph to replace the financial contributions of Cross City, and the technical expertise of Andy and me, it would have been more difficult for us to replace what he could offer.

The school budgeting data are public information, but obtaining that information can be difficult in practice. Even as the main developer on this project, it was challenging for me to get comprehensive listings of data necessary for the Budget Builder. It would likely be more difficult, or effectively impossible, for people outside the system to get access to all of that information. Furthermore, Geri Lim, the budget office manager, and later CFO, repeatedly defined data and budgets which were being developed by principals as planning documents not public information. In her (and the district's lawyers') opinion, budget data was not public until it had been formally adopted by the school board. However, it is difficult, if not impossible, to make significant changes to budgets at that point, defeating the purpose of citizen access.

A few principals gave us another reason why current, accurate data is important: although principals have nominal control over hundreds of thousands or even millions of dollars in their school budgets, after accounting for staff and other relatively fixed costs, they effectively control only several hundred or thousand dollars. Principals repeatedly told us that those small amounts of discretionary money made a large difference in the educational experience they could offer their students. Inaccurate or out-of-date information could easily throw off a budget by more than the amount that principals normally have to play with.

Technical Knowledge

A third source of power was held by Andy and me as the technical developers of the Budget Builder. Our technical expertise allowed us to make suggestions and give technical guidance to everyone in the project, and gave us the responsibility for implementing the ideas agreed upon in design meetings. Because we were often the ones generating ideas about how the Budget Builder should work, we were able to proactively add community-friendly ideas to the site and put other people in the position of reacting to them. Although our ideas weren't always accepted, by coming from a position of technical knowledge, we were often able to set the parameters for the discussions.

For example, we were able to implement many community-focused features that likely would not have existed otherwise, such as: putting the Budget Builder-including the budget data, forms, and guidelines-on the Internet instead of keeping it internal to the district; providing a context of school-based budgeting explanations and examples, so that the site was not strictly about the mechanics of budgeting; and keeping the site open to the public, without a need for password access.

Power Distribution

This project was unusual because of the separation of responsibilities and power between three organizations with different agendas. In many technology development projects, the responsibilities lie within the same organization, which ostensibly has goals that all of its members adhere to (or are at least aware of). In many other cases, the technical work may be contracted out to a different organization, but the contractors are under the direction of the original organization. In this case, each organization had various goals that were not in complete alignment, which prevented a simple, unidirectional approach to a shared goal. Each group was able to derive power from assets that the other groups needed but did not have themselves.

From one point of view, the Budget Builder project can be looked at as an example of potential community participation being stymied by an organization

Table 1: Power Relationships During Budget Builder Development

	Goals	Responsibilities	Assets	Power
SPS	system for principals	describing system, assembling data	system knowledge, data	could withhold difficult- to-obtain data
CCC	expand school-based budgeting & community control	raising money, school-based budgeting expertise	money	could withhold money
GSPA	community control, research project	technical development, community input	technical skills	could withhold expertise, could add features

looking to minimize "troublesome" community involvement. Because the Seattle Schools had data that were not easily accessible without their cooperation, we were forced to back down somewhat on our demands to include community in the development process. From a different point of view, the project can be seen as a success for community access, although a first step with much more work to come. The Seattle Schools were going to develop some sort of system to aid principals in their new task of developing school budgets. Because Cross City and GSPA brought a strong community orientation as well as money and technical expertise, we were able to turn the Budget Builder into a tool that is accessible to a much wider audience than the Seattle Schools ever envisioned.

The Missing Community

In the analysis above, an important fourth group was left out because it effectively had no power in this situation: community members outside the development team. Although citizens are regularly claimed to have ultimate power over public institutions through voting, there are many important day-to-day activities of those institutions that are far removed from that power. The development of the Budget Builder was one of those activities. Even though it was created in a public institution, it was developed mostly out of the public eye.

System Design Issues

In addition to the community participation and power relationships involved in the Budget Builder's development, some important system design issues surfaced. A recent article about lessons learned in Web development presents four pitfalls that developers of Web-based systems should be aware of (Dennis, 1998). Unfortunately, we learned nearly the same lessons they did, avoiding only one of the problem areas they outline. Their four conclusions (Don't chase new technologies, Users aren't techies, Training is critical, Web information systems are not just collections of Web pages) are listed as the following headings, while the explanations are from the Budget Builder project.

Don't Chase New Technologies

We were originally misled by the hype and promise of using the new programming language, Java, to develop the budget building module of our site. It wasn't until several months into the project that we learned that Java was in a rudimentary state, without support for many common features such as scrolling and printing that were essential to the project. There were good reasons for trying to use Java, most notably that it would have allowed a more interactive version of the tool. However, pursuing a Java version of the site used up money that might have been better spent on other tasks, and slowed down development of the site by taking up meeting and work time on features that were never completed. Now that Java supports some of the earlier missing features, has matured as a programming language, and has more widespread support, we may want to revisit the decision to abandon Java for HTML.

Users Aren't Techies

This was the one pitfall we were aware of from the beginning and did a good job of avoiding. However, there are many ways in which this trap can disguise itself.

"Don't assume that users are technical" is a maxim in the human-computer interaction field, but it can be hard to always keep in mind, especially when one is focused on a myriad of technical details. As a designer, it is very easy to get caught up in the technical sophistication of a particular feature, such as linking the Web site to databases or generating charts on-the-fly. The more difficult a feature is to implement, the more pride a designer takes in its success. However, the usefulness or usability of a particular feature is not necessarily linked to its technical complexity, and many times, the reverse is true. While designing, it is important to be aware of that distinction and hold the user's needs and abilities in the forefront.

From the beginning of the project, non-technical users were assumed to be the major users of the site, so we designed its features accordingly. We included navigation aids, on-line help and glossary information, and avenues for users to send us feedback. In addition, the Seattle Public Schools developed a training manual and offered training classes for their principals.

One important and often overlooked aspect of the technical design process concerns the hardware and software used by the designer. To develop an Internet site most easily, the designer wants to use the fastest computer hardware, speediest Internet connection, and latest software possible. This allows rapid creation of new versions and quick testing of changes. As a designer, it is very easy to get used to such a high level of capability. However, very few end users are likely to have equipment at the same level. The designer must constantly keep in mind the realities of the users and design accordingly.

In our case, we assumed that users would have slow connections to the Internet, older computers, small screens, and older versions of Web-browsing software. While the university computer on which I did most of the work was relatively fast, and had a fast Internet connection, my home computer was substantially slower, had a smaller screen, and relied on a slow modem to connect to the Internet—precisely the conditions that we were designing for. Regularly testing the Budget Builder with my home computer helped to keep our wider audience in mind.

Training is Critical

One of the important shortcomings in this project was our failing to provide adequate training for many of the site's potential users. Because we agreed to focus our work on the principals' version of the Budget Builder, there was no training of people outside of the district. The public meetings we held were demonstrations more than training sessions. For the principals as well, training was minimal, although at a somewhat higher level. Jay Iman of the Seattle Public Schools developed a training manual and offered training classes for the principals. The classes were poorly attended, however, since the principals were extremely busy. It is likely that other methods for training principals will need to be tried, such as sending experienced users out to the schools for one-on-one consultations.

Another crucial piece of the training link was left out as well. The budget analysts in the Seattle Schools budget office were assigned the task of assisting the principals with the Budget Builder, as well as their normal task of assisting with budget development. However, the budget analysts received no training in the use of the Budget Builder and received less formal support than the principals. It was assumed that the since the analysts were familiar with the budgeting process and forms, and were computer literate, they would have no problems using the Budget Builder. While I didn't hear about any problems directly from the analysts, they never gave any feedback during the design process outside of development meetings.

Training is crucial, although fostering it was challenging in this situation, particularly since anyone with access to the Internet was a possible user. Since it is impossible to train all potential users of the Budget Builder, more effort needs to go into providing a clear and simple user interface.

In addition to the technical training, Cross City is planning to sponsor school-based budgeting training sessions for anyone in the community who wishes to learn more about the topic. The most important piece of this training will not involve direct use of the Budget Builder, but will instead be focused on school-based budgeting principles and grassroots activism. Use of the Budget Builder as one tool to achieve school-based budgeting goals will occur later.

It is certain that with its limited resources, Cross City will not be able to directly train every community member interested in using the Budget Builder. For that reason, training will likely take the form of individuals who are familiar with the Budget Builder training each other informally. That likelihood reinforces the necessity of improving the Budget Builder interface so that it can be used with minimal training and support.

Web Information Systems Are Not Just Collections of Web Pages

My failure to recognize that the Budget Builder was a rather complicated software program, rather than a collection of Web pages, was a primary reason that there was no structured user testing of the site. Although user testing for readability is always a good idea (Morkes and Nielsen, 1997, 1998; Nielsen 1997, 1998), a site editor with a good knowledge of technical communication principles can be reasonably confident that a site will be understandable with standard editing methods. However, "understanding requirements is more difficult for Web information systems than for traditional Web development because users interact with the system much more than with simple Web pages. Unlike Web page development, in which users play small roles, user participation is as critical to Web system development as it is for traditional system development" (Dennis, 1998). Future improvements to the Budget Builder will be accompanied by more structured user testing, to give clearer guidance to the development process earlier on.

One reason why I didn't fully appreciate the systemic nature of the project was that the district budget staff was unable to describe their budgeting system for many months. In many early design meetings, I asked, "What does a complete budget consist of?" and usually got different, overlapping answers. We decided at the end of 1996 that the Budget Builder wouldn't be ready to use for the Spring of 1997. That had two effects: firstly, the route we took for the Budget Builder involved a highly simplified view of school budgets. This continued the impression that the Budget

Builder was a simpler project than it turned out to be. Second, because we weren't able to figure it out for them, the Seattle Schools budget office had to determine what a complete budget was made of. They eventually produced a set of budget forms and guidelines (known affectionately as "The Green Book," after the color of its cover) that described everything a principal needed to develop a complete budget.

A result of more clearly defining the existing SPS budgeting system was that the longer we worked on the project, the more the Budget Builder grew to look like their internal systems. Although we started out working from a clean slate, the site got more and more complex over time—partly out of necessity, because it's a complicated system—but also because SPS didn't attempt to simplify their systems. In the second year, we tried to make the budgeting section of the Budget Builder become an electronic version of the Green Book. Eventually, it approached the Green Book's complexity enough that Chris Warden of Cross City said, "The Budget Builder has become a data entry tool for the Seattle Schools!" She recognized that by increasing the complexity, we had lost the ability to do simple, straightforward budgets that had existed in our earlier version. We ended up restoring the simplified version and allowed people to work on whichever version they choose: the simplified version for planning or the Green Book version for submitting budget office-approved budgets.

Next Steps

The Budget Builder is a work in progress that is likely to continue for many years and in many forms. The Seattle Public Schools have committed to eventually using the Budget Builder as their preferred method for submitting school budgets. In addition, the Cross City Campaign has talked about expanding the Budget Builder to other cities.

Licensing the Budget Builder to the Seattle Public Schools

The Seattle School District wants to continue modifying the Budget Builder in future years to more closely mimic their internal systems, as well as to add features that principals have requested, such as real-time budget monitoring and modification throughout the school year. Because Cross City owns the Budget Builder, the two organizations recently negotiated a license to use and modify future versions.

The licensing was important because it gave Cross City a chance to make some demands of the Seattle Schools with regard to community access. For example, the agreement specifies that the Budget Builder must remain publicly available. Less successful from a community perspective, but still important, was a request by Diana Lauber of Cross City that all budgets be made public once submitted via the Budget Builder. Geri Lim, the current chief financial officer of the Seattle Schools, resisted that request, claiming that submitted budgets were "district planning documents," and that the district budget wasn't a public document until the district's combined budget was approved by the School Board. The final agreement says that principals are encouraged to make their school's budget available as soon as possible but has no firm requirement. The timing of when documents become "public" is crucial to how much input communities can have over their school's budget.

Expanding the Budget Builder to Other Cities

Less clear are the plans for expanding the Budget Builder's reach. Cross City works in several cities throughout North America, and developed the Budget Builder as a pilot project that would be moved to other cities. For much of the development, Cross City staff were adamant about funding only work that was general and applicable to any city where the Budget Builder might be used. District-specific modifications would have to be paid for by each district. Other cities have expressed interest in the system, particularly Los Angeles and Denver, but there are no specific plans at this time.

Moving the Budget Builder to a city like Los Angeles would present difficult obstacles. The Los Angeles school district is more than 10 times larger than Seattle's (680,000 students vs. 47,000; $5.3 billion annual budget vs. $350 million). Scaling up the system would not be trivial. A good part of the reason why the Budget Builder was able to be developed relatively quickly and inexpensively is that the Seattle school district is moderately sized. Moving to another city also presents a radical reconceptualizing of the Web site. Currently the Budget Builder is one set of Web pages on one single computer in one city. If Cross City decides to expand the project to other cities, what will the Budget Builder be and where will it be? How much of the site will Cross City maintain? Currently, the organization has a simple Web site and no one in their main office with the technical skills to maintain it. What parts of the site will be specific to each city and who will develop and maintain them? If each district sets up and pays for its own site, what's to keep them from excluding community participation? These questions are only beginning to be addressed.

Another factor that Cross City must consider as it expands is what level of resources it has available to support community organizing in each city. To increase the level of citizen participation in future Budget Builder development, Cross City needs to step up its school-based budgeting awareness and training campaign. That has already been difficult, and may be even more so if it turns its attention to other cities before community awareness in Seattle has crystallized.

Future Technical Work

Because of the complexity of the Budget Builder project and the project's limited funding, there is still much work that needs to be done on the site. Technical improvements will range from the mundane to the dramatic:

- Many graphic design enhancements will occur, to further unify the look of the site and decrease confusion. "In process" pages will be finished or cleaned out. Navigation assistance will be clarified.
- The help system must be dramatically improved, both because it is currently dismal, and because it is unlikely we will be able to provide training to all future users of the site. However, training for known users, such as budget office staff and principals, must be incorporated.

- Performance of the budget building section must be increased, as users complained most vocally about the system being slow.
- Processes must be designed to address problem areas that surfaced in past years, such as testing modifications to the site, and transferring data from SPS systems to the Budget Builder quickly and accurately.
- The Seattle Schools technical staff should be encouraged to think about how their data might be used publicly, especially as they perform major upgrades to new computer systems. Instead of using ambiguous codes (ex. EQUIP/MOVE/OTHER SR-DR) that may only have meaning within the district administration, they could use intelligible English (ex. Equipment/Moving/Other Services) to describe budget items.

Future Community Participation

For the Budget Builder to remain a tool that is accessible to people outside of the Seattle School District, we must widen participation in the development of future versions. It is unlikely that a community voice will remain on the technical team beyond my work this year. The district is likely to keep the Budget Builder technically separate from its computer systems for one more year—and therefore somewhat accessible to direct change—but the district's technical staff is increasingly anxious about bringing the Budget Builder into its grasp. Once that technical transition occurs, the only way that citizens will notice changes is if they are aware that the Budget Builder exists and are invested in having it remain publicly accessible. Some options for increased participation are:

- Incorporating a Community Steering Committee along the lines of the Principals Steering Committee. This would empower a body of citizens to meet regularly and suggest features and modifications to the existing Budget Builder.
- Community participation in usability tests. Tests for text readability and usability of the budget building section are necessary, and others will likely be too. Usability participants can be drawn from outside the district as well as users from inside the district.
- Cross City-sponsored training. Cross City has committed to a multi-year program of education about the importance of local control of school budgets and training in use of the Budget Builder. Part of the educational process can be opportunities for community members to comment on the Budget Builder.
- Piggybacking off of existing school activism groups. Although we have called on local educational activists for our previous public meetings, more could be done. Cross City can arrange to demonstrate the Budget Builder at meetings of the local groups. We can work with those groups to solicit feedback on how the Budget Builder could be used to achieve their goals.

Adopting these strategies will increase the chances that the Budget Builder will improve as a valuable public resource.

Conclusion

In this chapter, I have examined the development of the Seattle Public Schools' Budget Builder World Wide Web site with a particular eye towards the ways in which community participation affected the site design. There were both barriers to participation and successes due to its inclusion.

Barriers to Participation

Participation by potential users during the creation of technical systems is extremely important because the final form of technologies is highly dependent on the input received in early stages of their development. In this case, due to the power held by Joseph Olchefske and the Seattle Schools, the district administration was able to reduce the early inclusion of community views. Joseph had his own agenda that didn't exclude community participation, but didn't actively include it. However, his agenda effectively excluded it, since his priorities were only to get principals online and up to speed, and to have the system finished "enough" before including other parties in the process. In addition, he continually pushed the definition of "enough" further into the future, so that community participation was never a realistic option until the system was too far along to get much outside community input in the early stages.

However, Joseph provided a valuable community service by making the district's budgeting data available on the Internet. Although it was never discussed explicitly at the time, Cross City, Andy, and I felt that pushing Joseph too hard on the issue of community participation would have caused him to decide not to work with us. Our assets (funding and technical expertise) were more replaceable than his. Many other factors also contributed to the lack of community participation:

- Anne Hallett's desire to remain diplomatic,
- Cross City's inexperience working cooperatively with a school administration,
- Andy's and my failure to make clear how important early input was to the eventual design,
- my increasing technical workload and unwillingness to transfer much of it to others,
- complications of working long distance,
- the early nature of school-based budgeting in Seattle, and
- difficulties of organizing and holding community meetings.

Participatory Successes

Despite all of the reasons why community participation was problematic in this project, there are many ways in which it succeeded. From the point of view of the school district, the primary audience for the Budget Builder was school principals and budget staff. Using that definition, participation was very good, as principals and budget staff were regularly involved throughout the Budget Builder development.

Once the definition of community is broadened to include other school staff, parents, students, or even people not directly involved with a school, our participatory record looks weak. However, we did hold a few meetings in early stages with interested parents and teachers union representatives in which we gathered quite a bit of feedback about important principles. Unfortunately, the relationships lacked continuity, since with few exceptions we never met multiple times with the same people to verify our design against their original ideas.

Most importantly, the members of the design team qualify as community participants. The Cross City staff, although not residents of Seattle, held positions of district outsider, interested parent, educational activist, and others in their home community of Chicago. Much of their experience was directly transferable to the Budget Builder project. They were regularly involved in site design meetings and evaluations.

Andy and I, in addition to being district outsiders and residents of Seattle, were both sensitized to the need for community input into technical development. We constantly used our own experience as community members dealing with government bureaucracies, as well as the input we had gathered from others, to guide us in the technical work. Our community-focused work resulted in a product that:

- was accessible via the Internet, instead of being distributed on disk or via an internal district system;
- was publicly available to anyone, instead of only to principals and district staff;
- had more data visible than the district originally envisioned; and
- situated the mechanical budget-building tools in a context of explanations, examples, and assistance for users unfamiliar with the district's procedures.

Overall, the community participation we were able to incorporate had a noticeable effect on the final outcome of the project. Increasing participation in the next rounds of the design and addressing many of the problems outlined above, including simplifying the use of the site, offering more training, and involving users in testing the site, should all help to produce progressively more community-friendly versions of the Budget Builder as time goes on.

To see the current Budget Builder Web site, visit: http://bb.seattleschools.org. For a fuller description of the Budget Builder project, including an annotated version of the Budget Builder Web site, see Halaska (1998) or visit: http://www.socialdesign.org/bbanalysis/.

References

Dennis, A.R. (1998). "Lessons from Three Years of Web Development." *Communications of the ACM*, 41(7): 112-13.

Halaska, R.C. (1998). "Engaging Community in the Technical Design Process: An Analysis of the Development of the Seattle Public Schools' Budget Builder World Wide Web Site." PhD dissertation, University of Washington.

Morkes, J., and Nielsen, J. (1997). "Concise, SCANNABLE, and Objective: How to Write for the Web." http://www.useit.com/papers/webwriting/writing.html.

Morkes, J., and Nielsen, J. (1998). "Applying Writing Guidelines to Web Pages." http://www.useit.com/papers/webwriting/rewriting.html.

Nielsen, J. (1997). "How Users Read on the Web." Jakob Nielsen's Alertbox, 1 October. http://www.useit.com/alertbox/9710.html.

Nielsen, J. (1998). "Testing Whether Web Page Templates are Helpful." Jakob Nielsen's Alertbox, 17 May. http://www.useit.com/alertbox/980517.html.

Chapter XXIV

Discussions and Decisions: Enabling Participation in Design in Geographical Communities

Volkmar Pipek
University of Bonn, Germany

Oliver Märker, Claus Rinner and Barbara Schmidt-Belz
GMD—German National Research Center for Information Technology

Introduction

Communities frequently encounter design problems they have to deal with, be it their own representation in different media (e.g., the World Wide Web) or a design problem from the "real world" (e.g., urban planning). Major connotations with the term "community" are that there may be a great number of members and that the members of a community may differ according to their knowledge, experiences, abilities and intentions with great variety. This is especially true for geographic communities, where the only thing all members have in common is the place where they live, and which may have up to several million members.

Of course, geographical communities always have had to organise their design processes. As a part of political decision making in democracies, most "design decisions" are delegated to elected representatives in a city council, in a state parliament or in equivalent institutions in a federation of states. However, in some cases the integration of direct participation of community members into planning processes was intended or turned out to be helpful. This is especially true for urban and regional planning in Germany, where citizens are invited to participate in the planning process.

We refer to this procedure as an example for design processes in communities. We will discuss its background and its weaknesses regarding the aim of "good participation," describe options for information and communication technology

(ICT) support of this procedure, present an idealised participatory design system for urban and land-use planning and finally describe experiences with the GeoMed system, which is an approach to tackle the problem.

The Community Design Problem

Whenever in Germany a new street or railway track is being built, whenever new housing, industrial or commercial zones are planned, planning procedures are applied which are supposed to assure that the interests of the public community are not violated. Those planning procedures incorporate several participatory opportunities for citizens: the construction plans have to be made available to the public for at least one month. They are supposed to be explained in a way that makes them comprehensible to all citizens. Authorities are to actively invite those known to be affected by the planned measures, as well as relevant stakeholder organisations (e.g., environmental protection organisations) and there have to be opportunities to influence the decision through written objections or at hearings. The procedures and their aims are prescribed by federal laws with varying levels of detail. The responsible authorities then decide on the project. They base their decision on the results of the participation procedures.

The practice of those procedures shows deficits with regard to the aims. "Making plans available" usually means presenting them in the town halls. The comprehensiveness of explanations is often questionable. Invitations to stakeholder organisations sometimes have been selective. And, finally, the level of participation activities among citizens can be quite low. The citizens involved often feel that the authorities did not consider their opinions to a satisfactory degree.

Apart from political and ethical considerations on democratic practice this sometimes causes serious problems for the administrations. Projects that "successfully" went through the participation procedures have been hindered, delayed or even stopped afterwards by trials and public pressure, initiated by discontent citizens and causing financial losses to the community. So, the improvement of participation procedures is a point of interest for the authorities.

Planning Paradigms and Participation

That special community design problem of planning is subject to several academic disciplines which also tried to develop a theoretical foundation for the problem. Planning theory in Geography and Urban Planning now tries to capture the dynamics of a planning process based on philosophical or sociological work on human communication. Healey (1992) spoke of "the communicative turn in planning theory" in the 1980s. Several views on planning have been developed:

Planning as cooperative action: Based on the observation of innovations in the practice of urban planning that have been induced by new forms of work and organisation, Selle (1993, 1996) developed a cooperative perspective on the planning process and aligned it to the "disjointed incrementalism" approach in the problem-solving model of Braybrooke and Lindblom (1963). Its nucleus is that problems cannot be solved with extensive, general plans. Fragmentation of organisations and of society leads to very different perspectives and strategies on how to deal with problems. Self-referential autonomous parts of the social systems define and solve their problems on their own. In planning practice, observations confirmed this theoretical view, and strengthened the role of cooperation. According to Selle this new role of cooperation should be reflected in planning procedures in five ways:

- Designers, stakeholders and citizens should relate their thoughts and their acting to the thoughts and acting of other designers, stakeholders and citizens.
- Cooperation takes place in the "world between the worlds," i.e., in the public media space, where the values, ways of acting, and order principles of the different spheres are brought together. So, the notion of cooperation should not only refer to two or more actors, but always incorporate all actors.
- Cooperation is a dialogue-oriented, discursive process. Fixed procedures should be replaced by methods which offer participants to share and discuss different perspectives.
- Cooperation means that planning problems should not only be discussed, but even be solved in the public media space. Therefore, decision power has to be transferred from the political-administrative sphere to this space. Here, cooperation adds a new quality to the planning process, far beyond what participation means.
- The principle of cooperation demands openness according to processes, discussions and results. Communication processes have to be designed to give all participants the opportunity to actively take part in the development of a common understanding of the problem and its solution.

Planning as argumentation: Rittel (1972) introduced the term "wicked problem" for a class of problems with certain dynamics (no complete, stable problem description possible, infinite number of possible solutions, solutions not "right/ wrong" but "good/bad," etc.). He regarded planning to be such a "wicked problem." For the problem-solving processes of those problems ("Second Generation Approach") he states the following requirements:

- Participation of as many and as different people (i.e., experts and citizens) as possible has to be assured. They all should be guaranteed equal rights in the participation process.
- Since there is no objectivity in assessments and judgements, it is important to be able and willing to explain assessments to others.
- Argumentation has to be the basic mechanism of the process.

Rittel (Kunz and Rittel, 1970) developed a model of argumentation called "Issue-Based Information Systems" (IBIS). An IBIS model provides a limited number of types of speech acts, namely *issue, position,* and *argument,* together with a set of legal rhetorical moves. These define the possible relations between types of speech acts: for example, a position may respond to an issue, an argument may support or object to a position, and so on. Brewka and Gordon (1994) interpreted issues in a rather general manner as the "questions to be decided or goals to be achieved." Indeed, Rittel (1972) promotes IBIS as a general approach to solving wicked problems in community planning.

Planning as communicative action: Forester (1985) developed his "Critical theory of planning practice" on the basis of Jürgen Habermas' "Theory of communicative action." So, in contrast to the other approaches, he induces a theory of planning from a general sociological theory. His view on planning processes can be best described by showing his view on the role of their participants:

> "They [the planners] warn others of problems, present information, suggest new ideas, agree to perform certain tasks or to meet at certain times, argue for or against particular efforts, report relevant events, offer opinions and advice, and comment on ideas and proposals for action. These are only a few of the minute, essentially pragmatic, communicative actions that planners perform all the time, the 'atoms' out of which any bureaucratic, social, or political action is constructed. We can call these acts 'speech acts'" (Forester 1989, p. 142).

From that point of view he formulates communicative requirements for an innovative, democratic planning practice:

- Supply information to citizens, offer them the opportunity for a qualified, informed participation
- Include independent third-party expertise
- Listen carefully to judge on interests and obstacles as good and as early as possible
- Notify less-organised participants
- Make the decisions participatory (not only the preceding discussions)
- Encourage groups to demand for information
- Cultivate community networks (social networks, coalitions)
- Educate citizens on issues of planning processes and practice

All of the approaches are calling for the inclusion of as many and as different participants as possible, for an openness concerning planning process and its results and the notion of associated open communication processes as learning processes to understand each others' views. These aspects have to be observed when developing a computer-based planning process.

Another, more pragmatic approach, the "Planning Cell," developed by the German sociologist Peter C. Dienel (1978), has been tested for design processes in city communities successfully. A number of randomly chosen community citizens create a "citizen report" on a planning issue in a workshop-like discussion setting. During this workshop, which might last up to a week, they do not work in their usual jobs but concentrate on the planning problem. The whole workshop is supported by mediation measures to avoid communicative problems like opinion leadership, shyness, etc. The final report shows the main results of the discussion as well as minority opinions. It usually plays an advisory role in decision making, but may be backed by strong public pressure.

These approaches to a new culture of planning are not necessarily based on new technology. We believe, however, that ICT have a potential to support this new culture and should be developed with respect to these concepts. Of course, computer-based solutions have to compete with pragmatic approaches like the planning cell.

New Options through New Technologies

A computer offers numerous ways for distributing, visualising, exploring, and discussing community-related information. Many tasks in community design involve cooperative working. Groupware tools allow exchange of documents between members of arbitrary groups. The study of Computer-Supported Cooperative Work (CSCW) includes issues in user interface design for groupware tools, as well as document exchange through shared virtual workspaces, secure transactions, hardware and network access, virtual societies, and many others. When exchange of *geographically referenced information* is required, as in urban planning, Geographic Information Systems (GISs) are suitable tools to provide access to plans, maps, and additional statistical data. GIS researchers have introduced the notion of *Collaborative Spatial Decision-Making* (Densham et al., 1995) to label GIS that were augmented by groupware tools. In the following, innovations in both CSCW and GIS that are relevant for participation in community planning are described; the section concludes with two examples of promising combinations of GIS and groupware modules.

The development of the Internet gave many people unidirectional access to a global information space, organised in the WWW hyperlink structure, and bidirectional access to communication facilities, such as e-mail, newsgroups, chat, and HTML forms. Many groupware tools use the Web and related technologies for information exchange, as does, for example, GMD's *Basic Support for Cooperative Work*, BSCW (http://bscw.gmd.de, see Bentley et al., 1995).

In the community design scenario described above, an important requirement is computer support of participants in discussion procedures. Threaded newsgroups provide a simple mechanism for structuring asynchronous question-answer sequences. But discussions in planning are characterised by a *multiple-user, multiple-goal situation*; for example, in a public debate, participants form a very heterogeneous group, concerning their (political) opinions, their principle goals, their role in the planning process. Community discussion forums may also attract several hundreds or thousands of participants.

To cope with such large, conflictive discussions, *models of argumentation* that are more powerful than the newsgroup model have been considered for recording and structuring discussion procedures. The IBIS model as described above has been implemented in a number of computer tools, two of which are *gIBIS* (Conklin and Begeman, 1988) and *HyperIBIS* (Isenmann, 1993). More information on so-called "computer-supported collaborative argumentation" tools is available at http://kmi.open.ac.uk/sbs/csca/. The appeal of argumentation models for organising planning discussions is the natural language representation of a design process that keeps it comprehensible to community members.

A second improvement of discussion servers, additional to using powerful argumentation models, was to design *mediation* systems. They provide moderated discussions where a human facilitator supervises the discussion process. This mediator helps participants in using the discussion forum, e.g., when they have difficulties in incorporating their arguments in an ongoing debate. Thus, meditation systems cope with diverging capabilities of discussants. GMD's Zeno is an example of a groupware tool that combines an IBIS-like argumentation model with support for a mediator.

Advancements of Geographic Information Systems include two topics that are relevant in a cooperative design setting. On the one hand, access to GIS servers via the Internet enables distributed users to work on a single map or geo-referenced data set. On the other hand, modularization of the traditionally monolithic GIS packages, e.g., ESRI's *Arc/Info*, allows developers to create small tools that are centered around a well-defined service, thus simplifying the user interface and handling of the tools, and reducing bandwidth required for download.

Modularization began when the GIS vendors developed desktop mapping software, e.g., *ArcView* in the ESRI example, for those customers who did not need the complex functionality of their main products. Desktop GIS, in general, provide functions for layer management, zooming, panning, map layout, feature selection, and access to attribute data for features. But most of these tools lack GIS analysis functions like buffering and intersection of polygon layers, or network or 3D slope analysis. Splitting GIS into useful components is accelerated with a shift to the object-oriented programming paradigm. The OpenGIS consortium (http://www.opengis.org) organises the specification of geo-processing components, which could interact with each other in order to provide distributed GIS services.

As a matter of fact, it is true that current *Internet GIS* (also known as Online GIS, WebGIS, Internet Map Servers) are far from realising the vision of interoperable mapping functions. They rather emulate desktop GIS (*Webtop GIS*), insofar as they provide mapping functions for distributed geographic data available on the Internet. This helps to offer participants in spatial planning projects basic information for argumentation and decision support.

Three technical directions can be distinguished in WebGIS implementations:

- Simple *HTML*-based solutions consist of imagemaps, which provide links that can be followed by clicking on sensitive map areas. Some of these solutions are enhanced by HTML forms and server-side CGI scripts, which allow for multiple mapping functions (e.g., zooming, panning, and attribute retrieval) instead of simple browsing.
- Commercial solutions like ESRI's *ArcView Internet Map Server* rest upon an existing GIS that is linked with a Web server. Through a thin client software (HTML or Java), users can access the functions of the server GIS and get results in the form of imagemaps back to their computer.
- In pure *Java* applications, clients are thick, that is, all GIS functionality and most of the data is transferred to the client computer, when accessing the application.

Examples for the above types of WebGIS can be found in the EXSE project that aims at "the experimental evaluation of the technical and practical potentials of GIS applications in the World-Wide Web" (http://www.giub.uni-bonn.de/exse/). For example, the WebSD subproject provides a classical map server solution based on SICAD SD as GIS server, CGI scripts (Perl), and either a thin, HTML forms and JavaScript-based client interface, or a Java applet. For a deeper understanding of Internet GIS techniques and applications, see Plewe (1997) or Fitzke et al. (1997).

The *Descartes* system (http://borneo.gmd.de/and/, Andrienko et al. 1999) is an example of an Internet GIS that can provide highly specific services in a community environment. On top of a simple geodata model, Descartes offers automatic generation of cartographically sound thematic maps. Through the interactivity of the resulting Java maps, citizens can explore their region and achieve information to base their discussion contributions upon. Other authors promote the use of three-dimensional representations and of multimedia data like sound files, to provide a realistic insight in planning projects. Concerning their *Spatial Understanding and Decision Support System* (SUDSS), Jankowski and Stasik (1997) emphasise users' ability to explore plan alternatives, to perform newsgroup discussions, and finally, to evaluate plan alternatives through a formalised voting mechanism. The GIS-based video-conferencing tool, presented by Cowen et al. (1998), includes a collaborative whiteboard for annotating plan proposals. With the video-conferencing module, the aspect of synchronous communication between stakeholders in planning is well developed.

The main limitation of the tools described here is the fact that they are tailored for specific contexts. Some are network-enabled while others are limited to running on stand-alone computers; some support asynchronous discussion while others are designed for synchronous communication, as in public meetings or experts' negotiations. In the near future, we will probably see more and more combinations of single GIS and groupware applications that will build commercial, off-the-shelf solutions to support participants in community design procedures.

Requirements for a Community Design Tool

Bringing together the options information technology offers for participation support and the recommendations which evolved from the discussion in planning theory, computers can open up new horizons for design processes in geographical communities. We now want to integrate our ideas of computer support for participation into an architecture for community design support.

The starting point for that architecture is clearly the support of two aspects of participation: the representation of what is being designed and the option to interact with designers as well as other citizens in order to influence the design process. The representation can (and in our design problem should) be based on a model, which offers the option to generate different views on a subject (e.g., to combine geographical data with data on the estimated traffic flow).

Regarding the *necessary* functionality, this could be enough. But modern planning paradigms also introduce more informal requirements to participation procedures and thus participation tools. We sketch an architecture that combines all aspects discussed here in Figure 1.

Comprehensiveness: Plans should be presented in a way that makes them comprehensible for non-expert citizens. Of course, this requirement first calls for an appropriate graphical representation of plans, procedures and schedules. But since planning involves complex problem fields, interested participants should be supported in gaining competence in relevant fields of knowledge. Tools may also offer

Figure 1: Sketch of an Architecture for Participatory Community Design

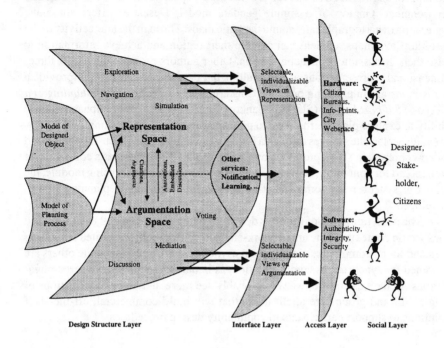

support for exploration, simulation and even distance learning. "Exploration" means giving citizens the opportunity to explore different plan alternatives from different perspectives or representations (two- or three-dimensional) in order to understand subsequently all aspects of the presented alternatives. This can be further supported with simulations based on acknowledged mathematical or formal models, for example, of the development of traffic noise or of pollution development for certain areas. Tools could even incorporate qualification modules which introduce citizens into relevant problem fields. All this may be supported with an expert network (maintained by stakeholder organisations) to support the learning process.

State of Discussions: Up to now the final decision on projects is still left to the city council and the authorities. Nevertheless it is important for all participants to have a reliable image of the current state of the discussion (in terms of "agree/disagree"). Voting mechanisms can be implemented, which might be used frequently. Weaker mechanisms may be applied to visualise the general support for certain opinions or persons.

Universal Public Access: This requirement is most important to respect, since even the partial transfer of participation procedures into virtual worlds is a substantial change to the accessibility and uniformity of documents and processes. The problem has to be tackled on different levels. On the hardware level, universal public access means having available the computers, networks and bandwidth to participate. On the software level, access means to have applications available that are easy to use even for casual users and that respect the requirements of minorities. As a special case this incorporates an integration of discussion and representation. It should be easy to relate a discussion contribution to the part of the representation it refers to (e.g., to cite it visually), and it should be easy to access those parts of a discussion that refer to a certain part of the representation.

Other requirements include:
- assuring the authenticity and integrity of each opinion expressed;
- support for moderation and mediation;
- support for usage patterns beyond the image of the "frequent, knowledgeable user" (e.g. notification services, storing discussion progress, offering different searching and retrieving techniques, individualisable views on the discussion structure as well as the currently favoured state of the design object and its alternatives); and
- support for socialisation and cooperation.

We have structured our architectural sketch into four layers: the design structure layer, which incorporates the modelled subjects and objects that play a role in the design (artefact to design, experts, design process, contributions to discussions, etc.) and their relations and representation. We further distinguish between the models of the design object and process (i.e., the pure data), the representations of the objects involved, and the argumentation space incorporating interactively established data like positions, evaluations, discussion contributions, decisions and relations. The interface layer covers all aspects of how the design structures can be visualised and manipulated. The access layer deals with hardware and software support for reliable and easily accessible systems for participation. The social layer refers to the "real" world, the participants, their interests and intentions.

In our opinion, for argumentation systems, the critical point to support in community design is an integration of information and interaction, which is attempted by the European project described in the following section.

GeoMed—A Joint European Project

GeoMed is partly funded in the European Union Telematics Applications Programme (http://ais.gmd.de/MS/geomed/overview.html). There are five developing project partners: the German National Research Center for Information Technology (GMD), TNO-FEL and TNO-Bouw (the Netherlands), VUB (Belgium), Intecs Sistemi (Italy), and Intrasoft (Greece). In order to ensure a user-driven development process, there are also four user partners in GeoMed: the City of Bonn (Germany), the City of Tilburg (the Netherlands), the Region of Tuscany (Italy), and the Technical Chamber of Greece.

The GeoMed system comprises the following integrated services:
- *Basic support for cooperative work (part of Zeno):* providing shared (virtual) workspaces.
- *Discussion forum (part of Zeno):* where users can discuss and deliberate planning issues and solutions, preferably structured according to the IBIS method, both direct or mediated discussions.
- *GIS viewer:* allowing to view GIS data via Internet. The users get a preview

of the whole map, they can pan, zoom and select layers. Users can also add new layers and edit simple graphics or annotations. This can also be seen as a step towards an integration on representation and argumentation.

- *GIS broker and payment broker:* enabling new services, where GIS data can be offered for sale, respectively retrieved by spatial and thematic queries, ordered and paid.
- *Services based on software agents:* performing notification and other services for users and organisers of workspaces and discussion forums.
- *Knowledge-based system applications:* which allow analysis of plans with respect to special regulations on the basis of a knowledge base.

For an efficient use of GeoMed it is important to integrate these components. For the shared workspace, the GIS viewer and the discussion forum, the functional integration also promises added values. A discussion forum can be created in a workspace like other types of resources, such as folders. The contributions to the discussion may be linked to other documents in the workspace, e.g., to plans, to complex statements or further explanations. In this way, a discussion becomes a natural part of a shared workspace.

GIS data are another special type of resource that may be kept and used in the shared workspace. If a user wants to see the GIS data, a special interface (implemented as a Java applet) is started, which provides functions like zoom, pan, or filter layers. The GeoMed GIS viewer also allows the user to create a new layer and edit some simple graphics or annotations. Note, however, that the GIS data, unlike other documents, are not to be uploaded to and downloaded from the workspace by its members, because of their special kind, their value, the heterogeneous formats, and their huge volume.

Figure 2: GeoMed Provides Integrated Services

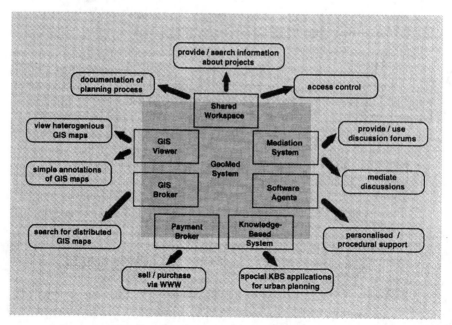

The most interesting integration is that of GIS maps and discussion forums. When the discussion concerns a plan accessible via the GIS viewer, a user can create a new layer and mark parts of the plan, and link the marks to his statements in the forum or illustrate his statements by sketching alternative solutions on the map (cf. collaborative whiteboard example above). In this way, it becomes easier and more natural to discuss features of a spatial plan. The ongoing work by Rinner (1999) is directed towards linking discussion contributions to map elements. The aim is to allow participants to explore the spatial distribution of contributions to a debate and to access these contributions via an intuitive map interface.

The prototype of the full GeoMed system shall be basis for demonstration at all four user sites. Demonstration means that the system is used and validated in a real planning project. In preparation of the demonstration, procedures of use are speci- fied, workspaces and forums are set up with appropriate structure and access rights. Some users, like project managers and mediators, shall have a special training. And last but not least, the system has to be filled with appropriate contents according to the project needs, like planning data and background information, so that users can start to work with it.

The ZENO Tool: First Experiences

The Zeno system (http://ais.gmd.de/MS/zeno/zenoSystem.html, Gordon et al., 1997) is an all-purpose system for computer-supported collaborative problem solving. The scientific background of Zeno is "computational dialectics" as a new subfield of computer science, whose subject matter is computational models of rational discourse (Gordon, 1994). In the GeoMed project, Zeno implements the mediation services and supports asynchronous discussions and argumentation processes within the framework of public participation in urban and regional planning.

The urban and regional planning domain is conceived as Collaborative Spatial Decision Making, its main ideas being:
- Collaboration of several planners and other actors is essential throughout the planning process.
- Representations of space (e.g., maps) play a central role in the planning process.
- Negotiation and decision making are crucial phases of each project.

The kernel of Zeno is an integration of shared virtual workspaces and asynchro- nous discussion forums.

The *shared virtual workspace* concept of Zeno was inspired by the BSCW system (http://bscw.gmd.de). It allows groups to provide and exchange their com- mon holdings of information and knowledge. A workspace is structured by folders and subfolders, and meaningful titles of resources help orientation. Different types of resources can be created or uploaded, like folders, documents of any format, or hypertext links. Meta data (like date of last change, author, size) and attributes (i.e., type-specific meta data like MIME Type of a document or URL of a link) of

Figure 3: Zeno Supports Heterogeneous User Groups in a Wide Range of Tasks

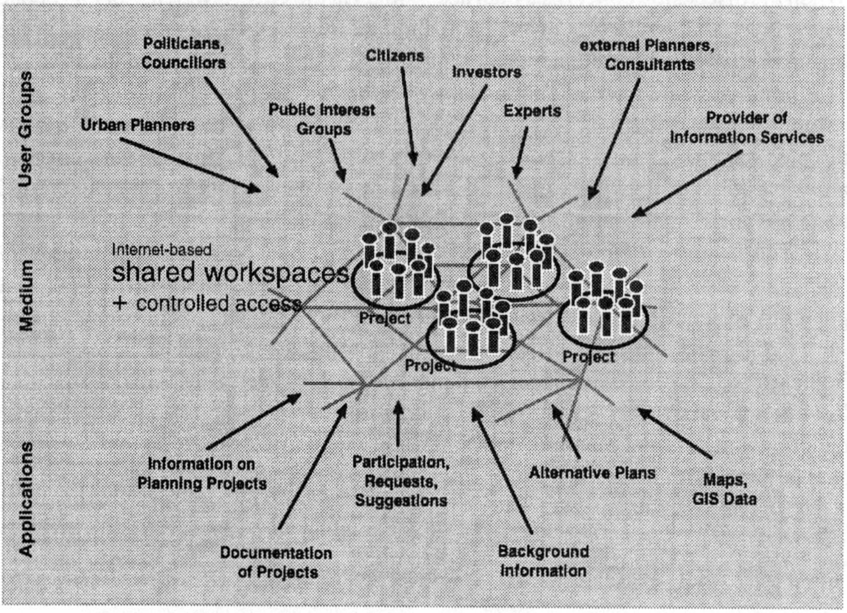

resources are further means of documentation and retrieval. Group members have access via the World Wide Web; a login procedure confines access to identified members. Insofar, a shared workspace is like a dynamically configured intranet. Access rights can be specified for all parts of a workspace and distinguish types of action which are allowed (like read, write, list or move). The concept corresponds to the requirements of public universal access, authenticity and security.

Anywhere in a workspace, a discussion forum can be created. It inherits the same central concepts like access control, resource handling and meta data. In a forum, the users can discuss any subject that arises in the course of cooperation. The documentation of a deliberation constitutes an essential part of the group's knowledge. Zeno supports structured discussions according to the IBIS argumentation framework (Kunz and Rittel, 1970, see above). Thus, a contribution to a Zeno discussion forum should be a new issue, or a position within an issue, or an argument supporting or contradicting a position. Additionally, comments and questions or answers are allowed at any level of the discussion. The forum may be moderated or not, special tools support a mediator's job. In Zeno, the IBIS model shall be augmented by rating and voting, and by means of moderation like closing a track of argumentation. Especially the support for rating discussion contributions may be a great help in discussions with a large number of participants, since with the computational approach for ratings in the Zeno concept (Gordon and Karacapilidis, 1997) controversial and non-controversial branches of the discussion can be distinguished and appropriately visualised.

Together with some more tools, this supports a wide range of tasks and user groups, as is illustrated in Figure 3.

The goals of Zeno with respect to the understanding of "planning as argumentation" (Kunz and Rittel, 1970) are:

- to make planning processes more transparent;
- to facilitate public participation;
- to help avoid or resolve conflicts;
- to facilitate and improve cooperation of planners, experts and communities; and
- to make urban planning more efficient, less time-consuming, and less expensive.

Figure 4: A Structured Zeno Discussion Forum on Traffic Issues

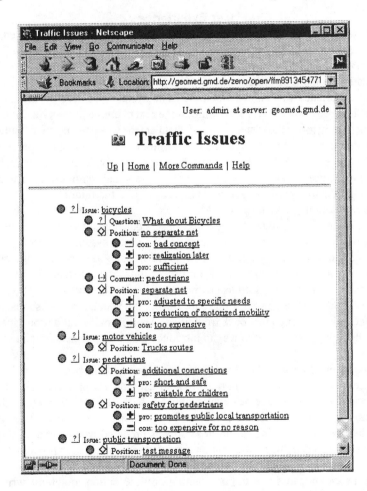

Zeno is Internet based, the clients can use Zeno via any up-to-date Web browser, the server side is implemented in Java and uses a relational database system. Two different user interfaces are available. One is a Java applet that communicates via the RMI protocol with the server. The applet is more compact and fast, but needs some installation effort on the client side. Sometimes, the applet encountered problems due to firewalls at a user's site, or would not work correctly due to incompatible Java interpretation by different Web browsers. Therefore the other user interface is based on pure HTML 3.2. It is simple, sufficiently fast and works without problems on any platform and browser version.

First User Experiences

Up to the end of 1998, experiences with the Zeno system were mainly based on laboratory experiments and the project team's internal use of the system. The experiments were set up as experimental games to validate the system and its use in the application domain of urban planning (see Schmidt-Belz et al., 1998, 1999). Additional experience was gained when the Zeno system was used during a real citizen participation procedure. Meanwhile, some real planning projects and initiatives are beginning to use the system. Among them are the German initiative "Future Cities," and international initiatives like the "Centre of International Cooperation, Bonn." By the end of 1999, the project shall also have gained some valuable experience in the use of Zeno for cooperation of geographical communities. The use for broad public participation is still to be investigated in the future.

As a validation event in the GeoMed project, a basic version of Zeno was submitted to a procedure, where two groups of users performed free planning tasks supported by the system. After a brief introduction to the system, the users were given a scenario, roles to play, and tasks to perform using the services of Zeno. Finally, on the basis of these experiences, the users validated the system with respect to ergonomic issues and discussed its prospective benefits and potentials. This was done in a two-day workshop for each group. During the validation workshops, the Zeno project got a thorough feedback from users. Apart from the validation aspect, this workshop also proved a good way to introduce and train planners in the use of Zeno.

As the second step of the GeoMed validation, Zeno (still the basic version) was used in a preliminary *public participation* of a land-use planning project (concerning plans for a new housing area to be developed in a suburb of Bonn). For a period of two weeks, citizens could find the plans, explanations and more information in a Zeno workspace. A discussion forum in Zeno was prepared where citizens could input their statements. In the public announcement of the participation process there was also a reference to Zeno. Of course, Zeno only was offered in addition to the traditional procedure, that is a public meeting and the option to submit written statements, or to talk to the planners during office hours.

A protocol of events in the workspace allows some cautious conclusions, though the event protocol does not reveal user names. In all, 760 events were registered, most of them "view" events, i.e., someone read a document. As the main plan document was viewed 109 times, this would be the maximum number of guests in the workspace. For comparison, a public meeting at this time was attended by about 100 citizens. Usually only citizens of the neighbourhood of the planning area are invited and only a small percentage of these people are really concerned. Taking into account that this was the very first offer for citizens of Bonn to participate via Internet, this is not a bad result. On the other hand, none of the guests in the workspace has sent a comment to the discussion forum. We speculate that the reasons for this are partly the novelty of the medium and partly difficulties with the old user interface. In spite of these slightly disappointing results, the effort was worthwhile, because planners and other actors in the administration of Bonn gained experience in the new technique and procedure.

A comparative study of Internet-based public participation in three European countries has been carried out by Burg (1999). Similar to our experiences, the author reports a very low degree of active on-line participation in formal zone planning procedures, with on-line contributions ranging between 0.1% and 6% of all contributions.

Summary of Experiences and Lessons Learned

Functionality and ergonomics. Some additional functions were asked for, but all in all, the functionality as specified for the full version of the Zeno system was estimated appropriate to support urban planning. The project was encouraged to implement the functionality as planned. The handling of the first version of the system proved inadequate and a substantial redesign of the user interface was required.

Issues in the organisational context of use. The first experiences made some issues obvious which would concern not only Zeno but also any system which supports group cooperation, Internet mapping, and public participation.

- *Initial difficulties.* Introducing a complex system like Zeno to an organisation certainly creates a need to restructure the traditional procedures. Regional and urban planning, however, are subject to many legal regulations, which set constraints to reorganisation. This means an extra effort for some pilot applications, until new model procedures have been worked out.
- *Threshold problems.* Many actors in many different roles take part in regional and urban planning. Whether these are individuals (like citizens) or organisations (like other departments, authorities or communities), each one is free to join in using Zeno or not. If a high percentage of actors does not (yet) use Zeno, its use will show rarely any benefits, because the development of a common understanding of a design problem as the principal criterion for cooperative planning (Selle, 1996) is lacking. Therefore, the intensity of use has to pass a certain threshold, in order to demonstrate the benefits of this new support.
- *Media diversity.* As long as some actors (like citizens) have no access to the technique required (or for some other reasons choose not to use Zeno), the traditional media for cooperation and participation, especially paper documents, will continue to be important. The problem of providing and handling heterogeneous media is not only an initial difficulty but will persist for many years to come. The coexistence of paper documents and electronic data brings some well-known problems for an administration. Such problems include the double effort of providing information both on paper and via Internet, to

archive and retrieve both paper and electronic documents in a planning process, or to update and control versions in redundant, distributed information bases. The coexistence of different communication media, however, is a chance. Face-to-face communication, meetings, or computer-mediated communication each have their special advantages and should be used in combination.

Aspects in the social and political context. Naturally, the first experiences described here do not yet allow a statement on how far Zeno meets the general goals mentioned above. Many discussions with users resulted in rather ambivalent assessments of the prospective benefits and risks of Zeno. Some examples of issues discussed by users are:

- Does Zeno facilitate citizen participation? Or will parts of the population be at a disadvantage because they have no access to Zeno or do not know how to use it?
- When the process and the diverging interests of participants become more transparent, will this avoid or produce conflicts?
- Will planning processes become more efficient, or, on the contrary, more complex?

Not surprisingly, users often came to the conclusion that the answers to these issues depend on the way the system is used (political layer) and on the cultural and socio-economic background of the use cases (social layer). For example, the modest degree of on-line contributions cited above may be explained by German or European particularities in community organisation. With respect to this realisation, it is very important to develop and validate models of use in the near future.

The Long and Winding Road to Computer-Supported Community Design Processes

The advantages of the new media seem to be just too obvious: different and explorable modes of representation, maybe even simulation of alternatives; the use of discussions as easily usable representations of the ongoing design activities; the reduced social context of text-based discussions helping for a more concentrated discussion; and the huge information pool provided by the WWW.

As the first experiences with Zeno, but also the results of earlier projects indicate, reality may look different. Computers and the Internet still have to be considered as highly selective because of affordability and usability issues. Media breaks between electronic and physical media still are necessary to incorporate all appropriate participants and potentially hinder communication. The missing social context can lead to more uncontrolled utterances like classical Usenet flames and can have other unwanted effects, as already encountered in socio-psychological research on e-mail systems (cf. Sproull and Kiesler, 1986). Additionally, the positive outcome of greater equality in communication between users that belong to different hierarchies has been doubted (cf. Weisband et al., 1993). A further problem results from the knowledge heterogeneity of the participants in a planning process. Participants range from naïve citizens to highly specialised planners and experts with many years' practice in the field of planning. In addition to simple linguistic problems (everyday usage vs. technical jargon), Kerner (1996) has identified a low comprehensive competence and insufficient consensual ability as important "expert deficits" with respect to problem-oriented discourses.

While these drawbacks refer to communication in general, respectively computer-mediated communication, IBIS and similar approaches face further problems. Isenmann and Reuter (1997) identified the following difficulties:

- IBIS requires the general agreement of all participants on argumentation and discourse as appropriate problem-solving methods. In practice this often is doubted on a problem-oriented ("IBIS is nice, but in our case...") as well as on a method-oriented ("I prefer to rely on facts and recognised methods rather than talk.") level.
- IBIS has to cope with high expectations since many users expect active problem-solving or knowledge structuring services from the system. Isenmann and Reuter recommend that "when using IBIS-like approaches, participants should be conscious of the support character of those systems, i.e., that these systems are primarily designed to facilitate communication and to stimulate the creative intelligence of their users." With the ZENO tool, no similar experiences have been made.
- IBIS requires users to decompose and classify accurately formulated knowledge items when they enter them into the s,ystem. This sometimes is difficult, and it might result in a loss of context.

All in all, IBIS is not a simple approach, requiring users to be trained in applying the method. Another speech-act-based approach to structured conversation, the Language/Action Perspective (Winograd, 1988) was heavily criticised on a more principal level for the restrictions it poses on communication. The main point was that the form and structure of a discussion can also influence the contents communicated (Suchman, 1993). Orlikowski (1995) hit the critical point in that discussion in clarifying that any categorisation stresses some aspects while suppressing other. What aspects are stressed and whether that should be considered as good or bad, she states as dependent on the context and the content of the categories. Especially when we think of the complexity of the planning problem and the possibly large number of participants, the benefit of structures and categorisations might outweigh the disadvantages. It is a problem with which empirical studies on IBIS have to cope, that IBIS might show its real benefit only in large discussions that rarely occur in experimental settings. So up to now most users only see the additional effort the system requires which may lead to a low acceptance. Here we should consider experiences made in groupware contexts which showed that the disparity in the distribution of cost and benefit among user groups and the lack of a critical mass of users can be a major obstacle for the success of cooperative systems (Grudin, 1989, 1994).

For the participants in the planning process, the facilities of influence would be of decisive importance. Regardless whether decisions to be taken at the end of the discourse are formally (e.g., in the case of a referendum) or informally (e.g., the citizen opinion as result of a planning cell—Dienel, 1978) bound to the participants' influence, the participants should be informed of where and how to exert real influence on planning ("qualification for participation," cf. Sell and Fuchs-Frohnhofen, 1994). The current planning practice does not always show a great willingness of the current decision makers to hand over decision competence.

If the application of network-based participation tools is not to miss its target, it should not be directed by technical feasibility, i.e. the available characteristics of network-based information and communication services, but it should consider primarily the requirements of a modern participation-oriented planning practice. Only if a network-based participation is implemented in the sense of the "new planning culture," an improvement of the participation situation can be expected. Otherwise, the application of network-based participation might degenerate to a symbolic and legitimatory measure.

Generalising from the planning problem, the question is whether and how the solution of design problems of a geographic community can be supported with ICT. Environmental issues regulations, communal education politics and laws can also be viewed as "design problems" in the context of geographic communities. The architectural sketch in Figure 1 could also be easily applied to other application fields. The limitations regarding access issues and computer-mediated communication issues of course still apply.

The access problem revisited. When lifting concepts made for participation in planning to a more general level of democratic decision finding, we have to have a closer look at access issues. Since participants may be very heterogeneous concerning education, knowledge, intentions, mental skills, etc., all participation procedures may be selective in one way or the other. Clearly, ICT support for participation processes adds to the selectivity of these processes by introducing a complex, expensive, high-tech device for communication and information. This effect could be alleviated, e.g., by offering free Internet access in citizen bureaus, designing the software according to software-ergonomical guidelines and choosing "lean" representations to allow an easy download via modem. Jones (1999) showed a method for a completely paper-based access to a (computer-based) community information system based on a "multifunctional device" (Scanner/Printer/Copier/Fax). The research initiative "Information Society for All" (Arnold and Vink, 1999) aims at enabling elderly or handicapped people to use computers, even for very special groups (Weber et al., 1999). On the other hand, ICT may lower access barriers, e.g., by offering multilingual support for immigrants (Lehtola, 1999), a problem which is normally ignored by authorities. We also observe that currently practised forms of communication and participation in geographic communities do not always match the criteria of a democratic participatory process.

We still have technological aspects to improve in computer-mediated discourse. The discussion of IBIS systems might have concentrated too much on the method instead of concentrating on the users needs. Surely IBIS systems address relevant problems (coping with chaotic, complex, large, discursive problem-solving processes), but do the users in our context feel that need?

We regard these questions as a guideline for further research effort:

How can users easily articulate their point of view? On the software-ergonomic side, we can closer relate the representation and the argumentation space. Support for citing an aspect of the representation, for evaluating it ("A is good, because..."), for establishing argumentative relations between two alternatives ("A is better than B, because...," "A and B do not fit together because..."), for standard utterances ("I agree") and for other "discussion acts" might also ease the use (for a corresponding approach for e-mail systems, cf. Camino et al., 1998).

Instead of typing messages, options for multimedia utterances should be explored. Of course, such tools had to be optimised for use in discourse settings regarding audio/video browsing, annotations (cf. Bargeron et al., 1998), and references to other discussion contributions.

A different approach to solve the problem is to support human mediators in a scenario of computer-mediated public participation, as is intended in the Zeno system. The idea is that only a moderator/mediator has to qualify in the refined possibilities of computer-mediated communication (e.g., the IBIS method). She transfers the utterances of the participants into an appropriate standardised notation. The public may concentrate on the issues, i.e., the design goals and the problem solution.

How can the qualification of users be supported? Users in computer-mediated discussion settings potentially need qualification support on the usage of computer tools, on the participation process itself (i.e., own options and influence on decisions, cf. Sell and Fuchs-Frohnhofen, 1994), and on the topics relevant for the design problem (e.g. urban planning, architecture, environmental sciences). Qualification in the latter case means to enable all users to actually understand what experts are talking about. In the design of community design tools, these issues could be respected by incorporating anchors for qualification modules. Other options could be drawn from the emerging research field of Computer-Supported Cooperative Learning (Koschman, 1996) and earlier work on learning with hypermedia systems (Jonassen and Mandl, 1990).

How can the socialisation be supported and used in the discussions? Providing computer-mediated communication creates a social space around the design problem. Tools may support socialisation by including social acts, e.g., allowing users to give recommendations of persons, design alternatives and discussion contributions, allowing users to give a mandate to someone who has similar opinions, allowing users to build coalitions and communities. That inclusion also means to represent the social aspects somehow, e.g., to bold print the names of persons with a lot of mandates at the interface. It would be interesting to see how those social options can be used in coping with large discussions (i.e., navigation).

But, remembering that we discuss design in geographical communities, we have to be aware that there is a whole reality outside the computer networks waiting to socialise people....

Again, as stated for the special case of planning, we want to stress that we have new technological options to support participation in design processes in geographical communities, but these options have to be used in new participation procedures that still have to be developed and evaluated — not only in a technical sense but with respect to criteria of the communicative planning theories.

References

(Translation of German titles by Volkmar Pipek)

Andrienko, G., Andrienko, N. and Voß, H. (1999). Thematic Internet Maps for Cities: Data Analysis and Publishing with Descartes. In *Proceedings of 21st Urban Data Management Symposium* (UDMS'99). Venice, April 21-23.

Arnold, A.G. & Vink, L.J. (1999). Broad Spectrum Approach and Information Society for All. Bullinger, H.J. and Ziegler, J. (Eds.). *Proc. of the Int. Conf. on Human-Computer Interaction 1999*. 2, 757-761, London: Lawrence-Earlbaum.

Bargeron, D., Gupta, A., Sanocksi, E., and Grudin, J. (1998). *Annotations for Streaming Video on the Web: System Design and Usage Studies*. Technical Report MSR-TR-98-60, Microsoft Research, Redmont, Washington. http://research.microsoft.com/pubs/.

Bentley, R., Horstmann, T., Sikkel, K. and Trevor, J.(1995). Supporting Collaborative Information Sharing with the World Wide Web: The BSCW Shared Workspace System. *The World Wide Web Journal: Proceedings of the 4th International WWW Conference,* 1, 63-74. O'Reilly, Sebastopol, CA. http://bscw.gmd.de.

Braybrooke, D. and Lindblom, C. (1963). *A Strategy of Decision*. New York, London.

Brewka, G. and Gordon, T.F. (1994). How to buy a Porsche: An approach to defeasible decision making. *Working Notes of the AAAI-94 Workshop on Computational Dialectics*. Seattle, Washington.

Burg, A. (1999). The influence of the Internet on public participation in land-use planning – the example of Germany, Great Britain and Sweden. *Proceedings of Computer-Supported Planning* (CORP 99). 1,141-146, Wien.

Camino, B. M., Milewski, A. E., Millen, D. R. and Smith, T. M. (1998). Replying to email with structured responses. *Int. Journal on Human-Computer Studies,* 48, 763-776.

Conklin, J. and Begemann, M. L. (1988). gIBIS: A Hypertext Tool for Exploratory Policy Discussion. Greif, I. and Suchman, L. (Eds.). *Conference on Computer Supported Cooperative Work*. Portland, Oregon: ACM.

Cowen, D.J., Shirley, W.L. and Jensen, J. (1998). Collaborative GIS. A Video-Conferencing GIS for Decision Makers. Proceedings GIS PlaNET, 7-11 September, Lissabon, Portugal.

Densham, P. J., Armstrong, M.P. and Kemp, K.K. (1995). Collaborative Spatial Decision-Making. Technical report 95-14, National Center for Geographic Information and Analysis (NCGIA).

Dienel, P.C. (1978). *Die Planungszelle - Eine Alternative zur Establishment-Demokratie* (The Planning Cell - An alternative to the establishment democracy). Opladen, Germany: Westdeutscher Verlag.

Fitzke, J., Rinner, C. and Schmidt, D. (1997). GIS-Anwendungen im Internet (Internet GIS applications). *GIS Geo-Informations-Systeme*. 10(6), S. 25-31.

Forester, J. (1985). Critical Theory and Planning Practice. *Critical Theory and Public Life*. Massachusetts, USA.

Forester, J. (1989). *Planning in the Face of Power*. Berkeley, Los Angeles & London.

Gordon, T. F. (1994). Computational dialectics. In *Workshop Kooperative Juristische Informationssysteme*. Number 241 in GMD Studien, 25-36, Sankt Augustin, Germany, September.

Gordon, T. F. and Karacapilidis, N. (1997). The Zeno argumentation framework. *Proceedings of the Sixth International Conference on Artificial Intelligence and Law*, S. 10-18. ACM.

Gordon, T. F., Karacapilidis, N., Voss, H. and Zauke, A. (1997). Computer-mediated cooperative spatial planning. Timmermans, H. (Ed.). *Decision Support Systems in Urban Planning*. London.

Grudin, J. (1989). Why groupware applications fail: Problems in design and evaluation. *Office: Technology and People*, 4(3), 245-264.

Grudin, J. (1994). Groupware and social dynamics: eight challenges for developers, *CACM*, 37(1), 93-105.

Healey, P. (1992). Planning through debate. The communicative turn in planning theory. *The Town Planning Review*, 63(2), 143-162, Liverpool, UK.

Isenmann, S. (1993). How to deal with wicked problems using a new type of information systems. Stowell, F.A., West, D., Howell, J.G. (Eds.). *System Science - Addressing global issues*, 367-372, Plenum Press, New York, USA.

Isenmann, S. and Reuter, W.D. (1997). IBIS - a convincing concept ... But a lousy instrument? Van der Veer, G., Henderson, A., Coles, S. (Eds.). *Designing Interactive Systems: Processes, Practices, Methods and Techniques* (DIS'97), Amsterdam, The Netherlands. New York: ACM Press.

Jankowski, P. & Stasik, M. (1997). Spatial understanding and decision support system: A prototype for public GIS. *Transactions in GIS*, Bd. 2, Nr. 1, S. 73-84.

Jonassen, D.H. & Mandl, H. (Eds.). (1990). *Designing Hypermedia for Learning*. Berlin, Germany: Springer.

Jones, R. M. (1999). Using the Paper User Interface to Support Community Involvement in Urban Development. Bullinger, H.J. and Ziegler, J. (Eds.). *Proc. of the Int. Conf. on Human-Computer Interaction*, 2, 427-431, London: Lawrence-Earlbaum.

Kerner, M. (1996). Ausblick. Kerner, M. (Ed.). Aufstand der Laien. Expertentum und Demokratie in einer technisierten Welt (Revolt of the laymen - Expertship and Democracy in a mechanised world). S. 293ff. Aachen.

Koschman, T. (Ed.). (1996). CSCL: *Theory and Practice of an Emerging Paradigm*, Hillsdale, New Jersey: Lawrence Earlbaum.

Kunz, W. and Rittel, H.W.J. (1970). Issues as elements of information systems. Center for Planning and Development Research, Institute of Urban and Regional Development Research. Working Paper 131. University of California, Berkeley. Also in IGP [Institut für Grundlagen der Planung, Universität Stuttgart], Technical Report H. S-78-2. Stuttgart, Germany.

Lehtola, A. (1999). Multilingual Information Services as a Goal. Bullinger, H.J. and Ziegler, J. (Eds.). *Proc. of the Int. Conf. on Human-Computer Interaction*, Vol. 2, 767-771. London: Lawrence-Earlbaum.

Orlikowski, W.J. (1995). Categories: Concept, Content, and Context. *Journal on Computer Supported Cooperative Work* (JCSCW). 3(1), 73-78.

Plewe, B. (1997). *GIS Online: Information Retrieval, Mapping, and the Internet*. Santa Fe, NM: OnWord Press.

Rinner, C. (1999). Argumaps for Spatial Planning. Laurini, R. (Ed.). *Proceedings of TeleGeo'99*, First International Workshop on Telegeoprocessing. May, 6-7, Lyon, France.

Rittel, H.W.J. (1972). On the planning crisis: Systems analysis of the first and second generation. IGP [Inst. for the foundations of planning, University of Stuttgart, Germany], Technical Report H. S-77-7. Stuttgart, Germany.

Schmidt-Belz, B., Rinner, C., and Gordon, T. F. (1998). GeoMed for Urban Planning—First User Experiences. Laurini, R., Makki, K. and Pissinou, N. (Eds.). ACM-GIS'98, *Proceedings of 6th International Symposium on Advances in Geographic Information Systems*. November 6-7, Washington DC, USA, S. 82-87.

Schmidt-Belz, B. and Fleischhauer, D. and Märker, O. (1999). Scenario-based System Validation by Users. Accepted for INTERACT '99, Edinburgh August 30-September 3.

Sell, R. and Fuchs-Frohnhofen, P. (1994). Gestaltung von Arbeit und Technik durch Beteiligungsqualifizierung ("The Design of Work and Technology by qualification for participation") [Schriftenreihe Sozialverträgliche Technikgestaltung, Materialien und Berichte Band 39]. Opladen, Germany.

Selle, K. (1993). Versuch über Planungskultur - Zustandsbeschreibungen und Einordnungen. (Experiment on Planning Culture - State descriptions and assessments). Bärsch, J. (Ed.). Das Ende der Normalität im Wohnungs- und Städtebau?: thematische Begegnungen mit Klaus Novy (The End of Normality in House Building and Urban Development? - Thematical Encounters with Klaus Novy). Darmstadt, Germany.

Selle, K. (1996). Was ist bloß mit der Planung los? Erkundungen auf dem Weg zum koooperativen Handeln (What's the matter with planning? Investigations on the way towards cooperative action). Dortmunder Beiträge zur Raumplanung, Vol. 69, Dortmund, Germany.

Sproull, L. and Kiesler, S. (1986). Reducing social context cues: Electronic mail in organizational communication. *Management Science*, 32(11), 1492-1512. Also Greif, I. (Ed.). (1998) *Computer-Supported Cooperative Work: A Book of Readings*. San Mateo, USA: Morgan Kaufman.

Suchman, L. (1993). Do Categories Have Politics? The Language/Action Perspective reconsidered. De Michelis, G.S. and Schmidt, K. (Eds.). Third Europ. Conf. on Computer Supported Cooperative Work (ECSCW'93). Milan, Italy, 1-14.

Weber, H., Leidermann, F. and Zink, H.J. (1999). Symbolic Telecommunication using the WWW. In: Bullinger, H.J. and Ziegler, J. (Eds.). *Proc. of the Int. Conf. on Human-Computer Interaction*, Vol. 2, London: Lawrence-Earlbaum.

Weisband, S., Schneider, S. and Connolly, T. (1993). Participation Equality and Influence: Cues and Status in Computer-Supported Cooperative Work Groups, De Michelis, G. S. and Schmidt, K. (Eds.). *Third Europ. Conf. on Computer Supported Cooperative Work* (ECSCW'93). Milan, Italy, 265-279.

Winograd, T. (1988). A language/action perspective on the design of cooperative work. *Journal on Human Computer Interaction*, 3(1), 3-30.

Chapter XXV

Radio B-92 in Belgrade Harnesses the Power of a Media Activist Community During the War to Keep Broadcasting Despite Terrestrial Ban

Robin Hamman
University of Westminster, UK

During the 1999 war between NATO and the former Yugoslavia, an opposition radio station in Belgrade used the Internet to continue to disseminate news and music despite having their terrestrial transmitting equipment confiscated by Serbian authorities. This article will discuss how Radio B-92 was able to do this through the close coordination of radio station staff in Serbia and their partners from within the European media activist community. This article will begin by setting the activities of Radio B-92 and its partners during the Spring of 1999 in a historical context by discussing the use of broadcast and other media during the wars and conflicts of the past.

Alternative Sources of Information at Times of War

During times of war and civil unrest, governments are often keen to control the flow of information. In occupied Europe during the Second World War, strict Nazi media policies gave rise to clandestine resistance newsletters and pirate radio stations. These new channels of information were used to bypass censorship and provided an independent source of news about the war for those who were otherwise bombarded with Nazi propaganda. Again during the May 1968 revolution in France, students and others involved in the general strike against the Gaullist government used pirate radio, art, and leaflets to further their cause. DeGaulle, realising the power of the media, placed heavy guards at government radio and television stations, and initiated strict policies limiting the freedom of the media, and forcing them to broadcast only pro-government viewpoints. It has been suggested that the failure of the May revolutionaries to gain political power is, in large part, due to their failure to wrest control of the media from DeGaulle (Barbrook, 1995).

For nearly a decade, activists working in war zones or areas of civil unrest have had access to advanced communications technologies, such as satellite telephones, fax machines and the Internet. Not only can they use these technologies to access news from the outside world, but also to transmit their own information and views. In recent years we have seen a number of examples of activists spontaneously utilising these new communications technologies.

In 1989, student protestors were seen running through Tiannenman Square with e-mail printouts and faxes of support from around the world. During the Gulf War, unfiltered news trickled out of Kuwait through e-mail and IRC chat rooms. During recent political turmoil in Malaysia, the government heavily censored the media. Western Web sites, according to news reports, became very popular during the crisis as they were "one of the few places where Malaysians could participate in no-holds-barred debate, or access uncensored information [about the trial of Anwar Ibrahim and the ruling regime]" (Denny, 1999, brackets added). There were even reports of newspaper vendors selling stacks of photocopies taken from printouts of Western Web sites while pro-regime newspapers went largely unsold.

In the age of the Internet, it is incredibly difficult for governments to effectively stop such information being made available across national borders short of cutting off all telephone services or through the large scale use of proxy servers which can be used to filter "objectionable" information. The ability to route around blockages or damage to the network was recognised by early pioneers of the ARPAnet, the precursor of today's Internet, who suggested that the Internet could potentially survive a nuclear attack (Hafner & Lyon, 1996). During Spring 1999, a group of on-line activists exploited the geographically distributed nature of the Internet to disseminate news and viewpoints about the war in Yugoslavia.

Radio B-92

Radio B-92 (http://www.b92.net) is an independent FM radio station based in Belgrade which has won a number of international press and media awards, including the prestigious "Free Your Mind" award presented to them by MTV Europe in 1998. Their broadcasts of music and uncensored news were, until the 2nd of April 1999, heard across Serbia through a network of local partner stations. Their signal was also picked up by the BBC World Service and retransmitted via satellite around the world. In December 1996, B-92 began using technology from Real Networks to stream live audio broadcasts and short video clips over the Internet.

From its start as a terrestrial broadcaster, B-92 has been a respected source of independent news in the Balkans. Its coverage of anti-government protests in Belgrade in 1996 and on recent events in Kosovo, against the wishes of the Milosevic government, has meant that it has long operated under the constant threat of closure. B-92 offices have been raided on numerous occasions and members of staff have been repeatedly harassed or arrested. On 23 March, with NATO bombardment imminent, the transmitter of Radio B-92 was confiscated yet again by the Serbian authorities and editor-in-chief Veran Matic was taken and held in custody at a police station for over eight hours. He was released soon after news agencies around the world began to report on his arrest after receiving urgent e-mail press releases from other members of B-92's staff.

B-92 stood its ground against the government by continuing to provide music and news, in both Serbian and English, over the Internet for 10 days following the confiscation of their radio transmitter. This was made possible through the coordinated efforts of B-92 staff based in their Belgrade studio and the support of media activists across Europe.

A Community of Media Activists

In early March, many of these activists met together for the Next Five Minutes Tactical Media Conference (http://www.dds.nl/~n5m) in Amsterdam. At this conference, discussions took place on how to best organise and run campaigns using the media, and plans were made for using the Internet to develop support networks and resource sharing among allied groups. While B-92 was not the sole topic of discussion at this large three-day conference which brought together around 300 media activists from around the world, B-92 did figure prominently there.

At the conference, Radio B-92 staff gave a presentation about their work and the various partnerships they had already set up. Their enthusiasm and hard-earned experience served as an inspiration for many of those in attendance, and representatives of B-92 were faced with a long stream of questions and utterances of support at the end of their session.

Partners Within the Media Activist Community

Launched on the day that the Milosevic regime closed the offices of Radio B-92 and confiscated their terrestrial transmitter, the Web savvy Help B-92 Campaign was a well-coordinated continuation of the support network started with 'Press Now!' In addition to providing technical support for B-92 and other independent news providers in the former Yugoslavia, campaign organisers set up a bank account for donations, and worked to publicise the plight of B-92 while also providing a contact point for journalists and others interested in the war.

According to Geert Lovink, a prominent Dutch media activist and one of the key organisers of Help B-92 Campaign, talks between B-92 and other groups throughout Europe started as early as 1992. In 1993, an organisation called 'Press Now!' was set up to support independent media production in the former Yugoslavia. This technical and financial support later proved vital to the success of B-92.

Because of the dangerous political situation and frequent crackdowns against the Serbian media, B-92 decided in 1996 to allow xs4all, an Internet service provider started by a group of Dutch media activists, to host its site from the Netherlands. By doing this, they hoped to keep the B-92 Web site out of the reach of Serbian officials. The Dutch ISP also provided the expertise and backbone needed for B-92 to create its own ISP in Serbia which they used to link independent media producers throughout the country.

During December 1996, B-92 supported large-scale political demonstrations against the Milosevic regime, in the process becoming the most listened to station in Belgrade, before Serbian officials banned B-92's broadcasts. In response, B-92 began streaming audio and video from its Web site, which the government was unable to stop since the server was based outside of Yugoslavian territory. These audio streams continued to be broadcast over the Internet from 1996 until 10 days into the NATO bombardment of Spring 1999.

Those involved in the Help B-92 Campaign were able help B-92 continue to provide Real Audio streams of music and news for 10 days following the government's ban on their terrestrial broadcasts. The campaign secured a pledge from Real Networks to provide an unlimited amount of audio and video stream connections to B-92. Anonymous e-mail lists were developed to protect the identity of those wishing to express their views about the war, and message-boards linking to the campaign site buzzed with information (see http://www.eGroups.com/list/kosovo). Encrypted e-mail services were provided for journalists and others in the former Yugoslavia who found themselves under threat. The campaign created a Web site banner in support of B-92, which is now displayed on hundreds of Web sites around the world, and over 15 million visitors are reported to have visited the B-92 site in the month following the start of NATO bombardment.

The Help B-92 Campaign Web site itself had around 15,000 visitors per day during the first two weeks it was online in April 1999 and messages of support and donations poured into the site from around the world. The campaign office worked with a team of translators all over Europe who, over the Internet, translate news from the former Yugoslavia into English, Dutch and half a dozen other languages including Chinese. Other campaign members then use e-mail and fax machines to distribute press releases and translations of news from Serbia while HTML coders place them on the campaign Web site.

Although their official headquarters is an attic space above De Balie, a cultural centre in central Amsterdam, the Help B-92 Campaign has people and organisations working for it throughout Europe. Micz Flor, a former University of Salford (UK) lecturer now based at Public Netbase to Vienna (http://akut.t0.or.at), worked with a team on one of many radio relays, recording and editing B-92's Real Audio broadcasts before rebroadcasting them on the Austrian public broadcasting network, ORF. This programme was broadcast for Austrian listeners, but according to Flor, could be received anywhere within a 1,000 km radius of Vienna.

The End of B-92's Webcasts

On 2 April, 10 days after they confiscated the transmitter of B-92, Serbian police entered and sealed their offices. All members of staff were sent home and a new general manager was appointed by Serbian officials. The former director of B-92, Sasa Mirkovic, issued a statement through the Web site vowing that B-92 will "find a solution how to continue broadcasting our signal and to inform all our audience all over the world. At the end, I would like to say that we all have to keep the faith."

The Serbian regime was not alone in recognising the power of the media during the war. Within a few days of the start of bombing, NATO pilots began targeting pro-regime television and radio stations with their bombs. The attack on one such television station led to over a dozen deaths.

Although the B-92 Web site remains online as of this writing in August 1999, visitors are still unable to access live news and music streams. Drazen Pantic, a Serbian mathematics professor who was awarded the Electronic Frontier Foundation's 1999 Pioneer Award (http://www.eff.org/promo/99pioneer.html) for his work developing OpenNet and other B-92 networking initiatives, confirms that "for now there is no jeopardy of the B-92 page being hijacked by regime." In April 1999, Pantic conceded that it appeared unlikely that B-92 would be able to continue to provide live content online in the short-term future due to the post-war political climate in Serbia.

The Help B-92 Campaign, however, continues to raise funds for the legal challenges which B-92 plans to lodge against the Serbian government. More importantly, some members of 'Press Now!' and the Help B-92 Campaign started a project called Open Channels for Kosovo (http://www.dds.nl/openchannels). According to Richard de Boer, who has worked as a translator with both Press Now! and the Help B-92 Campaign, Open Channels is more of an information service rather than just a support group. During the war they worked to translate and post the e-mails, messages, audio reports and other information coming from independent news sources inside Yugoslavia. These sources include over a dozen journalists in Yugoslavia and outlying countries as well as several individuals about which little is known other than their stories and views about the war.

The Open Channels Web site was not the only place to find independent news coming out of Serbia during the war. Many prominent newspapers, such as *The New York Times* and *The Guardian* in London, published reports from unnamed journalists living in Serbia, most of which transmitted their articles via e-mail. While most appealed for an end to the bombing of their country, few supported Milosevic or the atrocities in Kosovo. National Public Radio (NPR) in the United States began reading transcripts of correspondence between 16-year-old Finnegan Hamill, a high school student in Berkeley, California, and Adona* his 15-year-old e-mail pen-pal living in Kosovo (Conniff, 1999). Other Serbian voices were heard on the Internet. For example, Serbian Orthodox monk Jeromonah Sava from the Visoki Decani monastery near Pec (http://www.decani.yunet.com) started an e-mail list and created a Web page, both dedicated to the distribution of news from and about Kosovo at the beginning of the war.

By closing B-92, the Serbian regime may have succeeded in softening the voice of one independent news source in Serbia. However, because of the distributed nature of the Internet, and the well-organised support networks of activists using it, the regime has little chance of silencing the entire flood of independent news coming out of Yugoslavia. It could also be argued that NATO bombs probably did more to stop independent news coming out of Serbia than regime. Geert Lovink of the Help B-92 Campaign said during the war, "People will continue to send e-mail, as long as there are telephone switches. But if NATO bombs them [the switches], the telling of stories by independent sources in Yugoslavia will also end."

Lessons Learned from B-92

Radio B-92 in Belgrade is not the first, and certainly will not be the last, independent news provider to use the Internet during times of war. What is special about this case is that, by coordinating a community of supporters throughout Europe, Radio B-92 has been able to achieve more than they otherwise could have even imagined. They rose to prominence at the beginning of the war as the Western media received press releases and appeals for support via the Internet. News articles and reports followed, and many people around the World turned to B-92 to hear a different side of the story.

The community of activists, many of whom met at the Next Five Minutes Tactical Media Conference in Amsterdam shortly before the NATO bombardment of Serbia began, enabled B-92 to continue to reach audiences in Yugoslavia and around the World even when the odds were heavily stacked against them. In this case, community translated to power, even if Radio B-92 and the voices of individual Serbians did little to change the social or political structure of Yugoslavia.

It has been suggested that B-92 played an important role in the war because it, like the many individuals who sent news and personal diaries via email from Yugoslavia, helped put a human face on the population NATO was bombing. Many are now aware that there is a growing tide of opposition within Serbia to the Milosevic regime, and I feel that B-92 has played an important part in making people realise this. It can also be argued that, because they were being heard through the Internet, it became impossible for Milosevic to completely eliminate the opposition within Serbia for fear of an even stronger media backlash against him, both in Serbia and abroad. Either way, Radio B-92 and the group of European activists who supported their efforts during the war have demonstrated that the Internet can be a powerful tool for the organisation of campaigns in support of alternative political viewpoints, even during war.

References

Barbrook, R. (1995). *Media Freedom: The Contradictions of Communications in the Age of Modernity*. London: Pluto Press.

Denny, C. (1999). "Our Debt to the Web." *The Guardian, Online Supplement*. London. 11 July, 1999, pp. 12-13.

Conniff, K. (1999). "Bringing the Conflict Home." *Brill's Content*. June, 1999. New York.

Hafner, K. and Lyon, M. (1996). *Where Wizards Stay Up Late: The Origins of the Internet*. New York: Simon & Schuster.

O'Kane, M. (1999). "Playboy Son Begins to Feel the Heat" *The Guardian*. London. 30 April, 1999, p. 4.

Chapter XXVI

The Economics
of Community Networking:
Case Studies from the
Association for Progressive
Communications (APC)

Mark Surman
Commons Consulting, Canada

Introduction

It was a special moment. Non-profits were still figuring out the fax machine. No one had heard of the Internet. A few brave souls were stringing computers together, hanging modems and activists off the other end. The information — and the shifting political tide — were beginning to flow.

News and passion trickled from the ANC headquarters in London to every nook and cranny of South Africa. Meetings were planned and new social movements dreamed over a few modems and a 286 in Toronto. Lobbying tactics, grand visions and messages home all emanated from a little computer room as thousands of environmentalists converged on Rio.

At the center of all this was a band of computer activists calling themselves the Association for Progressive Communications (APC).

The APC is a global coalition of nonprofit organizations who supply Internet content and connectivity services to civil society. APC was founded by a group of seven organizations who had all been providing e-mail and on-line discussion forums to non-profits and non-governmental organizations (NGOs) since the mid-1980s. This group included Alternex in Brazil, GreenNet in the UK, Nicarao in Nicaragua, IGC (PeaceNet and EcoNet) in the U.S., NordNet in Sweden, Pegasus in Australia and Web Networks in Canada. APC now includes 25 member networks located on six continents.

From the beginning, these networks were driven by a clear political mission —

to help members of civil society get online and get their word out. They were also driven by the belief that creating self-sustaining nonprofit enterprises was the only way to make this happen. It is this mix of political vision and nonprofit entrepreneurship that sets APC members apart from many other pioneers in the area of on-line activism. In approaching the most common conundrum of alternative media — balancing mission and money — most APC members have tried to take the best from social movements and the business world. This has required a complex dance between internal democracy and customer responsiveness, low budgets and high quality technical services, political independence and private sector partnerships.

This article explores the mission/money dance by looking at how APC members have built valuable community networking projects that have been largely self-sustaining. This exploration includes:

- the early days of APC networks, when providing basic services like e-mail and discussion forums offered an excellent way to strike this balance;
- the difficulties that most APC members faced as they responded to the 'Internet explosion' of the mid-1990s;
- experiences with partnerships between APC members and private-sector Internet companies;
- the renewed focus on content, and on unique NGO Internet services, that have been emerging with many APC members during the late 1990s.

While the journey of most APC networks has been bumpy at times, the path they have taken points to a community communications model that at once supports political action and provides a financial base. This is a rare combination, and one well worth reviewing.

The primary source of information for this article is a series of interviews conducted with long-time staff at eight current or former APC members during May and June 1999. This included Alternex in Brazil, Econnect in the Czech Republic, Ecuanex in Ecuador, EDNA Internet in Senegal, GreenSpider in Hungary, Pegasus/C2O in Australia, SangoNet in South Africa and Web Networks in Canada. Additional information was gathered from other documents outlining the experiences of the seven founding APC members, and from a broader pool of literature about the Internet and communications.

Blazing the Trail For On-Line Community— APC in the Early Days of On-Line Activism

Since it was officially formed in 1990, the APC has grown to become a network of 25 organizations serving over 50,000 civil society users. Each member is an independent organization that provides Internet-based services to non-profits and activists within a particular country or region. In many ways, it is a loosely affiliated group that works more separately than it does together. Many members simply use APC to network content, find technical support and undertake the occasional joint project.

Despite this, there is a common spirit and an approach to 'doing business' which comes from the history that APC members share. They all started out with a dream of empowering non-profits by making expensive technology cheap. They all offered the same basic on-line services before the Internet came on the scene as a public medium. And, as the Internet matured, they all faced the challenge of transforming and remaining relevant to the communities they serve.

One of the most important common touchstones for older APC members is in the area of services. Even before coming together as a group, most of these organizations had already set up local or national BBS-style systems to support activists and social movements. These systems included basic text-only e-mail, discussion forums and user directories. Users would connect directly to the system using terminal software like ProComm or Kermit, chatting and sending messages to each other once they were online. In some countries, users outside of the city where the system was located could connect cheaply using telco-owned x.25 networks (this was well before Internet connections were common).

While it may seem trivial from today's perspective, offering these services to civil society in the late 1980s and early 1990s was a big leap forward. Few people had access to academic networks. Commercial networks like Compuserve and AOL were expensive and isolated from each other. Long distance phone and fax still relied on overpriced monopoly providers in most countries. The simple e-mail and conferences offered by APC members really did represent increased communications potential and decreased costs, especially for organizations working internationally. These services were core to the mission-based work of early APC members.

Of course, there was more to 'the mission' than cheap communications. The APC and other early on-line providers offered a new way of working, and a new approach to building communities. This was especially true of APC's international discussion forums — or conferences. Similar to the chat rooms made famous on the Well, APC conferences provided a place for like-minded people to gather, publish information and share ideas. These conferences covered every possible progressive interest area from women to the environment to peace. And, unlike many early on-line communities, the users of APC conferences came from all sections of the globe.

APC conferences quickly became a hub for information about social movements and important political events around the world. For example, conferences were a key communications platform for both the United Nations and NGOs before, during and after the 1992 Earth Summit (UNCED) in Rio. Huge numbers of official and unofficial documents were posted online and discussed on APC networks. At the same time, conferences were being used as a platform for organizing. Groups from around the world met online to prepare common strategies for Rio and to develop alternative policy approaches. Once on site, NGOs were able to maintain connections with colleagues at home through a series of computer centres operated by Alternex — APC's Brazilian member (Sallin, 1994).

As this illustrates, conferences lent themselves to a seamless connection between content and community. Both publishing and discussion happened in conferences. Granted, conferences were far from the best publishing medium. Uploads were painful and cumbersome, and there were no graphics or hyperlinks. But there was no better way at the time for the average activist to get his or her message out to the world. And, once a document was posted to a conference, discussion seemed natural. People would comment or add their own documents. Looking back, early APC conferences provided something that 'portal' builders can only dream of — a focussed collection of the best on-line content and opinion on a particular topic.

Building Sustainability—
The Early 'Mission-Driven Business Model'

For most early APC members, this mix of cheap communication services and progressive content and community made for the perfect mission-driven 'business model.' All on-line services at the time — commercial services, APC members, BBSs — were offered in the form of an 'account' which bundled dial-up access, e-mail and access to content. Content was limited to whatever was on the system you dialed in to, and e-mailing people outside your home system was either expensive or impossible.

In the case of APC members, the international e-mail connections and the content and community were valuable enough that many NGOs were willing to pay for accounts. Well before the Internet exploded on to the public scene, a number of APC members had between 5,000 and 10,000 paying users each. At the same time, the service offerings were much more modest and required less staff and capital outlay than today's Internet services. This combination of genuine user demand and reasonably low costs meant that many APC members were able to live up to the ideal of building financially self-sustaining systems.

Of course, there have been APC members who did not fit this model. Econnect in the Czech Republic started out by giving its accounts away, and ENDA in Senegal has always provided free or subsidized accounts to its existing project partners. There were also APC members who from the start tried to supplement the sale of accounts with other products and services. In its very early days, Web Networks sold computers, modems, desktop publishing, t-shirt design and handmade jewelry. Still, the 'dial-up e-mail and content account' was by far the most common business model for early APC members.

Despite the fact that this model was a perfect fit for the times, running a sustainable on-line service for non-profits was not as simple as plugging in the computer and selling accounts. At least some financing was needed as a foundation. Unfortunately, the idea of electronic communications was still a foreign concept to most of the donor community and grants were hard to come by. Web Networks, SangoNet in South Africa, and a few others were able to get small start-up grants that paid for basic equipment. Other early APC members were built on personal savings, volunteer labour and donated equipment. Like many non-profits, these networks were started by a few impassioned individuals with a lot of vision and very few resources.

In typical APC style, Pegasus in Australia came up with an entrepreneurial response to this lack of grant money. Hoping to create a network on the American EcoNet service model, they turned to the ethical investment community for

$AUS100,000 in start-up funding. A private corporation was created with the original investor and a small team of staff on the board. While the company was officially 'for-profit,' its aim was still to provide non-profits with on-line services.

In contrast, almost every other member of the APC is in some fashion a non-profit. Many started out as a project of a larger NGO (non-governmental organization). For example, Alternex was a part of IBASE and ENDA-Internet is a part of the larger ENDA Tiers Monde. Others, like Ecunaex in Ecuador, were created by a coalition of NGOs with a common need for communications services. Still others were started as independent non-profits driven by a small handful of committed individuals.

Looking back at this early period, it would seem that APC members had come up with a long-lasting approach for balancing mission and money. NGOs needed the services offered by APC members, and were willing to pay a modest sum to receive them. And the ways that NGOs were using these services were leading to concrete social impacts. New environmental coalitions were being formed in Hungary. Censored communication was circulating in South Africa. Educators interested in human rights were connecting their students and classrooms around the world. The model seemed almost perfect. But in the mid-1990s, as the WWW emerged and the Internet became a household word, everything began to change very rapidly.

When New Technologies Get Old—
The Internet Goes Mainstream

With the introduction of any new communications technology, there is a gradual move from utopian excitement to commodified stodginess. In the beginning, a new medium presents a great deal of opportunity and promise. For example, early 'broadcast' radio was a boon for labour and political activists in the U.S. The hand-built technology, lack of clear business models and lack of regulations meant that anyone could operate a radio station. But as a medium grows older, marginal players get pushed out and standardized commodities begin to dominate the market. Radio became driven by the sale of sets, and eventually by the sale of advertising slots. As this happened, RCA and regulators pushed amateurs — and activists — off the air (Marvin, 1988; McChesney, 1995).

While the details differ, a similar pattern emerged as the Internet began to grow older. At a mundane level, the hand-built e-mail and dial-up services of the early days have been replaced by low margin commodity IP (Internet Protocol) services offered by huge Internet Service Providers (ISPs) and huge, centralized WWW sites backed by venture capital and inflated IPOs (initial public offerings). This has put tremendous pressure on many small Internet endeavors — including some APC members.

APC members felt these changes most acutely in the area of connectivity. As commercial ISPs became mainstream and dial-up prices dropped during the mid-1990s, most APC members started losing users — and revenue. Not only was it almost impossible to compete on price, but it was also difficult to differentiate APC services from those of commercial providers. E-mail was e-mail. IP was IP. Why pay more to a non-profit when these services could be found cheaper down the street?

While this attitude did not represent a mass exodus, some APC members lost 10% to 20% of their users.

In addition to dealing with heavy competition in dial up, APC members were faced with the huge costs involved in becoming full service ISPs and in moving from text-only to graphically based services. Early APC systems had been based on "store and forward" technology that networked information en masse using overnight long distance phone calls. Users accessed these services using a plain text interface accessible through any terminal program. Keeping up with the competition in the ISP industry meant abandoning (or duplicating) these old services. This meant installing expensive dedicated Internet lines, setting up more local points of presence, hiring more support staff and building better billing infrastructure.

This situation put serious financial and operational pressure on many APC members. In some cases, this was in the form of huge debt loads. For example, Web Networks was over $CDN1,400,000 in debt by the middle of 1996 as a result of capital and staff costs related to ISP service expansion. Small potatoes in the private sector, but a big deal in the non-profit world. In other cases, APC members simply couldn't afford to become full ISPs until very late in the game. Ecuanex did not offer any kind of full Internet connection until 1997. Others had been providing these services in Ecuador since 1993.

While these pressures were significant, there was another issue that was at once bigger and more ethereal — was there still a 'mission-based' case for offering dial-up services? In the early days, activists couldn't get dial-up and e-mail from anywhere but an APC node. But by the mid-1990s large commercial ISPs had solved the 'access' issue in the big cities of most countries where APC exists (there are notable exceptions). Rural areas and smaller cities still needed connectivity, but this was beyond the reach and resources of APC members in most cases. Given this, why offer dial-up?

There have been many answers to this question, and they seem to be changing over time. In some places, the answer has been "let's keep selling dial-up to subsidize other NGO activities." In others it has been, "outsource dial-up so we can offer activists a one-stop shop." In still others, APC members have decided to get out of dial up altogether. Whatever the case, changes in the market forced the whole of APC to reconsider whether or how to be offering Internet connectivity.

At the same time, APC conferencing system — an even more clearly mission-driven aspect of the services — was also starting to crumble around the edges. The emergence of the WWW and more ubiquitous e-mail access quickly broke apart the traditional "one-stop shop for activist content + community all in one place" value of the conferences. People no longer posted long documents and policy papers to conferences. They posted them to WWW pages. People no longer set up their communities in conferences (which required all users to have an APC account). They set up mail lists. As a result, traffic declined dramatically in the vast majority of APC conferences during the mid- to late 1990s. There are still hundreds of conferences that are used for private group collaboration — something the conferences are excellent for. But there are very few public communities still active in APC's conferencing system.

Of course, there is nothing wrong with the fact that activists have moved onto the WWW and mailing lists. It makes total sense. The WWW is a better medium for publishing. E-mail lists are more accessible. But the feeling of community and collective intelligence which come from sharing the same on-line space have been lost as people have gone down their own path.

When combined, increased competition in dial-up and the decline of conferences meant that APC's 'almost perfect' mission-driven business model no longer worked. There was no longer a compelling reason to buy dial-up accounts from APC members, and the conferences didn't provide enough value for most people to keep their accounts open. By 1996, many APC members saw the writing on the wall and realized they needed to find a new way to deliver on the mission of "helping members of civil society communicate and get their word out." They also realized they needed new mission-driven business models if they were to stick to their original dream of self-sustaining non-profit communication networks.

When There's No Mission in the Money— Partnerships with the Private Sector

As APC members tried to survive the changing Internet marketplace in the mid-1990s, there were inevitable partnerships and contractual arrangements with private companies. Some of these were as simple as outsourcing dial-up — a straight trade of money for services. Others involved complex plans to use the thriving Internet commercial market to fund mission-driven work. Whatever the case, APC member relationships with the private sector have shown that there is still a need for community-based networks. The private sector does many things well. Social mission isn't one of them.

One of the most interesting APC private-sector partnerships emerged at Alternex. From its inception, Alternex had been a project of a well-known NGO called the Brazilian Institute for Social and Economic Analysis (IBASE). IBASE focuses research and advocacy around issues of social exclusion. Given the focus within IBASE, there have always been internal questions about whether or not Alternex was a proper fit. Information access was important, but was it a part of IBASE's core mission?

For almost 10 years, the answer to this question was 'yes.' But as IBASE's economic fortunes weakened and the Internet market took off, the situation changed. Against the recommendations of founder and IBASE director Carlos Afonso, Alternex was spun off into a private company in 1995. Majority control of the company was retained by IBASE, with a minority partner providing investment capital for expansion. The idea was that Alternex could continue to serve the NGO sector while at the same time making money from broader commercial demand for Internet services. The bulk of the profit would flow back to IBASE to underwrite its social research work.

Unfortunately, it didn't work out this way. As with the rest of the Internet industry, the profit didn't flow instantly. IBASE was disappointed with this and started to get anxious. In addition, the mission-based activities of Alternex began to weaken. Commercial clients who could pay more demanded resources and staff attention, often leaving NGO clients neglected. Afonso and others committed to serving NGOs did everything they could to fight this trend, but were outweighed by the imperatives that a bottom line-focussed business brings with it. Alternex was sold outright to its private-sector investors in late 1997.

As this illustrates, balancing mission and money by setting up a private sector 'front' is a tricky game. The rules change quickly, and it is tough to hold onto the original political dream. Moving on from this experience, Afonso has gone back to the nonprofit model to meet the Internet needs of NGOs in Brazil. In 1997, he started the Information Network for the Third Sector (RITS) to provide low-cost, cutting-edge on-line tools for civil society. RITS is not yet a member of the APC.

Although the circumstances were slightly different, the situation in Australia also resulted in the 'disappearance' of an APC member network. Growing out of the social investment movement and a lack of grant funding for nonprofit Internet endeavors, Pegasus had always been a private company. While the mission had always been key to Pegasus as a company, it could not avoid some of the trappings of private ownership. At a very basic level, the original investors — no matter how large their conscience and commitment — were at risk. With two investors having provided Pegasus $AUS100,000 each, and a number of smaller investors involved as well, the risk was not insignificant.

As the Internet market and competition grew in the mid-1990s, the risks started to seem especially acute. Other ISPs and Internet companies were receiving huge inflows of investment capital. In 1996, one Australian ISP was purportedly losing over $AUS500,000 per month just to build infrastructure and keep its market share. Another was swimming in paper wealth from an early Internet IPO. Pegasus just couldn't keep up on the investment front, and in turn could not pay for the management and technical expertise needed to compete. Given that Pegasus had decided to make a go of the commercial ISP game, this was a real problem.

Feeling these pressures, the shareholders of Pegasus decided to sell to a large Australian ISP called Microplex in early 1997. While the deal was a fairly straight-up commercial transaction, a great number of promises and assumptions were made about continuing with Pegasus' mission-based activities. In fact, many of the staff believed that Microplex saw market advantage in being able to effectively serve the non-profit community. Unfortunately, none of this was in writing. The more mission-focussed staff and many of the NGO clients were asked to leave Pegasus within a few months of the sale.

As in the case of Alternex, a new non-profit called Community Communications Online (C2O) was formed immediately. C2O retained relationships with many Pegasus users and has been involved in a number of interesting projects with organizations like IEARN. It has also taken Pegasus' spot as the Australian member of the APC. All this said, C2O has remained small and faces a good deal of competition from others serving the Internet needs of non-profits in Australia. It will grow, but it will take time.

A final example of APC member partnerships with the private sector can be found at Web Networks in Canada. In the mid-1990s, Web Networks faced a common business contradiction — high revenue and even higher costs driven by growth. New services were being introduced and revenue was expanding on a regular basis. But at the same time, paying a staff of 30 people and servicing the $CDN 1,400,00 debt were crippling the organization. Inspired in part by the not-yet-unsuccessful Alternex experiment, the management at Web Networks went out to seek private investors for a spin-off company.

Upon approaching one of the most promising investment prospects — Open Text Corporation of Waterloo, Ontario — Web Networks received a surprising response to its call. "We won't invest, but we'd be happy to buy you." Like many growing Internet companies, Open Text was searching for experienced programmers and technical managers. Web Networks had these.

The Open Text offer raised some obvious questions — how do you 'sell' a nonprofit? ... and keep its mission alive? The answer came in the form of a creative asset sale arrangement. Web Networks would sell over 90% of its assets and staff to Open Text in exchange for debt relief and ongoing service supply agreement. It would retain its customers and services but would refocus completely on content development with non-profits. Most technical and billing services required by Web Networks would be provided by Open Text on a cost-recovery basis. From the perspective of a non-profit interested in getting out of the Internet tech business and into the Internet content business, this seemed perfect. In mid-1996, the agreement was signed and the new Web Networks moved its remaining two staff into a small office next door to Open Text.

What seemed like a perfect deal in the beginning quickly became a bureaucratic nightmare. Web Networks had little control over its billing or customer service, and often found it hard to resolve technical problems. At the same time, Open Text was stuck providing a whole range of ISP-type services which were not part of its core business (Open Text makes intranet software). The relationship quickly soured, Web Networks started to be seen as a whiny charity case, and service quality plummeted. After only 16 months of a three-year service supply agreement, Web Networks and Open Text parted ways. Web Networks took all of its servers and billing infrastructure back in-house.

Interestingly enough, this was not the only major private sector partnership undertaken by Web Networks. At the same time as signing the deal with Open Text, Web Networks entered a connectivity outsourcing arrangement with PSINet. This was a more straightforward business relationship based on the bulk purchase of services from PSINet's core product line. Service was bumpy from time to time, and pleas to 'be nice to non-profits' with better prices didn't really work. But basically, it was a partnership that succeeded. Similarly, Web Network's e-commerce partnership with the Royal Bank (the second largest Canadian bank) has been based on the straight purchase of standard services. A significant bulk discount on Visa charges for Web Networks customers has been arranged, but the services are otherwise 'out of the box.'

All three of the major APC private-sector partnership experiences point to one lesson — expect business to act like business. There is nothing wrong with working hand in hand with a company to bring its services to the non-profit world at a lower cost in or in a better package. This is how business partnerships work. But expecting a company to provide services it doesn't offer (Open Text) or expect private-sector investors to be driven by anything but the bottom line (Alternex SA and Microplex) is just not reasonable. So, as long as both parties are getting something, and as long as non-profits are receiving a better service, private-sector partnerships fit well into the balance between mission and money. Where this is not the case, they can be very destructive.

From Pioneers to Platforms — Community Networking in Internet Everywhere Age

Amidst the chaos of competition in connectivity and experiments with the private sector, most APC members were starting to ask questions about their mission. What will non-profits really need as the Internet grows? What are my users asking for? What can we offer that is unique from the commercial Internet industry? For many, the answer was clear—content.

But as with connectivity, the world of Internet content has been changing rapidly over the last few years. Over 35% of the time people in the U.S. spend on the Internet is spent on WWW sites or other environments controlled by AOL or one of its brands (Hansell, 1999). Huge portal/cable deals along side an increased trend towards caching for high-speed Internet users has laid the groundwork for a 'mainstreaming' of content. Big money IPOs connected to aggressive branding and mergers have begun to consolidate 'category audiences' around the sites that were first to market. While there is still a great deal of space for new content models on the Internet, it is becoming tougher for small players — including NGOS — to get their message out.

Addressing this consolidation of Internet content and its implications for civil society has provided a new mission for some APC members. During the last few years, these organizations have shifted from 'getting non-profits on the Internet' to 'making sure their messages are heard.' Of course, taking this laudable goal of 'content enabling civil society' and making it work as self-sustaining nonprofit enterprise is not a simple task. In order to do this, APC members have tried everything from training to consulting to portal building to software development. Some of what has been tried has been useful. A little of it has been sustainable. And a lot of it can just be chalked up to learning. But as time inches on, some clear models are emerging.

The most obvious place to start on the 'content enabling' mission is with portals, or hubs that lead to information of interest to civil society and progressives. Of the seven networks interviewed for this article, all but one stressed that their most important next step was to create a non-profit information hub of some sort. Ecuanex and others in Latin America are working on a 'regional portal' dealing with key international development issues. GreenSpider is creating a service called Civil House which will provide support materials for fundraising, non-profit marketing, and strategic uses of the Internet. SangoNet is working with others to create an APC Africa site that will provide an overview of NGO news and donor information related to Africa. Common amongst all these projects is an impulse to stake a high profile claim in cyberspace for activists and civil society.

The original push to create 'content services' grew out of the declining relevance of dial-up and conferences. APC members were seeking new ways to 'provide value' — a.k.a. remain useful — to their NGO users. In response to this, a number of APC members started offering 'content only' accounts for a nominal monthly fee. These started out as a way for people with non-APC Internet connectivity accounts to access the conferences. Over time, WWW-based services like the Web Networks 'Community' grew to include events listings, news feeds, jobs and other information of interest to users.

Not surprisingly, this approach didn't work the first time out. For the most part, it was premised on the idea that users should pay for content and community. While this approach worked in both the commercial and nonprofit worlds during the days of direct-dial services, it just didn't wash in the Internet era. People weren't willing to pay for information, and rightly so. The Internet had been built on an 'information wants to be free' ethos.

Over the last two years, there has been a shift away from 'getting information to NGOs' and towards 'getting NGO information to the world.' Through 'portal' projects like the ones mentioned above, APC members in a number of countries are trying to find ways to drive more traffic to sites run by their activist users. This happens in a number of ways. High traffic APC member sites are used to highlight the best content of their users. User sites are 'aggregated' into focussed NGO directories and search engines. Users are provided with free listing services for events, jobs, press releases and other important information. Through this approach, APC, its members and its users are working collectively to break through the content consolidation on the Internet.

As more and more APC members try to move into this role, and as they bring more and more NGOs into the fold, two clear principles are emerging — aggregation and information sharing.

The 'traditional' wisdom in the Internet community over the past few years has been that aggregating — or bringing together content — is key to making anyone or anything seen on the Internet. To a great extent, this is true. You need to bring a site or a project to people's attention, or at least make sure they can look it up in a central directory, if you want it to be found. This need has driven the huge stock valuations of companies like Yahoo, and has encouraged others to created hundreds of smaller portal or aggregation sites focussed on a specific industry or issue. In many ways, the sites being developed by many APC members are simply applying this same wisdom to the NGO world.

But there is another principle in the emerging APC model — information sharing. Mainstream portal sites are about grabbing as much information as possible and organizing it within their own proprietary, siloed environment. Once they are set up, they do everything they can to win more market share than competing sites with a similar focus. While this model does produce useful directories, it misses one of the great lessons of the Internet — opening up and sharing information produces a more powerful end product.

Luckily, APC members have not been able to ignore this lesson. As they have moved towards creating aggregator sites, it has been clear that they cannot try to 'beat' other non-profit portals or hubs. Not only do these sites share similar political values, but more importantly, they are often run by APC users. Umbrella organizations, international federations and even loose coalitions all have a desire to aggregate information produced in their sector. Building on the mission of getting civil societies word out online, APC members need to drive traffic both to individual WWW sites and to aggregator sites within their user base.

This is where the principle of sharing comes in. Most of the technology behind aggregation sites is about 'pulling' information. The technologies emerging within the APC — and within other areas of the nonprofit sector — focus on a balance between push and pull. In an ideal world, this means not only that any APC user can feed their events listings to a central hub site with a lot of traffic, but also that they can grab events listings off the central site to list on their site. The result is a 'community of information' similar to what existed in the old conferencing days. People have the autonomy of their own WWW site and the power of a collective voice.

There are two approaches being played with in the creation of these shared publishing systems, or distributed portals.

The most straightforward approach simply allows a whole collection of APC users to share a single purpose database. For example, two dozen users might want a way to automatically publish and manage action alerts on their WWW site. Instead of setting up new software for each organization, the local APC member would simply give them the HTML code they needed to hook into an existing database. The basic result would be the same — action alerts would be published on each NGO's WWW site using their own graphics and design. But, through sharing the same database behind the scenes, action alerts can easily be aggregated on a central site or swapped between sites working on the same issue.

This approach was developed by Web Networks in Canada, and is being extended for use across the APC by Econnect in the Czech Republic. In the Canadian case, a number of unions, social service organizations and international development NGOs have found 'shared applications' incredibly useful. For example, the Canadian Council for International Cooperation (CCIC) has instituted a three-tier system to collect information for its In Common anti-poverty campaign. As a starting point, the development education organizations which belong to CCIC can post local In Common events or alerts by setting up the application on their own site or by posting to the main CCIC site. Information from all members is then aggregated on the central In Common site, creating a small anti-poverty portal. The information posted to In Common is also fed to Web Networks' content editors and then on to the main Web Networks Community site. As a result, information is presented in a number of different packages that will increase its exposure and impact.

This shared application approach fits well with the original APC ethos of self-sustainability for crucial non-profit communications systems. In the Canadian model, each database — alerts, events, resources, press releases — is offered as part of a monthly WWW hosting package. The price is lower than building the database from scratch, and the value is higher. At the same time, Web Networks is generating a ongoing, mission-driven revenue stream to support its activities. It is not yet clear whether this model has either staying power or international applicability. Nonetheless, this approach again shows that mission and self-sustainability don't need to be mutually exclusive.

The other model for information sharing and distributed portal is based on swapping data back and forth between sites using a technology called XML. Among other things, XML — or extensible markup language — allows WWW designers and database developers to embed information about the content of the pages that they are producing. For example, they may include information about the author, date, title and subject area. This makes it easy for other computers to grab selected pages and pull them back for local republishing.

The Institute for Global Communications (IGC) — which includes PeaceNet and EcoNet — is using XML to swap information with other non-profit WWW sites. For examples, Action Without Borders — a non-profit job and volunteer opportunity listing service — has agreed to allow peace- and environmental-related listings to be pulled down and republished on the IGC site. As a result, the selected jobs receive more traffic (because they are in two places), the IGC site builds a better profile (with people looking for jobs) and IGC's users get more exposure (they are listed on the IGC site right beside the jobs).

This approach to sharing and distributed portal building also has value in terms of connecting APC to other large NGO networks. For example, APC, One World Online and others involved in the distribution of civil society information have started to talk about standards for the exchange of data via XML. This type of collaboration holds a lot of promise in terms of the broader goal of increasing the voice of non-profits online.

Of course, experimenting with how to aggregate and how to share has not provided instant revenue for the whole of APC. In fact, the transition away from dial-up and towards content in the mission-driven business models of APC members has involved a complex set of strategies. Many members have relied heavily on developing WWW sites and on-line databases for NGOs. Others have focussed on training and capacity building to help NGOs build their own sites. Others have continued to rely on dial-up to keep them going.

SangoNet, Ecuanex and Web Networks have all moved heavily into the WWW site development side of things. Starting out with basic WWW sites, this area of work has tended to evolve towards database development and custom programming. In turn, it has also fed naturally into research and development for 'shared publishing' and other products. There is definitely a lot of mission-related satisfaction in doing this work — each project results in another NGO WWW site getting online. But the real value comes from taking these individual contracts and feeding them into the shared voice that is emerging on APC hub sites. At the revenue level, these services have been key to keeping a number of APC members alive.

Training, capacity building and support services have also been important in the transition away from dial-up. Almost all of the APC members interviewed for this article were engaged in some sort of capacity-building activity. For example, Econnect had been working over the years with environmental organizations opposing the construction of the Tamlin nuclear power station in South Bohemia. This included everything from providing e-mail accounts to training on how to get the message out. When the anti-power plant activists finally decided to blockade the construction site, Econnect was there with a local computer support center, a digital camera and help with getting news onto the Internet. This type of work has a clear impact in terms of helping NGOs get their word out online. While it is often reliant on donor money or subsidization from other services, most APC members remain committed to this type of work as a way to support and stay engaged with their users.

The other key service that has played a role in the transition away from dial-up has been ... dial-up. Revenue from Internet account sales has been so key to most APC members that simply dropping these services overnight would have been impossible. The most common strategy in transitioning away from providing Internet connectivity has been to outsource it to a larger commercial provider. While this is not a perfect approach, it has kept cash flow moving as APC members develop new service areas like those described above. Four of the six current APC members interviewed for this article had outsourced some or all of their connectivity services.

Unfortunately, these outsourcing arrangements have not always allowed APC members to have the lowest prices. Dial-up bulk buying packages from major ISPs tend to be designed for corporations with large sales forces or pools of mobile workers. Accounts are priced lower than retail but aren't really 'wholesale.' As a result, APC members like SangoNet, Web Networks and Econnect who have taken this approach are left with limited margins or high prices. This in turn impacts their ability to provide high-quality NGO support. Also, continuing to offer sales and support for dial-up is a significant distraction from the development of new content services. The distraction was serious enough in the U.S. that IGC sold the connectivity side of its business in mid-1999 in order to focus solely on content work.

A few APC members have remained squarely in the dial-up market and see it as key to their long-term strategy. For example, Ecuanex has been able to slowly build a sustainable and profitable connectivity user base. Connectivity prices in Ecuador are still almost twice what they are in North America and Europe, and thus Ecuanex has been able to compete on price and good service. At present, they have 200 users with a plan to grow slowly to a cap of 300 users once more phone lines are put in. ENDA also remains focussed on connectivity and is able to compete effectively with other local providers.

There are also one or two APC members who are still able to trade heavily on the conferences as a way to keep users. In Hungary, GreenSpider conferences were central to building the national alliance of environmental NGOs. Quite quickly, the conferences became a place where these groups met and discussed ideas. As a result, there is still a strong connection to conferences amongst Hungarian NGOs. GreenSpider's content services are able to build off conferences as a base and add other services over time.

Despite a bumpy and confusing period in the mid-1990s, it seems that most APC members have begun to effectively reinvent both their missions and their sustainability models. The focus is clearly on 'content enabling' civil society, on making sure the word gets out on-line. There are still challenges in figuring out how best to do this. And of course, there are the normal challenges of running a technically focussed non-profit enterprise –marketing, bill collection, finding qualified staff and staying relevant to users. But collectively, and sustainably, a good number of APC members seem headed back down the right path.

Learning From the APC—
New Social Models for New Media

One conclusion that could be drawn from the APC experience is that there is only a role for community-based communications efforts in the early days of a new medium. There is always a gap that needs to be filled—and opportunities to be taken—during this time. In APC's case, providing basic e-mail and connectivity to NGOs in the late 1980s and early 1990s was an obvious and important mission-based activity. But as the Internet market matured and these services became common-place, there was no real need for specialist non-profit providers of these services. Except in a few cases, continuing to focus primarily on these services would have been akin to running a community-based manufacturer of photocopiers or fax machines. Producing the basic tools of communication is simply something that the market can do better than non-profits.

But resting on this conclusion would miss the point — there is a real need for community-based institutions that can help civil society get its messages out. Traditional commercial and state media are notoriously bad at providing space for alternative voices. While there have always been alternative media outlets that address this problem, they have almost always remained on the margins.

The pioneers of community-based Internet services have a real opportunity to change this situation. The key to doing this is not to focus on infrastructure and technology, but rather on building social and economic models of communication that can provide sustainable, high-profile content platforms for civil society voices. Whether you call them portals, channels, shared Web sites or virtual communities, these platforms will be essential if alternative media is to move beyond the margins in the convergence age.

A number of APC members, along side organizations like One World Online and AMARC, are working hard to build these platforms. From these efforts, new methods of publishing, sharing and getting the attention of audiences are emerging. In some cases, these efforts have already produced great success stories. In others, they have fed back into the learning and new experiments which are popping up daily on the Internet.

One of the real tricks in making all of this succeed will be finding the right mission-driven business models to sustain the promotion of progressive voices on the Internet. The experience of APC clearly demonstrates that sustainable non-profit enterprise can work in this arena. As the broader economy related to Internet content becomes more clear, APC members and others will have the opportunity to build on this experience and create new models that work. In the meantime, the experiments—and the on-line activism—continue.

Acknowledgment

Original research for this article was undertaken with the support of the International Development Research Centre (www.idrc.gc.ca).

Works Cited

Hansell, S. (1999). "Now, AOL everywhere." *New York Times*. July 4, p. BU1

Marvin, C. (1988). *When Old Technologies Were New: Thinking About Electric Communication in the Late 19th Century*. New York: Oxford University Press.

McChesney, R.W. (1995). *Telecommunications, Mass Media, and Democracy: The Battle for the Control of U.S. Broadcasting, 1928-1935*. New York: Oxford University Press.

Sallin, S. (1994). *The Association for Progressive Communications: A Cooperative Effort to Meet the Information Needs of Non-Governmental Organizations*. New York: Harvard/CIESIN Project on Global Environmental Change.

Surman, M. (1995). *Wired Words: Utopia, Revolution and the History of Electronic Highways*. Toronto: Commons Consulting.

Interviews with:

Afonso, Carlos — Former Director, Alternex — July 1999

Esterhuysen, Anriette — Executive Director, SangoNet — June 1999

Fall, Moussa — Internet Manager, ENDA — October 1998 and June 1999

Hancherow, Tonya — Executive Director, Web Networks — June 1999

Haverkamp, Jan — Fundraising Director, Econnect — June 1999

Klinkera, Vasek — Executive Director, Econnect — May 1999

Nagy, Agoston — Partnership Director, Greenspider — June 1999

Roggerio, Roberto — Executive Director, Ecuanex — October 1998 and June 1999

Wilson, Paul — Former Director, Pegasus — June 1999

About the Authors

Alessandra Agostini (agostini@cootech.disco.unimib.it) graduated in computer science at the University of Milano in 1991. Since then she has been collaborating with the Cooperation Technologies Laboratory of the Department of Informatics of the University of Milano "Bicocca" (formerly University of Milano). Her experiences are in the CSCW field, and in particular in developing prototypes to support communication and cooperation between groups of people. Within this area, her interests range from workflow management systems and communication handlers, to supporting community work. She participated in various European-funded projects (Esprit ITHACA #2705; Esprit BRA COMIC #6225) and, at present, she is project manager of the Campiello project (Esprit LTR #25572).

Paul M.A. Baker is currently a visiting assistant professor at the School of Public Policy, Georgia Tech, and a research fellow at the Institute of Public Policy, George Mason University in Fairfax Virginia, teaching courses in the areas of American government, public administration, information policy, state and local government and research methodology. Baker is currently researching institutional issues involved in community information infrastructure development, municipal policymaking, and local govenment use of information and communication technologies. His research experience includes studies based on combined survey research and qualitative (focus and nominal group) approaches for both private and public sector clients. He completed a PhD in Public Policy at George Mason University and holds an MP in Urban Plannning from the University of Virginia.

Thomas Beale has been a practising software engineer since 1986, and has worked in Australia and the United Kingdom in supervisory control systems, software and document management, and currently works in health informatics. Since 1993 he has been studying how information technology can be used to address social and environmental problems, particularly with communities as a focal point. His professional work in health informatics, which now includes funded projects leading toward an electronic health record for primary care, including in developing countries and disadvantaged areas, has provided many insights into the use of IT outside the corporate sector. During the last two years, he has worked with members of an intentional village in Australia to develop an approach for the introduction of a village network.

Janice Brodman has 20 years of experience in designing, evaluating, and implementing organizational development programs in the U.S. and in developing countries. Her focus has been on effective technology cooperation, with specific attention to information and communications technologies (ICTs) adopted by non-profit organizations and medium/small enterprises. Her projects include development of multimedia training programs, formulation of ICT strategies, creation of WWW resources, and managing virtual conferences. She is the author of a number of seminal works on the factors contributing to successful use of new technologies in developing countries. She is also a frequent speaker at business and other conferences. During the past 10 years, she has conducted numerous assignments in technology transfer for a wide range of development agencies, including the United Nations, the World Bank, U.S. Agency for International Development (USAID) and country ministries. From 1983 to 1992, Dr. Brodman also worked with a major U.S. management consulting firm to develop information technology strategies for corporations. She has developed and taught short courses at Harvard University's John F. Kennedy School of Government, on effective use of information technology by public sector organizations. She holds a Ph.D. from Harvard University.

David Bruce is director of the Rural and Small Town Programme, Mount Allison University, where he has been employed since 1990. He has a BA in geography from Mount Allison University and an MA in geography from the University of British Columbia. David has knowledge and expertise in the field of rural community development and associated topics of housing, information technology adoption, home business issues, organizational growth and development, and community economic development. He has participated in more than 35 major research and outreach projects in most of Canada's provinces and territories. He is a certified trainer with the Canadian Housing and Renewal Association, and chair of the Editorial Board for its housing magazine, *Canadian Housing*. He also serves on Industry Canada's National Advisory Committee for its Community Access Program. David is a national CED technical provider through Carleton University's Community Economic Development Technical Assistance Program. David also teaches part time in the Department of Geography at Mount Allison University.

John Cawood is course leader of the BA/BSc information and communications degree at Manchester Metropolitan University. His research interests are in social informatics. He has published work on the social construction of network technologies and on-line teaching and learning.

Andrew Clement (clement@fis.utoronto.ca) is on the Faculty of Information Studies at the University of Toronto. His research interests are in the social implications of information technology and human-centred systems development. Current research focuses on public interest information policies for guiding the development of Canada's information infrastructure. He is co-editor of three books: *The Information Society: Evolving Landscapes* (Springer Verlag, 1990), *Information System, Work and Organization Design* (North-Holland, 1991), and *NetWORKING: Connecting Workers in and Between Organizations* (North-Holland, 1994). Recent articles have appeared in *Communications of the ACM, Computer Supported Cooperative Work (CSCW), Information Technology and People, AI & Society,* and *Canadian Journal of Information Science.* Prof. Clement is a member of ACM, CIPS and CPSR and is the Canadian representative to IFIP TC9 (Computers and Society). He serves as Chair of its Working Group 9.1 (Computers and Work).

Royal D. Colle has been a member of the Cornell University faculty for 33 years. Simultaneously he has been a consultant for the World Bank, FAO, WHO and others in applying communication technology to the challenges of social and economic development. In recent years he has worked on the potential of telecenters in developing nations. At Cornell, he was Chair of the Communication Department for 10 years and taught courses in communication planning and strategy. He has frequently conducted workshops on communication in Asia and Africa. Colle is a regular contributor to the Journal of Development Communication and serves on its Editorial Advisory Board.

Rosann Webb Collins is associate professor of information systems and decision sciences at the University of South Florida. Her research focuses on the impact of information technology on knowledge work, systems development, community informatics, and the legal and ethical issues in the use of computers. Her publications include a book on the deployment of global IT solutions and articles in *MIS Quarterly*, *The Information Society*, *Journal of the American Society for Information Science*, *Journal of Research on Computing in Education*, *Educational Technology*, and *International Library Review*.

Fiorella deCindio is associate professor of programming languages at the State University of Milano. Her research interests include Petri nets as concurrency theory, programming languages (namely, object-oriented and distributed programming languages) and the applications of the ICT to support life and work within social and office systems. In this last branch, in the '80 she did action research and education on workers participation in system design, then has been member of the team which conceived and developed one of the first CSCW prototypes (CHAOS, Commitment Handlig Active Office System). In 1994, she promoted the Civic Informatics Laboratory which launched several initiatives, including the Milano Community Network (Rcm) which is now a Participatory Foundation, and the Association for Informatics and Civic Networking of Lombardy (A.I.Re.C.) which groups the Community Networks in the Lombardy Region. Fiorella De Cindio is now President of both, and is among the promoters of the European Association for Community Networking (EACN).

Dave Eagle is co-director of CIRA and is a senior lecturer in the School of Computing & Mathematics. His research interests include the management and capacity planning of electronic networks, computer law and security, and community informatics. As the technical arm of CIRA he regularly acts as a consultant.

Morten Falch is associate professor at the Center for Tele-Information at the Technical University of Denmark. He received a masters in economics from University of Copenhagen in 1984 and a PhD from the Technical University of Denmark in 1993. He has formerly been employed as an economist in a Danish pension fund (pka) and at the UNIDO office in Tanzania. Since 1988 he has held various positions at the Institute of Social Sciences and Center for Tele-Information (CTI) at The Technical University of Denmark. He has participated in a large number of international research projects funded by RACE, ACTS and other international or national research programmes. He also participates in a joint research project on technology assessment at University of Ghana.

His major research area is "economic and social implications of advanced telecommunication services." This includes telecommunication economics, regulation, electronic commerce and financial sector applications. Morten Falch has also performed a large number of consultancies for the European Commission, UNCTAD and others. Among others he has contributed to a World Bank study on Multi-Media Multi-Purpose Communication Information Centres.

Susana Finquelievich completed her architectural degree at the National University of Rosario, Argentina, but she soon changed both her profession and her country. She graduated in urban and regional planning at the Polytechnic University of Szcsczecin, Poland, and later established herself in Paris. There, she finished a master's degree in urbanism and a doctorate in urban sociology, under Manuel Castells' direction. Dr. Finquelievich worked for several years as coordinator for Latin American Research in the Food/Energy Program of the United Nations University. In 1987 she returned to Buenos Aires, where as a senior researcher of the National Council of Scientific and Technical Research, she coordinates a team working on the social—and particularly the urban impacts—of the informational revolution. She also coordinates a research team at the University of La Plata.

She has published three books on the subject of social uses of ICT: *New Technologies in Town: Information and Communication in Everyday Life*, with J. Karol and A. Vidal, 1992; *Cybercities: Informatics and Local Management*, with J. Karol and G. Kisilevsky, 1996; and *The City and its ICTs*, with E. Schiavo, 1998. Nearly 100 of her papers have been published in specialized journals in U.S., Europe and Latin America.

Françios Fortier is a political economist researching the social implications of information and communication technologies. In 1997, he completed a doctorate in political science from York University in Toronto. Over the last decade, Fortier has done academic work and a number of consultations for non-governmental organisations, United Nation agencies, and research institutes. His most recent work includes the formulation of agricultural and rural development networking projects in several countries of Africa, Asia, and Latin America. He also teaches graduate courses on the political economy of the Internet.

Valeria Giannella (giannelv@brezza.iuav.it) lives and works in Venice where she obtained her degree in urban and regional planning and her PhD on planning processes for the city governance. She has collaborated for three years with the Venice School of Urban and Regional Planning and participated in several European projects focused on the mobilisation of local community subjects for environmental and economic restoration purposes. Her work for the EU-funded Campiello project is addressed to creating a meaningful inclusion of the Venice local communities in the building up of an information system which serves,not substitutes real life.

Antonietta Grasso (Antonietta.Grasso@xrce.xerox.com) studied computer science at the University of Milano where she got her degree in 1990. After that she joined the Cooperation Technologies Laboratory at the University of Milano where she worked on open architectures for process support until the end of 1995. Then she joined the Grenoble Laboratory of Xerox Research Centre Europe as research engineer in the Coordination

Technology group. Her work focused first on wide area inter-organizational process support systems, while her current research is on usage on multiple interfaces to support the creation of community memories and on collaborative information search. Her research interests are in the CSCW field, in particular the design of open architectures for process support and information sharing. A recent interest is around the design of user interfaces complementary to GUI ones. Her main current activity is in coordinating the activities in the European project, Campiello.

Michael Gurstein, Ph.D., a Canadian, completed a BA at the University of Saskatchewan and a PhD in the sociology at the University of Cambridge. Dr. Gurstein was a senior public servant in the Provinces of British Columbia and Saskatchewan. For a number of years Dr. Gurstein was the president of the consulting firm Socioscope Inc. in Ottawa, Canada, which specialized in the human aspects of advanced technologies. From 1992-95 Dr. Gurstein was a management advisor with the United Nations Secretariat in New York

Since 1995, Dr. Gurstein has been the ECBC/NSERC/SSHRC associate chair in the Management of Technological Change at the University College of Cape Breton and since 1996, the founding director of the Centre for Community and Enterprise Networking (C\CEN) of that Institution. C\CEN specializes in the application of information and communications technology to local economic development particularly rural development.

Dr. Gurstein has published widely in both scholarly and more popular journals and has contracted (with Idea Group Publishing) for this volume "Community Informatics: Enabling Communities With Information and Communication Technologies," and has the book "Burying Coal: Research and Development in a Marginal Community" (1999) by Collective Press. "The Net Working Locally: Information and Communications Technology in Support of Local Economic Development" is currently under review by several publishers.

Barry Hague is a senior lecturer in social policy and research coordinator for CIRA. He is reviews editor for the international journal *Information, Communication and Society* (Routledge) and co-editor of *Digital Democracy: Discourse and Decision Making in the Information Age* (Routledge, 1999).

Chris Halaska is a partner in Social Design Consulting, where he advances the practices of participatory design and democratic technology. In the last few years he has worked on projects for the Seattle Public Schools, the Cross City Campaign for Urban School Reform, Apple Computer, and Microsoft. He has a PhD in participatory design from the University of Washington. Most of his work involves encouraging user participation in the design of technical systems. In addition, he continues to do technical work, including designing web sites and database systems, electronic publishing, and providing computer support. And when that all gets to be a bit much, he plays African marimba.

Robin Hamman (robin@cybersoc.com) is a doctoral researcher at the Hypermedia Research Centre, University of Westinster (http://www.hrc.wmin.ac.uk). He is also a producer for the Communities Category of BBC Online and the editor of Cybersociology Magazine (http://www.cybersociology.com). Robin has been building and living in on-line communities since 1985 and has written about the Internet for a number of periodicals, journals, and edited collections.

Michael Koch (Michael.Koch@acm.org) is research scientist and assistant professor in the Informatics Department at the Technische Universitat Munchen in Munich, Germany. He has obtained a degree in computer science and a doctorate in the same topic from Technische Universitat Munchen. The work in the Campiello project took place during his 12-month post-doc stay at the Xerox Research Centre Europe in Grenoble in 1998. His main areas of interest are agent-based systems for community support and information/knowledge management, and the usage of Web technologies and on-line services supporting communities and groups.

Danny Krouk is a former consultant to the UCLA Advanced Policy Institute. He currently works for a software company.

John E.S. Lawrence is former Deputy Director of the Social Development Division in the Policy Bureau of the United Nations Development Program, and was for many years Principal Technical Adviser in Human Resources Development (HRD). He currently serves in a senior advisory capacity to UNDP's Health and Development Program, and has also worked closely with UN agencies, governments and NGOs preparing for the 2000 World Education Forum in Dakar, Senegal. He pioneered electronic networking in several aspects of UN functioning, and is especially interested in intersectoral policy space between education, health and employment sectors. His substantive specialty is in human development, with explicit focus on linking education/training to livelihoods.

Brian D. Loader is co-director of the Community Informatics Research & Applications Unit (CIRA) based at the University of Teesside, UK. He is also general editor of the international journal *Information, Communication & Society* (www.infosoc.co.uk) and is editor of *The Governance of Cyberspace* (Routledge, 1997); *Cyberspace Divide* (Routledge, 1998), with Barry Hague; *Digital Democracy* (Routledge, 1999), with Doug Thomas; *Cybercrime* (Routledge, 1999) with Bill Dutton, Nichole Ellisen and Nicholas Pleace; and *Key Concepts in Cyberculture* (Routledge, 1999). He has also published several articles and reports on technological change and social and political restructuring and is currently a member of the EU working group on ICTs, Social Movements & Citizens.

Oliver Märker is a PhD student at the German National Research Center for Information Technology. He has studied geography at the University of Bonn. His main research field is computer supported participation and tele-mediation in urban planning.

Peter Miller is a community technology consultant and public policy project coordinator for the Community Technology Centers' Network (CTCNet). He is former network director of CTCNet which supports community organizations that provide access to computers and related technologies, education programs and training to disadvantaged communities. Peter cochairs the Innovative Awards Committee for the OMB Watch Nonprofits' Policy and Technology project, was recently a panel chair for the federal Community Technology Centers grants program, and serves on the boards of the Ohio Community Computing Center Network, the Technology Education Council of Somerville, Massachusetts, the Civil Rights Forum on Communications Policy, and Computers In Our Future in CA. He is a member of Computer Professionals for Social Responsibility (CPSR), the Association for Community

Networking, and the Alliance for Community Media. He helped found the annual New England Computers and Social Change Conference and the Boston Computer Society's Nonprofit Assistance Program. Peter was previously the Executive Director of the Somerville (MA) Community Computing Center and Urban Planing Aid, Inc., the country's first advocacy planning agency. His articles on community technology have frequently appeared in the *Community Technology Center Review* which he founded and edits, and the *Community Media Review*, and the *CPSR Newsletter*.

Jo Pierson graduated in communication science at the Free University of Brussels (VUB) in 1996. Since then he has been researcher at the centre for Studies on Media, Information & Telecommunication (SMIT), headed by Prof. Dr. Jean-Claude Burgelman. His work includes research projects and consultancy assignments for private and public organisations with regards to ICT policy and user-oriented innovation studies. As research assistant for the Fund for Scientific Research—Flanders (Belgium) (FWO), he is preparing a PhD on the development and appropriation of ICT by small businesses.

Volkmar Pipek received a masters degree in computer science in 1996. Since 1997 he has worked in the Research Group Human-Computer-Interaction and Computer-Supported Cooperative Work (ProSEC) at the University of Bonn. He worked on organisational issues of collaborative computing, especially concepts for participative groupware development and introduction and user qualification and co-organised and taught in a distance learning project on "Computer Science & Society." Currently he leads a project on organisational learning in virtual organisations. His research interests span issues like knowledge management, CSCL, CSCW, electronic democracy, community networks and participatory design, and his focus lies on developing tools for collaborative discussion/decision support for design processes.

Bill Pitkin is a research associate in the UCLA Advanced Policy Institute and a doctoral student in the UCLA Department of Urban Planning. His research focuses on the role of information technology in community development and planning.

Agneta Ranerup (agneta@informatik.gu.se) is a lecturer in the Department of Informatics at the University of Göteborg in Sweden. She received her PhD in informatics from the University of Göteborg in 1996. Her main research interest is democratic aspects in informtion systems development in the public sector. More specifically, her research has focused how deliberative processes in local government can be supported by information technology. She is also involved in a project ("Access Kumla") where the TUC of Sweden, local governments, and civic associations are trying to use the Internet for providing local information as well as arenas for deliberation.

Neal Richman, PhD, is the associate director of the UCLA Advanced Policy Institute and is on the faculty of the UCLA Department of Urban Planning where he teaches courses in community development, real estate and professional practice.

Claus Rinner is working as a system developer at DIALOGIS Software & Services, Sankt Augustin, Germany. Among his tasks is designing interactive Internet mapping tools for exploratory data analysis. Claus holds a PhD in geography from the University of Bonn.

During his doctoral research at GMD, the German National Research Center for Information Technology, he was working on GIS-based discussion support for online planning. Claus graduated in applied systems sciences at the University of Osnabröck, where he studied environmental modeling and GIS.

Barbara Schmidt-Belz is a senior scientist at the German National Research Center for Information Technology. She has studied computer science at the University of Bonn and has since specialized in usability engineering. She has worked with users in different application domains, among them architecture, urban and regional planning, city administration and public participation. In the last few years, she has been involved in the development of the System Zeno.

Doug Schuler, as an activist and past chair of Computer Professionals for Social Responsibility (CPSR), has been writing and speaking about constructive uses of technology over 15 years. Doug is a co-founder of the Seattle Community Network (SCN) which, after three years, has nearly 13,000 registered users. He has written many articles on the "tapestry" of democratic and community-oriented communication technology and has co-edited three books on social implications of computing. Howard Rheingold called Doug's book *New Community Networks: Wired for Change* (Addison-Wesley, 1996) "an invaluable and deep sourcebook for grassroots activists." As an expert in the area of community computing, he has presented in Asia, Europe and North America. Doug holds a masters degree in computer science and one in software engineering and was a software engineer and researcher at Boeing's advanced technology center for 10 years. In addition to being a faculty member at the Evergreen State College on the technology and social implications of the Internet, Doug consults with organizations and institutions on how to use computer technology effectively. His web page on the "Community Networking Movement" is at http://www.scn.org/ip/commnet. He has most recently launched an international research+activism network that currently has members from more than 25 countries (http://www.scn.org/tech/the_network).

Leslie Regan Shade (shade@aix1.uottawa.ca) is at the Department of Communication, University of Ottawa. Her research interests focus on the social and policy issues surrounding information and communication technologies. Her work has been published in the *Canadian Journal of Communciation, Ethics and Information Technology*, *The Information Society*, and *Journal of Information Technology Impact*. She is working on a book, *Gender and Community in the social Construction of the Internet*, to be published by Peter Lang in 2000.

Seamus Simpson is a lecturer in the Department of Information and Communications at Manchester Metropolitan University. His research interests are in the public policy dimensions of ICTs and he has published work on telecoms policy and ICT convergence issues.

Roger S. Slack (PhD, Victoria University of Manchester, UK) is a research fellow at the University of Edinburgh Research Centre for Social Sciences. His interests include ethnomethodology, technology studies and the history/philosophy of artificial intelligence. He has co-edited two collections on the development and use of multimedia and authored a number of papers on community informatics, educational multimedia and ethnomethodological sociology.

Dave Snowdon (Dave.Snowdon@xrce.xerox.com) joined Xerox Research Centre Europe's Grenoble laboratory in February 1998. His research interests include novel user interfaces, information visualisation and the software architectures of collaborative virtual environments. He is currently working on the EU-funded project, Campiello, which is exploring novel user interfaces for sharing cultural information in communities. Before joining XRCE, Dave spent four years working in the Communications Research Group at the University of Nottingham during which time he built a number of prototype visualistion applications, worked on the design and implementation of a scalable collaborative virtual environment and was involved with the staging of two VR based public arts performances. He has also co-chaired two international conferences on the subject of collaborative virtual environments (CVE'96 & CVE'98) and is currently in the processing of organising CVE 2000.

James Stewart is a research fellow and doctoral student in the Research Centre for Social Sciences at the University of Edinburgh (http://www.rcss.ed.ac.uk/rcss_home.html). His doctoral work is on the appropriation of new media technologies in everyday life. He has researched and published on the innovation, development and use of new ICT systems in retailing, education and in other contexts, and is a writer and consultant on interactive television and the convergence of TV and the Internet. He edits Interactive Television News, an on-line industry newsletter (www.itvnews.com). This research was conducted as part of a major international study 'Social Learning Multimedia' funded by the European Commission. (http://www.rcss.ed.ac.uk/SLIM/SLIMhome.html). Acknowledgements for this research go to Stephanie McBride and Aphra Kerr of Dublin City University and Robin Williams at the University of Edinburgh.

Mark Surman has been helping others build on-line communities since the early days of the Web. He currently runs The Commons Group, a small communications strategy firm focused on on-line community and portal development. Recent Commons' projects have included product development workshops for community networks in 25 countries and on-line community vision documents for major government agencies. Mark has also served as the executive director of Web Networks, a Canadian Internet service provider serving the voluntary sector. Before moving on-line in 1994, Mark spent five years helping community organizations produce documentary film and video.

Kathryn Trees teaches in the School of Arts at Murdoch University. She is particularly interested in relationships between indigenous and non-indigenous people in Western Australia and has been involved in cross-cultural education of members of the judiciary for the past five years.

Andrew Turk teaches human factors aspects of information systems in the School of Information Technology at Murdoch University in Western Australia. His background in cartography and psychology has led to ongoing research interests in geographic information systems, visualisation and cognitive aspects of human-computer interaction. He has been working with Kathryn Trees for the last four years on the development of multimedia information systems to support education and negotiation within indigenous communities.

Index